Brahim Chaib-draa, Jörg P. Müller (Eds.)

Multiagent based Supply Chain Management

Studies in Computational Intelligence, Volume 28

Editor-in-chief
Prof. Janusz Kacprzyk
Systems Research Institute
Polish Academy of Sciences
ul. Newelska 6
01-447 Warsaw
Poland
E-mail: kacprzyk@ibspan.waw.pl

Further volumes of this series can be found on our homepage: springer.com

Vol. 11. Antoni Ligęza
Logical Foundations for Rule-Based Systems, 2006
ISBN 3-540-29117-2

Vol. 12. Jonathan Lawry
Modelling and Reasoning with Vague Concepts, 2006
ISBN 0-387-29056-7

Vol. 13. Nadia Nedjah, Ajith Abraham, Luiza de Macedo Mourelle (Eds.)
Genetic Systems Programming, 2006
ISBN 3-540-29849-5

Vol. 14. Spiros Sirmakessis (Ed.)
Adaptive and Personalized Semantic Web, 2006
ISBN 3-540-30605-6

Vol. 15. Lei Zhi Chen, Sing Kiong Nguang, Xiao Dong Chen
Modelling and Optimization of Biotechnological Processes, 2006
ISBN 3-540-30634-X

Vol. 16. Yaochu Jin (Ed.)
Multi-Objective Machine Learning, 2006
ISBN 3-540-30676-5

Vol. 17. Te-Ming Huang, Vojislav Kecman, Ivica Kopriva
Kernel Based Algorithms for Mining Huge Data Sets, 2006
ISBN 3-540-31681-7

Vol. 18. Chang Wook Ahn
Advances in Evolutionary Algorithms, 2006
ISBN 3-540-31758-9

Vol. 19. Ajita Ichalkaranje, Nikhil Ichalkaranje, Lakhmi C. Jain (Eds.)
Intelligent Paradigms for Assistive and Preventive Healthcare, 2006
ISBN 3-540-31762-7

Vol. 20. Wojciech Penczek, Agata Półrola
Advances in Verification of Time Petri Nets and Timed Automata, 2006
ISBN 3-540-32869-6

Vol. 21. Cândida Ferreira
Gene Expression on Programming: Mathematical Modeling by an Artificial Intelligence, 2006
ISBN 3-540-32796-7

Vol. 22. N. Nedjah, E. Alba, L. de Macedo Mourelle (Eds.)
Parallel Evolutionary Computations, 2006
ISBN 3-540-32837-8

Vol. 23. M. Last, Z. Volkovich, A. Kandel (Eds.)
Algorithmic Techniques for Data Mining, 2006
ISBN 3-540-33880-2

Vol. 24. Alakananda Bhattacharya, Amit Konar, Ajit K. Mandal
Parallel and Distributed Logic Programming, 2006
ISBN 3-540-33458-0

Vol. 25. Zoltán Ésik, Carlos Martín-Vide, Victor Mitrana (Eds.)
Recent Advances in Formal Languages and Applications, 2006
ISBN 3-540-33460-2

Vol. 26. Nadia Nedjah, Luiza de Macedo Mourelle (Eds.)
Swarm Intelligent Systems, 2006
ISBN 3-540-33868-3

Vol. 27. Vassilis G. Kaburlasos
Towards a Unified Modeling and Knowledge-Representation based on Lattice Theory, 2006
ISBN 3-540-34169-2

Vol. 28. Brahim Chaib-draa, Jörg P. Müller (Eds.)
Multiagent based Supply Chain Management, 2006
ISBN 3-540-33875-6

Brahim Chaib-draa
Jörg P. Müller

Multiagent based Supply Chain Management

With 149 Figures and 40 Tables

Springer

Prof. Dr. Brahim Chaib-draa
Computer Science/Software Engineering Department
Pouliot Building, Université Laval
Ste-Foy, PQ, Canada G1K 7P4
E-mail: brahim.chaib-draa@ift.ulaval.ca

Prof. Dr. Jörg P. Müller
Clausthal University of Technology
Julius-Albert-str. 4
D-38678 Clausthal-Zellerfeld
E-mail: mueller@in.tu-clausthal.de

Library of Congress Control Number: 2006925626

ISSN print edition: 1860-949X
ISSN electronic edition: 1860-9503
ISBN-10 3-540-33875-6 Springer Berlin Heidelberg New York
ISBN-13 978-3-540-33875-8 Springer Berlin Heidelberg New York

Springer is a part of Springer Science+Business Media
springer.com

Printed in The Netherlands

Cover design: deblik, Berlin
Typesetting by the authors and SPi
Printed on acid-free paper SPIN: 11679585 89/SPI 5 4 3 2 1 0

Multiagent based Supply Chain Management: A Brief Introduction

Brahim Chaib-draa[1] and Jörg P. Müller[2]

[1] Université Laval, Dépt. d'Informatique et de Génie Logiciel, DAMAS
Ville de Québec G1K 7P4 (PQ, Canada), chaib@ift.ulaval.ca
[2] Clausthal University of Technology, Germany, mueller@in.tu-clausthal.de

1 Introduction

Nowadays the agent technology is present in many industries including concurrent engineering, collaborative engineering design, manufacturing enterprise integration, supply chain management, manufacturing planning, scheduling and control, material handling, and holonic manufacturing systems. Supply chain management is concerned with planning and scheduling the different activities of organization across the supply chain, from raw material procurement to complete finished goods delivery. Such management is became vital today to the competitiveness of entreprises since it impacts their agility to meet changing market demands.

Agent technology and particularly multiagent systems was designed to capture many of the challenge involved in supporting changing supply chain practices. The rationale behind the choice of such technology in the context of supply chain management is as follows. Supply chains are made up of heterogeneous production subsystems gathered in vast dynamic and virtual coalitions, where each partner (or production subsystem) pursues individual goals, while satisfying both local and external constraints. In addition, the distributed manufacturing units have the following characteristics :

- *autonomy*: a company carries out tasks by itself without external intervention and has some kind of control over its action and internal state;
- *social ability*: a company in the supply chain interacts with other companies, e.g., by placing orders for products or services;
- *reactivity*: a company perceives its environment, i.e., the market and the other companies, and responds in a timely fashion to changes that occur in it. In particular, each firm modifies its behaviour to adapt to market and competition evolutions;
- *pro-activeness*: a company not only simply acts in response to its environment, it can also initiate new activities, e.g., launching new products on the market;

Agent technology and more specifically multiagent systems have been introduced for dealing with such characteristics. Moreover, multiagent systems offer a way to elaborate systems that are decentralized rather than centralized, emergent rather than planned, and concurrent rather than sequential; and this is what characterize supply chain systems in general.

The area intersecting supply chains management with multiagent systems has attracted in the last decade books, workshops, conferences, etc. Recently however, multiagent systems are became more formal borrowing from operations research, game theory and other facets from distributed decision making. We thought that this might be the right moment to edit a book that takes a close look at what has been done recently in the field of supply chain management using agent technology and more specifically multiagent systems. This book is intended as reference material for researchers, graduate students, and practitioners alike who are interested in pursuing research and development in this area, and who need for that, a comprehensive view of the existing literature, and some ideas for future directions.

The book contains sixteen chapters organized in four main parts as follows:

1. Introductionary Papers;
2. Multiagent based Supply Chain Modeling;
3. Collaboration and Coordination Between Agents in a Supply Chain;
4. Multiagent based Supply Chain Management: Applications.

2 Main Components of the Book

2.1 Introductionary Papers

Chapter 1 In this first chapter, Moyaux et al. introduce the topic of the book by first presenting supply chain management and then the multiagent systems. After that, they explain in detail how multiagent systems have been used in the context of supply chain management.

Chapter 2 In this chapter Ghenniwa et al. defend the idea that an electronic marketplace (*eMarketplace*) is a promising architectural model to develop collaborative supply chain management and integration platform. In their architecture, the *eMarketplace* exists as a collection of economically motivated software agents of service-oriented cooperative distributed systems. The authors discuss how coordination approaches as auctions and and multi-issues negotiation can be developed in the context of this *eMarketplace*. The objective here is to enable business entities to obtain efficient resource allocation while preserving long-term relationships.

Chapter 3 Müller and his colleagues investigate in this chapter the agent-based electronic business. More precisely, they argue how e-Business can benefit from methods, components and solutions based on software agent technology. To this end, they give an overview of the current e-business mainstreams and outline a number of important challenges e-Business has to tackle. Using three case studies, the authors investigate the required extent and the means to apply agent technology successfully in e-business applications in order to resolve some of these challenges.

2.2 Multiagent based Supply Chain Modeling

Chapter 4 In this chapter, Nagurney et al. develop a global supply chain network model in which both physical and electronic transactions are allowed and in which supply-side risk as well as demand-side risk are included in the formulation. The model consists of three tiers of decision-makers (i.e., agents): the manufacturers, the distributors, and the retailers who may be located in the same or in different countries and may conduct their transactions. The authors derive the governing equilibrium conditions sustaining the global supply chain network and provide qualitative properties of such equilibrium in terms of existence and uniqueness results and also establish conditions under which the proposed computational procedure is guaranteed to converge.

Chapter 5 Recently, researchers have introduced the Trading Agents Competition Supply Chain Management (TAC SCM) to capture many of the challenges involved in supporting dynamic supply chain practices. The game has been designed jointly by a team of researchers from the e-Supply Chain Management Lab at Carnegie Mellon University and the Swedish Institute of Computer Science (SICS). TAC SCM provides a challenging scenario for existing algorithms, due to the high dimensionality and the non-determinism of the environment, as well as the combinatorial nature of the problem. In this chapter, Doina et al. present RedAgent, the winner of the first TAC SCM competition (2003); an approach which is in fact based on multiagent design and in which many simple heuristics agents manage tasks such as fulfilling customer orders or procuring particular resources. The key idea here is to use internal markets as the main decision mechanism, in order to determine what products to focus on and how to allocate the existing resources.

Chapter 6 Lee and Chang propose in this chapter a framework of automatically formulating and modifying optimization models embedded in multiple software agents integrated along a supply chain. To implement the idea, the authors design an architecture of organization agents which are equipped with rule-based using backward and forward chain reasoning. The proposed framework is validated by applying a prototype to a cooperative delivery scheduling problem.

Chapter 7 In this chapter, Deshpande and his collegues propose a supply chain modeled as a multiagent system where the agents are located at every organization in the chain. They then put the emphasis on the real-time operations where it is desirable to re-optimize the schedules each time a new order is submitted. To this end, they exhibit a real-time scheduler that can manage new orders with soft real-time deadlines. Finally they explain in detail how their scheduler which uses fuzzy set approach has been implemented using a simulated environment for a supply chain.

Chapter 8 Here the author Kaihara focuses on the negotiation process in the supply chain formation under virtual entreprise environment to clarify the effective supply chain management. In this context, he defines each entreprise in virtual environment as an agent with multi-utilities and he propose a framework of multiagent programming with marketing science and N-person game theoretic approach as a negotiation algorithm amongst the agents. Finally, the author explains how he developed a computer simulation model to form supply chain management through multiple negotiations amongst several potential members in the negotiation domain, and clarifies the formation dynamism with the negotiation process.

2.3 Collaboration and Coordination Between Agents in a Supply Chain

Chapter 9 Smirnov et al. propose here an agent-based technological framework for dynamic configuration of a cooperative supply chain (CSC) as a flexible and agile system. In such CSC, agents negotiate and compromise on the optimal configuration in a sprit of cooperation, in order to maximize individual benefits and overall outcome of the system. The proposed approach considers configuring as a sequence of interrelated tasks, from which, partner selection among available enterprises and resource allocation are described in details. These tasks are solved within the framework applying different techniques of computational intelligence: (i) the theory of games with fuzzy coalitions; (ii) genetic algorithms, and (iii) constraint satisfaction problem solving.

Chapter 10 Moyaux and his colleagues describe here a simulation-based methodology which replaces the game theory usually used to analyze interactions in a formal way. They use such methodology to study the more-or-less selfishness of agents affecting their behaviour. The authors illustrate their methodology with the case study of a wood supply chain, in which each every company is seen as an agent which may use ordering strategy designed to reduce the fluctuations between the different participants of a supply chain (this phenomenon is called "bullwhip effect"). In this context, they assumed that each every agent utility can be split in two parts, a first part representing the direct utility of each agent (in practice, its inventory holding cost) and a second part representing agent social consciousness, i.e., its impact on the rest of the multiagent system (in practice its backorder cost).

Chapter 11 In this chapter Collins and Gini use auctions with temporal and precedence constraints for dealing with negotiation between agents. More precisely, they consider the problem of rational, self-interested, economic agents who must negotiate with each other in order to carry out their plans. Customer agents express their plans in the form of task networks with temporal and precedence constraints. The market runs a combinatorial reverse auction, in which supplier agents submit bids specifying prices for combinations of tasks, along with time windows and duration data that the customer may use to compose a work schedule. Two specific issues related to this problem have been studied by the authors. The first issue is the need for a market infrastructure to support decision processes and for which the authors have proposed a set of requirements for a market that can support this type of negotiation. The second issue is the determination of auction winners and for which the authors have explored and characterize a winner determination method.

Chapter 12 Babaioff and Walsh address the problem of coordinating the buying and selling of goods in multiple markets across a supply chain–as a mechanism design problem. The basic assumption which sustains that is that each agent has private information about its cost of providing a service, or about its value for purchasing a set of goods it desires. Because effective negotiation strategies in the presence of such private information can be difficult to design, they focus on incentive compatible mechanisms where the agent's dominant strategy is simply to report their private information truthfully. Unfortunately, with two-sided negotiation as it the case in supply chains, it is impossible to simultaneously achieve three desirable properties associated to incentive compatibility: perfect efficiency, budget instance, and individual rationality. In this chapter, Babaioff and Walsh address this impossibility and describe auctions that are incentive compatible, individually rational, budget balanced and where the efficiency is not perfect but just guarantee to be high.

Chapter 13 Fink addresses here the coordination problems between firms which collaborate within supply chains. To do that, he considers coordination problems at the level of operation management and assumes that potential decisions of the involved decision making units can be completely specified as formal contracts where the its elements are common knowledge. Fink explains in detail how he investigated an automated negotiation mechanism, with firms or decision making units represented as software agents. For taking into account asymmetric information and opportunistic behaviour, the author proposes a mediator agent supporting the negotiation process by randomly generating new candidate contracts. To this end, he introduces a verifiable mechanism which forces the involved agents to behave cooperatively to some extent. This mechanism and more generally the whole approach has been implemented and the results show that automated negotiations may provide an effective and efficient mechanism for coordinating decisions within supply chains.

2.4 Multiagent based Supply Chain Management: Applications

Chapter 14 Fischer and Gehring propose in this chapter a multi-agent based approach for supporting the planning of transshipment of imported vehicles via a seaport automobile terminal. First, the authors analyze the logistics supply chain for vehicle transport by focusing on the corresponding business process. On this basis, they subdivided the planning problem in three tasks and assign those planning tasks to different agent types, which are coordinated by a central coordinator agent. They then explain how their multiagent system has been subject to a test using practically relevant problem instances. Their derived test results show a robust behavior of such multiagent system with regard to varying problem data.

Chapter 15 Jeng and al. present in this chapter an agent-based framework for sensing, interpreting, predicting, automating and responding to business activities in the context of building manufacturing supply chain systems. By doing so the authors' goal is to minimize the latency (ideally, the should be almost zero) and to make well-informed and prompt business decisions. To achieve such goal, they developed a framework coined as CSR–commitment oriented sens-and respond framework which enables analysis across corporate business processes and then notifies the business of actionable recommendations.

Chapter 16 Srinivasan et al. describe, in this chapter, an agent-based model of a refinery's supply chain. The software agents emulate here the entities as procurement, sales, operations, storage and logistics departments of the refinery as well as the suppliers, logistics service providers and oil-exchanges. These agents model the embedded business policies and thus mimic the different business process of the entreprise. Uncertainty are captured by stochastic elements embedded in the agents; and the dynamics of the supply chain is emulated by discrete event simulation of the agent-based model. The authors have evaluated different business processes and supply chain configuration in regard to their effect on entity-specific as well as supply chain wide key performance indicators.

Contents

Supply Chain Management and Multiagent Systems: An Overview

Thierry Moyaux [1], Brahim Chaib-draa [1], and Sophie D'Amours [2]

[1] Universit Laval, Dpt. d'Informatique et de Gnie Logiciel, DAMAS & FOR@C
Ville de Qubec G1K 7P4 (Qubec, Canada), {moyaux, chaib }@iad.ift.ulaval.ca ,
[2] Universit Laval, Dpt. de Gnie Mcanique, FOR@C & CENTOR
Ville de Qubec G1K 7P4 (Qubec, Canada), sophie.damours@gmc.ulaval.ca ,

Abstract. This chapter introduces the topic of this book by presenting the fields of supply chain management, multiagent systems, and the merger of these two fields into multiagent-based supply chain management. More precisely, the problems encountered in supply chains and the techniques to address these problems are first presented. Multiagent systems are next broadly presented, before focusing on how agents can contribute to solving problems in supply chains.

1 Introduction

This chapter presents how multiagent systems are used to manage supply chains. To this end, supply chain management is first introduced as a business practice for solving some industrial problems by providing the concept of inter-company collaboration. Such a collaboration allows planning and synchronizing operations within a supply chain in order to solve the considered industrial problems, that is, within a network of firms producing and distributing products or services to end-customers. Instrial issues and some collaboration techniques are presented in Section 2.

The second area of interest in this book, namely, multiagent systems, is next introduced. The concept of "agent" is first defined, and next compared with another concept from Computer Science, the concept of "object". After that, the general agent architectures outline the different levels of agent sophistication. Then, we motivate the use of multiagent systems, and we compare these systems with some other scientific approaches. Finally, we illustrate this section with some examples involving multiagent systems in different fields or applications. This presentation of multiagent systems is developed in Section 3.

A synthesis of supply chain management and multiagent systems extends the previous illustrations of agents applied in different fields. For that purpose, agents are first introduced as a new information technology for supply chain management. The arguments pro agents outlined in Section 3 are next extended for the special case of agents in supply chains. Some projects applying agents to supply chains eventually illustrate this section. This synthesis of supply chain management and multiagent systems is detailed in Section 4.

T. Moyaux et al.: Supply Chain Management and Multiagent Systems: An Overview, Studies in Computational Intelligence (SCI) 28, 1–27 (2006)
www.springerlink.com

2 Supply Chain Management

We now introduce the concept of supply chain as a solution to some industrial issues. Specifically, industrial issues are first outlined next ,one of them called the bullwhip effect is detailed. Then, the concept of supply chain management is presented as a solution to these problems, as well as the collaboration implied by this concept. Finally, information technologies are presented as a tool supporting collaboration.

2.1 Industrial Problems in General

First, companies face a huge number of problems, such as how to make decisions concerning production planning, inventory management and vehicle routing. These three decisions are managed separately in most organizations because making each individual decision is very difficult, since many constraints have to be satisfied (production, shipping and inventory capacities, precedence order of activities, legal obligations, etc.) [1]. For instance, the multistage, multicommodity inventory management problem and the vehicle routing problem are both known to be NP-hard problems ([2] cited by [1]), i.e., very difficult.

Secondly, the problem is yet harder in reality because the decisions concerning production planning, inventory management and vehicle routing are *interdependent*. Hence, these three decisions should be taken together, which makes the planning problem harder.

Third, companies are not isolated, but impact on and are impacted by their partners. As a result, when a company maximizes its profits, it may disturb other companies, which may result in globally underoptimal decisions, because organizations may have different conflicting objectives [3, pp. 3]. The best solution would be to make the decisions together concerning production planning, inventory management and vehicle routing for several companies. As this planning problem is hard for a single company, synchronizing all companies decisions together is very hard.

The concept of supply chains was proposed to address this problem of minimization of total supply chain cost, while meeting fixed and given demand by points-of-sale, e.g. by retailers [4, pp. 8]. Before presenting supply chains, we focus on one particular example of industrial issues: the bullwhip effect.

2.2 A Particular Example of Industrial Problems: The Bullwhip Effect

We now present a phenomenon occurring in supply chains called the bullwhip effect. This effect consists in an amplification of the order variability. This variability is a problem because it makes demand (i.e., orders) more unpredictable. Figure 1 shows how this effect propagates in a simple supply chain with only three companies: a retailer, a wholesaler and a paper mill. In this figure, the retailer exclusively sells to the customer and buys from the wholesaler, the wholesaler sells to the retailer and buys from the paper mill, and the paper mill sells to

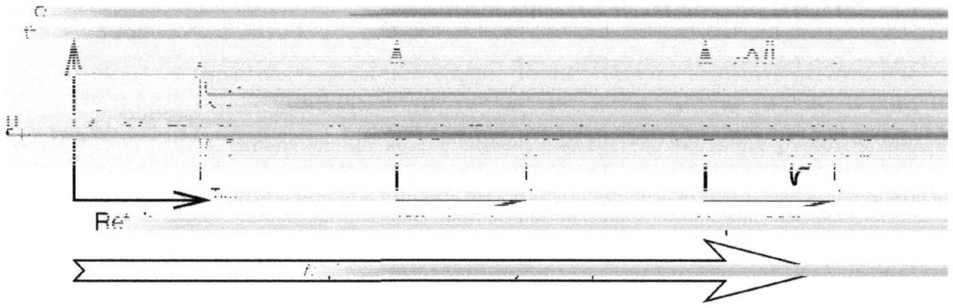

Fig. 1. The bullwhip effect [5, 6]

the retailer and buys from an unknown supplier. The ordering patterns of the three companies are similar in the way that the variabilities of an upstream site are always greater than those of the downstream site [5]. As a variability, the bullwhip effect is measured by the standard deviation σ of orders. Note that the means □ of orders are all equal in our example given in Figure 1.

There are several consequences of the bullwhip effect. It has been estimated that these consequences altogether would increase the costs by 200-300 MFIM (33-50 million euros) annually for a 300 kton paper mill [7] in Northern Europe. The consequences causing such costs are as follows:

1. Higher inventory levels : Every participant in the supply chain has to stockpile because of a high degree of demand uncertainties and variabilities induced by the bullwhip effect [6];
2. Supply chain agility [3] reduction : As inventory levels are high (cf. previous consequence "higher inventory levels"), the supply chain should sell products in inventory, before it sells the new products demanded by end-customers, which generates inertia in following end-customer demand. Moreover, demand uncertainties induced by the bullwhip effect make it more difficult for the supply chain to understand which product is demanded by end-customers;
3. Decrease of customer service levels : Demand variabilities may incur stock-outs, in which case, no products are available to be sold, and thus, no service can be given to customers;

The last two consequences of the bullwhip effect are related to the difficulties of planning under uncertainties, and consequently lead to:

4. *Ineffective transportation*: Transportation planification is made more difficult by demand uncertainties induced by the bullwhip effect;
5. Missed production schedules : Similarly to transportation, production planification is made more difficult by demand uncertainties induced by the bullwhip effect.

[3] The Iaccoca Institute [8] (cited by [9]) defines agility as the "ability of an organisation to thrive in a constantly changing, unpredictable business environment".

Several causes have been proposed to explain the appearance of the bullwhip effect, such as demand signal processing which uses forecasting methods not perfectly accurate, gaming among companies when demand exceeds supply, order batching which discretizes orders, and price variations which incite clients to over-order when price is low [6]. However, such a sharing of information is only possible when companies collaborate, because demand information is very important and should normally be kept secret Sharing demand information is often said to be the solution to the bullwhip effect (or at least a part of the solution) [3].

Notice that an interesting feature of the bullwhip effect is that it creates a link between the companies, which makes the concept of supply chain interesting. To see this link, consider the retailer in Figure 1 which does not suffer directly from the bullwhip effect, because it receives a quite constant demand from end-customers. As a consequence, the retailer has no direct incentives to reduce this phenomenon, while this would benefit the rest of the supply chain, and thus, the retailer itself indirectly because products may be cheaper, more available, etc.

We shall detail a thorough description of the bullwhip effect, a solution to it, and the incentives companies may have to use such a solution in Chapter 10. But at the moment, we present the concept of supply chain as a solution proposed for the industrial problems outlined in Subsections 2.1 and 2.2.

2.3 The Concept of Supply Chains as a Solution

We have not found the first definition of the term "supply chain", but we have found, for example, that Burns and Sivazlian [10] referred to it in the late 1970's. According to Muckstadt and his colleagues [11], there are many definitions and interpretations of the term "supply chain management". These authors defined a supply chain as "the set of firms acting to design, engineer, market, manufacture, and distribute products and services to end-consumers". In general, this set of firms is structured as a network, as illustrated in Figure 2 [12, 13] in which we can see a supply chain with five levels (raw material suppliers, tier suppliers, manufacturers, distribution centers and retailers). In the same context, Shapiro [4] noted that "supply chain management is a relatively new term that crystallizes concepts about integrated business planning that have been espoused by logistics experts, strategists, and operations research practitioners as far back as the 1950s". Similarly, Simchi-levi and his colleagues [3] defined this term as "a set of approaches utilized to efficiently integrate suppliers, manufacturers, warehouses, and stores, so that merchandise is produced and distributed in the right quantities, to the right locations, and at the right time, in order to minimize systemwide costs, while satisfying service level requirements". Poirier and Reiter [14] noted that the concept of supply chains improves the competitive position of collaborating companies, because it supports the creation of *synergies* among these companies. In particular, such synergies are due to the fact that a supply chain is a system, and as a consequence, this system is superior to the sum of the constituting companies. As previously explained, the concept of inter-company *collaboration* is a way to create such synergies in a supply chain.

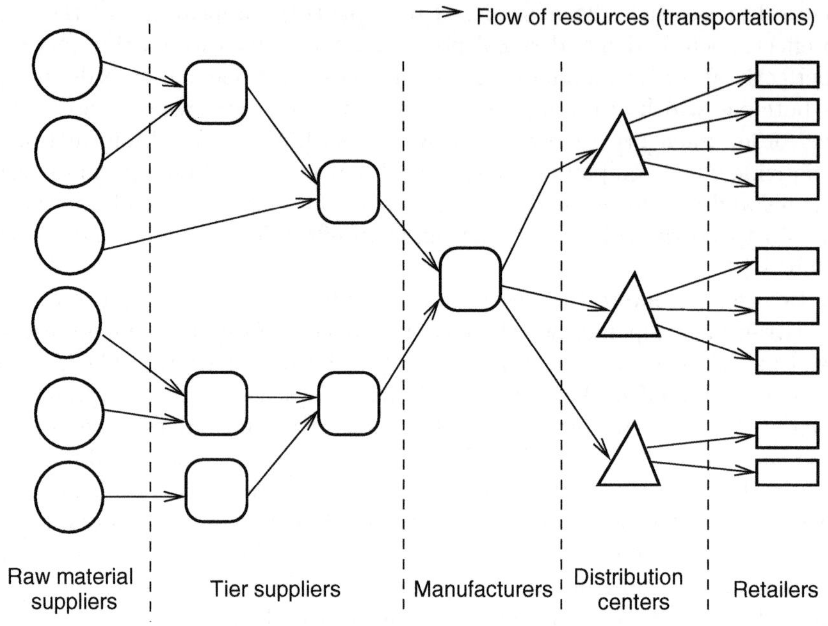

Fig. 2. An example of supply chain [12, 13]

2.4 Collaboration in Supply Chains

Muckstadt and his colleagues [11] noticed that the term *collaboration* is confusing because it has taken on several interpretations when used in the context of supply chain management. For example, various levels of collaboration techniques based on *information sharing* were set up in real supply chains. It is important to note that we refer in this chapter to collaboration as information sharing, even if collaboration is in general wider than only information sharing. We represent in Figure 3 how some of these information sharing techniques overlap. These techniques are essentially information centralization, Vendor Managed Inventory/Continuous Replenishement Program, and Collaborative Planning Forecasting and Replenishment. They are now reviewed in detail:

- *Information centralization:* This is the most basic technique of information sharing in which retailers broadcast the market consumption (approximated as their sales) to the rest of the supply chain. As we also refer to information centralization, it is necessary to distinguish information sharing from information centralization: the latter is a particular case of the former, because information centralization is the multi-casting in *real-time* and *instantaneously* of the market consumption information, while information sharing is only the sharing of the demand information between any companies. Moreover, several kinds of information may be shared, such as their available production capacity, their inventory level... and from this viewpoint, information sharing includes information centralization.

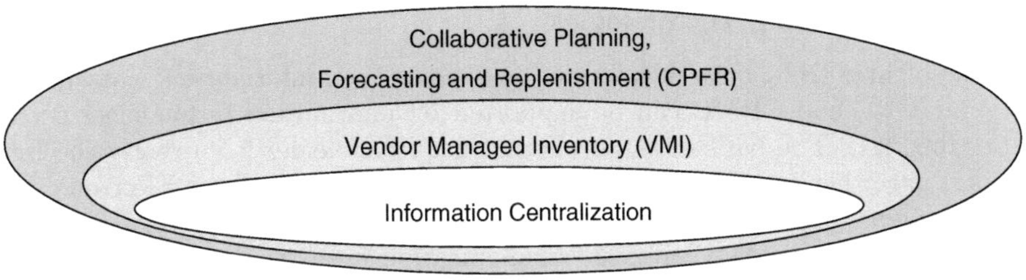

Fig. 3. Overlap of some collaboration techniques through information sharing

In the context of the bullwhip effect, Chen and his colleagues [15] formally showed, for two forecasting methods, that information centralization reduces this effect. In fact, information centralization reduces it, because each level of the chain can base its forecasts on the actual market consumption, instead of basing them on incoming orders, which can be much more variable than the actual market consumption [5, 6].

- *Vendor Managed Inventory (VMI)* [16] and *Continuous Replenishment Program (CRP):* These two collaboration techniques are very similar, but are used in different industries. The idea is that retailers do not need to place orders because wholesalers use information centralization to decide when to replenish them. Although these techniques could be extended to a whole supply chain, current implementations only work between two business partners. In fact, many customers are attracted to these techniques, because they mitigate uncertainty of demand, a consequence of the bullwhip effect. Moreover, the frequency of replenishment is usually increased from monthly to weekly (or even daily), which benefits both partners. These techniques were popularized in the late 1980's by Wal-Mart [17] and Procter & Gamble [18]. In particular, VMI has became one of the key elements of the quick response program in the grocery industry [19].
- *Collaborative Planning, Forecasting and Replenishment (CPFR):* This technique developed by the VICS Association [20] (Voluntary Interindustry Commerce Standards) is a standard that enhances VMI and CRP by incorporating joint forecasting. Like VMI and CRP, current implementations of CPFR only include two levels of a supply chain, i.e., retailers and their wholesalers. With CPFR, companies electronically exchange a series of written comments and supporting data, which include past sales trends, scheduled promotions, and forecasts. Conversely to the previous two techniques, thus CPFR shares more information than only demand information. This allows the participants to coordinate joint forecasts by focussing on differences in forecasts. Companies try to find the cause of such differences and agree on joint, improved forecasts. They also jointly define plans to follow when specific contingencies occur [3, pp.239].

2.5 Supporting Technologies

These three techniques of information sharing, i.e., information centralization, VMI/CRP and CPFR, can be supported by information technologies such as e-Hubs [21]. The basis of these information technologies is currently the Internet, but other technologies are also used, e.g. the protocol for Electronic Data Interchange (EDI). The first advantage of the Internet over every other technology is to provide a low-cost communication infrastructure available almost anywhere in the world. This first advantage allows companies to increase information streams, and more precisely in our context, to share more information. The second advantage of the Internet is to provide some standardized file formats (HTML, OWL, etc.), which reduces the cost of information technologies.

We shall develop the goals of information technologies and the way to achieve these goals during the presentation of the application of multiagent systems in supply chain management in Section 4.

3 Multi-Agent Systems

We now focus on the second field addressed in this book, i.e., multiagent systems. We first define the concept of agents, next we compare this concept with another concept from software engineering, i.e., the concept of object. Then, we outline some architectures of agents and some arguments in favour of the use of agents in general. Finally, we illustrate projects involving multiagent systems in different areas.

3.1 The Concept of Agents

Intelligent agents are a new paradigm of software system development. They are used in a broad and increasing variety of applications [22–24]. For a long time, there was no single definition of an agent and a multiagent system: several definitions cohabited in the past [25]. Nowadays, it seems that researchers agree on the following definition proposed by Wooldridge and Jennings [26]:

> The term "agent" denotes a hardware or (more usually) software-based computer system, that has the following characteristics:
>
> **Autonomy:** agents operate without the direct intervention of humans or others, and has some kind of control over its actions and internal state;
>
> **Social ability:** agents interact with other agents (and possibly humans) via some kind of agent-communication language;
>
> **Reactivity:** agents perceive their environment, (which may be the physical world, a user, a collection of other agents, the Internet, or perhaps all of these combined), and respond in a timely fashion to changes that occur in it;
>
> **Pro-activeness:** agents do not simply act in response to their environment, they are able to exhibit goal-directed behaviour by taking the initiative.

3.2 Comparison with Objects

Based on this concept of agent, Shoham [27] proposed a new programming paradigm called Agent-Oriented Programming (AOP) to replace the current Object-Oriented Programming (OOP). The difference between agents and objects is sometimes missed by programmers familiar with object-oriented languages, such as C++ [28] or Java [29]. The main difference between these two concepts is the *autonomy* of agents. In fact, while objects encapsulate some state on which their methods can perform actions, and in particular the action of invoking another object's method, an object has control over its behaviour. That is, if an object is asked to perform an action, it always does so, while an agent may refuse. Concerning this point, Wooldridge [30] recalls the slogan "Objects do it for free; agents do it because they want to".

Of course, some sophisticated objects may be very similar to agents. In fact, Wooldridge [31] noted that there are clear similarities, but obvious differences also exist. Let us consider the case of objects in Java that can easily be transformed into threads exhibiting some behaviour. Such active objects have some autonomy like agents, but their behaviour is only procedural in reaction to message requests. On the other hand, autonomy of agents makes them perform activities without external intervention [32]. In short, object-based concurrent programming has some relationships with distributed artificial intelligence [33].

But objects and agents also present differences. In particular, object state is much simpler than agent state. In fact, an object state is only a data structure, i.e., an aggregation of variables of different types (integers, booleans, character strings...) in a common structure, while an agent state consists of components such as beliefs, decisions, capabilities and obligations. As an agent state is more sophisticated, it is also referred to as a mental state [27].

Finally, it is important to note that agents have been programmed in C++ or Java, i.e., with an OOP language, but AOP languages have appeared. For example, JACK™ designed by the Agent Oriented Software Group [34] (Melbourne, Australia) is an AOP language. This language implements concepts from AOP upon an OOP language. That is, JACK™ provides an AOP compiler transforming JACK™ code into Java code. Next, the JACK™ compiler calls the Java compiler to transform the generated Java code into a runnable Java bytecode that works on any Java Virtual Machine.

3.3 Agent Architectures

In the same manner that there are several languages to implement agents, there are also different levels of complexity of this implementation. Such complexity depends on the task that agents have to carry out and on the environment surrounding them. Russell and Norvig [35] propose the following classification of agent architectures:

1. *Simple reflex agents*: This type of agent is the simplest, because percepts are directly related to actions via some condition-actions rules. What has occurred in the past is ignored, because these agents have no memory.

2. *Model-based reflex agents*: As agents cannot perceive their whole environment, model-based reflex agents keep track of the part of their environment they cannot currently observe. To achieve this, they have an internal representation of their environment, called a "model of the world", to guess the evolution of the environment and the impact of the agent's actions on this environment. Like simple reflex agents, model-based reflex agents select their action according to condition-action rules, but now, the condition only depends on the model of the world, and not on the current perception of the environment.
3. *Model-based, goal-based agents*: This type of agent has goal information describing desirable situations, because the current state of the model of the world is not always enough to select an action efficiently. That is, the model of the world is used to elaborate some predictions on how the world would be if the agent executes an action and what is the price to pay for that. The action to carry out is chosen so that the goals will be satisfied according to such predictions.
4. *Utility-based agents*: Goals just differentiate wishable states from non-wishable states, without further details, such as, the speed, the price or the safety to reach a wishable state. As a result, in order to improve the quality of agent behaviour, agents can be given a utility function mapping its state (or a sequence of states) in the model of the world, onto a real number describing the associated degree of agent's happiness. In comparison with goal-based agents, utility-based agents do not decide which action to do in order to achieve a goal, but which action to do to increase utility. This difference implies that both types of agents find which actions to do to achieve their goals, but utility-based agents find the best actions according to some given metrics. This agent architecture is hence much nearer to the definition of Economics agents (that only maximize their utility) than the previous three architectures.
5. *Learning agents*: Turing [36] has noted the huge amount of work it takes to program an intelligent machine, and has concluded that it would be easier to build learning machines and then to teach them. Another advantage of learning agents is their adaptability to unknown environments, and the improvement of their behaviour with time. Learning agents use the feedback from a critic to learn which perceptions of the environment are desirable, and in consequence, how to behave. Precisely, agents' learning consists in improving their future performance based on their past feedback from the critic, by optimizing their behaviour such as to maximize their utility when the world continues evolving as it has been. This kind of learning makes agents discover that some kind of (but not exactly) condition-action rules always do the same thing, based on their current knowledge.
   A problem arises here: after some learning time, agents are always going to do the same things because of these discovered rules, though the agents are not sure that these actions are optimal, while they might have a better performance if they had a wider knowledge of their environment. In fact, they should try to do very different actions than those prescribed by their learning

process. Some *exploration* of new actions should be carried out instead of only *exploiting* the learned knowledge.

3.4 Motivations for Multi-Agent Systems

Huhns and Stephens [37] noted that multiagent systems are generally less efficient than centralized solutions, because the distribution restrains optimization. But these authors also gave several advantages of multiagent systems. First, multiagent systems are easier to understand and implement, when the problem itself is distributed. This allows the multiagent system to give more flexibility when taking into account the modularity of the real, modelled system. Next, the distribution may force programmers to propose new algorithms to solve problems. In particular, the concurrency can be used to accelerate problem solving. Finally, a centralized solution may be impossible, because systems and data are in independent organizations. We develop this latter argument in Section 4, because it is the main one in favour of multiagent systems in supply chains.

Jennings [38] pointed out the flexible, high-level interactions of agents, that make the engineering of complex systems easier. This author recalls that *complex* systems are always distributed, and from his point of view, agent decomposition is very important to manage complexity. It follows from this, that designers need a means to reduce the complexity of the system control, in order to enhance their ability to model, design and build complex, distributed systems. Multiagent systems provide designers with this means through the decentralisation of control. In particular, the system complexity makes it very difficult to know every possible interaction in the system, because the system only has partial control and observability over its environment, and thus, this environment is highly unpredictable. Multiagent decentralisation takes this into account by letting each agent continuously coordinate its actions with other agents, instead of making this agent apply a behaviour prescribed at design-time. In short, some advantages of multiagent systems is the fact that modelling with agents:

- partitions the problem space of a complex system efficiently;
- is a natural way to modularise complex systems;
- focusses on the organizational relationships in complex systems.

Similarly, Wooldridge [30] says that *interaction* is now seen by most programmers as an important characteristic of complex softwares. For this reason, interactions, and thus multiagent systems, take a growing part in software engineering. Moreover, multiagent systems are an interdisciplinary field. For example, interactions in multiagent systems are also interesting to model dynamics in human societies.

We should note that there are also objections to multiagent systems. We now review such objections.

3.5 Differences between Multi-Agent Systems and Other Fields

In general, objections to multiagent systems are due to their similarity with other fields. To respond to these objections, Wooldridge [30] points out the difference between this field and some others:

Distributed/concurrent system

- *Similarity*: By definition, multiagent systems are a special case of distributed/concurrent systems. Therefore, experience in this field has to be kept by the multiagent system community, in particular to avoid discovering again how to manage mutual exclusion over shared resources, how to avoid dead- and livelocks...
- *Differences*: First, agents are autonomous, and therefore, synchronization and coordination are not structured at design-time, as they are in distributed/concurrent systems. In fact, agent synchronization and coordination is achieved at run-time. Secondly, agents are in general self-interested, while components in a distributed/concurrent system have the common goal of maximizing the overall system efficiency. For these two reasons, negotiation is important in multiagent systems, while it is unknown in distributed/concurrent systems.

Artificial intelligence

- *Similarity*: Historically, multiagent systems were born from Distributed Artificial Intelligence, which is a subfield of Artificial Intelligence [39].
- *Differences*: First, the main topic of artificial intelligence has been the study of components of intelligence (learning, planning, understanding images...), while the goal of research about agents is the integration of these elements. Therefore, during agent implementation, much more time is spent with computer science and software engineering, than with artificial intelligence. Secondly, social ability in systems has been ignored by artificial intelligence, while this is as important in an intelligent behaviour as learning or planning.

Economics/Game Theory

- *Similarity*: Like multiagent systems, Economics and Game Theory also deal with self-interested agents, and more precisely with their interactions. Some well-known reseachers have contributed to both computer science and economics/game theory, such as von Neumann and Turing. However, these two fields have been dissociated since these beginnings. Now, the situation is changing because game theory has more and more applications in multiagent systems, and economists are interested in multiagent simulations to understand inter-agent interactions.
- *Differences*: First, concepts in economics/game theory are descriptive, and thus, indicate nothing about how to compute a solution. Such computing is often very hard [40]. Secondly, game theory is built on the notion of rationality, but some debates are concerned to the question of its validity and/or utility for artificial agent societies. Thirdly, Boutilier [41] proposes another difference that is also related with rationality. This difference is about the assumption in economics/game theory that agents

are rational (the research questions concern the social consequences of this hypothesis), while programming this rationality is the problem itself in multiagent systems.

Social Science

- *Similarity*: Social sciences study the dynamics of human societies, while multiagent systems are concerned with artificial societies.
- *Difference*: It is not certain that the best way of building artificial societies is to base them on human societies. Moreover, other tools, such as the aformentioned game theory, also model human societies, and may thus be applied.

Because of similarities between multiagent systems and other fields (distributed/concurrent systems, artificial intelligence, economics/game theory and social science), agents have been applied in some of these fields. Furthermore, they have also been applied in many real-world applications, that are, in general, functionnally or geographically distributed. We now present some of these applications of multiagent systems.

3.6 Some Applications of Multi-Agent Systems

Multiagent systems have been used in many fields, as presented by Chaib-draa [24], Wooldridge [30] and Jennings [42], and Weiss [43]. As an illustration, we now outline some of these applications. Jennings, Sycara and Wooldridge [39] classify these applications in four classes:

Industrial applications: Industry was one of the earliest users of agent technology, especially in the following areas:

- *Manufacturing:* For example, the *Holonic Manufacturing Systems* (HMS) project [44–46] aims at standardizing architecture and technologies for open, distributed, intelligent, autonomous and cooperating systems in industry. Each component of these systems is controlled by agents, called "holons" for the combination of "holos" (the whole) and "on" (a particle) [46]. Each holon's goal is to work with the other holons, in order to control a production system in an efficient, scalable, open way. Applications of holons are, for instance, concurrent engineering, collaborative engineering design, and manufacturing enterprise integration [47].
- *Process control:* Process control is at a lower level than manufacturing, because manufacturing aims at controlling several workstations, while process control focusses on a single workstation. In fact, the complexity of a workstation may require the decomposition of its control into agents.
- *Telecommunications:* Telecommunication networks are geographically spread over a large area. Using agents to manage such networks is thus a natural metaphor. For instance, British Telecom [48] has developed the ZEUS Agent Building Toolkit for this purpose.

- *Air-traffic control* [42, 49]: OASIS is an air-traffic control system used at Sydney airport in Australia. Aircraft and the various air-traffic control systems are seen as agents. Agents are created when they approach Sydney airport. Their behaviour is both goal-directed ("I want to land"), and reactive to take real-time constraints into account. Some similar air-traffic control systems were designed for NASA [50], or by Cammarata and her colleagues [51].
- *Transportation systems:* Like telecommunication networks, the geographical distribution of transportation lead to the fact that agents are a natural metaphor. For example, *Automated Highway Systems* [52] unites several projects aiming at fully automatizing vehicle driving. Several goals are addressed, such as driving a vehicle without human intervention and collaborative driving. This second example consists of forming platoons of vehicles on roads, in order to improve the fluidity of traffic. Each vehicle is seen as an agent that tries to form a team with other vehicle-agents sharing the same part of trip.

Commercial applications: While agents for industry are quite often designed for a single, specific application depending on the company, commercial agents tend to be designed for a widespread diffusion. Among the areas of commercial agents, we can find:

- *Information management:* Since the users of Internet are more and more overloaded by information, agents can help them by filtering and gathering accurate information.
- *Electronic commerce:* Since Internet takes up a growing place in our everyday life, e-commerce promises to be more frequently used in the near future. In fact, agents can:
  - replace us to look for the products that best fit our needs;
  - bid for products on auctions sites, such as eBay [53], following a given strategy [54, 55];
  - try to form a coalition with agents buying a similar product, in order to have a price reduction due to the higher bought quantity [56–58].

  In particular, TAC (Trading Agent Competition [59]) aims at confronting agents to find the best buying strategy in situations close to real-life [60, 61]. TAC has several tracks, and one of them is about supply chain management.
- *Business process management:* Information systems are spread among the different departments in a company in order to bring information together. Using agents can make this information collection easier and more efficient. The collected information is useful for company managers when they make business decisions.

Entertainment applications: Although this industry is not seen as serious in computer science, it is currently growing. Specific areas of entertainment agents are:

- *Games:* For example, the concept of agents was applied in the game "Creatures" by Grand and Cliff [62] to build artificial pets living together in a simulated environment. These animals are built to resemble real-life animals, and in particular, their "brain" is a neural network.

- *Interactive theatre and cinema:* In these systems, users 'enter' the movie to play a role in this movie, and to interact with other characters played by artificial agents. Programming these agents so that they resemble real people is an issue, because they have to look like human beings, to behave like them.

Medical Applications: Agents are more and more used in medical applications, for instance:

- *Patient monitoring:* For instance, Hayes-Roth *et al.* [63]'s system Guardian is distributed to respect the fact that a team in a Surgical Intensive Care Unit is made up of people who have different expertise and who collaborate. Guardian has a hierarchical structure, in which a control agent controls perception/action agents and reasoning agents, in order to help manage patient care.
- *Rescue team management:* RoboCup Rescue [64, 65] is a competition involving the simulation of an earthquake similar to Kobe (Japan) in 1995. Agents model teams of firemen, policemen and ambulances, that have to be coordinated in order to minimize both the number of dead civilians and the number of destroyed buildings. The idea is that an earthquake scenario cannot be studied in real-life, and thus, has to be simulated in order to find which behaviour rescue teams should have.

Some other applications cannot be put in these four classes. For instance, *interface agents* assist users in software, like the paper clip in MS-Office [66], even though this is currently a mono-agent system. *Simulations of ecological and social* systems are another kind of application of multiagent systems. For example, Franchesquin and Espinasse [67] programmed a multiagent simulation, that takes into account both ecological and social dynamics, in order to study the hydraulic management of the Camargue (south of France).

Since we focus in this book on agents in industry, and more precisely, on supply chain management, we will describe additional multiagent systems in industry. It is worth noting here, that the HMS (holonic manufacturing) project, presented above, looks similar to multiagent systems for supply chain management, but it is indeed different. In fact, the HMS project addresses problems at a lower level, that is, intra-company, while supply chains are made up of several companies. As a consequence, assumptions about agents in supply chains (selfishness, available information, etc.) have to be a bit different than assumptions about holons. The other chapters in this book make such supply chain-related assumptions.

4 Multi-Agent Systems in Supply Chain Management

The first section of this chapter introduced supply chain management, and the second one multiagent systems. We now focus on the merging of these two fields into multiagent-based supply chain management. We first show how computers are currently used in supply chains, then we give some arguments justifying the use of multiagent systems in supply chains, and finally some examples illustrate this section.

4.1 Information Technologies in Supply Chain Management

According to Simchi-Levi and his colleagues [3], "information technologies is an important enabler of effective supply chain management. Much of the current interest in supply chain management is motivated by the possibilities that are introduced by the abundance of data and the savings inherent in sophisticated analysis of these data". It follows that information technologies in supply chains pursue three goals:

- *collecting* information on each product from production to delivery or purchase point, and providing complete visibility for all parties involved;
- *accessing* any data in the system from a single-point-of-contact, e.g. from a PDA linked to the company information system through a wireless link;
- *analyzing* data, planning activities, and making trade-offs based on information from the entire supply chain.

To achieve these activities, information technologies use certain means:

- information technology *infrastructure* (network, databases...);
- *e-commerce;*
- *supply chain components*, which are the various systems directly involved in supply chain planning, i.e., Decision Support Systems (DSS).

The *standards* gathering these three means are, for example, the protocol for Electronic Data exchange (EDI). Although regarded as a success because it is used by large corporations, EDI was never accepted by the majority of the communities of the business world as a means of trading electronically, because its is a barrier for small companies [68]. This explains why new Internet-based standards currently emerge. In particular, the eXtended Markup Language (XML) [69] is used in more and more applications on the Internet. But XML is too generic to enable collaboration in supply chains. Therefore, some XML-based standards were proposed, such as the Resource Description Framework (RDF) [70] to define a common vocabulary for describing resources, the Web Ontology Language (OWL) [71] to give semantics to Web pages, the Common Business Library (CBL) [72] for describing documents such as orders or catalogues, etc. Please refer to Singh and Huhns [73]'s book for an overview of these technologies and many others.

Concretely, information and decision technologies take the form of:

- *Enterprise Resource Planning* (ERP) is a class of software systems organizing and managing companies [74], e.g., PeopleSoft/Oracle [75], or SSA Global [76];
- *E-commerce*, and in particular marketplaces, such as Commerce One [77] and Ariba [78];
- *Advanced Planning and Scheduling* (APS) is a class of software for Decision Support System (DSS) in supply chains.

According to Shapiro's decomposition of information technologies [4], the first two applications (ERP and e-commerce) belong to "Transactional Information Technologies" because they are concerned with acquiring, processing and communicating raw data. On the other hand, APS and DSS belong to "Analytical Information Technologies" because they allow analyzing raw data in order to help managers, which is a task at a higher level. In practice, companies first install transactional tools, because analytical tools need them to be fed with raw data.

More and more, multiagent systems are seen as a new technology for improving or replacing technologies used in both transactional and analytical information technologies. We now explain why agent technology seems so promising in the context of supply chains.

4.2 Using Multi-Agent Systems in Supply Chain Management: Motivations

Some arguments in favor of using multiagent systems in supply chain management can be found in the literature. In fact, researchers have already applied agent technology in industry to concurrent engineering, collaborative engineering design, manufacturing enterprise integration, supply chain management, manufacturing planning, scheduling and control, material handling, and holonic manufacturing systems [47].

Concerning supply chains, Dodd and Kumara [79] think that Mark Fox (e.g. [80]) was probably the first to organize the supply chain as a network of intelligent agents. Indeed, supply chains are made up of heterogeneous production subsystems gathered in vast dynamic and virtual coalitions. Intelligent distributed systems, e.g. multiagent systems, enable increased autonomy of each member in the supply chain. Each partner (or production subsystem) pursues individual goals while satisfying both local and external constraints [81]. Therefore, one or several agents can be used to represent each partner in the supply chain (plant, workshop, etc.). Moreover, the agent paradigm is a natural metaphor for network organizations, since companies prefer maximizing their own profit than the profit of the supply chain [82]. In fact, the distributed manufacturing units have the same characteristics as agents [83] (based on Wooldridge [26]'s definition of agents, quoted previously):

- *autonomy*: a company carries out tasks by itself without external intervention and has some kind of control over its action and internal state;
- *social ability*: a company in the supply chain interacts with other companies, e.g. by placing orders for products or services;
- *reactivity*: a company perceives its environment, i.e., the market and the other companies, and responds in a timely fashion to changes that occur in it. In particular, each firm modifies its behaviour to adapt to market and competition evolutions;
- *pro-activeness*: a company not only simply acts in response to its environment, it can also initiate new activities, e.g. launching new products on the market;

Issue	Autonomous agents	Conventional systems
Model	Economics, biology	Military
•••••• ••••••• • ••••••••••••••••		
1 Theoretical optima?	No	Yes
2 Level of prediction	Aggregate	Individual
3 Computational stability	Low	High
•••••• ••••••• • •••••• ••• •••••		
4 Match to reality	High	Low
5 Requires central data?	No	Yes
6 Response to change	Robust	Fragile
7 System reconfigurability	Easy	Hard
8 Nature of software	Short, simple	Lengthy, complex
9 Time required to schedule	Real time	Slow

Table 1. Agent-based vs. conventional technologies [84].

Moreover, multiagent systems offer a way to elaborate production systems that are decentralized rather than centralized, emergent rather than planned, and concurrent rather than sequential. Therefore, they allow relaxing the constraints of centralized, planned, sequential control [84]. Unfortunately, an agent-based approach is not a panacea for industrial softwares. Like other technologies, this approach has advantages and disadvantages: it must be used for problems whose characteristics require its capacities. According to Parunak [45], five characteristics are particularly salient. In fact, agents are best suited for applications that are *modular, decentralized, changeable, ill-structured* and *complex.*

To judge relevance for supply chains of autonomous agents, Parunak [84] compares this approch with conventional technologies in Table 1, thus highlighting differences between these two philosophies. To this end, multiagent systems are identified as biological (ecosystems) and economical (markets) models, whereas traditional approaches are compared with military patterns of hierarchical organization. Table 1 summarizes the main disadvantages of multiagent systems:

1. theoretical optima cannot be guaranteed, because there is no global view of the system;
2. predictions for autonomous agents can usually be made only at the aggregate level;
3. in principle, systems of autonomous agents can become computationally unstable, since, according to System Dynamics, any system is potentially unstable.

But on the other hand, the autonomous, agent-based approach has some advantages too:

4. because each agent is close to the point of contact with the real world, the systems's computational state tracks the state of the world very closely...

5. ... and without need for a centralized database;
6. because overall system behaviour emerges from local decisions, the system readjusts itself automatically to environmental noise ...
7. ... or to the removal or addition of agents;
8. the software for each agent is much shorter and simpler than would be required for a centralized approach, and as a result is easier to write, debug and maintain.
9. because the system schedules itself as it runs, there is no separate scheduling phase of operation, and thus no need to wait for the scheduler to complete. Moreover, the optima computed by conventional systems may not be realizable in practice, and the more detailed predictions permitted by conventional approaches are often invalidated by the real world.

All these reasons show the relevance to use agents in supply chain management. In other words, thanks to their adaptability, their autonomy and their social ability, agent-based systems are a viable technology for the implementation of communication and decision-making in real-time. Each agent would represent a part of the decision-making process, hence creating a tight network of decision makers, who react in real-time to customer requirements, in opposition to the flood of current processes, which is decided before customers place an order [79].

4.3 Using Multi-Agent Systems in Supply Chains: Examples

We now illustrate the use of agents in supply chains by presenting various projects. These projects can be separated into two broad families: supply chain management projects [85] and supply chain design projects. Moreover, the manner of solving problems also differs depending on projects, e.g. the number and the role of agents vary considerably, depending on the particular point under study. To highlight these differences, Table 2 summarizes the projects which are now described:

1. DragonChain was implemented by Kimbrough's team [86] at the University of Pennsylvania (Philadelphia, PA, USA) to simulate supply chain management, and more particularly to reduce the bullwhip effect. For that, they base their simulation on two versions of the *Beer Game*, the *MIT Beer Game* (i.e., the original game [102]) and the *Columbia Beer Game* [103], and they use agents that look for the best ordering scheme with genetic algorithms. Note that Chapter 10 uses simulations based on a model similar to the Beer Game, while Chapter 9 also applies genetic algorithms.
2. Agent Building Shell at the University of Toronto (Ontario, Canada) is a library of software classes providing reusable elements for building agent systems. These agents have four layers: a layer for knowledge management, an ontology layer, a layer of cooperation and conflict solving, and a layer of communcation and coordination. This latter layer is insured by the COOrdination Language (COOL). This project has involved several researchers, such as Mark Fox, Teigen, Barbuceanu and Beck [87–90].

Project	Studied problem	Approach	Number and role of agents
1. DragonChain [86]	Management (Bullwhip effect)	Genetic algorithm seeking the best ordering scheme	1 agent/company
2. Agent Building Shell [87–90]	Management (Coordination)	COOrdination Language (COOL)	1 agent/company
3. MetaMorph 1 & 2 [81]	Management (Coordination)	Mediator-agents	1 agent/company +mediator-agents
4. NetMan [83, 91]	Management (Intra- and inter-company operations management)	Contract driven coordination in Convention, Agreement and Transaction (CAT) formalism	1 agent/workshop
5. BPMAT &SCL [12, 92, 93]	Modelization (Which elements are common to all supply chains?)	Comparison of three very different supply chains	BPMAT models companies & SCL intercompany streams
6. MASCOT [94]	Management (Agility increase)	Comparison of several coordination policies	1 agent/company
7. DASCh [95, 96]	Management (supply chain modelization techniques)	Delays and uncertainties on streams modelled as agents	2 agents/company + 1 agent/stream
8. Task dependency network [97–99]	Design & management (Partner selection)	Comparison of auction protocols	1 agent/company
9. MASC [100]	Design (Partner selection)	Auction-based protocol under constraints	1 agent/company + 2 directory agents
10. OCEAN [101]	Management (Global cooperation emerging from local competitions)	Negotiation system in a multiagent contract network	1 agent/company (1 agent = system of 6 agents)

Table 2. Some projects applying agents to supply chains.

3. MetaMorph II is an improvement of a first project called MetaMorph. Agents form a federation centered around mediators that have two roles: they allow agents to find each other, and they coordinate these agents. These two projects were developed at the University of Calgary (Alberta, Canada) by Maturana and others [81].
4. NetMan (NETworked MANufacturing) formalizes networked organizations and production operations in order to obtain agile manufacturing networks in a dynamic environment. Conversely to DragonChain, this multiagent system manages an actual supply chain, rather than the Beer Game. Each company is cut in NetMan centers, i.e., independent, collaborating business units. The NetMan centers of a company coordinate with each other and with other customers' and suppliers' NetMan centers. This coordination is based on contracts and conventions, which are formalized according to the model *Convention, Agreement, Transaction* (CAT). This work was carried out at Université Laval (Quebec City, Quebec, Canada) [83, 91]. Chapters 14, 15 and 16 detail additional industrial applications of agents.
5. BPMAT is a software library developed by IBM [93] to model company activities (note that Chapter 15 presents another project of IBM). SCL is an

addition to this library for modelling inter-company flows. BPMAT and SCL are based on Swaminathan, Smith and Sadeh [12, 92]'s work at Carnegie Mellon University (Pittsburgh, PA, USA), which sought elements common to any supply chain by comparing three chains from distinct industrial sectors. Chapter 8 also deals with supply chain modeling.

6. MASCOT (MultiAgent Supply Chain cOordination Tool) is a reconfigurable, multilevel, agent-based architecture for planning and scheduling aimed at improving supply chain agility. It coordinates production among multiple (internal or external) facilities, and evaluates new product/subcomponent designs and strategic business decisions (e.g., make-or-buy or supplier selection decisions) with regard to capacity and material requirements across the supply chain [94]. Like BPMAT and SCL, this work was also accomplished at Carnegie Mellon University (Pittsburgh, PA, USA).
7. DASCh was developed at ERIM (Ann Arbor, MI, USA) by Parunak and his colleagues [95, 96] to explore the modelling techniques of networks of suppliers and suppliers' suppliers. In particular, flows of products and information flows are viewed as agents to model imperfections in these flows. The approach of risk management in Chapter 4 may manage such imperfections as well.
8. The Task Dependency Network is an asynchronous, decentralized market protocol (auctions) for allocating and scheduling tasks among agents that contend for scarce resources, constrained by a hierarchical task dependency network [97, 99]. An additional paper [98] extends this protocol to model supply chain formation. This work is a Ph.D. thesis defended in 2001 by Walsh [97] (supervised by Wellman) at the University of Michigan (Ann Arbor, MI, USA). Walsh presents a more recent work with Babaioff in Chapter 12.

   In similar ways, other works use market mechanisms to coordinate supply chains [104, 105].
9. MASC studies coordination modes between companies in supply chains. These coordination modes are calls for submissions, which submitters answer according to their capacity and production load. Companies winning this auction next take part in the supply chain carrying products to the consumer. This work was completed at the Université d'Aix-Marseille 3 (Marseilles, France) [100]. Chapters 2, 3, 5, 11 and 13 (i.e., those about e-commerce and negotiation) also address these questions about where and how many items to buy.
10. OCEAN (Organization and Control Emergence with an Agent Network) is a control system with an open, decentralized and constraints-based architecture in which there is responsiveness, and distribution of production resources and technical data. This system was designed to react to environment dynamics in order to show that cooperation at the global level may emerge from competitions at the local level. This work was completed at INSA de Lyon (Lyon, France) and at the Université de Montpellier 2 (Montpellier, France) [101]. Chapters 7 and 8 also treat environment dynamics.

5 Conclusion

The literature review in this chapter has introduced the topic of this book by presenting the two areas involved and their merger. Specifically, the first area involved, i.e., supply chain management, was first introduced by outlining some industrial problems, and how the concept of supply chain as well as the collaboration it enables allow solving these problems.

After that, we turned to multiagent systems as a way to implement the supporting technologies required by the concept of supply chain. We described what an autonomous agent is, and how agents are different from objects. We also presented some agent architectures, the motivation for the use of agents in any field, and a comparison of multiagent systems with other areas. Some applications of multiagent systems illustrated what multiagent systems are, followed by a focus on one specific application, that is, on multiagent-based supply chain management.

This focus on the use of agents in supply chains first detailed the information technologies required by supply chains, next focussed the previous motivation for agents on their application to supply chains. It is important to note that what supply chains require is the main characteristic of agents, i.e., their autonomy. Finally, the summary of some projects using agents in supply chains illustrated this chapter.

Similarly to this illustration, the rest of this book contains additional applications of agents to supply chain management or to closely related areas, such as e-commerce and risk management in supply chains.

6 Acknowledgment

We would like to thank FOR@C, the Research Consortium in E-business in the forest products industry (Université Laval, Québec, Canada), for supporting this research. This work was also partially supported by the National Sciences and Engineering Research Council of Canada (NSERC).

References

1. Mallya, S.N.: Demand Diven Production and Distribution: Seamless Integration of Downstream Supply Chain Decisions using Electronic Data Interchange. Ph.D. thesis, Drexel University (Philadelphia, PA, USA) (1999)
2. Savelsbergh, M.W.P.: Local search in routing problems with time windows. Annuals of Operations Research **4** (1985) 265–305
3. Simchi-Levi, D., Kaminsky, P., Simchi-Levi, E.: Designing and managing the supply chain. McGraw-Hill Higher Education (2000)
4. Shapiro, J.F.: Modeling the Supply Chain. Duxbury (2000)
5. Lee, H.L., Padmanabhan, V., Whang, S.: The bullwhip effect in supply chain. Sloan Management Review **38** (1997) 93–102
6. Lee, H.L., Padmanabhan, V., Whang, S.: Information distortion in a supply chain: The bullwhip effect. Management Science **43** (1997) 546–558

7. Carlsson, C., Fullér, R.: Reducing the bullwhip effect by means of intelligent, soft computing methods. In: Proc. 34th Hawaii International Conference on System Sciences (HICSS-34), (Hawaii, USA) (2001)
8. Iaccoca Institute: 21st Century Enterprise Strategy: An Industry Led View (1991) Lehig University, Bethlehem (Pennsylvania, USA).
9. Rigby, C., Day, M., Forrester, P., Burnett, J.: Agile supply: Rethinking systems thinking, systems practice. International Journal of Agile Management Systems **2** (2000) 178–186
10. Burns, J.F., Sivazlian, B.D.: Dynamic analysis of multi-echelon supply systems. Computers & Industrial Engineering **2** (1978)
11. Muckstadt, J., Murray, D., Rappold, J., Collins, D.: Guidelines for collaborative supply chain system design and operation. Information Systems Frontiers **3** (2001) 427–435
12. Swaminathan, J.M., Smith, S.F., Sadeh, N.M.: Modeling supply chain dynamics: A multiagent approach. Decision Sciences **29** (1998) 607–632
13. Davidsson, P., Werstedt, F.: A multi-agent system architecture for coordination of just-in-time production and distribution. Knowledge Engineering Review **17** (2002) 317–329
14. Poirier, C.C., Reiter, S.E.: La Supply Chain - Optimiser la Chaîne Logistique et le Réseau Interentreprises. Dunod (Les Réalités de l'entreprise) (2001) Translation of "The Supply Chain Optimization: Building the Strongest Total Business Network" (1996) by Monique Sperry.
15. Chen, F., Drezner, Z., Ryan, J.K., Simchi-Levi, D.: The bullwhip effect: Managerial insights on the impact of forecasting and information on variability in a supply chain. In: Quantitative Models for Supply Chain Management. Sridhar Tayur, Ram Ganeshan and Michael Magazine (1994)
16. Taras, J.: VMI Web site (2003) `http://www.vendormanagedinventory.com/` (accessed Sept. 8, 2003).
17. Wal-Mart Stores, Inc.: Web site (2004) `http://www.walmart.com/` (accessed April 5, 2004).
18. Procter & Gamble Company: Web site (2004) `http://www.pg.com/` (accessed April 5, 2004).
19. Waller, M., Johnson, M.E., Davis, T.: Vendor-managed inventory in the retail supply-chain. Journal of business logistics **20** (1999) 183–203
20. VICS Association: Web site (2003) `http://www.cpfr.org` (accessed Sept. 10, 2003).
21. Lee, H.L.: E-business and supply chain integration. Stanford Global Supply Chain Forum `http://www.stanford.edu/group/scforum/Welcome/EB_SCI.pdf` (2001)
22. Chaib-draa, B., Moulin, B., Jarras, I.: Agent et systèmes multiagents. In J.P. Briot et Y Demazeau, ed.: Principes et Architecture des Systèmes Multi-Agents. Hermes, Lavoisier (2001)
23. Chaib-draa, B., Moulin, B., Mandiau, R., Millot, P.: Trends in distributed artificial intelligence. Artificial Intelligence Review **6** (1992) 35–66
24. Chaib-draa, B.: Industrial applications of Distributed AI. Communication of ACM **38** (1995) 49–53
25. Ferber, J.: Les systèmes multi-agents, vers une intelligence collective. InterEditions, Paris (France) (1995)
26. Wooldridge, M., Jennings, N.R.: Intelligent agents: Theory and practice. The Knowledge Engineering Review **10** (1995) 115–152
27. Shoham, Y.: Agent-oriented programming. Artificial Intelligence **60** (1993) 51–92

28. Free Software Foundation: Web site of GCC (GNU compiler collection) (2004) http://gcc.gnu.org/ (accessed April 14, 2004).
29. Sun microsystems: Java 2 platform (2003) http://java.sun.com/ (accessed Sept. 19, 2003).
30. Wooldridge, M.: An Introduction to Multi-Agent Systems. John Wiley & Sons, Inc. New York, NY, USA (2001)
31. Wooldridge, M.: Intelligent agents. In Weiss, G., ed.: Multiagent Systems - A Modern Approach to Distributed Artificial Intelligence. MIT Press, Cambridge (Massachusetts, USA) (1999) 28–77
32. Guessoum, Z., Briot, J.P.: From active objects to autonomous agents. IEEE Concurrency **7** (1999)
33. Gasser, L., Briot, J.P.: Object-based concurrent programming and distributed artificial intelligence. In Avouris, N., Gasser, L., eds.: Distributed Artificial Intelligence: Theory and Praxis. Kluwer Academic Publishers (1992)
34. Agent Oriented Software Group: JACK Intelligent Agent (TM) User Guide. Technical Report Release 4.1 (2003) http://www.agent-software.com/.
35. Russell, S., Norvig, P.: Artificial Intelligence - A modern approch. Englewood Cliffs, NJ: Prentice-Hall (2d Edition) (2003)
36. Turing, A.M.: Computing machinery and intelligence. Mind **59** (1950) 433–460
37. Huhns, M.N., Stephens, L.M.: Multi-agent systems and societies of agents. In Weiss, G., ed.: Multiagent Systems - A Modern Approach to Distributed Artificial Intelligence. MIT Press, Cambridge (Massachusetts, USA) (1999) 79–120
38. Jennings, N.: On agent-based software engineering. Artificial Intelligence **117** (2000)
39. Jennings, N., Sycara, K., Wooldridge, M.: A roadmap of agent research and development. International Journal of Autonomous Agents and Multi-Agent Systems **1** (1998) 7–38
40. Papadimitriou, C.H.: Algorithms, games, and the internet. In: Proc. STOC 2001 (extended abstract in ICALP 2001). (2001)
41. Boutilier, C., Shoham, Y., Wellman, M.P.: Economic principles of multi-agent systems (editorial). Artifical Intelligence **94** (1994)
42. Jennings, N., Wooldridge, M.: Applications of intelligent agents. In: Agent technology: Foundations, applications and markets. Springer-Verlag New York, Inc. (1998) 3–28
43. Weiss, G.: Multiagent systems - A modern approach to distributed artificial intelligence. The MIT Press (1999)
44. Gruver, W.A., Kotak, D.B., van Leeuwen, E.H., Norrie, D.H.: Holonic manufacturing systems: Phase 2. In: Holonic and Multi-Agent Systems for Manufacturing - 1st International Conference on Industrial Applications of Holonic and Multi-Agent Systems, HoloMAS 2003, Prague (Czech Republic) http://gerstner.felk.cvut.cz/HoloMAS/2003/ (2003) 1–14
45. Parunak, H.V.D.: Industrial and practical applications of DAI. In Weiss, G., ed.: Multiagent Systems - A Modern Approach to Distributed Artificial Intelligence. MIT Press, Cambridge (Massachusetts, USA) (1999) 377–424
46. Shen, W., Norrie, D.H.: Agent-based systems for intelligent manufacturing: A state-of-the-art survey. Knowledge and Information Systems **1** (1999) 129–156
47. Shen, W., Norrie, D.H., Barthès, J.P.A.: Multi-Agent Systems for Concurrent Intelligent Design and Manufacturing. Taylor & Francis (London and New York) (2001)
48. British Telecommunications plc: The ZEUS agent building toolkit (2004) http://more.btexact.com/projects/agents/zeus/ (accessed May 21, 2004).

49. Ljungberg, M., Lucas, A.: The OASIS air traffic management system. In: Proc. 2nd Pacific Rim International Conference on Artificial Intelligence (PRICAI '92). (1992)
50. Callantine, T.J.: Air traffic controller agents. In: Proc. 2nd int. joint conf. on Autonomous Agents and MultiAgent Systems (AAMAS), Melbourne (Victoria, Australia) (2003) 952–953
51. Cammarata, S., McArthur, D., Steeb, R.: Strategies of cooperation in distributed problem solving. In: Proc. 8th International Joint Conference on Artificial Intelligence (IJCAI-83). (1983) 767–770
52. Hallé, S., Chaib-draa, B., Laumonier, J.: Car platoons simulated as a multiagent system. In: Proc. 4th Workshop on Agent-Based Simulation, Montpellier (France), SCS (2003) 57–63
53. eBay, Inc.: Web site (2004) `http://www.ebay.com/` (accessed May 16, 2004).
54. Ben-Ameur, H., Chaib-draa, B., Kropf, P.: Multi-items auctions for automatic negotiation. Journal of Information and Software Technology **44** (2002) 291–301
55. Rosenschein, J.S., Zlotkin, G.: Rules of Encounter: Designing Conventions for Automated Negotiation Among Computers. MIT Press, Cambridge (Massachusetts, USA) (1994)
56. Asselin, F.: Formation de coalitions d'agents logiciels pour le groupement d'acheteurs. Master's thesis, Université Laval, Département d'Informatique et de Génie Logiciel, Quebec City (Quebec, Canada) (2002)
57. Asselin, F., Chaib-draa, B.: Towards a protocol for the formation of coalitions of buyers. In Crainic, T.G., Gavish, B., eds.: Proc. Fifth International Conference on Electronic Commerce Research (ICECR-5). (2002)
58. Breban, S., Vassileva, J.: A coalition formation mechanism based on inter-agent trust relationships. In: Proc. 1st int. joint conf. on Autonomous Agents and MultiAgent Systems (AAMAS), Bologna, Italy (2002) 306–307
59. Trading Agent Competition: Web site (2004) `http://www.sics.se/tac/` (accessed May 16, 2004).
60. Cheng, S.F., Leung, E., Lochner, K.M., O'Maley, K., Reeves, D.M., Schvartzman, L.J., Wellman, M.P.: Walverine: A walrasian trading agent. In: Proc. 2nd int. joint conf. on Autonomous Agents and MultiAgent Systems (AAMAS), Melbourne (Victoria, Australia) (2003) 465–472
61. Vetsikas, I.A., Selman, B.: A principled study of the design tradeoffs for autonomous trading agents. In: Proc. 2nd int. joint conf. on Autonomous Agents and MultiAgent Systems (AAMAS), Melbourne (Victoria, Australia) (2003) 473–480
62. Grand, S., Cliff, D.: Creatures: Entertainment software agents with artificial life. Autonomous Agents and Multi-Agent Systems **1** (1998)
63. Hayes-Roth, B., Hewitt, M., Washington, R., Seiver, A.: Distributing intelligence with an individual. In Gasser, L., Huhns, M., eds.: Distributed Artificial Intelligence, Vol. II. Kaufmann (San Mateo, California, USA) (1989) 385–412
64. Hiroaki, K.: Robocup rescue: A grand challenge for multi-agent systems. In: Proc. International Conference on Multi-Agent Systems (ICMAS 2000). (2000)
65. Paquet, S.: Learning Coordination in RoboCupRescue. In Xiang, Y., Chaib-draa, B., eds.: Lecture Notes in Artificial Intelligence. Volume 2671., Springer-Verlag (2003) 627–628 Proc. 16th Conference of the Canadian Society for Computational Studies of Intelligence (AI 2003).
66. Microsoft Corp.: Pack MS-Office 2000 (2004) `http://www.microsoft.com/office/` (accessed Jan. 29, 2004).

67. Franchesquin, N., Espinasse, B.: Analyse multiagent de la gestion hydraulique de la camargue: Considérations méthodologieques. In: Proc. Journées Francophones des SMA et de l'IAD, St Etienne (France) (2000) 325–342
68. Cingil, I., Dogac, A.: An architecture for supply chain integration and automation on the internet. Journal of Distributed and Parallel Databases **10** (2001) 59–102
69. World Wide Web Consortium: Web site (2003) `http://www.w3c.org` (accessed Sept. 10, 2003).
70. World Wide Web Consortium: Specification of the Resource Description Framework (RDF) 1.0 (2003) `www.w3.org/TR/rdf-schema/` & `www.w3.org/TR/REC-rdf-syntax` (accessed Sept. 10, 2003).
71. World Wide Web Consortium: Specification of the web ontology language (OWL) (2004) `http://www.w3.org/2004/OWL/` (accessed March 31, 2005).
72. OASIS: The XML cover pages - XML Common Business Library (xCBL) (2003) `www.oasis-open.org/cover/cbl.html` (accessed Sept. 19, 2003).
73. Singh, M., Huhns, M.N.: Service-Oriented Computing - Semantics, Processes, Agents. John Wiley and sons, Ltd. (2005)
74. Garé, B.: Entreprise, ouvre-toi ! Logiciels & systèmes **52** (2000) Dossier spécial chaîne logistique.
75. Oracle: Web site (2005) `http://www.oracle.com/peoplesoft/integration.html` (accessed Oct. 5, 2005).
76. SSA Global: Web site (2005) `http://www.ssaglobal.com/` (accessed Oct. 5, 2005).
77. Commerce One, LLC: Web site (2005) `http://www.commerceone.com/` (accessed Oct. 7, 2005).
78. Ariba, Inc.: Web site (2003) `http://www.ariba.com` (accessed Sept. 5, 2003).
79. Dodd, C., Kumara, S.R.T.: A distributed multi-agent model for value nets. IEA/AIE 2001 (2001) 718–727
80. Fox, M.S., Chionglo, J.F., Barbuceanu, M.: The integrated supply chain management (1993) Internal report of the Enterprise Integration Laboratory, Department of Industrial Engineering, University of Toronto (Ontario, Canada).
81. Maturana, F., Shen, W., Norrie, D.: Metamorph: an adaptive agent-based architecture for intelligent manufacturing. International Journal of Production Research **37** (1999) 2159–2173
82. Viswanathan, S., Piplani, R.: Coordinating supply chain inventories through common replenishment epochs. European Journal of Operational Research **129** (2001) 277–286
83. Cloutier, L., Frayret, J.M., D'Amours, S., Espinasse, B., Montreuil, B.: A commitment-oriented framework for networked manufacturing coordination. International Journal of Computer Integrated Manufacturing **14** (2001) 522–534
84. Parunak, H.V.D.: Applications of distributed artificial intelligence in industry. In O'Hare, G.M.P., Jennings, N.R., eds.: Foundations of Distributed Artificial Intelligence. John Wiley & Sons (1996) 71–76
85. Chen, Y., Peng, Y., Finin, T., Labrou, Y., Cost, S.R.: Negotiating agent for supply chain management (1999)
86. Kimbrough, S.O., Wu, D., Zhong, F.: Computers play the Beer Game: Can artificial agents manage supply chains? Decision Support Systems **33** (2002) 323–333
87. Teigen, R., Barbuceanu, M.: The supply chain demonstrator (1996) Technical report and user guide by the Enterprise Integration Laboratory, Department of Industrial Engineering, University of Toronto (Ontario, Canada).

88. Barbuceanu, M., Fox, M.: The architecture of an agent building shell. In: Intelligent Agents II, Springer-Verlag, LNAI 1037. (1995) 235–250
89. Barbuceanu, M., Fox, M.: Cool: A language for describing coordination in multiagent systems. In: Proc. First International Conference on Multi-Agent Systems. (1995)
90. Beck, J., Fox, M.: Supply chain coordination via mediated constraint relaxation. In: Proc. 1st Canadian Workshop on Distributed Artificial Intelligence, Banff (Alberta, Canada) (1994) 61–72
91. Lyonnais, P., Montreuil, B.: Un environnement distribué de simulation à base d'agents. In: 3e conférence francophone de MOdélisation et SIMulation (MOSIM) - conception, analyse et gestion des systèmes industriels, Troyes (France) (2001)
92. Swaminathan, J.M., Smith, S.F., Sadeh, N.M.: Modeling the dynamics of supply chains. In: Proc. AAAI-94 Workshop on reasoning about the shop floor, Seattle (Washington, USA) (1994)
93. IBM Research: BPMAT: A business process modeling and analysis tool (2003) `http://www.research.ibm.com/pdtr/bpmat.htm` (accessed Sept. 19, 2003).
94. Sadeh, N., Hildum, D., Kjenstad, D., Tseng, A.: MASCOT: An agent-based architecture for coordinated mixed-initiative supply chain planning and scheduling. In: Workshop notes, Agent-Based Decision Support for Managing the Internet-Enabled Supply Chain, 3rd International Conference on Autonomous Agents (Agents' 99). (1999)
95. Parunak, H.V.D., VanderBok, R.: Modeling the extended supply network. In: ISA-Tech'98 (Houston, Texas, USA), Industrial Technology Institute (1998)
96. Baumgaertel, H., Brueckner, S., Parunak, V., Vanderbok, R., Wilke, J.: Agent models of supply network dynamics. The Practice of Supply Chain Management (2001) URL http://www.anteaters.net/ sbrueckner/publications/2001/-AgentModelsOfSupplyNetworkDynamics_submitted_20010209.pdf.
97. Walsh, W.E.: Market protocols for decentralized supply chain formation. Ph.D. thesis, University of Michigan, Ann Arbor (Michigan, USA) (2001)
98. Walsh, W.E., Wellman, M.P.: Modeling supply chain formation in multiagent systems. In: Agent mediated electronic commerce (IJCAI workshop). (1999) 94–101
99. Walsh, W.E., Wellman, M.P.: A market protocol for decentralized task allocation. In: Proc. 3rd International Conference on Multi-Agent Systems (ICMAS-98), Department of computer science and information systems, University of Hong Kong (1998)
100. Labarthe, O.: Modélisation multi-agent de chaînes logistiques. Mémoire de Diplôme d'Études Approfondies, Université d'Aix-Marseille 3, Marseilles (France) (2000)
101. Bournez, C., Gutknecht, O.: Modèle de contrôle par émergence de coordinations dans un réseau de contrats multiagents. In: Journées Francophones de l'Intelligence Artificielle Distribuée et des Systèmes MultiAgents, Montreal (Quebec, Canada) (2001)
102. Sterman, J.D.: Modeling managerial behavior: Misperceptions of feedback in a dynamic decision making experiment. Management Science **35** (1989) 321–339
103. Chen, F.: Decentralized supply chains subject to information delays. Management Science **45** (1999) 1076–1090
104. Eymann, T.: Markets without makers - A framework for decentralized economic coordination in multiagent systems. In: Lecture Notes in Computer Science 2232. (2001) 63–74 Proc. 2nd International Workshop WELCOM.

105. Fan, M., Stallaert, J., Whinston, A.B.: Decentralized mechanism design for supply chain organizations using an auction market. Information Systems Research **14** (2003) 1–22

eMarketplace Model: An Architecture for Collaborative Supply Chain Management and Integration

Hamada Ghenniwa[1], Jiangbo Dang[2], Michael Huhns[2], Weiming Shen[3]

[1] Dept. of Electrical and Computer Engineering
University of Western Ontario, London, Ontario Canada
hghenniwa@eng.uwo.ca
[2] Dept. of Computer Science and Engineering
University of South Carolina, Columbia, SC, USA
dangj@sc.edu, huhns@sc.edu
[3] Integrated Manufacturing Technologies Institute
National Research Council, London, Ontario, Canada
weiming.shen@nrc.gc.ca

Abstract. The current economic climate forces businesses to collaborate more frequently and build efficient organizations and supply chains that reduce time-to-market and costs. This chapter argues that an electronic marketplace (*eMarketplace*) is a promising architectural model to develop collaborative supply chain management and integration platform. It supports coordination mechanisms and integration at the business and systems levels of the enterprise and the supply chain. In this architecture, the *eMarketplace* exists as a collection of economically motivated software agents of service-oriented cooperative distributed systems. It enables and supports common integration and economic services between market participants. This chapter presents an agent-oriented dynamic trading mechanism that produces an integrated supply chain for the *eMarketplace*. Future *eMarketplaces* need to support both market-based and relationship-based supply chains. To this end, this chapter discusses coordination approaches based on market mechanisms such as auctions and multi-issue negotiation. The objective is to enable business entities to obtain efficient resource allocation while preserving long-term relationships.

1 Introduction

Information technology has enabled, and in some cases has forced, companies and organizations to redefine their business models and to reorient their internal capabilities to exploit electronic business (*eBusiness*) techniques. They are finding it necessary to collaborate to build more efficient operations and supply chains that reduce times-to-market and costs. *eBusiness* is the use of the Internet along with other electronic means and technologies to conduct such collaborations within businesses, from businesses to consumers, among businesses, and from businesses to government.

Traditional models of *eBusiness* integration, such as those based on EDI (Electronic Data Interchange) and enterprise-centric views, are useful for businesses with

H. Ghenniwa et al.: *eMarketplace Model: An Architecture for Collaborative Supply Chain Management and Integration*, Studies in Computational Intelligence (SCI) **28**, 29–62 (2006)
www.springerlink.com

well-defined trading relationships, but not enough for the rapidly growing and changing global marketplace. In these models, point-to-point interfaces are created to support transactions involving replenishment orders for the goods of a previously negotiated contract. In the sell-side model, either a single distributor is responsible for aggregating all the suppliers, or the customer is responsible for comparison-shopping between suppliers. This makes it inefficient and expensive for both customers and suppliers. In the buy-side model, the buying organizations are responsible for setting-up and maintaining catalogs of their suppliers, and hence it is costly and technically demanding.

An electronic marketplace (*eMarketplace*), however, appears to be a promising integration model for *eBusiness*. *eMarketplaces* provide an integrated and efficient environment for consumers, who depend on a variety of products and services that can spread across several suppliers or marketplaces. Likewise, they provide suppliers with the ability to reach, discover, and develop new customers within a single *eMarketplace* or across various *eMarketplaces* quickly with low cost. In general, *eMarketplaces* offer businesses the chance to develop and enhance their most important relationships—those with customers and suppliers. It enables the creation and leveraging of services and supply operations in a way that seamlessly integrates business entities (customers, suppliers, partners, and competitors) in a dynamic trading community. In this work, we view an *eMarketplace* as a cooperative distributed system that integrates participating business entities, including consumers, suppliers, and other intermediaries. This architecture enables and facilitates common economic services and commerce transactions between the consumers and suppliers, such as brokering, pricing, and negotiation, as well as cross-enterprise integration and cooperation in an electronic supply chain. In this architecture, the *eMarketplace* exists as a collection of economically motivated software agents.

The rest of the chapter is organized as follows. First, it reviews some of the business models related to *eBusiness* applications with a brief analysis of the main architectural design issues for *eMarketplaces*. After that, it briefly describes an architecture for a cooperative distributed system, Business-Centric Knowledge-Oriented architecture (BCKOA), for *eMarketplace* integration. This is followed by a description of a layered BCKOA implementation for an *eMarketplace*. Then the main components of an agent-oriented BCKOA for an *eMarketplace* are presented, including a supply chain automation system for integration and management using a group of cooperating software agents. Unlike traditional buyer-centric approaches, the proposed architecture emphasizes on the supplier perspective as well through the enablement of relationship-based supply chain management. A short description of an ongoing implementation of the proposed model for virtual enterprise *eMarketplace* is described next. It next describes a multi-issue negotiation model and protocol that considers both functional and quality attributes. Then it discusses some of the related work in both the academic and industrial communities. Finally, it summarizes the main contributions of this chapter.

2 *eBusiness* Models

One of the most frequently mentioned barriers to successful *eBusiness* applications is the lack of an appropriate business model. In simple words, it is "an architecture for the product, services, and information flows, including a description of the various business actors and their roles; and a description of the potential benefits for the various business actors; and a description of the sources of revenues" [47]. Possible architectures for business models can be constructed by combining interaction patterns of its components with value-chain integration [4][15]. These can be fully open, with an arbitrary number of business participants (customers and suppliers), or semi-open, with one customer and multiple suppliers or vice-versa. In principle, several architectures can be conceived for *eBusiness* applications; in practice, however, only a limited number can be realized [47]. The following are the most widely realized models [47].

A basic model is *eShop*. It is based on providing a self-service storefront with the company's catalogs and product offerings on the Web. The business objective is to lower the sales cost. A major concern with this model is making customers responsible for surfing a large number of *eShop* sites for comparisons among the products from different suppliers. An *eProcurement* model, however, focuses on the buying aspect of the business. A typical architecture for *eProcurement* consists of a browser-based self-service interface to the corporate purchasing system or its ERP. The supplier catalogs are presented to end-users through a single unified catalog, thereby facilitating a corporate-wide standard procurement process. In addition, *eProcurement* might support calls for tender through the Web, which might be accompanied by an electronic submission of bids. Nonetheless, an *eProcurement* model does not support dynamic trading. The business objective of this model is cost savings on purchasing operations. Online auction models have also received much attention for automating dynamic trading. The primary business objective is to increase efficiency, reduce waste, and minimize overall cost. Other models are based on creating value-chain businesses. One model describes service provisioning of specific functions, such as electronic payments or logistics. Other approaches are also emerging in production and stock management, where new intermediary service providers are formed to provide specialized expertise to analyze and fine-tune production. The business objective of this model is to generate revenue based on fee or revenue percentage.

Although each of the above models attempts to provide an *eBusiness* solution, none addresses the creation and leveraging of services and supply operations in a way that seamlessly integrates business entities (customers, suppliers, partners, and competitors) in a dynamic trading community. A very promising business model that can effectively deal with this challenge is *eMarketplace.* This model supports value-chain integration and provisioning in its structure and services. It combines the advantages of the sell-side, the buy-side, and the value-chain models. The business objective of the *eMarketplace* model can be based on a combination of subscription fees, transaction fees, and service fees.

The next section lays the engineering foundation for developing an architectural framework for *eBusiness,* with special attention on an *eMarketplace* model. The specification of an *eMarketplace* as a cooperative distributed system describes the architecture of an ontology driven *eBusiness* environment that deals with technological and business issues.

3 *eMarketplaces*: Requirements Analysis and Design Issues

Early attention to *eMarketplaces* focused on lowering the business operation costs. Garciano and Kaplan [18] suggested that the transaction cost savings alone from *eBusiness* exchanges could be a significant portion of the total cost of production and order fulfillment. However, as *eBusiness* grows and becomes viable in the real world, its corresponding *eMarketplaces* must expand to support a broader base of services ranging from baseline interaction and directory services to specialty market services, such as dynamic trading, supply chain integration and management. By automating and lowering the cost of searching and matchmaking between consumers and suppliers, *eMarketplace* becomes an appropriate solution for businesses to conduct large volumes of transactions using dynamic trading approaches such as auctions. Also, through the facilitation of collaboration and information-sharing services, *eMarketplaces* enable and support sharing of supply chain information such as forecasts and inventory levels. *eMarketplaces* can also improve the efficiency of the supply chain by automating business processes such as procurement, order management, and fulfillment. In addition, an *eMarketplace* should enable and strengthening the relationship between business participants and their supporting systems.

To this end, a fundamental aspect that our proposed *eMarketplace* architecture supports is to maintain various relationships between customers and suppliers. This enables both customers and suppliers to leverage economies of scale in their trading relationships and provide them with an access to various liquid marketplaces. This in turn allows the use of dynamic pricing models[1], such as exchanges and auctions. The following subsections provide a detailed analysis of the aspects that sets the foundation for the proposed architecture.

3.1 Market Structure and Economy Model

The market structure governs the trading process and defines the formal rules for market access, traders' interactions, price determination, and trade generations. Its behavior restricts the set of message sequences that traders may exchange and determines the trading outcome. Therefore, a market institution [32] is the specification of the set of admissible messages (i.e., traders' actions, usually price and/or quantity offers), and the final goods/services allocation given any combination of messages chosen by the participants and any initial allocation. In classical economic theory, there are several market models for specific trading situations and structural behaviors. In the commodity market model, various suppliers and consumers participate to

[1] We use the term dynamic pricing broadly to refer to short-term flexibility of prices to respond to changing supply and demand conditions.

trade goods/services (commodity) of the same type. The market price is publicly agreed upon for each commodity independent of a particular supplier. The challenge in this market structure is to deploy a pricing methodology that produces price adjustments that bring about market equilibrium (i.e., equalize supply and demand).

In an auction-based market, each participant (consumer and supplier) acts independently and contracts to buy or sell at a price agreed upon privately. An auction-based *eMarketplace* is a form of centralized facility, or clearinghouse, by which costumers and suppliers execute trades in an open and competitive bidding process. In open auctions, bidders can know the bid value of the others and will iteratively have an opportunity to offer competitive bids. However, in open distributed environments where an auction can be distributed over space and/or time, an iterative mechanism might not be feasible. A standard form of one-shot auctions is the first-price sealed-bid auction. It avoids iterations but introduces another computational problem of counter-speculation of the other agents' valuations and might not achieve the highest price. The Vickrey auction eliminates the computational cost of both the iterative valuations and counter-speculations overhead.

The two market structures above are not appropriate for bargaining situations where few participants try to reach an agreement that will leave them at least as well off as they could be if they reached no agreement. Most of these situations cannot be entirely determined by the market forces. In bargaining, both customers and suppliers have their own objective functions and they negotiate with each other as long as their objectives are met. The participants can engage in direct negotiations with each other using their respective bargaining strategies to arrive at a "fair" price for a particular item. This market structure does not support a specific negotiation protocol; rather the participants will use an unrestricted bidding protocol. A major challenge in this structure is how to enable any participant to determine the "fair" price.

3.2 Supply Chain Management and Integration

To provide smooth and effective integration at the business level, the *eMarketplace* architecture accommodates and supports interfaces to the existing business models of the participant entities through cooperative supply chain integration and management. An *eMarketplace* can be treated as a physically and logically distributed system of interacting autonomous business entities. Yet, there is a need for well-accepted interoperability standards, which must be meshed for supply chain integration to meet business demands. Conceptually, a supply chain manages coordinated information and material flows, production operations, and logistics of the *eMarketplace*. It provides the *eMarketplace* with flexibility and agility in responding to customer demand shifts without conflicts in resource utilization. The fundamental objective is to improve coordination within and between various participant business entities in the supply chain. Effective coordination can lead to a reduction in lead times and costs, alignment of interdependent decision-making processes, improvement in the overall performance of each participant in the chain, as well as the supply chain itself. In an *eMarketplace* setting, supply chain management can be viewed as a cooperative distributed problem-solving activity among a society or group formed by autonomous business entities that work together to solve a common problem [45]. The decision-making process is centralized for the group, but decentralized for the local decisions of each member. Therefore, the problem of supply chain design in an *eMarketplace*, as discussed later, can be solved by the design of a structure and mechanism for coordination and integration in a distributed system.

The choice of coordination mechanism depends on the setting of the interaction between consumers and suppliers. There are four possible settings between them (i.e., consumer-to-supplier): many-to-many, one-to-one, one-to-many, and many-to-one. Market transactions are appropriate for a many-to-many setting. Since adequate liquidity is critical to the success of an *eMarketplace*, only commodities, near commodities, or other highly standardized products are likely to attract adequate trading volumes to support many-to-many interactions. At the other side of the spectrum is the one-to-one setting of negotiation and partnerships, where prices often vary by customer in relation to different non-price attributes, such as purchasing volumes and service requirements. Coordination is based on one-to-one negotiations that are influenced by the long-term relationship between the consumer and the supplier. In one-to-many and many-to-one settings, participants can have more flexibility to select the most beneficial coordination mechanism. In this context *eMarketplaces* can provide flexible structure for mechanisms that improve supply chain coordination. When a single consumer is interacting with multiple suppliers, the consumer can use either a portfolio of long-term contracts, or market-based approaches such as reverse auctions. When a single supplier is interacting with multiple consumers, the supplier has a number of choices, including revenue management, "forward" auctions, dynamic pricing, and long-term contracting.

Using market-based mechanisms to match supply and demand prevents consumers with low valuations from receiving limited goods and prevents suppliers with high production costs from supplying limited demand. Nevertheless, there are a number of challenges that need to be addressed for *eMarketplaces* to be successful, including adequate market liquidity and establishing approaches for realizing the benefits of market mechanisms without undermining existing supply chain relationships [22]. In relationship-based supply chains, long-term relationships have higher value than that of the efficient resource allocation. In these settings, prices are usually negotiated rather than determined by the market, and efficient allocation is sacrificed in favor of the other benefits of relationship-based negotiations. Usually supply and demand are balanced by non-price mechanisms. In this context, allocation is treated as a function of negotiation and relationship rather than as allocation efficiency in the economic sense. For example, to deal with an oversupply, suppliers may negotiate special deals on forward buys or inventory buys to "borrow" demand from the future, or even build-up excess inventory. However, in the absence of market-clearing prices, supply and demand usually are not in equilibrium. For example, supply shocks or unanticipated demand increases can lead to shortages due to inability of the prices to change in a sufficient rate to dampen demand or stimulate production. The same for weak demand or excess production which can result in overstocked inventory while prices are not able to fall in a rate to equilibrate supply and demand. Furthermore, contract prices that lag true market-clearing prices can cause poor capacity investment decisions, leading to an undesirable cycle of oversupply and undersupply. *eMarketplaces* can address these challenges by providing a platform that combines both relationship-based and market-based coordination mechanisms. Therefore, relationship-based supply chain participants can use the market-based mechanism as a spot market to buffer supply and demand shocks [9][10]. For example, suppliers can use it to offload excess inventory. Also, consumers can use it to deal with periodic shortages. In addition, through spot markets contract prices can be adjusted in response to shifts in supply and demand by providing benchmarks for contract negotiations.

Dynamic pricing also provides an important allocation mechanism for highly differentiated goods and services. Transactions can be complex and often require evaluation and negotiation along multiple attributes and different factors. They can affect the purchasing volumes between a given set of supply chain partners. Transactions may also involve bundles or combinations of possibly complementary goods and services. Auctions and multi-issue negotiations can provide supply chain participants with the adequate decision support tools to efficiently carry out complex multi-dimensional transactions.

3.3 Foundation Architecture for Integration

It is also important that the architecture of an *eMarketplace* supports and leverages the participants' legacy environments with minimum overhead. The support can take place, as will be described later, over technology-independent cooperative distributed system architecture. Another key factor for the foundation of an *eMarketplace* is the ability to operate in an open environment. This is driven by the fact that in many cases a customer's needs may go beyond the specialist capabilities of any single *eMarketplace*. The architecture of the *eMarketplace* provides the foundation to integrate and leverage the participants' resources, such as applications and databases. Traditionally, the foundation technology that enables enterprises to connect resources together is known as *middleware*. Mainstream middleware solutions focus on integration at the data-level, such as those based on OMG CORBA™ (Object Management Group, Inc. 1995) and J2EE™ (Java™ 2 Platform, Enterprise Edition). Enterprise application integration (EAI) has emerged as middleware technology with an objective to ease the burden and lower the costs of application integration. However, different EAI tools are developed to accommodate different levels of integration requirements, including Object-level, business process-level, and cross-enterprise process-level. However, there are currently very few EAI solutions specifically designed for cross-enterprise integration. While EAI tools focus on technology-centered integration, other complementary approaches focus on integration as an architectural aspect. One approach is a mediator-based architecture [52], which comprises a layer of "intelligent" middleware services to link data resources and applications. Another approach is the facilitator [18], in which integration is based on the principle that any system (software or hardware) can interoperate with any other system without the intervention of human users or their developers. This level of automation depends on supporting ontologies to describe the resources. Facilitators use meta-level information in converting, translating, or routing data and information.

In the proposed *eMarketplace* environment there are significant interactions between the systems deployed by the participating business units, their customers, and other businesses. Therefore, designing *eMarketplaces* requires embodying greater levels of business knowledge within the *eMarketplace* transactions, activities, and service definitions. Additionally, it requires a greater degree of communication, coordination, and cooperation within and among the business entities and their systems in the *eMarketplace*. In other words, the *eMarketplace* architecture represents an integrated body of people, systems, information, processes, services, and products. Several attempts in business-process reengineering addressed structural integration only. The focus was on reorganizing enterprise units along critical business processes, such as supply chain and the product life cycle [22]. However, in this chapter, the focus will be on structural, behavioral, and informational integration of the participant business entities. The following sections address these aspects in more details in order to set the foundation for the proposed architecture.

4 Business-Centric Knowledge-Oriented Architecture

The *eMarketplace* architecture must be semantically rich and describe the organization and the interconnection among the software components, business services, and business ontologies of the *eMarketplace*. In this work, we deal with both the *fundamental* and the *practical* issues of integration.

Fundamentally, we view integration as an abstraction level at which a distributed system environment can be described as collective coherent universe of cooperating entities. Here we describe a business-centric knowledge-oriented architecture (BCKOA) for cooperative distributed systems. BCKOA specifications provide the abstraction to support the domain entities and applications independent of any specific technology. The main elements of BCKOA include domain services, integration services, and domain ontology. A key to BCKOA is a service-oriented model in which the overall connectivity of the system supports a "virtual" point-to-point integration mechanism. BCKOA recognizes the separations among functionalities supported by its services. Yet, they can be ubiquitously integrated in an *ad hoc* structure to fulfill a complex business service or a market structure.

To support heterogeneity and technology-independent properties at the system level, the boundaries between the layers correspond to standardized interfaces. Additionally, BCKOA includes domain ontology to capture and implement the conceptualization of an application domain at the knowledge level. BCKOA provides three families of integration services. (1) Ontology and semantic integration services support the semantic manipulations needed when integrating and transforming information or knowledge to satisfy a BCKOA task; also when capabilities require re-using components. (2) Coordination and cooperation services support ad hoc and automated BCKOA configurations. This includes locating and discovering domain and BCKOA services that are potentially relevant to a domain or BCKOA service. (3) Wrapping services make different applications, components, objects, or modules comply with internal or external standards. Such standards may involve the interface to the software system or its behavior.

The specifications of BCKOA services are independent of any component framework, but their implementation can be based on the services provided by the target framework. Here, the concept of service is viewed as a computational model that enables a designer to capture and represent complex applications in open environments, such as *eMarketplaces*, as a software artifact independent of the target framework. To this end, we proposed a meta-model for services description based on DAML-S constructs [9] and BCKOA services. The meta-model is treated at two levels: (i) UML-based model for service capabilities and processes, and (ii) agent-oriented model for service interactions (cooperative or competitive) [30].

A key challenge in putting BCKOA into a practical context is the transformation or the mapping of its abstract description into the specification of the target component framework. To deal with this issue, BCKOA requires that business-object implementations be obligated to conform to the domain ontology. The business-object specification itself in the domain ontology becomes the reusable component that can be configured and assembled into multiple solutions (business-objects), independent of technology implementation. Therefore, the domain ontology in BCKOA governs the structural and the behavioral semantics of the business-objects in a way that is consistent across all implementations, and is accessible from any implementation. The BCKOA framework, shown in Fig. 1, provides an integrated execution environment for integrated business object implementations. Mapping a BCKOA description to an implementation framework is driven by three specifications: domain ontology description, maps, and a profile. Technology mapping specifications include a map to specify a transformation from the BCKOA domain ontology and services to the implementation components and service extensions for the target component framework. The mapping of each business concept representation to its implementation is managed by a *profile*, as a set of properties that defines the environment for a mapping. This mechanism enables an automated transformation from a relatively stable domain ontology and service description to different component technologies. This framework has been supported by an integration development environment (SOAStudio) [31]. It also includes development workspaces and mechanisms as well as basic application programming interfaces. It enables designers to utilize the proposed framework effectively and transparently to develop agent-based services and further build business systems like *eMarketplace*. They are based on the proposed meta-model, and have agent constructs in software entity and DAML-S description to build ontologies. The extracted WSDL interfaces are published through UDDI-compliant registrar, which is also an agent-based service.

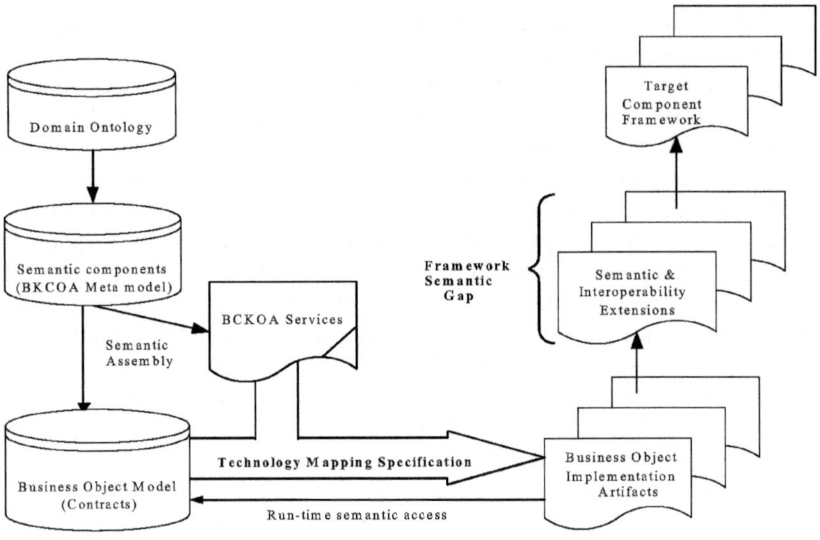

Fig. 1: BCKOA Framework

5 BCKOA-based *eMarketplace*

This section describes the application of BCKOA to develop an *eMarketplace* as a cooperative distributed system. The objective is to provide an automated framework that enable businesses (suppliers, customers, and intermediaries) to effectively engage in complex and diverse collaborative activities. The proposed BCKOA-based *eMarketplace* is shown in Fig. 2(b), which builds upon the abstraction architecture of the *eMarketplace* in Fig. 2(a) [20]. The lower layer of the *eMarketplace* architecture in Fig. 2(a) is the infrastructure that represents one or more physical network-based environments in which *eBusiness* systems can exist. The BCKOA model, in Fig. 2(b), supports the *eMarketplace* infrastructure using two layers: the distributed-computing layer and the integration-services layer. The assumption is that this infrastructure can support various markets for providing or obtaining specific goods and services. Yet, each *eMarketplace* may be independent and may support its own rules, procedures, and protocols as described by the market layer.

The market layer may support several business domains as described by the business layer. BCKOA, in Fig. 2(b), provides the integration between the business context of the market, and the services provided by the participant entities. A business-entity may participate in multiple *eMarketplaces*. A bank, for example, could participate with different roles in investment management market, mutual fund management market, and financial advisory market.

In BCKOA, the business-service layer is supported by three types of services, depicted in Fig. 2(b), (1) Business-specific services, (2) Business-entity services, which represent the implementation of the business services by specific business entities, and (3) Market services, which are categorized further into core, such as dynamic trading and supply chain services, and value-added, such as procurement process and workflow services. Here we focus on the core services. Ideally, the market services should be able to offer a wide variety of coordination and trade mechanisms to fit with multiple business models.

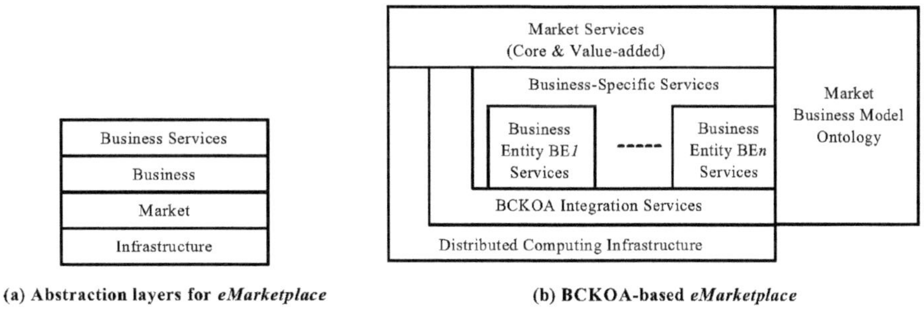

Fig. 2: Use of BCKOA for the Architecture of an *eMarketplace*

Based on the success of applying economic theories in the real world as a sustainable model for exchanging and regulating resources, goods and services, we propose to apply a flexible computational economy framework for market services. Therefore, a BCKOA-based *eMarketplace* incorporates mechanisms for different types of market structures, such as auctions and bilateral negotiation. Each of which is viewed as a separate market session. In this chapter we will focus on multi-issue negotiation and auction based market sessions. Each participant (consumer or supplier) acts independently and contracts to buy or sell at a price agreed upon privately. Here we focus on coalition-based model for multi-issues negotiation [13] and private-value auctions, such as the Vickrey mechanism [51]. As it will be discussed later that coalition-based model for multi-issue negotiation is more effective and computational efficient than traditional issue-by-issue and package deal approaches. Also, Vickrey auction provides a market mechanism that is simpler, but more efficient and more stable than open auction mechanisms and classical sealed bid auctions [49]. While it is a simple yet powerful mechanism, it is important to mention that the Vickrey mechanism may not be appropriate in all domains. For example, truthful bidding is not necessarily the dominant strategy for domains where an agent's marginal costs (and thus its reservation price) are determined by other agents' valuations, such as the case with *public-value auctions* in the stock market [43].

In BCKOA-*eMarketplace*, supply chain management is treated as a coordination methodology that manages information and material flows, production operations, and logistics. The objective is to provide an automated coordination mechanism for the participants in a supply chain. The proposed solution combines both the multi-issue negotiation as coordination mechanism for relationship-based supply chain and Vickrey based auctions as coordination market-based supply chain. The adopted integration framework for the supply chain makes use of the methodologies reported in [41][46]. In this work, we particularly extended Singh's application to supply chain integration [24]. The methodology, as described later, promotes the interchange of standard business documents and compensate for exceptions that might occur during execution. This methodology requires that the participant business entities in a cooperative supply chain only describe their supply processes using Open Applications Group (OAG) standard business documents and UML interaction diagrams. These are converted automatically into roles and specifications of the software agents for the corresponding business entities.

A combination of the market services and the business-entity services can be used to generate different business models of an *eMarketplace* as desired by the participating business entities. This structure enables a business-entity to integrate and describe the types of business services offered and the information needed to use a particular service offering within the *eMarketplace*. The details of each service type and the required information might vary among business entities, although the description of the service type is based on some common conventions described for an *eMarketplace* using service meta-model [30].

BCKOA recognizes the integration services as separate functionalities. Yet, they can be ubiquitously integrated in an *ad hoc* structure to fulfill a complex business service or a market structure. The interaction mechanisms supported by the integration layer describe both the pattern and protocol of exchanging messages between the services, such as brokering, resource discovery and ontology mapping

6 Agent-Oriented *eMarketplace* Model

All services (business, market, and integration) in a BCKOA-based *eMarketplace* usually involve complex and nondeterministic interactions, often producing results that are ambiguous and incomplete. Auctions and *ad hoc* service integrations are some examples. In addition, the dynamic nature of the environment requires that the components of the system be able to change their configuration to participate in different, often simultaneous roles in *eMarketplaces*. These requirements could not be accomplished using traditional ways of manually configuring software. Agent-orientation is a very promising design paradigm for integration. In fact, such a paradigm is essential to model an open environment, such as an *eMarketplace*, especially considering the multiple dynamic and simultaneous roles a single business-entity may need to participate in given *eMarketplace* sessions (a financial services organization may have representatives acting on its behalf simultaneously within the context of brokering, service provisioning, and marketing).

Software agent technology provides the next generation in the evolution of computational modeling, programming methodologies, and software engineering paradigms. Here, we view "agent" as a metaphorical conceptualization tool at a high level of abstraction (knowledge level) that captures, supports, and implements features that are useful for distributed computation in open environments. The first feature of an agent is that it should be able to operate as a part of a community of cooperative distributed systems, including human users. In our view, an agent can be described as a collection of primitive components that provide a focused and cohesive set of capabilities. Fig. 3 depicts the Coordinated Intelligent and Rational, Agent (CIR-Agent) model [21]. The basic components include a problem solver, interactions, and communication, as shown in Fig. 3(b). A particular arrangement or interconnection of the agent's components is required to constitute an agent, as shown in Fig. 3(a). This arrangement reflects the pattern of an agent's mental state as related to its reasoning to achieve a goal.

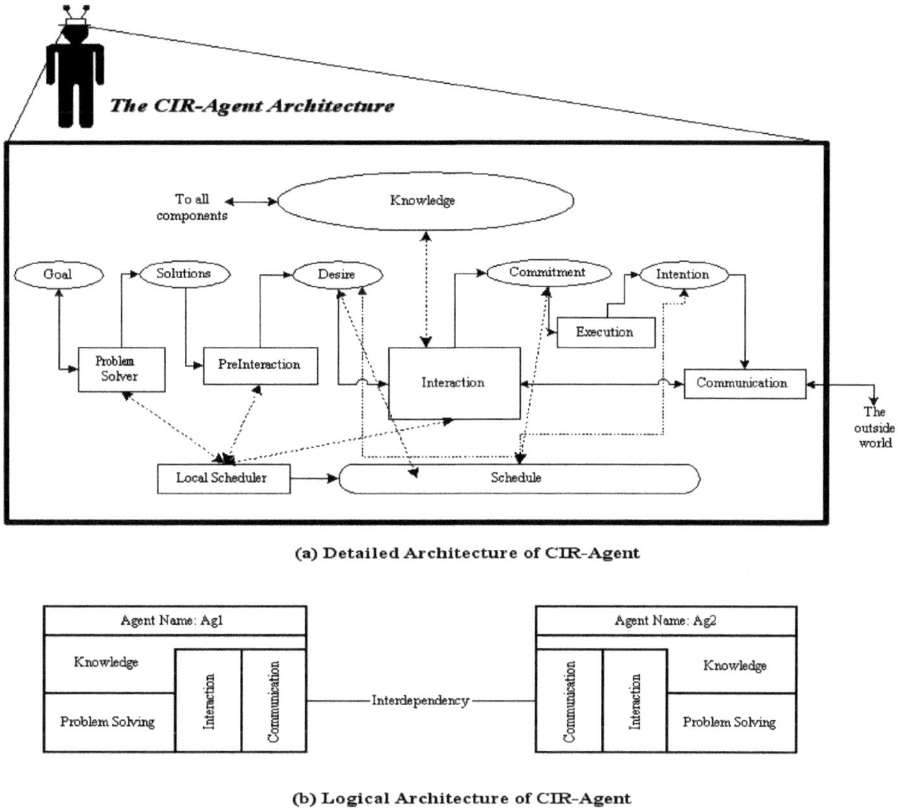

Fig. 3: The CIR-Agent Architecture

However, no specific assumption is made on the detailed design of the agent's components. Therefore, the internal structure of the components can be designed and implemented using object-oriented or any other technology, provided that a developer conceptualizes the specified architecture of the agent as described in Fig. 3(b). A CIR-Agent model provides software engineers with features at a higher level of abstraction that are useful for cooperative environments. It supports flexibility at different levels of the design: system architecture, agent architecture, and agent component architecture. These degrees of flexibility allow information systems to adapt to changes with minimum requirements for redesign. An agent within the context of a BCKOA-based *eMarketplace* might play several roles and should be able to coordinate, cooperatively or competitively, with the other agents, including humans. Therefore, as shown in Fig. 4, an agent's role can be categorized as user-interface, business-specific service, business-entity service, market service, or integration service.

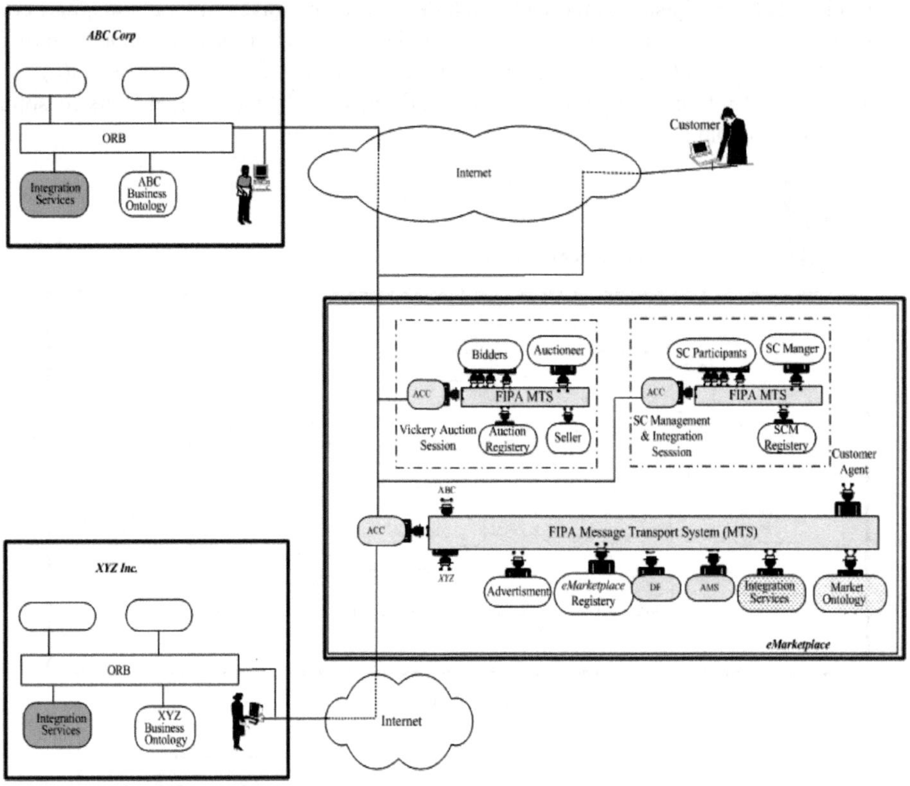

Fig. 4. The architecture of the *eAuction* Market

User interface agents play an important and interesting role in many applications. The main functionality of user interface agents is to support and collaborate with users in the same work environment to achieve the users' goals.

Business-specific service agents are specialists that provide a collection of business-services available in the *eMarketplace*. Performing the functionality of a business service is typically the cooperative integration of several agents including business-specific service agents and market service agents. A business-specific service agent may be a representative in the *eMarketplace* for some functionality that is based on legacy applications or libraries, such as a product catalogue Web site.

Market service agents are specialists that provide a collection of functions for generic *eBusinesses* in *eMarketplace* environments in which a single entity (usually an agent) can perform its tasks in the *eMarketplace*. Market services (value-added and core services) are horizontal, i.e., services that are used in several business domains by several business entities. Here the focus is on core services, particularly dynamic trading services using coalition-based formation of multiple issue negotiation Vickrey auctions and supply chain integration, which will be discussed further in the following sections.

Integration service agents are specialists that provide a collection of integration functions for a cooperative distributed system in which a single entity (agent, component, object, etc.) can perform its tasks. Integration services are used by several distributed entities. For example, a brokering service provides a capability-based integration in the *eMarketplace*. The brokering agent allows agents (for integration, market, or business services) to describe the properties of a requested service. Then, on behalf of the requester, it establishes interactions with service providers to fulfill the requests. The brokering agent is responsible for identifying and interacting with other integration services, such as resource discovery services and ontology manager services to accomplish its tasks. Another type of integration agent provides view-integration, which is a service to merge and map the description of business-objects (e.g., source schemas) in the *eMarketplace* supported by the business ontology into an integrated view or schema. For instance, a catalogue service might require information provided by several business entities supporting different product schemas. A view integration service provides the integration into a common definition language (e.g., XML-based), which is in turn mapped into a target representation language by a specialized language mapping service. View integration is responsible for identifying and interacting with several services to fulfill its functionality, including brokering, source-schema, ontology, and language-mapping services.

7 Multi-Attribute Negotiation Service: Coalition Deal Negotiation Model

Many researchers have investigated multiple issue negotiation [16][28][36]. One approach [16] developed an optimal agenda and procedure for two-issue negotiation. It introduced two negotiation procedures: *issue-by-issue negotiation* and *package deal.* However, for n-issue negotiation, over $n>2$ *issues*, which is a common setting in many *eMarketplace* applications, the computational cost to reach a package deal might exceed the benefits obtained by optimizing the participants' utilities and becomes impractical. Furthermore, these models often assume that private information of the business entities (software agents herein) is common knowledge while neglecting the associated computational cost. Therefore, these models do not fit typical dynamic and competitive environments. To deal with these challenges, namely utility optimization and computational efficiency, we propose an approach based on the *coalition deal* for multiple issue negotiation [13].

To this end, our approach of a coalition deal negotiation is to extend the properties of issue-by-issue negotiation and the package deal procedure with the flexibility to balance between time and utility.

Definition: *For a coalition deal, all negotiation issues are partitioned into disjoint partitions and each partition is negotiated independently of other partitions. Like the package deal, issues inside the same partition are negotiated as a whole package and an offer includes a value for each issue in this partition.*

Clearly, from the above definition, issue-by-issue negotiation can be treated as a specific case of a coalition deal with one issue per partition. The package deal can be viewed as a coalition deal with one partition for all issues. However, coalition-deal negotiation provides (a) better utility than issue-by-issue negotiation, (b) less computational cost than package deal negotiation, (c) more flexible negotiation, and (d) better management for negotiation. To discuss this formally, consider multiple-issue negotiation with set I of k issues, where $I=\{I_1,I_2,...,I_k\}$. Let IP be the set of all partitions of size s over I, where $IP=\{IP_j|1\leq j\leq s\}$, *where* IP satisfies the constraint: $\forall 1\leq m\leq s$, $1\leq n\leq s$, $m\neq n$, such that $IP_m \cap IP_n = \phi$ and $\cup_{j\in IP}\cup_{i\in j} i = I$. For two agents a and b, respectively they can be defined as a 4-tuple of a parameters:

$$S_a = \langle \mathrm{P}_a^{IP}, U_a^{IP}, T_a, \delta_a \rangle$$

$$S_b = \langle \mathrm{P}_b^{IP}, U_b^{IP}, T_b, \delta_b \rangle$$

where $\mathrm{P}_a^{IP} = \{P_a^i \mid i \in j, j \in IP\}$ denotes agent a's set of reserve prices over I and P_a^i denotes a's reserve price over issue i, which belongs to partition j, $U_a^{IP} = \{U_a^i \mid i \in IP\}$ denotes agent a's utility functions over IP where U_a^i denotes agent a's utility function over one partition I from IP, T_a and δ_a denote agent a's bargaining deadline and discount factor. Agent b's negotiation parameters are defined analogously. An agent's utility from IP of I is the sum of its utilities from all partitions. For a coalition deal, each partition is negotiated separately and independently of other partitions. An agreement can take place either on some of the partitions or all of them. For each partition, an offer includes a value for each issue inside this partition that would be the same as the package deal for this partition. This allows trade-offs to be made between issues inside this partition. An agreement has to take place either on all the issues inside the partition or none of them. For each partition, we assume that the agents use the same protocol as for the package deal, but instead of making a set of offers over I, an agent offers a set of offers over issues from this partition. An agent can make trade-offs only across issues in the same partition, resulting in a set of offer sets, all of which give it equal utility.

At time t, Agent b generates a set of offers over a partition IP_i of k_i issues that give itself the equal utility. We define $P_{b,t}^{IP_i} = \langle P_{b,t}^{IP_i(1)}, P_{b,t}^{IP_i(2)}, ... P_{b,t}^{IP_i(k_i)} \rangle$ as agent b's current utility optimal offer over partition IP_i for agent a if it gives a the maximum utility. For a coalition deal, each partition is considered using the package deal negotiation protocol. In this context, agent a's action $Ac_{a,t}$ for the coalition deal procedure is defined as follows:

$$Ac_{a,t} = \begin{cases} \mathit{Quit} & \mathit{if} \quad t \geq T_a \\ \mathit{Accept\ Package\ deal} \quad \mathit{for} \quad IP_i & \mathit{if} \quad U_a^{IP_i}(P_{b,t}^{IP_i}) \geq U_{a,t+1}^{IP_i}{}' \\ \mathit{Offer} \quad \mathrm{P}_{t+1}(U_{a,t+1}^{IP_i}{}') \quad \mathit{for} \quad IP_i \in IP & \mathit{otherwise} \end{cases}$$

Where $U_{a,t+1'}^{I}$ is the utility value for an agent to generate its count-offer at time $t+1$ over partition IP_i. Similarly, we define agent a as playing its equilibrium strategy for the package deal over a partition if $U_{a,t+1}^{IP_i}{}' = (1 - y_{a,t+1}^{IP_i})U_{\max,a}^{IP_i}$, where $U_{\max,a}^{IP_i}$ is the maximum possible cumulative utility agent a can get from partition IP_i. The equilibrium strategy for agent a and agent b over other partitions is defined analogously.

Coalition Deal Utility and Efficiency

For a given set of issues, $I=\{I_1,I_2,...,I_n\}$, and a partition $IP=\{IP_1,IP_2,...,IP_k\}$, we assume generating a price value for any issue require the same unit of computational cost. Furthermore, we assume issue-by-issue negotiation can be performed in parallel for each issue and the same for every partition in a coalition deal. Therefore, to compare the computational efficiency, we need to compare the computational cost of generating an offer in each round only. If each negotiation type approach requires the same number of rounds to reach an agreement, then we can compare their computational costs by comparing the cost of generating an offer in each round.

An n-issue negotiation can be viewed as a distributed search through an n-dimensional space. In issue-by-issue negotiation, each issue is negotiated separately. Therefore, this lead to a computational complexity of $O(m^n)$, where n is the size of the issue set and each issue may have m possible values. The complexity gets worst when we have to solve this problem each round using the package deal negotiation procedure. In coalition deal negotiation, however, issues are partitioned into k disjoint partitions and each partition is settled independently of the other partitions. Like the package deal, issues inside a partition are negotiated as a whole and an offer includes a value for each issue in the partition. Therefore, the computation problem is reduced to the sum of k independent search problems where the i-th search is in an n_i-dimensional space, where $n_i << n$ and $\sum_{i=1}^{k} n_i = n$. For sequential implementation the corresponding computational complexity is $O(km^{n_s})$, which can be reduced to (m^{ns}) using parallel implementation, where $n_s = \arg\max n_i$. Moreover, we can limit the maximum size of a partition to a constant C without losing value due to interdependency as discussed below. In this case, the computational cost of a coalition deal reduces to $O(nm^C)$. The computational complexity of the parallel implementation will be $O(m^C)$ to generate a coalition deal. Now we discuss some basic ideas for partitioning the set of issues:

- If agent's utility from issues is additive, the issue set could be partitioned arbitrarily and independently, agent can choose different partition sizes corresponding its computation capacity and time pressure.
- In the situation where some issues could be dependent and their utilities are in form of one multi-variable function, one variable for each issue. We can put the dependent issues into the same partitions and assume a more general form of additive utility in which each addend is a multi-variable function for dependent issues.
- It is always possible to form a better utility frontier for one partition by putting issues with distinct comparative interests into the same partition. In real world, people usually put their most interested issues with their opponents' most in-

terested issues into one basket. If the opponent's preference is unknown, people can put their most interested issues together with the issues that they most likely to compromise.

Therefore, the proposed coalition deal provides more flexibility to balance between computation costs and utility gains:

- A business entity can adjust the partition size based on its current time pressure and computation resource.
- Most distributed automated negotiation systems encounter a message congestion problem and the coalition deal can mitigate it. A message congestion problem occurs when an agent cannot process its received messages as fast as they arrive.
- If new issues are introduced by adding new partitions the coalition deal provides better scalability without affecting the existing issues.

Using negotiation mechanism to match supply and demand prevents consumers with low valuations from receiving limited goods and prevents suppliers with high production costs from supplying the limited demand.

8 *eAuction* Market Service

Auctions are dynamic and often efficient mechanisms for selling items in complex *eMarketplaces*. The proposed auction market here incorporates a Vickrey auction mechanism. In private-value Vickrey auctions, an agent's valuation is determined locally and independent from other agents' valuations. It provides incentives to promote truthful bidding among self-interested agents and avoid the computational cost of counter-speculations. Hence, there is no need for iterative negotiation strategies, dynamic strategic behavior to reveal other bidders beliefs, or high degree of security. Furthermore, it has been shown that this mechanism can achieve the same utility for the participants as other less direct mechanisms where bids might not be the true valuation [49].

8.1 The Auction Market System Architecture

Our proposed auction market is a collection of auction sessions along with some generic services required by all markets such as the market registration, advertisement and search engines. Each auction session has an initiator agent represented by the auctioneer-agent to sell or buy items or services under particular auction rules. The term "session" emphasizes the temporally extended nature of these auction events or the market and enables the formation of spot markets. A forward auction session (regular or a spot market) extends from the initial call-for-bids through the negotiation, awards, and final closing.

Generically, each session mechanism should ensure the continuity of competitive bidding and deliver dynamic tracking of the negotiation status. A multilateral market can also be applied for supply chain management, in which both supply and demand curves are constructed to determine a market-clearing price. In fact, several market mechanisms can be applied to model the multilateral market. For example, Wilson [53] initiated the study of double auction as a means to model multilateral trading. In such a double auction, all trades are made at a single market clearing price. McAfee [33] proposed a double auction model that explicitly considers the role of an auctioneer who intervenes in the trade and keeps track of supply and demand at asked and bid prices. Babaioff and Nisan [4] proposed protocols for exchange of information between multilateral markets along a single supply chain. Each market form a link in the supply chain operates as a double auction, where the bids on one side of the double auction come from bidders in the corresponding segment of the industry, and the bids on the other side are synthetically generated by the protocol to express the combined information from all other links in the chain.

One of the common services provided by an auction market structure is the "Registry", through which business-entities, suppliers and consumers, register with the *eMarketplace* and corresponding representative agents are created. The entries in the agent's registry database would include the identity of business-entity agent, its role as a supplier or consumer, its list of preferences, such as product, price range, and specific supplier brand. Potential suppliers advertise their goods or services with the *advertise agent*. The *search agent* represents the first interaction with the trading agents after they join the *eMarketplace*. It is deployed to expand the supplier and buyer agents' awareness by recommending a certain auction session based upon their reported preferences and/or product or service specifications.

Auction sessions can be seller centric "forward" auction or buyer (consumer) centric "reverse" auction. A seller-centric auction helps a supplier-agent in maximizing its revenue at equilibrium, whereas a buyer-centric auction helps a buyer-agent in identifying the lowest price at equilibrium. In bilateral markets a combination of double auction (both forward and reverse auctions) and supply-demand profile integration help a market-based supply chain determine efficient allocation.

Any potential supplier or consumer (buyer) agent that wishes to join an auction session is required to register once again with the auction's *registry service* a priori to be able to participate in the bidding process. Finally, the *auction session agent* can be viewed as a repository of auction protocols and auction initialization parameters. Once an auction event is decided upon, the auctioneer-agent is able to extract the auction parameters such as start bid, ask price, and increment bid from the *auction session agent*. In addition, the *auction session agent* will provide the auctioneer-agent with the specifications of interaction rules such as the way it will be performed either open or sealed, the direction of bidding progression either ascending or descending and the stopping criteria.

8.2 Market Session: Vickrey Auction

The basis of work presented here is limited to the Generalized Vickrey Auction (GVA). We restrict our discussion on the forward auction protocol. In a single item forward auction, each agent, A_i, has a "maximum willingness to pay" or "value" for the item that is assumed to be private information known only by the agent A_i, denoted by v_i. Then, the objective is to award the item to A^*_i with the highest bid value. Although the Vickrey auction ensures that the item is awarded to the highest bidder-agent, but it does not maximize the supplier's revenue unless "reserve prices" are imposed [25]. Given a set of combinatorial bids, the supplier-agent (through auctioneer-agent) then decides how best to allocate individual goods to those bundles for which bids were placed, with the objective to maximize revenue. Groves' mechanism in *combinatorial auctions* is a combinatorial allocation problem (CAP). It involves allocating bundles of items that may overlap to *n* bidders. CAP is an optimization problem equivalent to the weighted set packing problem [40]. The proposed auction market is comprised mainly of three agents: the *Auctioneer-agent, the Bidder-agent* and the *Supplier-agent.* Under the general Vickrey mechanism, it is in the interest (the dominant strategy) of the bidder to report its true valuation function. Then, the auctioneer-agent

- calculates the allocation (x_i^*) that maximizes the sum of the bids subject to the items constraint;
- calculates the allocation $(x_{\sim i}^*)$ that maximizes the sum of the bids other than that of bidder agent *i* such that it excludes all items allocated to agent *I*;
- announces the winners and their payment given by

$$p_i = \sum_{j \neq i} v_j(x_{\sim i}^*) - \sum_{j \neq i} v_j(x^*)$$

Under the assumption of quasi-linear preferences, each bidder-agent calculates its utility. For bidder-agent *i* the utility will be $u_i(x^*) = v_i(x^*) - p_i = v_i(x^*) - \sum_{j \neq i} v_j(x_{\sim i}^*) - \sum_{j \neq i} v_j(x^*)$

From the computational perspective, the above GVA introduces several challenges at two distinct levels. At the auctioneer-agent level, the winner determination is an *NP-hard* optimization problem in which the auctioneer needs to solve the problem twice: first with all participating agents reporting their preferences and then with each agent removed from the system to compute payments. At the bidder-agent level, agents must compute and communicate their complete valuations for an exponential number of bundles of items for different outcomes, each of which might involve solving a *hard local optimization problem.* These computational issues need to be resolved without losing the main game-theoretic properties, especially allocative-efficiency and strategy-proofness. Firstly, the GVA requires complete information revelation from each agent. Secondly, the valuation problem for a single bundle can be hard [40], and in combinatorial domains there are an exponential number of bundles to consider. Thirdly, agents must communicate that information to the auctioneer, which might be costly in network resources in addition to security problems

Fortunately, in sealed-bid auctions, as with our case, the bidding rules amount to no more than a bidding language, i.e., the syntax and semantics of bids. Agents would require a representation language, such as OR* [37], with expressive and compact methods to accurately express their preferences, make the bid structure more explicit and reduce the intractability of the valuation complexity. In other words, rather than requiring the agent to solve valuation problem locally, it simply sends the local problem specification directly to the auctioneer. In this manner, strategy-proofness is not reserved, given that the auctioneer can interpret the bidding language faithfully.

The second challenge concerning the computational complexity of the auctioneer's *winner determination* problem can be handled by several optimal and approximations algorithms [17]. In our case, once all agents have placed their bids in the OR* bidding language for any arbitrary set of bids, and if these bids conform linear-order bids, nested-hierarchical bids or a combination of both structures, then the linear program guarantees an exact optimal solution. This linear formulation can utilize standard algorithms and hence can be run directly on standard commercially available optimization software, such as CPLEX [26]. In other cases, the linear programming relaxation may not yield non-integral-solution. Thereby, heuristics can be used to somehow find an approximation to the optimal solution. A *greedy* algorithm that provably runs in polynomial time, however does not guarantee to always produce an optimal solution. Alternatively, a recursive *branch-and-bound* algorithm based on an upper bound LP solution can be used. This technique trims the search space in exponential searches and can provably produce an optimal solution, however, does not guarantee to always run in polynomial time.

8.2.1 Auction Session Module

In this section, we focus on the Vickrey auction market components and functionalities as an integral *eBusiness* model in a BCKOA-based *eMarketplace*. The auction market mainly recognizes three types of agents representing the auctioneer, suppliers, and consumers. These agents represent specialist market service agents that provide a collection of functions to achieve the intended business objective of auctions. However, the trading process mainly involves the auctioneer and consumers (or bidders) agents. Initially, both supplier- and consumer-agents register with the auction session. The supplier-agent then formulates a sell-order and assigns it to an auctioneer-agent that has the capability to carry out the selling task on its behalf in an auction session. The auctioneer-agent receives a request from the supplier-agent to process the sell order. Consequently, the auctioneer-agent formulates an "announcement" including all the auction parameters supplied by the supplier-agent such as the item details, quantity, minimum price and desired selling deadline. This "announcement" is encapsulated inside a "call-for-bid" message that is sent out to all potential bidder-agents registered with the current auction session.

At a certain time, the auctioneer-agent sends "propose" messages to all bidder-agents indicating the start of the bidding process. Bidder-agents express their preferences in OR* bid format and sends "bid" messages back to the auctioneer-agent. When the selling deadline reaches, the auctioneer-agent ceases to accept any messages and send a "reject" message for late arriving messages. Immediately, the bid evaluation process commences by comparing the structure of aggregated bids against certain classes of bids that indicate an optimal allocation of winners and calculation of payments is ultimately guaranteed. Otherwise, the auctioneer-agent runs a heuristic algorithm in order to arrive at a computationally feasible approximate winner allocation. Finally, the auctioneer-agent sends "inform" messages to all the bidder-agents as well as the supplier-agent to notify them of the winner and his payment. Each bidder-agent is responsible for notifying the consumer of the auction outcome. The supplier-agent must also inform the human supplier (or the supplier's and consumer's representative-agents in this case) of the auction outcome at the end of the auction session.

8.2.2 Agent Architecture

Each agent's architecture is based on the CIR-agent model [21]. The primitive components of the CIR-agent are arranged in a particular order to represent each agent's role and capability. Basically, each agent consists of *knowledge* and *capability* packages, each of which is tailored according to the agent's specific role. An auction session mainly recognizes three types of agents, namely, supplier-agent, auctioneer-agent and bidder-agent. The architecture of each agent type is described in detail below.

An auctioneer-agent is composed of two packages: *knowledge* and *capability*. The *knowledge* package contains the information in the agent's memory about the environment and the expected world. This includes the agent self-model, other agents' model, goals that need to be satisfied, possible solutions generated to satisfy each goal, and the local history of the world that consists of all possible local views for an agent at any given time. The agent's knowledge also includes the agent's desires, commitments and intentions toward achieving each goal.

The *capability* package includes the *reasoning* component, the *domain actions* component which contains the possible set of domain actions that when executed the state of the world will be changed, and the *communication* component where the agent sends and receives messages to and from other agents and the outside world. The *reasoning* component is further decomposed into the *problem-solver* component and *coordination* component. The *problem-solving* component maps the agent's goal into a solution. For example, the auctioneer's goal is to determine a market price for an item of unknown value. The problem-solver involves aggregating bids and evaluating them against a certain criteria. The *coordination* component includes several interaction devices that deal with various interdependency problems, such as capability dependency and conflict of interest [21].

Fig. 5 depicts the logical structure of the *reasoning* component representing the transformation process of the mental state of the auctioneer-agent, once a goal pops-up. Therefore, the arrival of OR* bids via agents communication messages coming from the bidder-agents through the *communication* component assigns a goal-state. The problem-solver component is responsible for the collection of bids and the winner determination process to allocate the item or bundles and their respective prices. Through the *interaction* component the auctioneer-agent sends out an "accept" message at the beginning of auction session informing the supplier-agent that the auction will be carried on its behalf. Once the auction ends, the outcome must be reported to all participants. This information is encapsulated in outgoing messages and sent to the outside world through the *communication* component.

Similarly, as shown in Fig. 6, the bidder-agent is designed using the same CIR-Agent architectural principles. The exception is the classes that constitute the *problem solver* component, which are tailored according to its specific role and functionality as described in Vickrey auction session module. The bidder-agent interacts with the consumer via a user-interface designed for the OR* language to express the atomic bids as well as constraints that signify which item/bundle is mutually exclusive. The formulated bid represents the body of the agent's communication message that will be sent to the auctioneer-agent via the communication module.

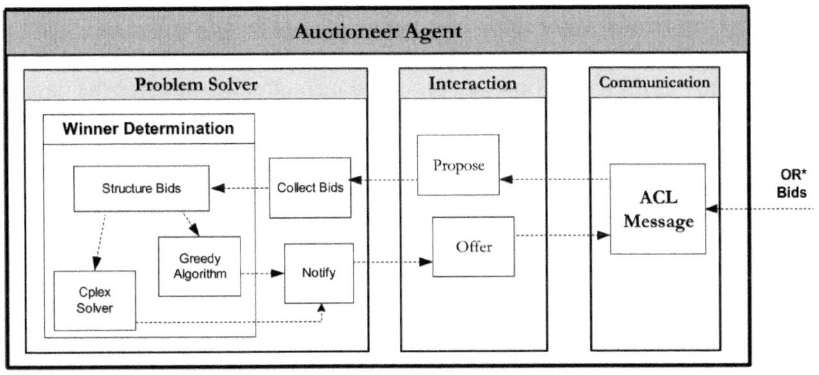

Fig. 5 Architecture of the auctioneer agent's reasoning component

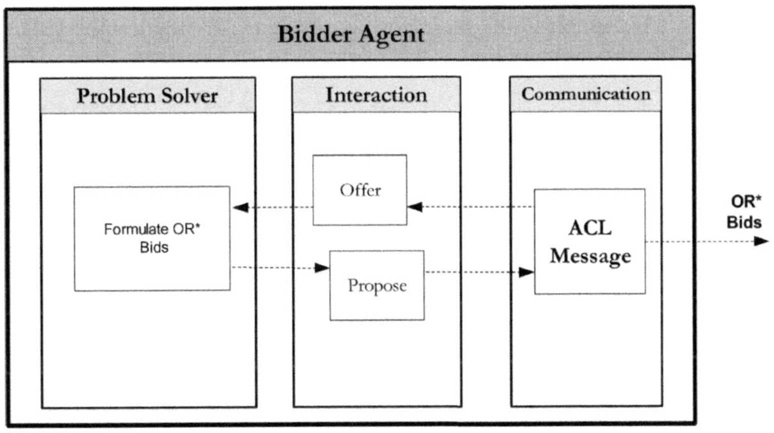

Fig. 6 Architecture of the bidder agent's reasoning component

9 Agent-Based Supply Chain Integration Service

An *eMarketplace* enables the creation and leveraging of services and supply operations in a way that seamlessly integrates business participants in a dynamic trading community. Negotiations and auction-based markets provide efficient coordination mechanisms for the supply chain. As described in Section 3, in relationship-based supply chains, long-term relationships usually have higher value than that of the efficient resource allocations. BCKOA *eMarketplace* addressed these challenges by providing a platform that combines both relationship-based and market-based coordination mechanisms. Utilizing multi-issue negotiation services the agents can create a spot market, in which the set of issues can include both price and non-price attributes. According to different context, agents may adopt different mechanisms to balance between long-term relationships and prices in a negotiation by assigning different weights to relationship issues and price issues in a utility function.

For supply chain integration, our proposed methodology requires the description of the supply processes involved between participating business entities using OAG standard business documents and UML interaction diagrams. The methodology begins with capturing a supply chain scenario and its associated UML interaction diagrams, exemplified in Fig. 7. The interactions, as shown in Fig. 7, consist of the exchange of structured documents, such as the OAG business-object documents (BODs). For B2B interactions, a ProcessPO BOD is a *directive* that carries the composite semantics of *request* and *inform*, in which the sender requests that the recipient evaluates a purchase order and informs the sender of the results. The informal semantics is that ProcessPO will be followed by a response from the recipient and that the response will be either an AckPO or a DeclinePO. Using the semantics of each document, the messages in the interaction diagram are converted into a bipartite conversation graph (not shown here), which delineates each participant's conversations. A bipartite conversation graph is used to identify the roles of the participants in B2B transactions. This graph is the basis for constructing Dooley graphs [14], shown in Fig. 8 as collaboration diagrams.

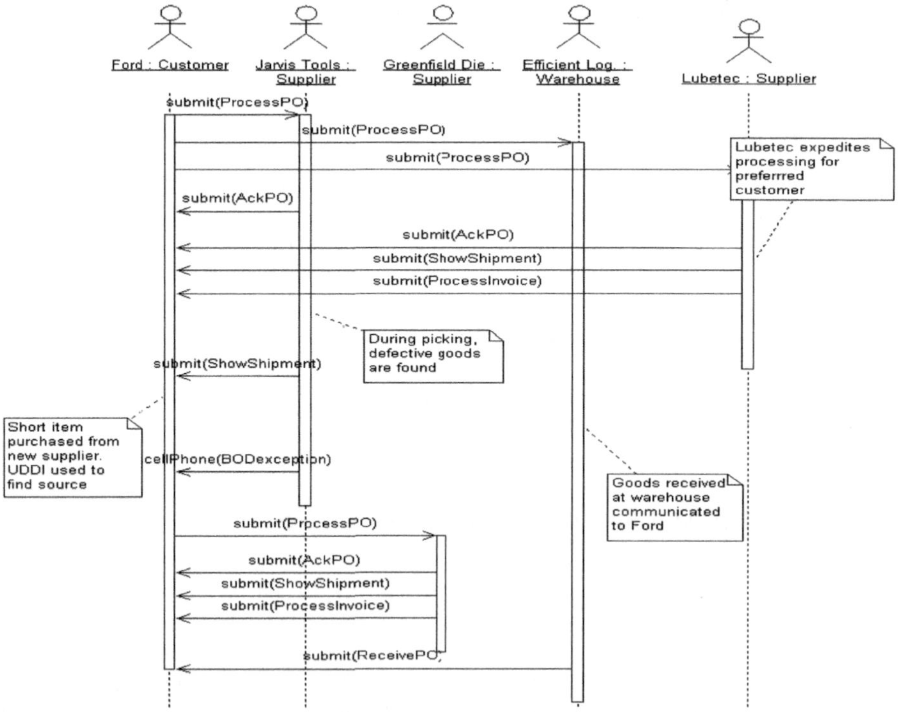

Fig. 7. Interaction Diagram for the OAG Scenario Involving Ford and its Suppliers

Note that collaboration participants can fill different roles at different times, and thus can be involved in many conversations simultaneously. Each of the roles identified in the collaboration diagram can be assigned to an agent in the supply chain. Moreover, the diagram for each role is mapped directly into a state-machine description for the agent's behavior. The corresponding agents then carry on the roles and for the affiliated business entities, and then collectively they manage the supply chain process.

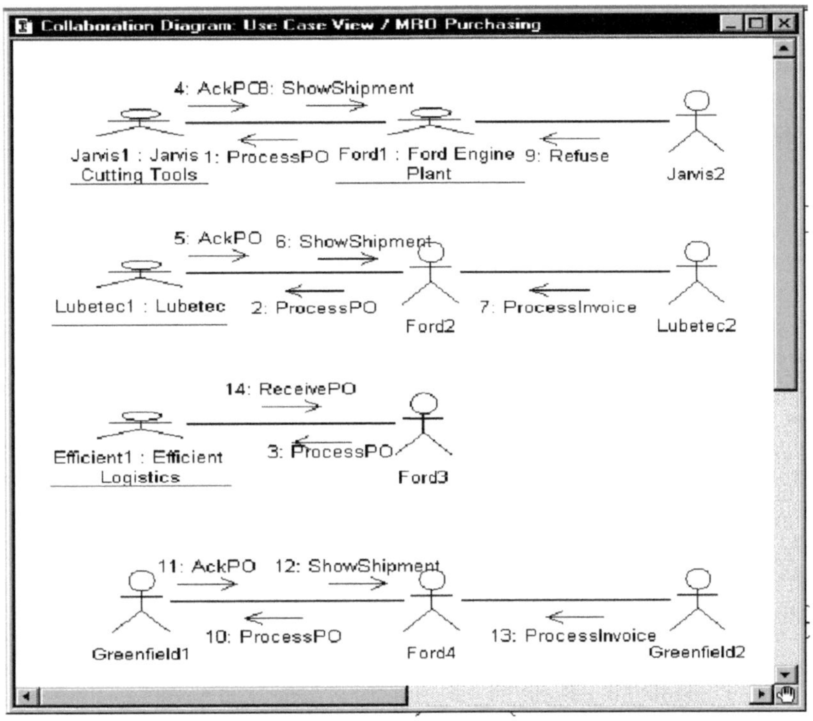

Fig. 8. Collaboration Diagram with Participant Roles for Ford Interoperability Scenario

The methodology, summarized in Fig. 9, uses—and begins to formalize—the BODs that OAG and RosettaNet are standardizing. It provides a basis for the convergence of multiple standards for supply chain management, which could become ready-to-use technology for different participant business entities in the *eMarketplace.*

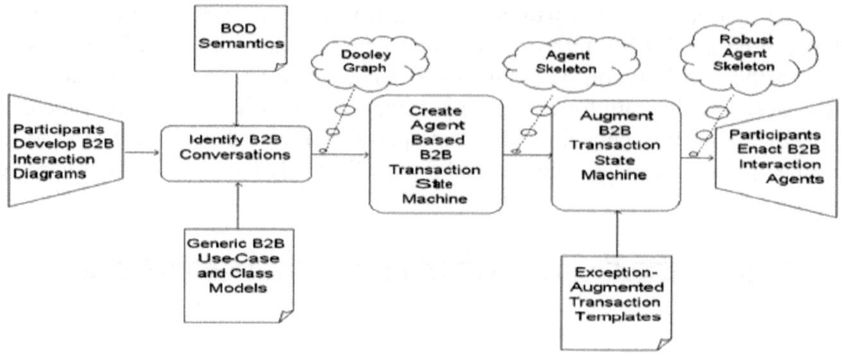

Fig. 9. Agent-Based Coordination Methodology for B2B Automation

10 Prototype Implementation

To validate and experiment with our analyses and foundations described in the previous sections, we have developed a prototype of an agent-oriented BCKOA for an *eMarketplace,* as shown in Fig. 4. *ABC* Corp and *XYZ* Inc. are virtual business entities registered with the *eMarketplace* for both purchase and sales services. Both organizations use a BCKOA-based computation environment. Individual customers or business-entity personnel in the *eMarketplace* can participate in the market through their user interface agents. Similarly, an agent in the eMarketplace represents each business-entity service. These agents provide thin, intelligent, highly autonomous interfaces for the business-entity services that might be based on legacy applications. For example, the *ABC* purchasing-service agent represents the implementation of the business-specific purchases by *ABC* in the *eMarketplace.* Each user interface and business-entity service agent is registered in the *eMarketplace.* Thus, a user interface agent can benefit from the market, business-specific, and business-entity services by interacting with their representative agents. Each business-entity service must also be registered with a registry agent for the corresponding business-specific service. Each layer, and its registry services, are intended to provide some aspect of information about the *eBusiness* environment and enable an interested party to obtain information to potentially use offered services, or to join the *eMarketplace* and either provide new services or interoperate as a trading partner with other business-entities in the *eMarketplace.*

In the current prototype, we experiment with auction market structures, as shown in Fig. 4, where customers and suppliers are brought together to trade with each other and prices are set by the selected market structure. An individual customer is able to participate in the market through a dedicated user interface agent possibly assigned by the *eMarketplace.* Similarly, each participating business entity is assigned to a team of CIR-Agents for the registered services and representative personnel who might have a direct contact with the market as well as with the customers. The market ontology provides a conceptualization of the domain at the knowledge level.

The implementation utilizes the JADE platform [6], which is a software framework to develop agent applications in compliance with the FIPA specifications (The Foundation for Intelligent Physical Agents 1998) for multi-agent systems. JADE deals with all aspects that are external to agents and independent of their applications. These include message transport, encoding, parsing and agent lifecycles. JADE supports a distributed environment of agent containers, which provide a run-time environment optimized to allow several agents to execute concurrently. This feature has been utilized to create several concurrent market sessions, such as commodity and auction sessions. JADE provides support for standard FIPA ontologies and user-defined ontologies. Although our implementation takes advantage of the JADE platform and its supporting agents, such as a directory facilitator, the architecture of the application agents is based on the CIR-Agent model (shown in Fig. 3). Java features, such as portability, dynamic loading, multithreading, and synchronization support make it appropriate to implement the inherent complexity and concurrency in an *eMarketplace*. These features are also instrumental for executing the CIR-Agents concurrently. The design of each agent is described in terms of its knowledge and capabilities. The agent's knowledge includes the agent's self-model, goals, and local history of the world, as well as a model of its acquaintances. The agent's knowledge also includes its desires, commitments, and intentions as related to its goals.

The main capabilities of the CIR-Agent include communication, reasoning, and domain actions. Implementation of the communication component takes advantage of JADE messaging capabilities. It is equipped with an incoming message inbox, whereby message polling can be both blocking and non-blocking, and with an optional timeout mechanism. Messages between agents are based on the FIPA ACL. The agent's reasoning capabilities include problem solving and interaction devices. The problem solving of an agent is implemented through the use of complex behaviors. Behaviors can be considered as logical execution threads that can be suspended and spawned. The agent keeps a task list, containing active behaviors. The problem-solving component varies from one agent to another as will be the in the following subsections. The agent behaviors can be classified as follows: behaviors that are concerned with market services, such as a market-registry and auction services; and behaviors that are concerned with providing business-specific services, such as selling and purchasing.

10.1 Auctioneer-Agent

The role of the auctioneer agent is to formulate the auction proposal, sends it out to all registered bidder-agents and collects the bids. At the end of the bidding-time, it evaluates the bids, determines the winner bidder-agent along with the allocation and payment and notifies both the supplier-agent and bidder-agents of the outcome. As described before, the auctioneer-agent is a collection of *knowledge* and *capabilities* components. The knowledge component includes the agent's self-model, model of other agents, and the local history. The main capabilities of the CIR-Auctioneer agent include *communication, reasoning* and *domain actions.* The *communication* component consists of JADE *ACLMessage* class that is implemented by the problem-solver or the coordination components to construct messages via several FIPA performatives such as REQUEST, INFORM, QUERY-IF. The auctioneer-agent's *reasoning* component consists of the *problem-solver* and *coordination* components, which are described below in detail.

The auctioneer-agent's *problem-solver* component contains a set of *Behavior* classes that represent the auctioneer-agent's specific tasks. These classes implement all the agent's platform tasks such as registration with the DF service. Also, they represent *simple behaviors* and *cyclic behaviors* to handle the incoming messages from both the supplier-agent and the bidder-agents. A basic set of methods that can be called to implement the application tasks of the auctioneer-agent such as the "*callForProposal*" method that formulates the "*announcement*" of an auction event, the "*updateBidList*" method that checks the auction's start time, and collects the bids and the "*informAuctionResults*" method for notification purposes. The winner determination class is described in detail below.

In case of combinatorial auctions, the auctioneer-agent has to solve the winner determination problem as an optimization problem expressed as an integer-program that is further relaxed into a linear-program. The "*structureBid*" method in BidEvaluation class determines whether a tractable or intractable solution is necessary according to certain restrictions on the size of OR*-bids. Interfacing to standard algorithms such as the CPLEX solver solves the tractable cases. AMPL (A Modeling Language for Mathematical Programming) was used to formulate a model of the optimization problem, as shown in Fig. 10, in order to be input to the CPLEX Solver. AMPL is a comprehensive and powerful algebraic modeling language for linear and nonlinear optimization problems, in discrete or continuous variables. AMPL's familiar algebraic notation makes it ideal for rapid prototyping and development of models as well as it can communicate with a wide variety of solvers. For intractable cases we use a greedy allocation algorithm [37] to find an approximate allocation.

The coordination component contains a class that extends a JADE behavior class namely the FIPA *ContractNetInitiatorBehavior* through which the auctioneer agent sends an INFORM message, which contains the auction specifications and the sell parameters to a set registered bidder-agents. The bidder agents can respond either by a PROPOSE-message (followed later by a BID), or a REFUSE message to reject participation in the auction and finally and NOT-UNDERSATND message in which case the auctioneer agent has to resend an INFORM message with the auction specifications and the sell parameters.

```
set n;                              # n is set of atomic bids
param p {i in n};          # p is bundle price
var x {i in n};                     # x is the winning bundle
maximize Total_Price: sum {i in n}x[i] * p[i];
# objective total of winning prices is maximum
subject to sum_Bundle: sum {i in n} x[i] <= 1;
#constraint: without allocating a bundle more than once
subject to Bundle_value {i in n}: 0 <= x[i]; # non-negative bundle
values
```

Fig. 10 Combinatorial Auction model in AMPL

10.2 Bidder-Agent

The role of the bidder-agent is to express its preferences in OR* bid format and sends "bid" messages out to the auctioneer-agent. The *problem-solver* component contains a Bid class that implements a *cyclic behaviour* in order to respond to incoming messages from the auctioneer-agent that requests bids. This class implements all the bidder-agent's tasks such as registration with the DF service as well as a single method that formulates preferences and bid valuations in OR* language. The *coordination* component contains a class that extends the JADE behavior class *ContractNetReponder Behavior* through which the bidder-agent prepares the PROPOSE message that is later followed by the formulated OR* Bids or the REFUSE message to refuse the proposal or the NOT-UNDERSTAND message to request for resending the information of the announcement again.

10.3 Supplier-Agent

The supplier-agent's role is to formulate the sell-order and assigns it to the auctioneer-agent. The supplier-agent's reasoning component consists of the following: The implementation of the supplier agent's *problem-solver* component, includes the "*SellOrder*" class that implements a method called *processSellOrder* that displays the user-interface through which the human supplier enters the item descriptions. The *SellOrder* class implements a *cyclic behaviour* that responds to incoming messages from the auctioneer-agent that either notifies the supplier of the auction outcome or requests the supplier-agent to leave the auction once the auction session has ended. This component contains the "*SellOffer*" class. It formulates a REQUEST message to ask the auctioneer-agent to carry out the selling task on the behalf of the supplier-agent. It also implements a simple behavior to respond to auctioneer-agent ACCEPT OR REJECT messages.

11 Related Work and Discussion

There have been several recent attempts to promote *eMarketplace* models by the academic and industrial communities. For example, the electronic market-place (EMP) [8] was an attempt to develop a business-to-business system architecture. It is viewed as a DBMS solution to support many-to-many relationships between customers and suppliers. The Global Electronic Market (GEM) [42] attempted to develop a logical market framework and infrastructure. A main objective was to separate system-related and market-related design issues. In GEM, the market provides trading mechanisms that include bids and offers. A more complex architecture for *eMarketplace* is MAGMA [48], with its special focus on the infrastructure required for conducting commerce on the Internet. In MAGMA, the *eMarketplace* has been viewed in terms of three main functionalities, namely, traders, advertising, and banking. Alternatively, OFFER [7] proposed a brokering-based marketplace. In OFFER the *eMarketplace* is viewed as a collection of suppliers, customers, and brokers. A customer can search for a service either directly in the e-catalog of the supplier or use the e-broker to search all the e-catalogs of the suppliers that are registered with this broker. E-brokers employ a simple auction mechanism. In a different approach, MOPPET [3] proposed an *eMarketplace* system as agent-oriented workflows. MOPPET viewed the market as a workflow management system carried out by several types of agents: task, scheduling, facilitator, and recovery agents.

Another approach was driven by the bottom-up modeling of market processes with self-organizing capabilities [2]. The objective was to develop a computational study of economies modeled as evolving systems of autonomous interacting agents, and known as agent-based computational economics (ACE) [29][46]. The ACE researchers relied on computational laboratories [33] to study the evolution of decentralized market economies under controlled experimental conditions. The goal was to develop analysis tools that enable an economist to test economic theories developed using standard modeling approaches

Several companies have emerged to automate logistics and re-supply within specific industrial segments. For example, Ariba [1] developed a marketplace based on procurement portals and dynamic exchanges for horizontal marketplaces. Ariba Dynamic Trade, for instance, attempts to provide dynamic trade mechanisms, such as auctions, reverse auctions, and bid/ask exchanges and negotiations. SAP Service Marketplace (SAP Services Marketplace, SAP AG) is an Internet portal for the SAP community. It provides basic online services such as catalog browsing, matchmaking, and ordering from SAP and its partners. Other approaches were directed to support vertical marketplaces, such as PaperExchange [39], that enables customers and suppliers to negotiate pricing and transact directly with one another. PaperExchange also attempts to provide several supporting services, such as logistics and clearing services, industry-specific job listings, industry events, news headlines, and a resource directory. VerticalNet [50] also built a set of Web-based marketplaces for specific industrial segments, such as financial services, healthcare, and energy. Each Web site forms a community of vendors and customers in a specific area. Vertical trade communities are introduced in segments with a substantial number of customers and suppliers, fragmentation on both the supply and demand sides, and significant on-line access

Another direction adopted by many major software vendors is to develop Internet-based commerce platforms. Examples are IBM CommercePOINT [25], Microsoft Site Server Commerce Edition [35], Oracle Internet Commerce Server INTERSHOP [27], and Sun JavaSoft JECF (Java Electronic Commerce Framework) (Sun Microsystems). These proprietary attempts focus on providing infrastructure services such as security payment directories and catalogs to be integrated with existing systems and the Web.

The proposed agent-oriented BCKOA *eMarketplace*, however, provides a framework of enterprise integration that deals with several systems and business issues. Unlike the above-mentioned attempts, it is fundamentally based on business integration rather than systems integration. The objective is to develop an architecture that is semantically rich in describing an organization and the interconnection among all elements of the *eMarketplace*, including people, business services, software components, and business ontologies. Technologically, BCKOA is service-oriented in the sense that it enables business entities and their supporting systems to join the market at the highest abstraction level (service level) with minimum overhead and independent of specific technology. Utilizing BCKOA, the business-object implementations of the participating services can be integrated in the execution environment. In addition, BCKOA provides an appropriate architecture for an *eMarketplace*. The form of the architecture supports *eMarketplace* functions that are inherited from real-world marketplaces. They are complex and nondeterministic, yet they characterize a real business environment. A BCKOA *eMarketplace* provides an integration environment for the broad-based services that are required for interaction and directory services, dynamic trading, cooperative supply chain integration and management. Therefore, we believe that the BCKOA-based *eMarketplace* is appropriate for integrating horizontal business services, vertical business services, specific business functionalities, and leveraging legacy systems in a way that supports end-to-end integration. Furthermore, a BCKOA *eMarketplace* provides a wide variety of coordinating and trade mechanisms to fit multiple business models. In our research we have applied a flexible computational economy model for the market services layer. Therefore, a BCKOA-based *eMarketplace* incorporates mechanisms for different types of market structures.

We believe that agent-orientation is an adequate paradigm for producing the information architecture of next-generation *eBusiness* systems, especially *eMarketplaces*. Agent technology richly enables and supports the automation of complex tasks and yields systems that are reliable and able to assume the responsibilities of the *eMarketplace* in which they compete. The components of agent-based BCKOA, namely, *eMarketplaces*, business entities (products, suppliers, customers, etc.), and the foundation (integration) architecture and services that glue them together, are essential to building robust many-to-many value chains in emerging *eBusiness*.

12 Conclusions

This chapter presents our ongoing research on developing an agent-oriented architecture for an *eMarketplace* that provides intelligent integration of collaborative supply chains. The objective is to develop an engineering foundation for the *eMarketplace*. To this end, several *eMarketplace* design and integration issues have been addressed. The focus is on developing a foundation architecture that supports services ranging from baseline integration services to specialty business-related services, and to integrate business organizations in an open market environment.

This chapter presents an agent-based Business-Centric Knowledge-Oriented Architecture (BCKOA) for an *eMarketplace* model. BCKOA is a service-oriented architecture for cooperative distributed systems independent of any specific technology. In the BCKOA based *eMarketplace* several types of agent roles are identified: user-interface, business-specific services, market services, and integration services. This chapter proposed an agent-oriented dynamic trading mechanism (Vickrey auction) and multi-issue negotiation mechanisms for supply chain. Also it proposed a methodology based on OAG standard business documents and UML interaction diagrams for supply chain integration. A prototype of BCKOA using CIR-Agent model and a

FIPA-compliant platform, JADE, is developed with special focus on an auction market structure.

Currently, the objective is to demonstrate the feasibility and the effectiveness of the proposed architecture as a design and integration model for *eMarketplaces* platform for supply chain management and integration. In continuing our research, the computational effectiveness of the architecture will be the main concern. Also, we will expand the application and the implementation of our prototype *eMarketplace* to investigate the most appropriate techniques to support secure, reliable, and effective transactions.

Acknowledgements

This work has been partially supported by Mackenzie Financial Corporation and National Sciences and Engineering Research Council of Canada (NSERC). The authors would like to thank Mr. Hussam Dafaalla and Dr. Yinsheng Li for their help in the implementation of the prototype and the SOAStudio.

References

1. Ariba, B2B Marketplaces in the New Economy, (n.d). from http://www.commerce.net/other/research/ebusiness-strategies/2000/00_07_r.html, (October 19, 2000).
2. W.B. Arthur, J. Holland, B. LeBaron, R. Palmer, and P. Tayler. Asset pricing under endogenous expectations in an artificial stock market model. Proceedings on the Economy as an Evolving Complex System. W.B. Arthur, S.N. Durlauf, and D.A. Lane (Eds.). (1997).
3. S. Arpinar, S. A. Dogac, N. Tatbul, An Open Electronic Marketplace through Agent-based Workflows: MOPPET, International Journal on Digital Library, 3 (1) (July,2000).
4. M. Babaioff and N. Nisan, "Concurrent auctions across the supply chain", Proceedings of the 3rd ACM conference on Electronic Commerce, 1-10. ACM Press, 2001
5. A. Bartelt, W. Lamersdorf, A Multi-Criteria Taxonomy of Business Models in Electronic Commerce, Proceedings of the IFIP/ACM International Conference on Distributed Systems Platforms, (2001).
6. F. Bellifemine, A. Poggi, and G. Rimassa, JADE – A FIPA-compliant agent framework, Proceedings of PAAM'99, London, (April 1999) 97-108.
7. M. Bichler, C. Beam, and A. Segev, OFFER: A Broker-centered Object Framework for Electronic Requisitioning, IFIP Conference Trends in Electronic Commerce, (1998).
8. S. Boll, A. Gruner, A. Haaf, and W. Klas, EMP-a database driven electronic marketplace for business-to-business commerce on the Internet, Journal of Distributed and Parallel Databases, Special Issue on Internet Electronic Commerce 7 (2) (1999, April).
9. G. Cachon and M. Lariviere, "An Equilibrium Analysis of Linear, Proportional, and Uniform Allocation of Scarce Capacity," IIE Transactions 31, No. 9, 835-850 (1999).
10. D. Carlton, "The Theory of Allocation and Its Implications for Marketing and Industrial Structure: Why Rationing is Efficient," Journal of Law & Economics 34, No. 2, 231-262 (1991).
11. DAML-S 0.7 Draft Release, (n.d). http://www.daml.org/services/daml-s/0.7

12. J. Dang, D. Shrotri, and M. N. Huhns, Distributed Coordination of an Agent Society Based on Obligations and Commitments to Negotiated Agreements," In Challenges in the Coordination of Large Scale Multiagent Systems, P. Scerri, Ed.: Springer Verlag, Lecture Notes in Computer Science, 2005.
13. J. Dang and M. N. Huhns, "Coalition Deal Negotiation for Services," presented at 1st International Workshop on Rational, Robust, and Secure Negotiations in Multi-Agent Systems, AAMAS 2005, Amsterdam, Netherland, 2005.
14. R.A. Dooley, Repartee as a Graph. In R.E, Longacre, (Ed.) An Anatomy of Speech Notions, (Lisse: Peter de Ridder Press, 1976) 348-358.
15. M. Dubosson, A. Osterwalder, and Y. Pigneur, eBusiness Model Design, Classification and Measurements. Thunderbird International Business Review, 44 (1) (2002).
16. S. S. Fatima, M. Wooldridge, and N. Jennings, "Optimal negotiation of multiple issues in incomplete information settings," presented at Third International Joint Conference on Autonomous Agents and MultiAgent Systems (AAMAS'04), New York, USA, 2004.
17. Y. Fujisima, K. Leyton-Brown, and Y. Shoham "Taming the computational complexity of combinatorial auctions", In Proceedings of the Sixteenth International Joint Conference on Artificial Intelligence, Stockholm, (1999) 548-553.
18. L. Garciano and L. Kaplan, "The Effects of Business-to-Business E-Commerce on Transaction Costs," Journal of Industry Economy 49, No. 4, 463-485 (2001).
19. M. Genesereth, An Agent-Based Approach to Software Interoperability, In Proceedings of the DARPA Software Technology Conference, (1992).
20. H. Ghenniwa, eMarketplace: Cooperative Distributed Systems Architecture, 4th International Conference on Electronic Commerce Research, Dallas, Texas, USA. (November 2001).
21. H. Ghenniwa, and M. Kamel, Interaction Devices for Coordinating Cooperative Distributed, Intelligent Automation and Soft Computing, 6 (2) (2000) 173-184.
22. W. Grey, T. Olavson, and D. Shi, "The role of e-marketplaces in relationship-based supply chains: A survey" IBM Systems Journal, Vol. 44, No 1, 109-123 (2005)
23. M. Hammer, and J. Champy, Reengineering the Corporation, (Harper Collins 1993).
24. N. Ivezic, M. Barbacci, D. Libes, T. Potok, and J. Robort, An Analysis of a Supply chain Management Agent Architecture, Proceedings of the Fourth International Conference on Multiagent Systems, (IEEE Computer Society Press, Los Alamitos, Calif., July 2000) 401-402.
25. IBM Corporation CommercePOINT Payment. (n.d). From http://www.internet.ibm.com.commercepoint.payment
26. ILOP CPLEX Suite, (n.d.), From www.cplex.com.
27. Intershop Communications, Inc. Intershop 3, from http://www.intershop.com. (1998).
28. C. M. Jonker and V. Robu, "Automated Multi-Attribute Negotiation with Efficient Use of Incomplete Preference Information," presented at Third International Joint Conference on Autonomous Agents and MultiAgent Systems (AAMAS'04), New York,USA, 2004.
29. B. LeBaron, Agent-based computational finance: Suggested readings and early research, Journal of Economic Dynamics and Control, 24 (2000) 679-702.
30. Y. Li, H. Ghenniwa, and W. Shen, Integrated Description for Agent-Oriented Web Services in eMarketplaces, Proceedings of AI'2003 Workshop on Business Agents & the Semantic Web, Halifax, NS, (2003) 11-17.
31. Y. Li, H. Ghenniwa, and W. Shen, Web Service-Oriented Agents and Integration Description, to appear in Computational Intelligence a special Issue on Business Agents and the Semantic Web, (2004).
32. K. McCabe, S. Rassenti, and V. Smith, Institutional Design for Electronic Trading, Conference on Global Equity Markets, N.Y. University, Salomon Center, (1992).

33. R.P.A McAfee, “dominant strategy double auction”, Journal of Economic Theory, 56(2):434–450, (1992).
34. D. McFadzean, D. Stewart, and L. Tesfatsion, A computational laboratory for evolutionary trade networks, IEEE Transactions on Evolutionary Computation, 5 (2001) 546-560.
35. Microsoft Corporation, Internet Commerce, from http://www.microsoft.com. (1998).
36. T. D. Nguyen and N. Jennings, "Coordinating multiple concurrent negotiations," presented at Third International Joint Conference on Autonomous Agents and MultiAgent Systems (AAMAS'04), New York,USA, 2004.
37. N. Nisan, Bidding and allocations in combinatorial auctions, In ACM Conference on Electronic Commerce, (2000).
38. M. J. Osborne and A. Rubinstein, A Course in Game Theory: The MIT Press, 1994.
39. Paperexchange Marketplace. (n.d). From http://www.paperexchange.com
40. D. Parkes, iBundle: An Efficient Ascending Price Bundle Auction, In Proceedings of the first International Conference on Electronic Commerce, (ACM Press 1999) 148-157.
41. H. Van Dyke Parunak, Visualizing Agent Conversations: Using Enhanced Dooley Graphs for Agent Design and Analysis, Proceedings of the Second International Conference on Multiagent Systems, AAAI Press, Menlo Park, Calif., (1996) 275-282.
42. B. Rachlevsky-Reich, and I. Ben-Shaul, et al., GEM: A Global Electronic Market System, Information Systems Journal, Special Issue on Electronic Commerce, 24 (6) (1999).
43. T. Sandholm and V. Lesser, "Issues in Automated Negotiation and Electronic Commerce: Extending the Contract Net Framework," presented at First International Conference on Multi-Agent Systems (ICMAS'95), San Francisco, CA, USA, 1995.
44. M.P. Singh, Synthesizing Coordination Requirements for Heterogeneous Autonomous Agents, Journal of Autonomous Agents and Multi-Agent Systems, 3 (2) (2000) 107-132.
45. A. Smirnov, and C. Chandra, Ontology-based Knowledge Management for Cooperative Supply Chain Configuration, Proceedings of AAAI Spring Symposium Bringing Knowledge to Business Processes, Stanford, California, (AAAI Press, March 20-22, 2000).
46. L. Tesfatsion, (Ed.) Special Double Issue on Agent-Based Computational Economics, Journal of Economic Dynamics and Control, 25 (3-4) (2001).
47. P. Timmers, Electronic Commerce: Strategies and Models for Business-to-Business Trading, ., (Wiley & Sons Ltd., 1999).
48. M. Tsvetovatyy, M. Gini, B. Mobasher, Z. Wieckowski, MAGMA: An Agent-Based Virtual Market for Electronic Commerce, Journal of Applied Artificial Intelligence, special issue on Intelligent Agents, 11 (6) (September1997) 501-523.
49. H. R. Varian, Mechanism Design For Computerized Agents, The First USENIX Workshop on Electronic Commerce, New York, 11 (19) (July 1995) 13-21.
50. VerticalNet® Marketplaces. (n.d). From http://www.VerticalNet.com
51. V. Vickrey, Counter Speculation, Auctions, and Competitive Sealed Tenders, Journal of Finance, 16 (1961) 8-37.
52. G. Wiederhold, Mediators in the Architecture of Future Information Systems, IEEE Computer, 25 (3) (March 1992) 38–49.
53. R.B. Wilson, “On equilibria of bid-ask markets”, in Arrow and the ascent of modern economic theory, New York Univ. Press, New York 375–414. (1987).

Software Agents for Electronic Business: Opportunities and Challenges (2005 Re-mix)

Jörg P. Müller[1], Bernhard Bauer[2], Thomas Friese[3], Stephan Roser[2], Roland Zimmermann[4]

[1] Clausthal University of Technology, Germany
[2] University of Augsburg, Germany
[3] University of Marburg, Germany
[4] University of Erlangen-Nuremberg, Germany
Email: mueller@in.tu-clausthal.de

Abstract. Despite the consolidation of expectations related to information technology that we have seen over the past three years, Electronic business (e-Business) remains a major driver of information technology. In this paper, we investigate how e-Business can benefit from methods, components and solutions based on software agent technology. We give an overview of the current electronic business mainstreams and outline a number of important challenges e-Business has to tackle. By means of three example case studies, we investigate the required extent and the means to apply agent technology successfully in e-Business applications in order to resolve some of these challenges.

1 Introduction

When the first version of this paper was written about four years ago, electronic business was on the rise towards the top of the hype curve. The rapid emergence of the World Wide Web and the first seemingly sensationally successful examples of Electronic Commerce / Electronic Business (e-Business) solutions and products created an image of electronic business as a potential revenue generator and technology driver beyond rationality.

Just four years later, the situation has changed dramatically. The main keywords characterizing this development are terms such as globalization, total cost of ownership, mergers and acquisition, and enterprise interoperability. There is a strong pressure on IT strategy and operation departments to cut costs (corporate IT spending has continually shrunk from 2001 to 2004, and is expected to grow only very moderately over the next five years [3]). Thus, e-Business is no longer mainly an instrument for increasing profit, but for reducing cost, speeding up business transactions, and supporting (as opposed to hindering) change and transformation processes. Thus, e-Business has matured and the initial expectations to it have been cut back to a more realistic scope.

On the other hand, we have been witnessing a number of important social and technological developments over the past years that create new challenges in the e-Business context:

J.P. Müller et al.: *Software Agents for Electronic Business: Opportunities and Challenges (2005 Re-mix)*, Studies in Computational Intelligence (SCI) **28,** 63–102 (2006)
www.springerlink.com

- **Globalization** creates global competition but also new opportunities for cooperation and revenue generation.
- Enterprises aim at **optimizing** the **total cost of ownership** for IT assets, and to optimize their product and service portfolios, leading to intensive mergers and acquisitions and outsourcing activities.
- **Legal requirements** of globally operating companies and the pressures of changing markets and competition put new requirements on governance, i.e. the ability to define processes, key performance indicators, and metrics for process execution, and ensure the appropriate "end-to-end" transformation of process models from a business layer down to implemented systems.

These developments create a greater than ever need for IT systems and solutions to support adaptability and self-organization. They also put enterprise interoperability into the focus of technological considerations in a sense that methodologies, architectures, and methods that support the design of interoperable cross-enterprise systems and applications have to be developed. There is an increased need for support of distributed collaborative and model-driven engineering in a cross-enterprise context, e.g. in product-related business. Furthermore, models and implementations of security and trust are required to support the applicability and application of technology innovations (such as (semi-)automated process / service composition based on a service-oriented architecture).

Thus, to summarize, while the view of e-Business has dramatically changed, leading some technological approaches of the past *ad absurdum*, there are new (and partially not so new) research and technology challenges to be tackled.

Looking at the development of agent technology within the past years provides a picture which reveals some parallels to the situation of e-Business described above: agents companies focusing on aspects such as personal assistant agents, recommending agents, interface agents for electronic commerce, matchmakers, etc. have vanished from the markets, while some of the features they supported have been silently incorporated into electronic business solutions (e.g. Amazon) or search engines (e.g. Google). At the business-to-business side, some companies focusing on agent-based logistics and marketplaces solutions have gone out of business or were acquired by other companies. Those who were successful do often no longer market their products and solutions as being agent-based. At the platform side, FIPA is still struggling and trying to find its unique value proposition. It is doubtful whether FIPA's move to become an IEEE-supported body will pay off given that the premier venues for e-Business-related standardization are elsewhere (e.g. OMG, OASIS, W3C). Thus, similar to electronic business as a whole, we have witnessed a phase of disillusioning and redefinition of the field of agents and multiagent systems. Also, similar to e-Business, the field of agent technology has matured in terms of its theoretical foundations, as well as the available software engineering tools and methodologies. It has benefited from aligning itself and exploiting synergies with technological innovations such as web services, Semantic Web, Ambient Intelligence, autonomic/organic computing, P2P and grid computing, and service-oriented architectures (as opposed to trying to compete with these technologies, as has often happened in the past). In addition, there are signs that the agent and multiagent research community is on a good way to refocus on fields where the essence and original benefits of agent technologies lie: mechanisms for modeling, engineering, understanding, and analyzing decentral systems consisting of autonomous and (possibly) self-interested systems, and the interactions among the elements of these systems (agents).

Before the background of these changes, in this paper we shall address the following questions:

- What are the technological challenges faced by e-Business today?
- How can software agent technologies help enterprises tackle these challenges of e-Business?
- What are promising or required areas of research, integration, and development of agent technology to make it applicable to the challenges?

To illustrate our findings, we look at three case studies in which agent- and related technologies are applied:

- agent-enabled supply chain event management;
- decentralized management of business resources (document, objects and services) using a peer-to-peer approach;
- application of model-driven development and agent-oriented software engineering for the modeling of interoperable enterprises and business processes.

Section 2 outlines a number of essential challenges in the area of electronic business. In Section 3 we discuss agent technology and outline benefits of agent technology with respect to e-Business. Section 4 investigates three case studies in the domains of supply chain event management, decentral business resource management, and model-driven development of business systems. Section 5 discusses findings from the case studies, concludes, and outlines areas for future work.

2 Areas and Challenges of e-Business

In this section, we will outline several areas of e-Business, define a model-driven architectural view on e-Business and have look at fields where we believe agent technology can play an important role over the next few years.

2.1 Areas of e-Business

Figure 1 shows the classification of business activities used by a large international electronics company. This classification provides one definition of areas of electronic business (out of many possible) based on business process types. Three general types of processes are distinguished: Management processes, core (value-creating) business processes, and support processes. In terms of the core business processes, three areas are distinguished:

- **Customer relationship management:** Includes all activities to plan customer interaction, understanding market needs and customer requirements, acquiring customers and managing offers, and deal with customer contacts efficiently and effectively.

- **Supply chain management:** Considers all activities to enable efficient and effective coordination along the value chain, including planning, supplier relationship management, manufacturing, order delivery, and returns management.
- **Product lifecycle management:** Covers customer-oriented planning, management, monitoring and controlling of the whole product lifecycle. The product lifecycle includes all processes in the product's lifecycle from the initial idea right through to phase-out. The goal of *Product Lifecycle Management (PLM)* is to increase product lifecycle profitability based on creation of value for the customer.

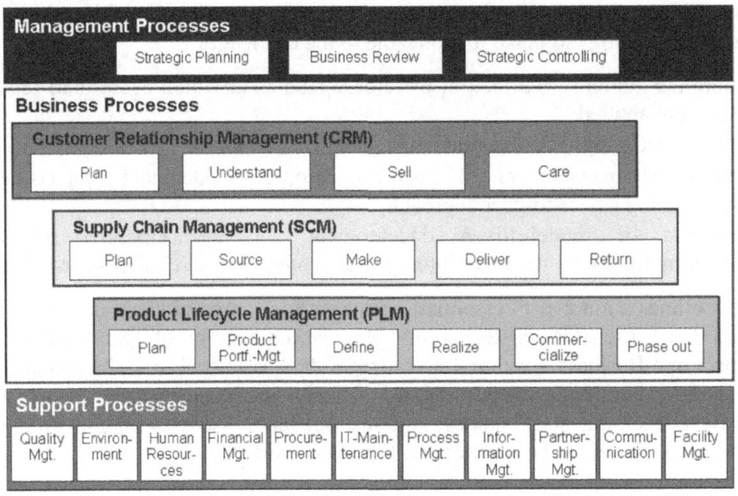

Figure 1. Business process types

2.2 e-Business Architecture

Business processes model enterprises' specific process flows in describing those activities which are conducted in pursuing enterprises' objectives and for generating revenue. In implementing value-added chains business processes interact with partners. Those interactions can be organization internal as well as cross-enterprise interactions. In order to enable business processes to collaborate with partners and to facilitate the composition of business processes, the paradigm of service-oriented architecture is applied to business process modeling (see [10]). Business processes and activities are treated as components which provide services to and consume services from other business process components. Interacting business processes form a network of interconnected processes where conversations are conducted (e.g. agent interaction protocols for the *Contract Net Protocol* or auctions). The same process can appear in multiple solutions and can be connected to different partners in each case. The service-oriented approach to business process modeling is appealing mainly because it provides a natural logical view to distributed systems.

In the context of the collaboration of business process components **orchestration** and **choreography** is distinguished. [29] defines **orchestration** and **choreography** for the collaboration of business process components as follows:

- **Orchestration:** Orchestration refers to an executable process that may interact both with **internal** and **external services**. Orchestration describes the interactions between services/agents, including the business logic and the execution order of the interactions. Within an orchestration, the process is always controlled form the perspective of one of the business parties.
- **Choreography**: Choreography describes processes in a more collaborative way, where each party/agent involved in the process describes the part it plays in the interaction. Choreography tracks the sequence of messages exchanged among multiple business parties, e.g. in interaction protocols. Often, choreography is associated with the public message exchange that occurs between multiple services.

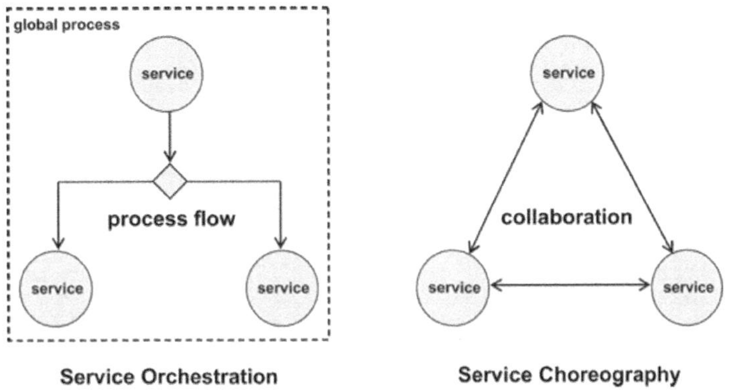

Figure 2: Orchestration and choreography

[10] makes a distinction between an internal and an external view on business processes. Depending on the viewpoint, a process is described either as an executable, abstract, or collaborative process (see Figure 3).

- **Executable process:** The **internal view** models the 'how' of a business process to a level of detail the modeler knows. In [17], processes, which model process flows as a set of partially ordered **tasks**, are called **executable processes**. As the flow of the processes' interactions is described from the point of view of a single process, which coordinates participating sub-process, this kind of process composition is referred to as process orchestration.
- **Abstract process:** The **external view** models the 'what' of a business process. Each process specifies its roles, which it realizes in the collaboration with other processes, but doesn't provide any indication about its own realization. The interfaces of such business processes components are called **abstract processes**. Abstract processes additionally describe their public interactions they perform in relation to their roles in collaborations.

- **Collaborative process:** In the case of process choreography the collaboration between abstract processes is described in **collaborative processes**. Collaborative processes use abstract processes to model the message exchange between processes and the sequence of the message exchange from the viewpoint of an external observer. The collaborations between the involved parties are modeled as interaction patterns between their roles they assume. Collaborative processes are also termed **business protocols** [17], since they model the external message exchange, which takes place among the participating parties such as interaction patterns in communication protocols.

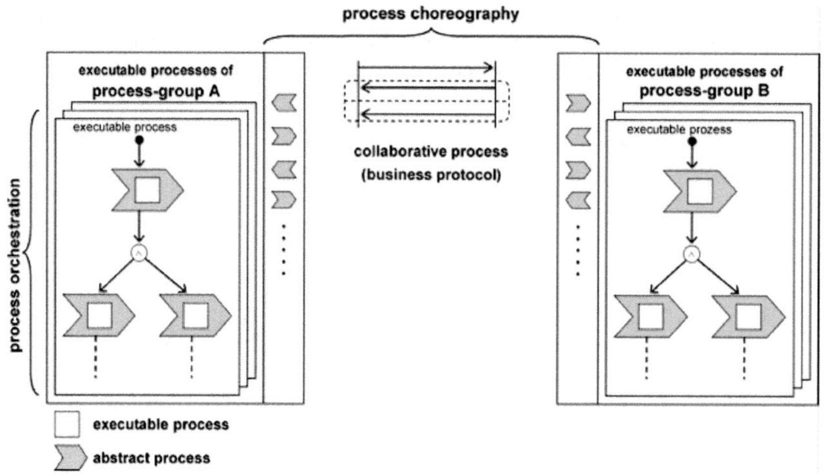

Figure 3. Modeling collaborative business processes

Collaborative processes and business protocols respectively can be realized in multiple ways, which differ in how the business protocols' conversation flow is coordinated. The two most promising approaches are discussed in the following. In the first approach, the collaborative process describes a logical view on the cross-enterprise process and on a process spanning multiple process groups, respectively.

The collaborative process determines the sequencing of the message exchange between communication partners in defining links between the abstract processes of message senders and message receptors (see Figure 4). The description of the collaborative processes does not contain any logical expression (e.g. case distinctions), which could control the process spanning multiple process groups. Such decisions are made in the executable processes, which implement the abstract processes. This is a problem since the control of collaborative process's control flow is distributed to multiple group internal executable processes. The change of the collaborative protocol generally results in the change of multiple executable processes or the introduction of wrapper processes.

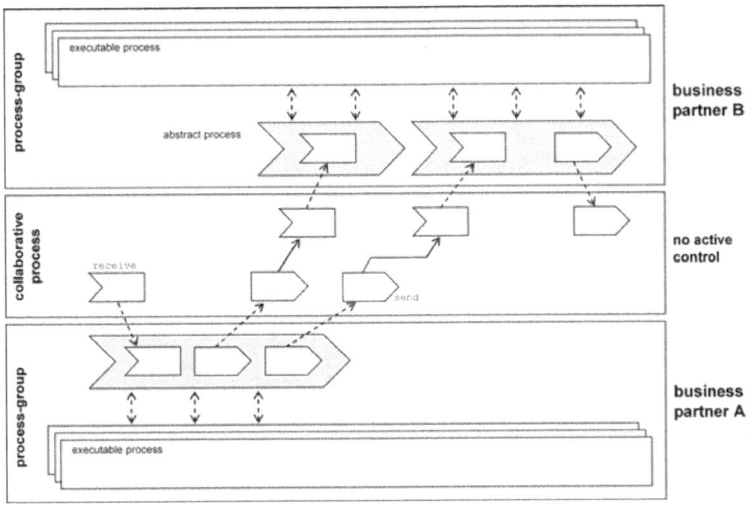

Figure 4. Collaborative process with no active control

In the second approach (see Figure 5), a collaborative process also models a view on a cross-enterprise process and on a process spanning multiple process groups. In contrast to the prior alternative the collaborative process acts as an intermediary and broker respectively, which controls the control flow between the communication partners actively. In doing so it is decided by control elements (e.g. the OR in Figure 5), to which abstract process a message is routed.

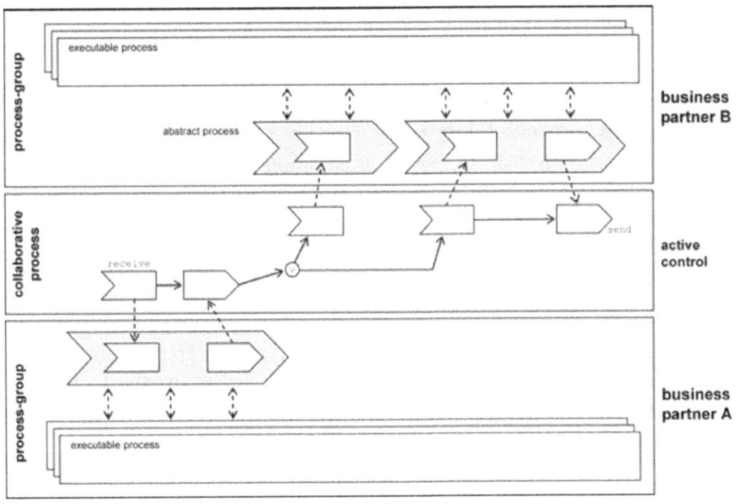

Figure 5: Collaborative process with active control

This decision could be reached for example on the basis of the state of the exchanged message. A change of the business protocol would merely result in the change of the collaborative process. The internal executable processes of the communication partners should remain unchanged.

2.3 Challenges in e-Business

In this section, we will outline some of what we believe are challenges and opportunities for research on e-Business. We believe that in order for a technology to succeed in e-Business it will be necessary to take an incremental approach, thus adding value on top of existing platforms and services.

2.3.1 Challenge: Semantic Interoperability

How can machines better understand the content of communication with humans or other machines? How can e-Business consumers find supply matching their demand? How can suppliers find potential customers? How can applications created by different companies interoperate more smoothly even if there is no explicit standard for the interaction? How can techniques for achieving interoperability at the level of semantics be used to add value to existing platforms and standards in e-Business?

We believe that semantic interoperability is maybe the hardest of all challenges to be overcome, because it requires the understanding of semantics by machines, which is still in its beginnings.

The vision that enterprises will be able to seamlessly interoperate with others is an important aspect and can be formulated for e-Business more concise as a vision of on-demand, plug-and-do-business where networked organizations can easily find, communicate and do business with other organizations. Characteristic properties and behaviours of networked (future) e-Business solutions will be:

- Smart operations using the "plug & do business" principle (for standard eCommerce processes, such as billing).
- "Model & configure" instead of "Write integration code" (using innovative, visual knowledge model integration).
- Model-generating customisable computing solutions built on dynamic architectures and intelligent infrastructures.
- Allocate competent resources or train to qualify, and then perform "just-in-time" task assignment and monitoring of critical tasks.

Customisable computing solutions will be built form dynamic architecture models and intelligent infrastructure services, making computing simpler, more effective and user-friendly, integrating all legacy systems and thinking, and augmenting human creativity and learning [18].

2.3.2 Challenge: Support for Flexible Organization Structures and Collaborative Business Processes

How can e-Business infrastructures keep up with ever changing structures of enterprises? When departments change their responsibility or structure, when a partner becomes a competitor or vice versa, how can communication and security policies be maintained smoothly? How can the participants in the value chain communicate and collaborate securely and efficiently over firewalls and enterprise boundaries? How do suppliers know which events in their internal supply chain are of relevance and need to be communicated to their (downstream) partners? How can an existing IT infrastructure migrate towards this functionality?

How can distributed individuals, teams, departments, and enterprises collaborate and coordinate their activity? How can the information necessary for this be exchanged efficiently? How can tasks be formulated, associated with skills, responsibilities, and permissions, and be allocated? How can humans collaborate with applications acting on their behalf or on behalf of other humans? While various theoretical approaches exist, supporting flexible organization structures in practice is a huge challenge.

In particular in a corporate context, huge efforts are made to set up infrastructures that allow humans to share information, resources, applications, and to work towards a common goal. Still, the problem remains hard due to its dynamic nature and through increased mobility of individuals.

2.3.3 Challenge: Pro-active, Adaptive Processes and Agent Grid Services

How can complex, distributed processes be modeled as active, situated entities that can interoperate with other processes that can monitor their environments and change their behavior on demand? How can these processes deal with change? What needs to be done to implement (pro-)active processes based on existing process models, e.g., in distributed, flexible manufacturing and supply chain management.

While many approaches to process modeling exist, automating them still remains a challenge due to the required level of introspection (i.e., a process needs to be able to judge a situation with respect to its competence, e.g., to recognize unforeseen conditions) and due to incomplete sensor-actor loops (i.e., decisions relevant to the process are often made by entities outside the control (and visibility) of that process).

One important emerging trend is the proliferation of electronic services (eServices). Grid technology is a technology which promises to offer a global platform to support on-demand virtual organizations for coordinated sharing of distributed data, applications and processes.

Service orientation of an Agent Service Grid makes it a promising platform for seamless and dynamic development, integration and deployment of service-oriented applications. The application components can be discovered, composed and delivered within a Grid of agent services. Dynamic business processes can be created to link eServices in a loosely coupled architecture to create agile applications spanning organizations and computing platforms.

2.3.4 Challenge: Dynamic IT

How can eServices be deployed on demand? How can big times in business be handled? How can the utilization of the infrastructure of an enterprise be optimized? How can the network react to load balances on eService access?

Dynamic IT refers to an emerging framework that untangles the isolated, under-leveraged infrastructure, data and applications to create a set of services to support business planning and execution. It also encompasses an IT infrastructure and service composition that can be fully leveraged, while also being cost-effectively managed and provisioned. According the analyst firm IDC, the newly emerging dynamic information technology (IT) model will change the way in which IT vendors view, manage and support all types of alliance and channel partners. Moreover IDC believes that dynamic IT will demand greater interoperability and flexibility between its participants. It is about creating a high-performance IT capability that can support the rapid pace of business change. In particular the growing role of IT is an operational component that underpins business strategy and success. The key findings include:

- The growing dependence on IT by business requires organizations to continuously re-architect their IT capabilities for speed and efficiency. Thus, many enterprises today are looking to evolve their overall computing frameworks to not only decrease costs but also to better address their organizations' goals with less disruption.
- Unique priorities are driving organizations to pursue dynamic IT transformation on two major (and complementary) paths: business process enablement and IT operational improvement. Practicality will necessitate a step-at-a-time approach for many, but, to be competitive, organizations will ultimately have to address both areas.
- Suppliers of IT products and services must address a critical new set of capabilities to be successful in the transition to dynamic IT.

The most pressing needs driving the shift to dynamic IT are:

- **Speed and Flexibility**: Respond faster to changing business needs
- **Performance**: Provide better IT service levels in support of business activities
- **Cost**: Continuously drive down IT costs

Therefore such a system should aim at freeing system administrators from most system operation and maintenance tasks, while providing users with a system that runs at peak performances on a 24/7 basis. This allows maintenance and adjustment for operations in the face of changing components, workloads, demands, and external conditions and in the face of hardware or software failures, both innocent and malicious.

2.3.5 Challenge: Security, Privacy, and Trust

The questions addressed by this challenge include the following: How can e-Business systems provide individualized content and services while at the same time ensuring adequate (and user-defined) levels of privacy? How can user profiles be created, maintained, and adapted to cope with multiple threads of a user's activity, with disruptive events, and with longer term changes in the user's preference and activity structure? How can teams and organizations be profiled? How can humans delegate complex tasks to machines? How can they be sure the machine has understood the task correctly? How can they specify the permissions and limitations of authorization of machines during task execution? How can they monitor the progress of task execution? How can a flexible hand-over between humans and machines be implemented to guarantee human and machine abilities are used for their best? How can transactions done by machines be verified and liabilities be enforced? What are legal aspects to be observed?

The importance of this question has been recognized. P3P is an attempt to provide a uniform representation of privacy preferences for an individual or company. However, it only covers privacy related preferences, and not the individualization and context-awareness (see Section 2.3.7) of presentation, content, and services.

While many researchers investigate secure protocols and efficient strategies e.g., for automating negotiation, it is our firm belief that the most difficult part of the problem is at the user interface layer in developing appropriate models for efficient delegation and transparent collaboration between humans and machines.

Moreover different levels of trust are a key issue making business in ad-hoc manner successful. Only with the necessary trust cooperations between business partners are possible.

2.3.6 Challenge: Adaptive Decision Making for Evaluation and Selection of Products and Services

Before the background of emerging dynamic service-oriented infrastructures, the questions posed here include: How can configuration and purchasing processes be automated? How can machines support users in finding products and services matching a demand profile? How can they evaluate different configuration and product offers? How can they interoperate with existing e-Business platforms? How can transactions made by machines be tracked down to the users? How can fraud be prevented? How can the risk for users or enterprises be minimized that faulty software makes wrong suggestions or buys the wrong items at too high a price? How can these models be defined? How can machines be effectively instructed for these decision-making assistance tasks? How can decisions (or suggestions for decisions) by machines be made transparent to humans?

Finding answers to these questions will be a precondition to any automation of matchmaking, contracting, or shopping / procurement processes in e-Business. As machines change their role from slaves to assistants, complex domain models are required to enable them to provide assistance in complex decision situations such as contract negotiations.

2.3.7 Challenge: Mobility Support and Context Awareness

While Ubiquitous Computing and Ambient Intelligence have reached the status of research buzzwords, we are still waiting for the promises of these technologies to show impact on enterprises. How can mobile users be ensured effective access to corporate processes and knowledge? How can they find and communicate with team members remotely? How can location and general context information be used to make services more useful and intelligent? How can information be routed, transported and presented in a way that accommodates available networks and devices? What infrastructure is necessary to leverage the potential of ubiquitous computing for enterprises and individuals? How can RFID technology reduce costs and make production / logistics more efficient?

In a society where mobility plays a more and more important role, providing answers to these questions is crucial. Moreover virtual teams all over the word make it necessary to support the notion of virtual organizations and their support for working collaboratively together. RFID tags and other mobile devices support smart and ubiquitous access to all kinds of information and providing information of all kinds of products and services. Based on sensors and other context information, like calendar entries, context aware behavior of systems in e.g. manufacturing as well as intelligent management of state information throughout the whole lifecycle of a product or plant can be obtained (see e.g., [38]).

3 Agent Technology for e-Business

In this section we define our notion of agent technology and then elaborate opportunities for agent technology to tackle the e-Business challenges defined in Section 2.

3.1 Agents Definitions and Characteristics

In [22], an agent is defined as

... a computer system, situated in some environment, that is capable of flexible, autonomous action in order to meet its design objectives.

In an e-Business context, we might want to phrase the definition in a slightly different way:

an agent is an intelligent autonomous computer system that does something useful on behalf of a human or an organization.

With software agents usually special properties are associated, see e.g., [26] for an overview. It becomes clear that agent technology is based on and needs to be developed in the context of existing basic technologies, such as software technologies. Content is described using standards like XML or RDF, OWL; existing communication mechanisms, like SOAP or WAP, are applied within agent technology. In the case of e-Business we also view platforms as base technology.

Software agents add a number of additional features to these technologies:

Agents enable **collaboration** between humans and machines, thus providing a new metaphor for multimodal *user interaction.* Users instruct agents to act on their behalf instead of having to program them. Two core areas within the discipline of intelligent user interfaces include intelligent dialog systems, and believable virtual personalities. At the same time, agents introduce the notion of *machine-machine collaboration.*

Negotiation and communication on a semantics level or goal-oriented level allow agents to coordinate their tasks and resources. In order to support the communication at a semantics level, *ontologies* are applied to assign meaning to symbols and expressions within a given domain language. Based on negotiation, communication, and ontologies, *cooperation* between agents can take place. *Coalition formation* is useful in eCommerce where virtual purchasing cooperative societies can be built to obtain quantity discount. *Organizational models* can be applied for the mapping of real organizations within a company to software organizations based on agents.

The **intelligence** of an agent is manifested by its ability to adapt its behavior to changing environmental conditions, and to employ *planning and decision-making*, e.g., for process scheduling and resource planning based on negotiation techniques. In particular planning can be performed by a single agent or in cooperation with several agents. *Learning* can be applied in several dimensions, e.g. learning of user preferences, learning of negotiation strategies of other agents up to adapting processes in the context of e-Business. *Matchmaking* is the process of mediating demand and supply based on profile information. It plays a crucial role, e.g. in agent based electronic marketplaces: the task to be solved is to find the most appropriate agents, products, or services for a task, negotiation, or market transaction.

Agent technology supports **mobility** at different levels: mobility of users, devices, and code. In the latter case we speak of *mobile agents*.

Multi-agent systems are **distributed** systems; therefore distribution is an inherent feature of agent-based systems. In this context both agents and resources can be distributed. The agent infrastructure enables transparent access to distributed resources and transparent interaction among distributed agents.

These pillars of agent technology result in individualized added value services. They support a potentially nomadic user with intelligent assistance, based on search, integration and presentation of distributed information and knowledge, advanced process control support, and e-Business and enterprise applications. To conclude, agent technology provides

- integration of distributed heterogeneous systems
- mobility of persons and software
- coordination of heterogeneous components and services
- support of users for routine tasks
- personalization of tasks and the exchange of personalized information and
- easy integration of additional components and frameworks

Therefore the promises of agent technology in e-Business are:

- To achieve enriched, higher level communication
- To enable more intelligence service provision, and process management e.g. by personalization and integration of different services to value-added
- To deal with the enlarging amount of information and functions, and
- To allow self-organizing of processes.

To realize this potential, agents need to communicate to discover their peers, to negotiate and to co-operate in open environments where everybody can add their contribution when and how it is deemed appropriate. Most importantly, agent systems will need to build on and interface with a variety of existing and upcoming developments and standards, like OWL, ebXML or WebServices. This includes support for a wide range of devices, but also integration of telecommunication and internet, and the integration with ERP, supply chain management systems, m/eCommerce platforms up to application servers and WebServices.

Thus, agents will only be able to fulfill their potential if they provide a standardized, open and generic infrastructure. Another important requirement for the success of agent technology is the availability of generic services, like for virtual team coordination, self-organization of processes, recommendation engines, mobility support and location-aware services. Such services are designed for multiple reuses and cover areas where higher level of intelligence is needed and agents seem more relevant than ever. In addition successfully marketing agent technology requires presenting it as an incremental extension of known and trusted technologies (e.g., object-oriented programming), and providing explicit engineering tools to support proven methods of technology deployment.

3.2 Challenges

In the following section we shall investigate where and how agent technology can provide means for dealing with the above challenges.

3.2.1 Challenge: Semantic Interoperability

Usually agents cooperate with each other to solve their tasks. The important issue concerning the cooperation among agents is that they should be domain independent and cooperate dynamically with unknown partners. Two levels of cooperation can be considered: The first level - a meta-level - supports mechanisms for searching for cooperating agents, to analyze their features and to start cooperating with them depending on their domain specific behavior. At the second level of cooperation the agents need a language to solve their domain specific tasks.

Semantic interoperability can be obtained on three levels: Use of ontologies – common vocabulary, interaction protocols – common interaction pattern - and semantic based model-driven software development – common development models.

In such an open cooperation between software modules one problem arises, namely the usage of a common language between the partners with identical syntax and semantics. Agents can have different terms for the same concept and identical terms for different concepts. A common *ontology*, then, is required for representing the knowledge from various domains of discourse.

As a first approximation, an ontology is a collection of well-defined terms of a specific domain, i.e. an ontology is a specification of the objects, concepts, and relationships in an area of interest. Moreover within an ontology, semantic constraints can be defined.

The *knowledge model* is a specification of the set of primitives used by a certain class of representation languages. As such, a knowledge model can be considered a meta-ontology, i.e. an ontology can be defined using some knowledge model. The *ontology sharing* problem is the problem of ensuring that two agents who wish to converse do share a common ontology for the domain of discourse. Minimally, agents should be able to discover whether or not they share a mutual understanding of the domain constants. Currently, quite a few web ontology languages are available, including SHOE, OML, XPL, ONTOBROKER, RDF, and RDF Schemas, DAML, and OIL. The most influential one is the standardization of DAML-OIL by W3C, called OWL.

In e-Business semantic interoperability is a crucial point. The standardization efforts can help to manage part of these problems. Semantic Web activities and the ontology work in the area of agent-based systems can help to overcome additional interoperability issues.

Also for agent-based systems and their development a certain integration has to be achieved. Agents are especially suitable for the realization of interoperable e-Business applications. They support 'plug & do business' since they support the meta-search of agents and can allocate resources to perform 'just-in-time' task assignment and monitoring of critical tasks by comprising autonomic / self-x properties. Solutions requiring dynamic architecture models and intelligent infrastructure services can be built by the means of agents.

In addition agent technology supports standardized interaction protocols as pattern of communication which are also in the focus of W3C's Protocol working group and of WSCL. These protocols are a necessity for the interoperability of standardized automated negotiations.

Nevertheless to achieve interoperability in e-Business the 'model & configure' principle instead of "write integration code" applies to agent-based systems. Thus agent-based systems have to support interoperability frameworks (like the Interoperability Framework presented in [6], see Figure 6) describing integration of model-driven software development, service-oriented software development and execution environments as well as methodologies, standards and domain models.

The reference model for conceptual integration (left side of Figure 6) has a MDD point of view focusing on the enterprise applications and software system. The relationships between the different levels of abstraction are shown with respect to an enterprise system. MDA and ADM could be perceived as a "top-down" and a "bottom-up" approach to software development and integration.

The reference model for **technical integration** (right side of Figure 6) has a service-/agent-oriented point of view where a software system (agent) provides a set of services required by the businesses and users of the enterprise. The architecture of enterprise applications and software systems (agents) is described according to a 4-tier reference architecture where each tier provides different software services required by the enterprise. The software system itself is coupled to an ICT infrastructure illustrated by a service bus that provides the necessary communication infrastructure and is also used for integrating two (or more) enterprise systems. The service bus will make use of infrastructure services, and registry and repository.

In the reference model for applicative integration (Figure 6, mid part) enterprise and software models are related in a holistic view, regardless of modelling language formalisms, by the use of metamodels. This provides a good starting point for the development of agent-based systems. Enterprise models describe enterprise aspects, including business models capturing actors and stakeholders, business services, business information and business processes. These business models provide a context for the software solutions that need to be developed and integrated. Software models describe how software systems are used to support the businesses of an enterprise. The software models further refine the business models in terms of software realisation models. The resulting system should be an agent-based system.

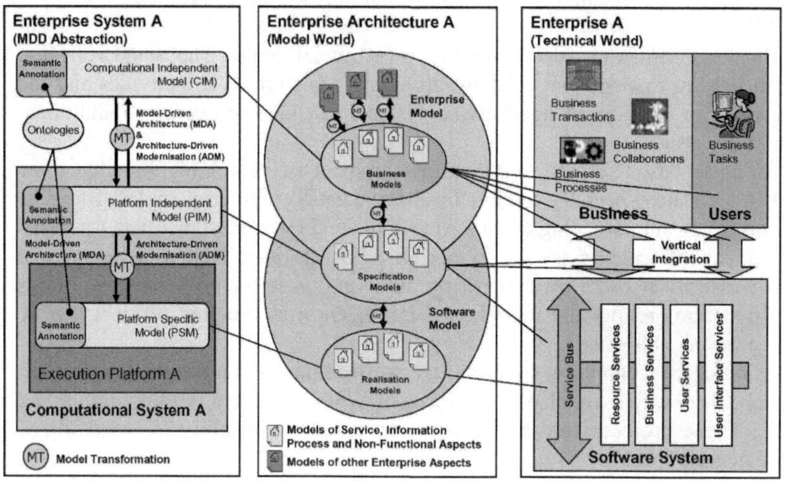

Figure 6. The ATHENA Interoperability Framework

MDD methodologies need to follow a structured approach where interoperability requirements from business operations in a networked enterprise drive the development of software solutions. This means that MDD methodologies need to be related to enterprise architectures. Figure 6 shows how the model world, reflecting the applicative integration, is related to the reference models for conceptual and technical integration. Enterprise and software models can be built to understand and analyse the interoperability requirements of an enterprise.

Thus agent technology can result in solving this challenge. We refer to Section 4.3 where we will have a closer look on such a model-driven development approach to a service/agent-oriented implementation of e-Business systems.

Approaches based on such an interoperability framework provide an opportunity to build a multi-agent based distributed workflow systems starting from a business (process) model narrowing the gap between the high level requirement and system specification in the development of multi-agent system and connects the business workflow community and multi-agent community. Thus, business users can produce their own business (process) models that can be grounded in multi-agent systems.

3.2.2 Challenge: Support for Flexible Organization Structures and Collaborative Business Processes

Organizational models and their representation and support are a core topic in e-Business research. By an organization we understand all regulations, which provide for a co-ordination of the enterprise and for its adjustment at the company target. Current research in organizational models focuses on techniques to analyze organizations and their processes.

Due to the nature of agents as decentralized, autonomous entities, organizational modeling technique can be transferred to multi-agent systems (for details see e.g. [34]; parts of the section are based on this reference). As already stated not all agents can perform all activities, but some agents are specialized for certain tasks. Organizational models are a structuring mechanism to obtain results which are not achievable by an individual agent, but need coordination of the individual agents. One rationale for the existence of organizations is to overcome the limitations of individuals. Transferring this to agents we have four basic limitations [34]:

- **cognitive limitations:** agents have cognitive limitations and need to cooperate to achieve higher levels of performance;
- **physical limitations:** agents have only limited resources and therefore must coordinate their actions, especially accessing non-local resources;
- **temporal limitations:** agents are temporally limited and therefore must join together to achieve goals which transcend the lifetime of a single agent;
- **institutional limitations:** agents are legally or politically limited and therefore must attain organizational status to act as a corporate actor rather than as an individual actor.

Research in the field of computational organizational theory uses computational and mathematical methods to study both human and automated organizations as computational entities, in particular necessary in the e-Business domain and can be solved using agent technology:

- Human organizational models are characterized by acquiring, manipulating and producing information through joint, interlocked activities of human beings and automated information processing.
- Automated organizational models are characterized by multiple distributed agents, which exhibit collective organizational properties.
- For the modeling of an organization usually the following points have to be considered:
  - agents - humans as well as artificial - comprising the organization by modeling their organizational roles and decision making influenced by their capabilities and knowledge;
  - the organization's design or structure specifying the aspects like tasks, roles, skills, collaborative teams, agents hierarchies, and resources;
  - tasks the organization carries out, especially taking into consideration temporal constraints or other dependencies on the single tasks, and similarities of tasks;
  - any environment of the organization, and the organization is involved in;
  - the organizational material transformation and / or information processing technology;
  - any stressors, like time pressure, deadlines, and turnover, on the organization.

Thus, virtual organizations and flexible organizational structures show many of the characteristics that make an agent-based approach seem adequate. However, any agent-based solution must build on established standards, especially in the area of process modeling and enterprise application integration.

Communication is a key feature in multi-agent systems. Communication between agents is necessary to exchange information, to distribute tasks, plans and goals, to coordinate actions, to negotiate prices and resources, to manage shared resources and to recognize, avoid and manage conflicts. To enable negotiation between different agents, e.g. for task delegation, communicative acts are used, grounding on speech act theory. Examples of communicative acts are *cfp* (call for proposals) starting a contract negotiation between agents, or *request* to request an agent to perform some task. These communicative acts and additional information, like sender or receiver of a message, are covered by agent communication languages (ACLs). The main representatives of ACLs are FIPA-ACL [8] from the standardization committee FIPA, and KQML [7]. Within ACL messages the content languages are used to code the real content of the message, e.g. the starting of an auction or the proposed price of a good.

Agents can interact in various patterns called *interaction protocols* also known as *conversation* or *communication* protocols which are built on top of the communicative acts. Agent interaction protocols as defined e.g. by FIPA can be seen as pattern for interaction between different agents that is formally defined and abstracted from any particular sequence of execution steps. Whichever kind of conversation is chosen, the pattern or protocol of that conversation must be understood by the participating agents. The behavior of an agent how to participate in some auction, for example, is fixed by the agent interaction strategies.

In e-Business, the communicative act, interaction protocol and interaction strategy layer are important for the automation of business processes and automated supply chain management.

3.2.3 Challenge: Pro-active, Adaptive Processes and Agent Grid Services

Agent-oriented systems are based on small units with negotiation facilities as shown above which can coordinate the tasks and goals independent of some central process, self-organization is one of the key issues agent-based systems can support. Moreover, the pro-active behavior of agents based on planning and scheduling together with autonomous behavior allow agents to organize complex tasks and processes in agents' inherent distributed environments on their own. In particular, agents can support pro-active processes with organizational concepts. Applying learning techniques can improve the process formation.

A good example to study these aspects is the application of agents to e-Manufacturing. According to [20], the scheduling and manufacturing control structure used within the MABES system is a distributed autonomous agent framework. Each agent is responsible for monitoring and acting on a part of the manufacturing process. A component may be a process, such as a press; or a stack of pre-processed components. The agents interact to control the flow of parts through either a traditional push or a lean pull or takt system. Within the adopted approach, the overall desired behavior for a manufacturing line emerges from individual behaviors of, and interactions among, distributed agents. For other research on agent-based manufacturing we refer to [5], [9], [28].

The upcoming technology of electronic services enables the development and deployment of loosely coupled, interoperable, distributed, heterogeneous systems. These will help establish relationships amongst service providers, customers and intermediaries more rapidly and with reduced setup time and cost. Nevertheless, many issues remain largely unresolved. In particular convergence of convergence of e-Service enabling technology like grids, web services, and self-management software, are yet to be reached.

Since the Grid community has started exercising the fruitful adoption of Web services mechanisms and protocols, a reference, generic Grid Services architecture has indeed been produced: the Open Grid Services Architecture (OGSA). OGSA defines Grid middleware in terms of its services, the interfaces these services expose, the individual and collective state of resources belonging to these services, and the interaction between these services within a service-oriented architecture (SOA). Services are not layered, but loosely coupled peers that, either singly or as part of an interacting group of services, realize the capabilities of OGSA through implementation, composition, or interaction with other services.

A key part of agents' appeal back in 2001 was the service-orientation: agents are producers, consumers and indeed brokers of services. Web and Grid Services have gone some way to provide this, and bring also a promise of take-up in the enterprise that perhaps was not evident through agent deployments. But agents also provide some other properties, and it could well be that the Grid community is beginning to discover it needs these. Grid is now fundamentally service-oriented and the emphasis has shifted to the notion of the Grid as being about virtual organizations (VOs). In a large scale service-oriented Grid world including VOs we will have to focus on what will happen when this world exists - when there are large numbers of services, with variable availability, qualities of service and cost. This is when the other aspects of agency become valuable.

The first of these is *negotiation*. In Grid infrastructure and applications we cannot make assumptions about constant availability of resources. In a service-oriented world it is necessary to organize sets of available services on demand in response to dynamic requirements and circumstances. The agents' research community has acquired a wealth of expertise in this area. In the Grid, people have just started exploring this area, for example in WS-Agreement and WS-Negotiation. The grid community is also initiating research on Grid economies which again could be better informed by results from the agents community.

The second is *autonomy* - a behavior not provided by Web Services but absolutely a given when working with agents. Automation is fundamental to the Grid but currently is not handled very flexibly. In contrast to Grid Services, autonomous agents work to achieve their individual objectives and they interact to meet their objectives in their common environment - techniques which directly address the Grid requirements.

The third point is *architecture*. In agent applications the agents often work with content which is not itself communicated in the agent communication language, as exemplified by information agents on the Web. This separation of the coordination architecture is also a Grid characteristic - in Grid applications, the high throughput requirements are still met by established Grid technologies while the Web services are used to control them rather than replace them.

3.2.4 Challenge: Dynamic IT

Dynamic IT as described in Section 2.3.4 needs:

- Support for flexible organization structures as well as
- Pro-active, adaptive processes and Agent Grid Services

Thus Dynamic IT shares many concepts with common trends such as resource virtualization, Service Oriented Architecture (SOA), utility computing and Grid Computing. Large IT vendors are developing architectures and products following the dynamic IT model. Main common properties of dynamic IT solutions are.

- **Virtualization** - Simplifies and improving use of computing resources by hiding the details of the underlying hardware and system software, allowing for consolidation and the ability to adapt to changing demand.
- **Provisioning** – Automation of software and hardware installation and configuration for new business service development. Provisioning focuses on self-configuration and dynamic allocation of individual elements of the IT infrastructure so that identities, storage, or servers are provisioned as business needs dictate.
- **Policy-based Orchestration** – Policy-based orchestration helps to automatically control availability, security, optimization and provisioning so that the entire IT infrastructure responds dynamically to changing conditions according to defined business policies. Orchestration meets the demand for resources from users, by implementing dynamic provisioning.

Through automation of policy-based orchestration and provisioning costs (for software installation, storage provisioning, etc.) shall be decreased, while resource utilization and response time is improved.

A key feature of dynamic IT system will be its **self-managing**, **self-provisioning** and **self-adaptation** orientation starting from grid infrastructure over protocols and middleware abstracting from an underlying grid. E-Services can be developed, making use of the underlying adaptive grid's functionality like load-balancing, policy-based orchestration and intelligent grid protocols without any knowledge of the underlying infrastructure, protocols and resources since this is abstracted and virtualized. The deployment of the e-Service instances even on runtime for adjustment operations (in the face of changing components, workloads, demands, or external conditions) is automated by the dynamic IT-system, making system administrators free from most of system operation and maintenance tasks. These features of such systems impact on most reliability indicators e.g. infrastructural changes response time, lessening of human configuration errors and a significant reduction of a number of typical cost items like network set up costs, maintenance costs, deployment etc.

A large scale distributed system like the Grid demands configuration and its components inevitably exhibit failures. People talk of the need for the grid to be self-healing, self-managing or self-organizing - to exhibit emergent behavior which can be achieved applying agent technology. The need for this increases as grids become interconnected and decision-making becomes decentralized. Autonomy is part of this picture but there is more - it's also about giving up control, about self-organization. It's fair to say that the agents' community has yet to establish best practice in large scale decentralized self-organizing multi-agent systems. The Autonomic Grid is perhaps a job of work for both communities and beyond, drawing on biologically-inspired computing, complex adaptive systems and cybernetics.

3.2.5 Challenge: Security, Privacy, and Trust

Agent technology can support different kinds of personalization and individualization. First of all agent technology can support end-to-end processes from the end-user to some service and their scheduling using planning and scheduling mechanisms based on market-based resource management. These processes can be optimized according to the users' preferences like preferred working times.

Tracking the user, i.e. taking the actual time, the actual position and the actual tasks of the user into consideration and comparing these information with the planned activities can result in *situation awareness and reaction*, i.e. depending on the current situation, e.g. time, position, tasks to be performed in the future; specific action options are triggered in order to help to fulfill the schedule of a user if some extraordinary events occur like a traffic jam on the way to the next customer or if tasks take longer than expected. In particular *automating standard procedures*, e.g. appointment scheduling, using knowledge about the working and travel times, the position of a user, the performed work or the consumption of some material, e.g. semi-automatic ordering of new material could be included in the supply chain.

At the service level, user preferences like preferred jobs, hotels or brands, can be taken into consideration and adapted depending on user behavior. User agents can select what information to present and how to present it depending on the user's profile.

Further important topics are *privacy* and *trust*. Using agent technology ensures adequate user privacy, since user-side agents hold the private data and decide (based on instructions by the user) which information are given to others. The same holds for user agents on trusted server which can guarantee that user information is kept private. The user agent is the instance which selects the appropriate information for the user depending on their preferences

3.2.6 Challenge: Adaptive Decision Making for Evaluation and Selection of Products and Services

Helping individuals and organizations to find what they need in distributed electronic marketplaces and vast heterogeneous content repositories requires a number of tasks to be accomplished well: (1) individualization of requests; (2) automation of search; and (3) interpretation and evaluation of candidates. Agents have the potential to contribute to each of these challenges. By maintaining and learning user profiles, agents can construct appropriate demand profiles. By using brokering services (e.g., yellow pages) and by doing query planning to combine multiple services, agents can find possible matches for a demand profile. In this context, they make use of existing services as described e.g., by [2]. Finally, matchmaking combined with the ability to negotiate provides the means for evaluating possible candidates and helping the user to get the best deal.

The interoperability with existing e-Business platforms is largely based on the usage of standards and enabling technologies. We have observed that e-Business software companies enter (and, occasionally, leave) the market with products the features of which contain agent support for personalization and marketplace interaction (e.g., [25], [24]). While it is not always easy to see how much of this is marketing and how much is technology, we believe that this is an encouraging sign that agents will play a considerable role to tackle this challenge within the next three to five years, and that the big players in the market might include agent features into their products and solutions once the basic foundations to deal with the topic of adaptive decision-making assistance for e-Business has been established in enterprises. Matchmaking is the process of mediating demand and supply based on profile information. The task to be solved is to find the most appropriate agents, products, or services for a task, negotiation, or market transaction. Most real-world problems require multi-attribute matchmaking, i.e., the ability to combine various dimensions of decision-making to define an overall solution to a matchmaking problem, requiring the interplay of multiple matchmaking algorithms. In addition, in order to be applicable for real-world applications, the matchmaking component must be easily integrated into standard industrial marketplace platforms. The essential requirements to be met by a matchmaking framework are the following:

- Demand and supply profiles in electronic marketplaces are often complex (e.g., matching job profiles against applicant profiles) and require multi-dimensional matchmaking. Often, a combination of existing methods is adequate to deal with different aspects of matchmaking.
- Demand and supply information is distributed and heterogeneous. Thus, distributed search and ontology mapping may be required to achieve comparable profiles.
- Demand and supply profiles differ depending on the application domain; also the underlying business logic to determine the quality of a match (distance functions) is very domain specific in part. A framework must support this variety.
- A framework should restrict the effort of developing new marketplace solutions by enabling reuse of existing profiles and business logic. In particular, developing a matchmaking solution should require only little coding, and should be done mainly through customization.
- The matchmaking framework should assist marketplace developers by supporting a clear process for building matchmaking solutions; appropriate tools should support the enforcement of this process.

In order to be able to accomplish tasks (e.g., negotiations) automatically, user agents have to be equipped with knowledge (e.g., negotiation strategies). In addition, models of user preferences and profiles can be developed, to better seize dependencies between individual dimensions by mapping domain-specific attributes (e.g. price, material, extras) to general preference dimensions (security, comfort, thriftiness, achievement). This allows users to express their preferences via these general dimensions.

3.2.7 Challenge: Mobility Support and Context Awareness

Agent technology supports human mobility at different levels. Code mobility is provided by mobile agents, which can migrate from one machine to another. Because of their migration capabilities mobile agents can (1) benefit from the capabilities of a visited computer. This can be data of a special database, or services, such as online banking or the connection to wireless networks which are only accessible on the visited computer; (2) take part in auctions on a user's behalf while the user is offline, for instance because of areas with no reception. Moreover in this scenario the response times are shorter than when using mobile communication; (3) profit from the resources of different servers to calculate solutions in parallel. The performance of the client machine can be low, however the server machines have to be powerful; (4) be supported by the client as well as by the server and the dynamic adaptation to the requirements of the user. This proceeding allows installing the necessary software (dynamic software update); (5) allow independent, off-line operation, e.g., after agents have migrated to some server to perform their tasks a notebook can be switched off. This fact results again in higher availability and robustness, since the agents change their environment system and in contrast to typical client-server approaches no permanent connection to the server has to be established; (6) save bandwidth if the agent as a software program and all its data needs less resources than the data which have to be processed by the agent; (7) communicate fast because of the local communication via e.g. shared memory; (8) support load balancing.

As already stated situation-aware services can be supported by software agents. Moreover the upcoming lightweight agent platforms deal with mobile communication and mobile collaboration supporting agents on small (mobile) devices.

Having sensor information and automatic analysis of them, together with other context information, agents can operate on the behalf of the user to some extend.

4 Agent-Enabled e-Business Applications: Case Studies

In this section we describe some case studies for the use of intelligent decentralized technologies in e-Business applications. The examples include agent-based supply-chain event management (SCEM), decentralized business resource management, and model-driven development.

4.1 Agent-Enabled Supply Chain Event Management

In this section an information logistics solution is presented which provides event-related information in supply chains to facilitate effective reactions to disruptive events in fulfilment processes (production, transportation, warehousing). The concept relies on agent technology to support the autonomy of supply chain partners while at the same time it guarantees flexible interaction among supply chain partners.

4.1.1 Problem Description

During execution of processes stochastic problems occur which have a significant impact on the performance of supply chains: Timeliness of fulfillment, quality measurements, costs and revenues of supply chain partners are affected negatively. Even small events in suppliers' processes result in deviations from globally planned and optimized schedules with serious impacts on supply chain performance. Negative consequences can be reduced, if event-related information is provided to supply chain partners shortly after such events have occurred and corrective actions can be taken. However, a lack of reliable and accurate information on events and insufficient communication of event-related data between supply chain partners is observed and the supply chain control loop of planning, execution and control is effectively blocked regarding the *control* function. An information deficit at supply chain partners regarding event-related information is the consequence. Improved information management in supply chains is required to reduce this deficit and increase the time for taking appropriate reactions to disruptive events. To realize this objective supply chain partners ought to act proactive: First, partners in the supply chain have to "sense" what kind of information they might need in the future and act proactive by pulling information from all available data sources including related partners. Second, information on disruptive events identified by a partner should be communicated to potentially interested supply chain partners proactively (information push). An SCEM (Supply Chain Event Management) solution has to enable and support both types of pro-activeness.

Agent technology is selected since it provides features for SCEM such as autonomy, reactivity and proactiveness of behavior and social ability to communicate with other agents (or humans). Proactiveness as the primary requirement for a SCEM solution is inherently satisfied by agent technology: An agent is endowed with a goal it pursues and its advanced capabilities to engage in dialogs with other software agents enables it to proactively gather SCEM data and to decide on the necessity of proactively generating alert messages. A second aspect is agent technology's support for designing distributed systems. Jennings generalizes the applicability of agent technology to all those problems that are characterized by independent roles of different actors where these have to interact in order to solve the problem [21]. This characterization applies to the SCEM problem where actors try to satisfy their implicit demand for SCEM information on disruptive events by interacting with each other.

4.1.2 Technology Used

To provide event-related information in a proactive fashion to supply chain partners and thus reduce the information deficit, various technologies are employed in the agent-based SCEM concept (for details see [35]).

An ontology designed for SCEM purposes ensures semantic interoperability among supply chain partners [36]. It is used by all agents types associated with SCEM in a supply chain. The ontology is based on the Ontology Web Language OWL. This representation guarantees reasoning support since it is based on concepts derived from description logic. A conversion of the SCEM ontology into a Java Beans representation is used for a prototype implementation. Java Beans permit Java-based agents to create knowledge instances in terms of the SCEM ontology and access these directly through method invocation [35]. However, agent applications begin to support direct use of OWL for message content and knowledge representation thus obliterating additional conversions (e.g. Whitestein LS/TS[1]).

All agent interactions are based on FIPA standards, in especially the FIPA ACL message format and the FIPA *Request Interaction* protocol. These standards facilitate

[1] http://www.whitestein.com/pages/solutions/ls_ts.html

integration of new partners into the SCEM agent society of a supply chain. Within every enterprise, several agent types perform activities such as data gathering, data analysis and alert generation. Coordination among these agents is solely achieved through message exchange. The inter-organizational perspective of the SCEM problem represented by the information deficit is challenged through proactive data gathering for event information on suborders and alert generation aimed at downstream customers in a supply chain. Complex interaction patterns evolve from the autonomous behaviours of the different agent types across a supply chain thereby realizing inter-organizational event management.

One agent type uses a rule-based system in combination with a learning component for focussing monitoring efforts on potentially critical orders [4]. The objective is to improve efficiency of the SCEM concept through reduced monitoring efforts while at the same time automatically adjusting monitoring activities to evolving focus-points in supply chain processes. Rules which hint to potentially critical orders are termed critical profiles. They consider e.g. the destination of an order or other order characteristics. All critical profiles are continuously evaluated by the agent as to their ability to identify critical orders.

Analysis of gathered SCEM data is based on Fuzzy Logic to imitate human heuristics behavior in complex situations which are characterized by multiple suborders and a variety of indicators to be considered at once [37]. A human actor initially configures an agent for data analysis by providing example assessments of various situations which are characterized by certain indicators. These data sets are used as learning sets for training of the Fuzzy Logic system and training results are used in configuration files which are directly accessed by software agents.

Four agent types are designed which realize different roles in event management. The associated activities result in one SCEM process which is realized by these agents (see Figure 7 and [35]). A discourse agent manages all external communication while a coordination agent is responsible for decisions on monitoring of orders and generation of alerts. Surveillance agents gather data from internal data sources and information on related suborders of suppliers. Wrapper agents for proprietary data sources (e.g. data bases) provide data in the format of the SCEM ontology to other agent types if requested. The SCEM process is replicated in every enterprise of a supply chain thus assuring distributed inter-organizational event management.

Both, coordination and surveillance agents exhibit proactive behaviour by starting monitoring activities, initiating data gathering rounds and preparing alerts for customers. These activities result in agent messages to other agent types which act upon the received content in their own fashion. Complex interaction patterns evolve, one of which is depicted in Figure 8. In the example the discourse agent of an enterprise receives either a request for data from a customer or an alert of one of its suppliers regarding a suborder.

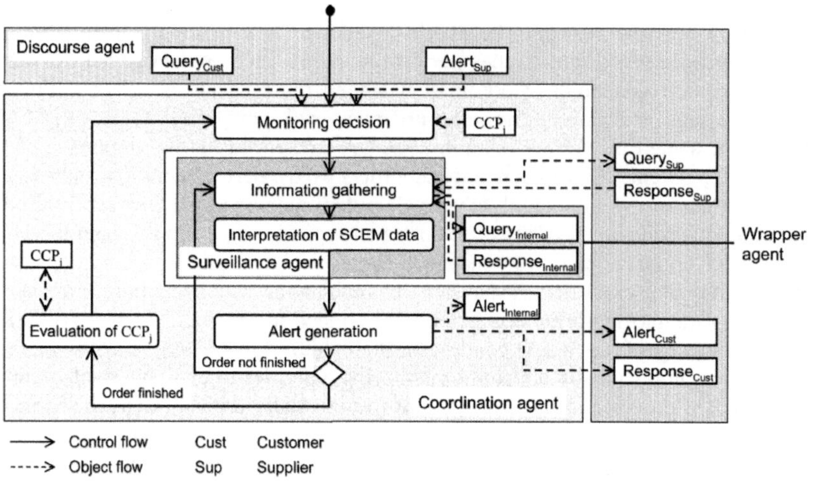

Figure 7. SCEM process and associated agent types

These incoming messages are verified and forwarded to the coordination agent, which decides to request data from a surveillance order. In case no surveillance agent is yet responsible for a certain order a new surveillance agent is initialized. This agent gathers data from internal data sources and requests data on related suborders, if any exist. Based on the Fuzzy Logic mechanism the agent analyzes all received data and forwards this information to the coordination agent which decides on generation of an alert or at least creates a response to a pending customer request.

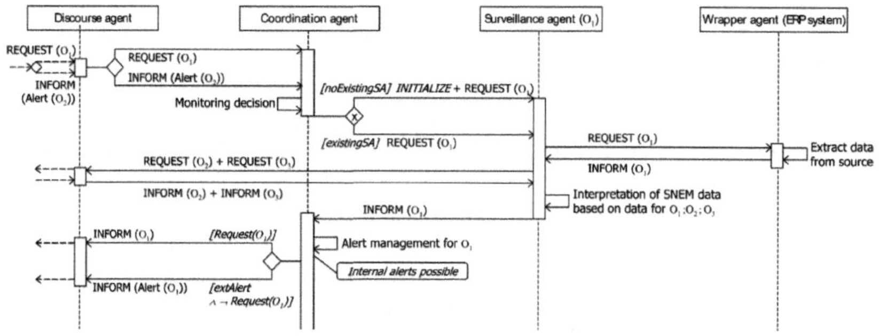

Figure 8. Evolving agent interaction pattern

Variants of interaction patterns evolve, if for instance a monitoring decision based on critical profiles is generated by a coordination agent. In addition, cascading data requests in a supply chain from customers to suppliers to suppliers of suppliers and so on evolve, depending on the complexity of order relationships in a supply chain.

4.1.3 Application Context/Validation

The agent-based concept is modelled in detail based on AUML patterns which describe structural features such as knowledge assets and agent behaviours as well as resulting interaction patterns [35]. Two prototypes based on the agent-concept are realized: First, an implementation in a laboratory environment enables controlled experiments and second, an industry showcase illustrates integration of the agent-based SCEM approach into existing IT environments. Both prototypes are based on FIPA conform platforms (JADE, FIPA-OS) and are detailed in [35]. Integration of the laboratory prototype with other multi-agent-systems (MAS) is demonstrated in the supply chain scenario *Agent.Enterprise* where other MAS are responsible for production planning and a multi-level supply chain is simulated (www.agententerprise.net, [11]).

An evaluation of the agent-based SCEM approach indicates significant monetary benefits not realized by common SCEM systems. Evaluation is guided by the question: *By how much can follow-up costs of disruptive events be reduced by a SCEM system?* A theoretical cost-benefit-model presented in [35] provides hypotheses on the specific benefits of agent-based event management in multi-level supply chains. They indicate that superior performance results from inter-organizational communication among agents in a supply chain and from focused monitoring efforts based on critical profiles. Depending on cost parameters and environmental conditions, the model predicts additional reductions of follow-up costs for disruptive events compared to existing SCEM concepts of up to 50% compared to conventional SCEM approaches [35]. These hypotheses are substantiated in experiments with the laboratory prototype where significant reductions of delays are realized. Rating these results with costs indicates similar reductions of follow-up costs as predicted by the theoretical cost-benefit-model [37].

Finally, assessment of the industry showcase shows significant acceleration of data gathering processes compared to the status-quo of a logistics service provider. An additional process analysis also provides insight into realistic cost reduction potentials for the logistics service provider. Compared to the status quo a decrease of up to 98,000 Euros per quarter is estimated for a middle-sized logistics service provider assuming severe disruptive events in about 11% of its international orders [35].

4.2 Decentral Business Resource Management

4.2.1 Problem Description

The objective of the research reported on in this section is to employ and further develop concepts for distributed and decentral resource management to increase the adaptability of business systems and business IT architectures. This higher degree of adaptability is required to flexibly plan and enact distributed business processes across multiple partners, and to cope with the complexity of and need for collaboration in decentralized business relationships. The ability of enterprises to manage business resources in a decentralized fashion is required in order to deal with the modeling, enactment and management of cross-enterprise business processes, to obtain communication and co-ordination in virtual enterprises and distributed engineering teams, and to support distributed information management and event-based workflow support, e.g., in supply chain event management scenarios as described in Section 4.1.

We start by defining the term Decentralized Business Resource Management. We use the notion of *Resources* in very general terms. Resources are considered to be data or metadata descriptions of entities existing in the real or electronic world, for example:

- database entries, data structures;
- documents, web pages, or service descriptions;
- people, workers, doctors, patients, patient information, medications;
- relationships, associations, hyperlinks;
- software components or business objects;
- tasks, activities, or processes;
- storage space, CPU capacity, goods, or materials.

A *Business Resource* is a resource that is relevant in a business context, i.e. to manage, perform or support value creation processes. In particular, we envisage business documents, business objects, skills, business services, and business objects to be the types of resources that will be primarily considered under the term Business Resource.

By *Decentralized Business Resource Management*, we understand the extension and application of decentralized architectures and coordination concepts to enable business users and business applications to create, publish, find, access, observe, modify, control, and manage business resources, and to work on/with them collaboratively, even in cases where no central infrastructure or facility is available.

4.2.2 Technology Used

In the beginning of the project, we were investigating several methods and architectures for dealing with the management of distributed resources in a decentral fashion. While agent technology appears to be a natural candidate for dealing with decentralization, we found existing agent platforms of little help, since they do not support for essential decentral functions such as decentral lookup, routing, search, and replication protocols. Rather, we found that Peer-to-Peer Computing provided us with the base protocols we were looking for to establish basic decentralized resource management features. Thus, the decision was made to build and use a peer-to-peer platform for decentralized business resource management instead of building it on an established agent platform such as Jade, FIPA-OS, or COUGAAR.

The Business Resource Management Framework (BRMF) [12] provides components both for the management and transmission of static resources such as business documents, and for the management, integration and runtime support for active components (actors for short) in the web service based business application domain. Figure 9 illustrates the architecture of the BRMF.

It is built on top of the RMF [13] that offers message based communication between all nodes in the P2P network, as well as fault tolerant publishing, discovery and retrieval of general XML resources and an additional publish/subscribe mechanism. The RMF implements a distributed hash table - referred to as the P2P information space - that is collaboratively created using the aggregated resources of all peers in the network.

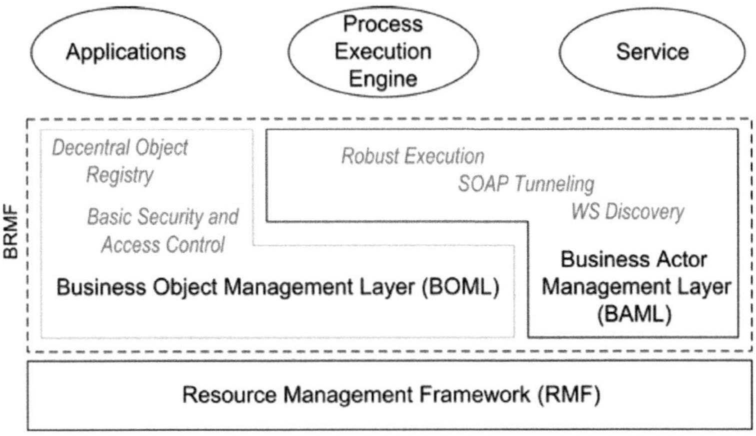

Figure 9. Architecture of the Business Resource Management Framework

4.2.2.1 Business Object Management

A central requirement for the realization of a business application distributed within a single enterprise or across different companies is the ability to manage and transmit business objects. In the context of the BRMF we use the term business object to refer to a collection of properties exchanged during business interactions. Often, these objects will adhere to a schema that can be defined and expressed using XML Schema or an interface definition language.

It is the purpose of the BOML within the BRMF to present an easy to use interface for application developers that allows creating and manipulating such business objects while hiding the complexity of the underlying communication infrastructure. Functions are provided that allow publishing, retrieving and altering the objects without the explicit need for end users to deal with the underlying P2P infrastructure in a safe and secure way. The ability of the P2P infrastructure to configure itself and adapt to changes and failures in an autonomous manner is used to build a middleware solution that also eases its application in different scenarios, both for application developers and system administrators. Even though the basic entity that is managed by the RMF is an XML resource, it is not possible to publish arbitrary XML documents right into the RMF. Two possible ways of publishing such a business object in the RMF exist:

- If the object is an XML document or can be serialized as an XML document, we can construct an RMF resource as an envelope containing the business object. This resource document has all the required elements needed by the RMF to manage the resource (i.e. the unique resource id and a set of keywords associated with the resource) in addition to the original business object XML document.
- The second option is to publish a reference resource into the RMF instead of the entire business object as an XML document. This resource contains the necessary

information to discover the object and retrieve it from an external object management system.

Management of the business objects in the P2P information space is performed by the BRMF runtime components together with a specialized business object binding. This object binding is a Java class hierarchy, created by a binding compiler from the business resource schema and a binding profile. Application developers are aided in the construction of the binding profile by a binding editor and an analysis component that helps to improve the constructed object binding by analyzing a reference set of business objects or information collected from the P2P infrastructure during runtime of the application.

4.2.2.2 Business Actor Management

While the BOML is intended for the management of static business resources, the management of active components (actors for short) in the business application domain (i.e. web services and business process execution engines) is the functionality provided by the Business Actor Management Layer (BAML). The BAML consists of components offering both end user management applications and APIs (e.g. for business process execution engine developers). The requirement of actor discovery is met by using the functionality of the underlying BOML. A corresponding resource binding for the static descriptions of the actors can be generated and exposed via a specialized interface. The BAML would in this way provide functionality to register services using their WSDL interface descriptions to an application developer. If an end user is not an application developer himself, a generic management application should expose the corresponding functionality. For example, a service provider can use a generic service registration application that takes the URIs of WSDL service descriptions as input and registers the services for use in the BRMF.

More sophisticated functionality is provided by specialized components collected in the BAML. For instance, the Robust Execution Layer (REL) [14] offers self-healing execution of business processes without requiring a change to either the process specification or the process execution engine. In the light of a service failure, the REL uses the service lookup functionality of the BAML to locate alternative invocation targets and redirects the service call. In a similar fashion, the BAML can expose an interface to tunnel service invocations through the underlying P2P communication infrastructure, offering the ability to implement a load balancing and scheduling system in the BAML.

4.2.3 Application Context/Validation

Within the European project ATHENA [1], we have been investigating the usage of mechanisms for decentralized business resource management in three application domains: Collaborative product development in the automotive industry, procurement for the furniture industry, and change management processes in the aerospace industry. In this section, we focus on the automotive application.

Outsourcing activities in the automotive industry lead to increasingly complex business processes spanning organizational borders between original equipment manufacturers and a multitude of suppliers. Those suppliers themselves have to deal with increasing demands regarding quality and cost of their products as well as their internal business processes and automated integration with the OEMs systems. In [15], the basic concept of a service-oriented architecture supporting essential workflows in virtual prototyping for product innovation is presented.

In this scenario, automotive suppliers consume computational resources - mainly CPU cycles used to run finite element simulation and optimization software for sheet metal forming - offered by computational resource providers. While the automotive suppliers need to offer their services (e.g. receive inquiry for quotation) to OEM partners, a number of computational resource providers offer their services (e.g. finite element simulation) to the suppliers.

All partners can use the service registrar component of the BAML to register their services within the P2P information space. Seamless communication between these services can be easily achieved despite existing firewalls or the fact that hosts offering web services reside within a private network address space behind a Nework Adress Translation (NAT) router, because the SOAP tunnelling mechanism included with the BAML is configured for the corresponding service containers. Every partner in the scenario has a vital interest in non-interrupted execution of its internal business processes as well as the availability of its services. Making services highly availably requires the use of redundant service hosts, registering redundant services with the P2P information space. Since the partners will configure their process execution engines to use the process execution support mechanism, the engines will resolve the redundant partner services in case of individual service failure.

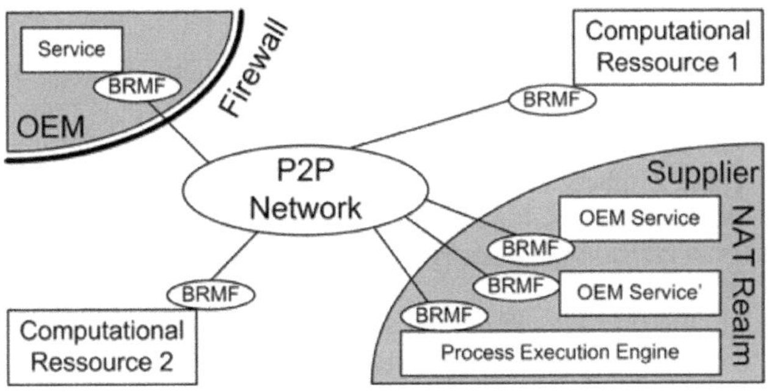

Figure 10. Sample application, connecting OEM, Supplier and Computational Resource Providers via the BRMF P2P Network

The sample scenario is shown in Figure 10. Even though the partner systems at the OEM as well as the supplier site are behind firewalls or NAT routers, they can seamlessly communicate with all partners using the underlying P2P network. No manual intervention by administrators is needed to re-configure either the OEM firewall or the supplier NAT routers. Redundant service instances (OEM Service and OEM Service*) are published to the BRMF by the supplier in order to be available to the OEM's internal processes even in case of single service failure. Similarly, the suppliers may protect their internal business processes from fault by automatically switching to another computational service provider.

4.3 Model-Driven Development and Integration

4.3.1 Problem Description

To enable enterprises to keep up with the constant change in business relationships and cross-organizational value chains, it is crucial to develop adaptive business systems and value chains. In order to achieve this, methodologies, methods, and infrastructures to support end-to-end modeling of cross-organizational business processes are required, allowing changes to business processes being defined at the business level and providing well-defined (and possibly largely automated) model transformations and refinements down to the level of information and communication technology systems.

The focus of the case study presented in this section is on business process modeling, as opposed to run-time business process management. It contributes at various/all level of detail for model driven development. Given the ultimate goal to support end-to-end business processes, by providing a process of gradual transition from abstract conceptual descriptions of business process to concrete, executable business processes, requires the consideration of mappings to the information and communication technology level.

4.3.2 Technology Used

The case study makes use of *model-driven development and architecture* (MDA™) as promoted by the Object Management Group (OMG).

Within MDA the software development process is driven by the activity of modeling the business software system. One of the major differences to traditional development processes lies in the nature of the artifacts that are created during the development process. These artifacts are formal models, i.e. models that can be understood by computers and finally be transformed into a representation that lends itself to execution which can effectively supported e.g., by a web services information and communication infrastructure.

The Model Driven Architecture (see [23]; this subsection is also based on this reference) is a framework for software development driven by the Object Management Group (OMG). The following three models are at the core of the MDA:

- **Computation Independent Model (CIM):** This is the most abstract model within the MDA. It describes business logic, business processes and workflows, and business requirements to a software system independent of computational technology.
- **Platform Independent Model (PIM):** This model is defined at a high level of abstraction; it is independent of any implementation technology. It describes a software system that supports some business. Within a PIM, the system is modelled from the viewpoint of how it best supports the business.
- **Platform Specific Model (PSM):** In the next step, the PIM is tailored to specify a system in terms of the implementation constructs available in one specific

implementation technology. A PIM is transformed into one or more PSMs. For each specific technology platform a separate PSM is generated. Most systems today span several technologies; therefore it is common to have many PSMs with one PIM. The final step in the development is the transformation of each PSM to code. Because a PSM fits its technology rather closely, this transformation is relatively straightforward.

4.3.3 Application Context/Validation

Figure 11 gives an overview of the case studies results and shows how the case study fits to e-Business systems architecture presented in Section 2.2. The focus of the case study is model-driven development of services-oriented, collaborative solutions for agents-based systems.

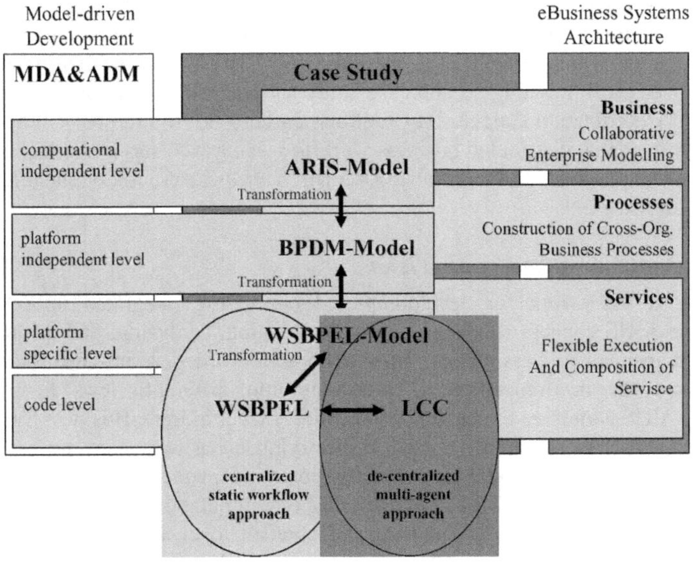

Figure 11. Model-driven development and e-Business architectures

The artifacts to be found in the case study can be assigned to the various levels of abstraction (computational independent, platform independent, platform specific and code level) provided by the MDA and also considered by the recently emerging architecture driven modernization (ADM) of the OMG. The points addressed by the e-Business Systems Architecture, collaborative enterprise modeling at business level, construction of cross- organizational business processes at process level, and flexible execution and composition of services at service level, can be found in the various models of the case study. Starting from the business viewpoint of business analysts models are to a huge extend automatically refined to specification views of system architects and finally to the realization view of software developers (viewpoint definition see [6]). Changes of viewpoints and modeling methods are a very crucial point in the development of an information and communication system, since mistakes in here are rarely found before the system is deployed. Thus a main issue addressed in this case study is that business models are consistent with the information and communication technology systems models which a grounded on service-oriented architecture. Finally the case study provides an opportunity to build multi-agent based distributed workflow systems starting from a business process model rather from an interaction protocol, narrowing the gap between the high level requirement and system specification in the development of multi-agent system. [16]

The case study starts with a transformation from Architecture of Integrated Information Systems (ARIS) models at computational independent level to platform independent models based on the Business Process Definition Metamodel (BPDM). This transformation encodes reusable knowledge about the platform, i.e. service-orientation. After the result of the transformation has been manually refined to a fully detailed BPDM model, another transformation produces platform specific workflow models based on a UML-Profile for the Web Service Business Process Execution Language (WSBPEL). In most cases only slight refinements on these models are necessary to produce code. Beside the generation of WSBPEL code, which can be run by a centralized workflow engine, the case study shows how a multi-agent protocol (Lightweight Coordination Calculus (LCC [30])) ca be used to interpret a WSBPEL specification to enable distributed business workflow using web services composition a the multi-agent platform. In the following we will have a closer look on the different abstraction levels:

4.3.3.1 Computational independent level

ARIS forms a framework for developing and optimizing integrated information systems. The ARIS concept serves as model for creating, analyzing, and evaluating business management process chains. Thus ARIS allows the description of business processes and the decomposition of processes into different views to reduce complexity. ARIS addresses the business and process level of the e-Business Systems Architecture. ARIS is commonly used for specifying a business view on business processes and therefore for modeling at computational independent level. [31]

In a model transformation, where ARIS models of computational independent level are mapped to BPDM models of platform independent level, knowledge about the target models' IT-solution architecture is encoded. In the case of the case study knowledge about how to realize process description in a service-oriented environment is encoded. Supporting a transformation from ARIS into BPDM takes the fact into account that ARIS is the leading modeling business process methodology in industry and tying to make it easier for enterprises to migrate their business process models into a model-driven framework. Finally, choosing BPDM as a vendor-independent representation will leverage the support for modeling collaboration which will be provided in BPDM and makes the approach particularly suitable for the modeling of decentralized, cross-enterprise processes.

4.3.3.2 Platform independent level

Resulting from an OMG Request for Submission, the primary objective of the Business Process Definition Metamodel (BPDM) initiative is to provide an abstract model for the definition of business processes (see [19]). BPDM is specified as a

UML 2.0 profile enabling generic UML tools to both author or consume business models. As BPDM provides basic concepts from business process modeling, such as processes, tasks, rules, transactions, workers, and organizations, as first-class citizens, and additionally provides support for the modeling of collaboration, it appears a promising approach to combine the openness and genericity of UML with the expressiveness and vocabulary required for business process modeling.

BPDM as a metamodel describing logical relationships focuses on two aspects of business processes modeling: the internal behavior realized by operational processes, using a flow model for describing how a process is performed, and the external interactions, used to connect processes with other components in a service-oriented architecture. The external view specifies partner roles describing external process interactions called abstract processes as well as interaction protocols between those partner roles. Interaction protocols are modeled as collaborative processes between partner roles implemented by operational processes.

BPDM models generated by the model transformation from computational independent to platform independent level only contain information for which it was possible to be derived from the ARIS models. Furthermore BPDM provides concepts to represent more detailed information about information and communication systems. Thus service and collaborations of the generated BPDM models have to be enriched with additional information, e.g. about event and document correlation, to fully detailed platform independent models. This fully detailed platform independent model can be transformed to platform specific models by model transformations comprising platform specific knowledge. In the case study BPDM models are transformed to an UML profile for the WSBPEL process execution platform. Like described in the model transformation is mainly straight forward, since WSBPEL is similar to BPDM but platform specific. WSBPEL is based on the service-oriented approach too.

4.3.3.3 Platform specific level

The UML profile used at platform specific level in the case study enables automated business processes to be modeled using UML tools such as Rational Rose and Rational XDE. The UML 1.5 profile for automated business processes (business process definition PSM) supports concepts that are based on those supported by the Web Services Business Process Execution Language (WSBPEL).

WSBPEL has its origins in the join of IBM's Web Service Flow Language (WSFL) and Microsoft's XLANG to one web service centered process execution language; it is maintained by the Organization for the Advancement of Structured Information Standards (OASIS). Defining a notation to specify business process behavior based on Web Services, executable business processes model actual behavior of a participant in a business interaction, while abstract processes specify the visible message exchange of each of the parties involved in a business protocol. WSBPEL is meant to be used to model the behavior of both executable and abstract processes. [27]

Since UML 1.5 profile for automated business processes (business process definition PSM) supports concepts that are based on those supported by WSBPEL, the platform specific models can easily be transformed to WSBPEL code (see UML profile and transformation specification [19]).

4.3.3.4 Code generation for web service based process execution

Composition of web services has received much interest as a means of supporting business-to-business or enterprise application integration. Currently, there are two main approaches for web services composition: one is a static, workflow based approach, for example on WSBPEL as described above, which is a de facto standard for distributed workflow systems using web services composition. A workflow engine is used to run the whole business process model, web services thus can be invoked as the business process executes. However, the downside of this approach is that, although the workflow engine can execute these invocations asynchronously, the process is still centralized, which means it suffers from the single point-of-failure weakness that plague centralized designs [32] and in some environments, centralization is nor possible, for example, in a peer to peer mobile devices based environment. In addition, centralized design may require heavyweight servers. [16]

An alternative approach is to apply a multi-agent system for web services coordination ([30], [33]). With this approach, each agent in the multi-agent system is associated with a web service which contains the necessary external behaviors for the participants (agent). The flow control logic is defined with in the multi-agent system protocol which is passed between all the agents together with the messages to tell each agent what to do next to enable their coordination. The infrastructure of the system is depicted in Figure 12.

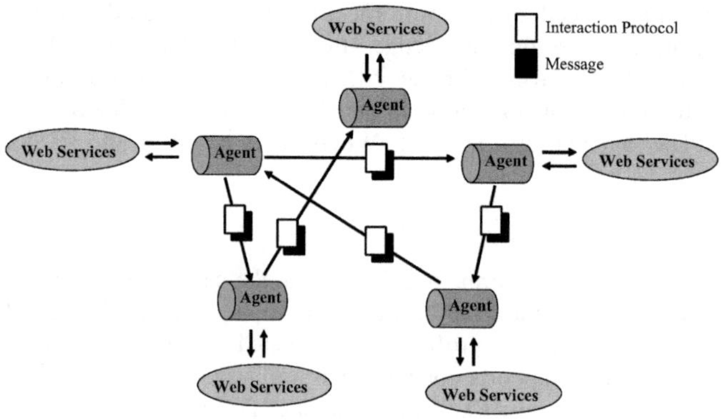

Figure 12. Decentralized MAS infrastructure

Although the problem of centralization is overcome by using the approach, a shortcoming of it is that the interaction protocol (system specification) is at a very low level of system design. It specifies the message passing that takes place between different participants at implementation level, mixing both the business and technical requirements. Therefore, huge effort on the validation and verification is required for the interaction protocol production in order to make sure that the protocol is strictly consistent with the high level requirements of the business process model [16].

In our case study the approach described in [16] is applied with which WSBPEL specification can be used to parameterize a generic multi-agent interaction protocol, thus all existing WSBPEL specification and available tools can be exploited when we try to enact a distributed business workflow using web services composition on a multi-agent platform. The distributed multi-agent workflow enactment mechanism is developed using the WSBPEL specification. A multi-agent protocol (Lightweight Coordination Calculus) is used to interpret a WSBPEL specification to enable distributed business workflow using web services composition on the multi-agent platform. WSBPEL can be adopted directly in the multi-agent system, thus reducing the effort on the validation and verification of the interaction protocol. This approach also provides a lightweight way of re-design of large component based systems. Thus, we see how agent technology can contribute to the development of e-Business systems from a model-driven perspective.

5 Discussion, Conclusions, and Outlook

The contribution of this paper is twofold: Firstly, we outlined what we believe are the key challenges in today's e-Business. Secondly, we investigated how agent technology can help tackling some of these challenges. We demonstrated possible approaches in using agent technology for e-Business solutions as well as main advantages in building agent based e-Business systems. This chapter shows that agent technology is well suited to develop highly dynamic, generic and intelligent integration frameworks for e-Business based on standard vertical and horizontal services on top of existing standard platforms. However the agent community has to take upcoming Standards and work in other areas into serious account, like Semantic Web, Web Services, Web Service Choreography, Grid Technology as well as Autonomic Computing initiatives.

- *Agent-based SCEM*: It is shown, that a society of agents is able to cooperate autonomously and provide benefits not yet realized by common SCEM system architectures. Semantic Interoperability is solved by using an ontology which is OWL based and might e.g. be directly integrated into industry-strength agent applications such as Whitestein's LS/TS.
- *Decentralized Business Resource Management*: This use case is an example how decentral lookup and routing functionality provided by peer-to-peer protocols enables a new class of robust, scalable, an self-organizing business applications with no need for central information repositories or control instances. From an agents perspective, our observation that, in contrast to the speech act related work resulting from multiagent systems research along the lines of KQML and FIPA, peer-to-peer computing managed to create an impact in a very short time, due to fairly simple protocols with a clear, value creating functionality: the virtualization of resources. Yet, we expect agent technology to contribute to making peer-to-peer systems more flexible (e.g., by adding negotiation functionality) and situation-aware.
- *MDA-Agents*: The case study shows, that agent technology can be combined with standard software engineering approaches and the Web Service community and add the potential of agents to the Web Service area.

We believe that certain functions of agent technology will play an important role in the development of next generation e-Business systems in coming years. Agent technology assumes a key role in flexibly combining existing heterogeneous e-Business solutions, adding smart functionality and automating standard processes. This hypothesis is solidified, if we look at the main abilities agents provide as compared to other technologies. Three main abilities are identified:

- Enriched, high-level communication (agent communication languages based on existing transport encoding and underlying networking protocols, co-ordination of tasks, collaboration based on semantics and ontology's);
- Provision of increasingly intelligent services and process management e.g. through personalization and integration of different services into value-added services (service wrapping, brokering, matchmaking, negotiation, auctioning, preference modeling, adaptive behavior through learning mechanisms);
- Self-organization of processes (autonomous, flexible and pro-active behavior through planning, scheduling, and learning functionality) by adding these functionality to the emerging peer-to-peer and grid service platforms.

One point this paper makes is to show that agent technology has much to offer to next-generation e-Business solutions. We believe that the main benefit of agents in e-Business will be reached in the following application areas:

- Support semantic interoperability of enterprises, processes, and applications;.
- Exploit model-driven software engineering of e-Business applications and their implementation on an agent-related middleware;
- Support Dynamic IT infrastructures and protocols by (quality of) service negotiation and semantic models;
- Provide services pro-actively and adaptively service provisioning based on automated planning methods.
- Leverage service-oriented middleware by integrating (and mediating between) heterogeneous platform paradigms such as Peer-to-Peer, Grid, Semantic Web (Service), and Web Services.

However, we should end with a word of caution. For agents to satisfy the expectations there are some important preconditions. First, the research community needs to establish a focus on pragmatic solutions that build on existing standards. Second, existing solutions and technologies need to be used efficiently and enhanced incrementally instead of re-inventing the wheel. Third, while agents are an important enabling technology, they do not liberate software engineers and system developers from making a careful requirements analysis and design of processes and systems; Fourth, for some areas, such as advanced supply-chain management, automated negotiations and collaborations, new architectures and methods will be required to achieve the necessary level of scalability, flexibility, as well as trust and security. Self-organizing manufacturing and supply-chains whose nodes consist of autonomous collaborating agents seem to be appropriate building blocks for these solutions.

References

[1] ATHENA IP European Project, FP6-IST-507849, http://www.athena-ip.org.

[2] Bauer, B.; Berger, M.: Agent-Based Personal Travel Assistance, Proceedings of International ICSC Symposium on Multi-Agents and Mobile Agents in Virtual Organizations and E-Commerce (MAMA 2000), Wollongong, Australia, 2000

[3] Bartels, A. IT Spending Outlook: 2004 To 2008 And Beyond. Forrester Market Overview. Forrester Research, August 24, 2004.

[4] Bodendorf, F., Zimmermann, R., "Proactive Supply Chain Event Management with Agent Technology" *Int. Journal of Electronic Commerce*, 9 (2005) 4, 57-90.

[5] Bussmann, S., Jennings, N.R., Wooldridge, M.J. Multiagent Systems for Manufacturing Control: A Design Methodology. Springer-Verlag, 2004.

[6] Elvesæter, B., Hahn, A., Berre, A-J., Neple, T. Towards an Interoperability Framework for Model-Driven Development of Software Systems, First International Conference on Interoperability of Enterprise Software and Applications, Geneva Switzerland, 2004.

[7] Finin, T. and Fritzson, R.: KQML - a language and protocol for knowledge and information exchange, In Proc. 13th Intl. DAI Workshop, Seattle, WA, USA, 1994.

[8] FIPA 98 specification: http://www.fipa.org/repository/fipa98.html

[9] Fischer, K.: Holonic Multiagent Systems - Theory and Applications. In Proceedings Portuguese Conference on Artificial Intelligence (EPIA) 1999: 34-48.

[10] Frank, J., Gardner, T., and Johnston, S. Business Process Definition Metamodel – Concepts and Overview, IBM, 2004.

[11] Frey, D.; Stockheim, T.; Woelk, P.; Zimmermann, R. Integrated Multi-agent-based Supply Chain Management. In: Proc. 12th IEEE International Workshop on Enabling Technologies: Infrastructure for Collaborative Enterprises (WETICE-2003). pp. 24-29.

[12] Friese, T., Müller, J. P., Smith, M., Freisleben, B.: A Robust Business Resource Management Framework Based on a Peer-to-Peer Infrastructure. In: Proc. 7th International IEEE Conference on E-Commerce Technology, 2005

[13] Friese, T., Freisleben, B., Rusitschka, S., Southall, A.: A Framework for Resource Management in Peer-to-Peer Networks. In: Proceedings of NetObjectDays 2002. Volume 2591 of Lecture Notes in Computer Science., Erfurt, Germany, Springer (2002), pp. 4–21.

[14] Friese, T., Müller, J.P., Freisleben, B.: Self-Healing Execution of Business Processes Based on a Peer-to-Peer Service Architecture. In: Proc. 18th Int. Conference on Architecture of Computing Systems. Vol. 3432 of LNCS, Springer (2005), pp. 108–123.

[15] Grauer, M., Barth, T., Gerdes, M., Reichwald, J.: On distributed virtual prototyping in the automotive supplier industry using service-oriented computing. In: Proceedings of the GI Tagung Produktdatenmanagement. (2005).

[16] Guo, L., Robertson, D., Chen-Burger, Y-H. Enacting the Distributed Business Workflows Using BPEL4WS on the Multi-agent Platform, Conference on Multiagent System Technologies, LNAI, Koblenz Germany, 2005.

[17] IBM: Business Process Execution Language for Web Services Version 1.1, http://www.ibm.com/developerworks/library/ws-bpel/, 2003.

[18] IDEAS. The Vision for 2010, www.ideas-roapmap.net, 2003.

[19] Iyengar et al. Business Process Definition Metamodel. Revised Submission to BEI RFP bei/2003-01-06. 2004.

[20] Ivezic N., Tom Potok, and Line Pouchard. Multiagent Transitional Operations, Autonomous Agents 2000

[21] Jennings, N.R., "An Agent-Based Approach for Building Complex Software Systems", *Communications of the ACM* 44 (2001) 4, 35-41.

[22] Jennings N. R., K. Sycara, and M.J.Wooldridge. A Roadmap of Agent Research and Development. JAAMAS. 1(1):1998, pp. 7-36.

[23] Kleppe M., Warmer J., Bast W. MDA Explained – The Model Driven Architecture: Practice and Promise, Addison Wesley, 2003
[24] LivingSystems: www.livingsystems.com
[25] LostWax: www.lostwax.com
[26] Müller, J. P. The design of intelligent agents. Lecture Notes of Artificial Intelligence, Vol. 1077. Springer-Verlag, 1997.
[27] OASIS: Web Service Business Process Execution Language Version 2.0, Working Draft December 2004, 2004.
[28] Parunak, H. Van Dyke: A Practitioners Review of Industrial Agent Applications. Autonomous Agents and Multi-Agent Systems 3(4): 389-407 (2000)
[29] Peltz, C. Web Service Orchestration and Choreography, Web Services Journal Volume3 Issue 7, 2003.
[30] Robertson, D. A Lightweight Method for Coordination of Agent Oriented Web Services, AAAI Spring Symposium of Semantic Web Services, 2004.
[31] Roser, S., Bauer, B.: A Categorization of Collaborative Business Process Modeling Techniques, SoS4CO Workshop in 7th CEC Conference, IEEE, Munich Germany, 2005.
[32] Vidal, J.M., Buhler, P., Stahl, C. Multiagent systems with workflows, IEEE Internet Computing, 2004.
[33] Walton, C.D. Model Checking Multi-Agent Web Services, AAAI Symposium of Semantic Web Services, 2004.
[34] Weiss G., Multiagent Systems - A modern approach to DAI. Cambridge: MIT Press, 1999.
[35] Zimmermann, R.: Agent-based Supply Network Event Management, Dissertation, Nürnberg, Juli 2005. To appear in: Whitestein Series on Agent Technologies, 2005.
[36] Zimmermann, R., Käs, S., Butscher, R., Bodendorf, F., "An Ontology for Agent-Based Monitoring of Fulfillment Processes". Tamma, V., Cranefield, S., Finin, T.W.; Willmott, S. (eds.), Ontologies for Agents: Theory and Experiences, Whitestein Series in Software Agent Technologies, Birkhäuser, Basel, 2005, S. 323-345.
[37] Zimmermann, R.; Winkler, S.; Meyreiss, C.; Bodendorf, F.: Assessment of Supply Chain Event Management Data by an Agent-based System with Fuzzy Logic. Proceedings of: Agent-based Simulation 2005, Erlangen, 2005.
[38] R. Zimmermann, J. P. Müller: An Ambient Intelligence Information Infrastructure for Production-to-Maintenance Processes. In Proc. of the 16th FAC World Congress, International Federation of Automatic Control, Prag, July 2005.

Global Supply Chain Networks and Risk Management: A Multi-Agent Framework

Anna Nagurney [1], Jose M. Cruz [2], and June Dong [3]

[1] Radcliffe Institute Fellow,
Radcliffe Institute for Advanced Study, 34 Concord Avenue,
Harvard University,
Cambridge, Massachusetts 02138
and
Department of Finance and Operations Management, Isenberg School of Management, University of Massachusetts, Amherst, Massachusetts 01003
nagurney@gbfin.umass.edu
[2] Department of Operations and Information Management,
School of Business, University of Connecticut,
Storrs, Connecticut 06269
jcruz@business.uconn.edu
[3] Department of Marketing and Management,
School of Business, State University of New York at Oswego,
Oswego, New York 13126
dong@Oswego.edu

Abstract. In this paper, we develop a global supply chain network model in which both physical and electronic transactions are allowed and in which supply-side risk as well as demand-side risk are included in the formulation. The model consists of three tiers of decision-makers/agents: the manufacturers, the distributors, and the retailers who may be located in the same or in different countries and may conduct their transactions in distinct currencies. We model the optimizing behavior of the various decision-makers, with the manufacturers and the distributors being multicriteria decision-makers, and concerned with both profit maximization and risk minimization. The retailers, in turn, are faced with random demands for the product. We derive the governing equilibrium conditions and establish the finite-dimensional variational inequality formulation. We provide qualitative properties of the equilibrium product flow and price pattern in terms of existence and uniqueness results and also establish conditions under which the proposed computational procedure is guaranteed to converge. Finally, we illustrate the global supply chain network model and the computational procedure through several numerical examples. This research illustrates the modeling, analysis, and computation of solutions to decentralized supply chain networks with multiple agents in the presence of electronic commerce and risk management in the global arena. Moreover, it highlights and applies some of the theoretical tools that are now available for such multi-agent problems.

A. Nagurney et al.: Global Supply Chain Networks and Risk Management: A Multi-Agent Framework,
Studies in Computational Intelligence (SCI) 28, 103–134 (2006)
www.springerlink.com

1 Introduction

Growing competition and emphasis on efficiency and cost reduction, as well as the satisfaction of consumer demands, have brought new challenges for businesses in the global marketplace. At the same time that businesses and, in particular, supply chains have become increasingly globalized, the world environment has become filled with uncertainty. For example, recently, the threat of illness in the form of SARS (see [1]) has disrupted supply chains, as have terrorist threats (cf. [2]). On the other hand, innovations in technology and especially the availability of electronic commerce in which the physical ordering of goods (and supplies) (and, is some cases, even delivery) is replaced by electronic orders, offers the potential for reducing risks associated with physical transportation due to potential threats and disruptions in supply chains.

Indeed, the introduction of electronic commerce (e-commerce) has unveiled new opportunities for the management of supply chain networks (cf. [3]) and has had an immense effect on the manner in which businesses order goods and have them transported. According to Mullaney et al. [4] gains from electronic commerce could reach $450 billion a year by 2005, with consumer e-commerce in the United States alone expected to come close to the $108 billion predicted, despite a recession, terrorism, and war.

The importance of global issues in supply chain management and analysis has been emphasized in several papers (cf. [5], [6], [7]). Moreover, earlier surveys on supply chain analysis indicate that the research interest is growing rapidly (see [8], [9]). Nevertheless, the topic of supply chain risk management is fairly new and novel methodological approaches that capture both the operations as well as the financial aspects of such decision-making are sorely needed. In particular, the need to incorporate both supply-side and demand-side risk in supply chain decisionmaking and modeling is well-documented in the literature (see, e.g., [10], [11], [12], [13], [14]).

Frameworks for risk management in a global supply chain context with a focus on centralized decision-making and optimization have been proposed by Huchzermeier and Cohen [15] and Cohen and Mallik [6] (see also the references therein). In this paper, in contrast, we build upon the recent work of Nagurney, Cruz, and Matsypura [7] in the modeling of global supply chain networks with electronic commerce and that of Nagurney et al. [16] and Dong, Zhang, and Nagurney [17, 18, 19] who introduced random demands in a decentralized supply chain network. In particular, we develop a global supply chain network model with both supply-side risk (handled as a multicriteria decision-making problem) and demand-side risk (formulated through the use of random demands).

The paper is organized as follows. In Section 2, we develop the global supply chain network model and derive the optimality conditions of the various decision-makers/agents. The model can handle as many manufacturers, countries, currencies, as well as distributors, and retailers, as required by the specific product application. Moreover, we establish that the governing equilibrium conditions can be formulated as a finite-dimensional variational inequality problem. We emphasize here that the concept of equilibrium, first explored in a general setting for supply chains by Nagurney, Dong, and Zhang [20], provides a valuable benchmark against which prices of the product at the various tiers of the network as well as product flows between the tiers can be compared.

In Section 3, we provide qualitative properties of the equilibrium pattern, and, under reasonable conditions, establish existence and uniqueness results. We also establish properties of the function that enters the variational inequality that allows us to establish convergence of the proposed algorithmic scheme in Section 4. In Section 5, we apply the algorithm to several global supply chain network examples for the computation of the equilibrium prices and shipments. The paper concludes with Section 6, in which we summarize our results and present suggestions for future research.

2 The Global Supply Chain Network Model with Risk Management

In this Section, we develop the global supply chain network model consisting of three tiers of decision-makers/agents. The multicriteria decision-makers on the supply side in the form of manufacturers and distributors are concerned not only with profit maximization but also with risk minimization. The demand-side risk, in turn, is represented by the uncertainty surrounding the random demands at the retailers. The model allows for not only physical transactions but also for electronic ones. For the structure of the global supply chain network, see Figure 1.

In particular, we consider L different countries with a typical country denoted by l, $\hat{l}$, $\bar{l}$ (since we need to distinguish a given country in a tier). There are I manufacturers in each country with a typical manufacturer i in country l denoted by il and associated with node il in the top tier of nodes in the global supply chain network depicted in Figure 1. Also, we consider J distributors in each country with a typical distributor j in country $\hat{l}$ being denoted by $j\hat{l}$ and associated with second tier node $j\hat{l}$ in the network. There are a total of JL distributors in the global supply chain network. A typical retailer k in country $\bar{l}$ dealing in currency h is denoted by $kh\bar{l}$ and is associated with the corresponding node in the bottom tier of the network. There are a total of KHL retailers in the global supply chain.

We assume a homogeneous product economy meaning that all manufacturers produce the same product which is then shipped to the distributors, who, in turn, distribute the product to the retailers. In order to include the influence of the Internet, we allow the manufacturers to transact either physically with the distributors, or directly, in an electronic manner, with the retailers. Hence, the links connecting the top and the bottom tiers of nodes in Figure 1 represent electronic links. Moreover, the retailers at the bottom tier of nodes of the global supply chain network can be either physical or virtual retailers.

The behavior of the various supply chain decision-makers/agents represented by the three tiers of nodes in Figure 1 is now described. We first focus on the manufacturers. We then turn to the distributors, and, subsequently, to the retailers.

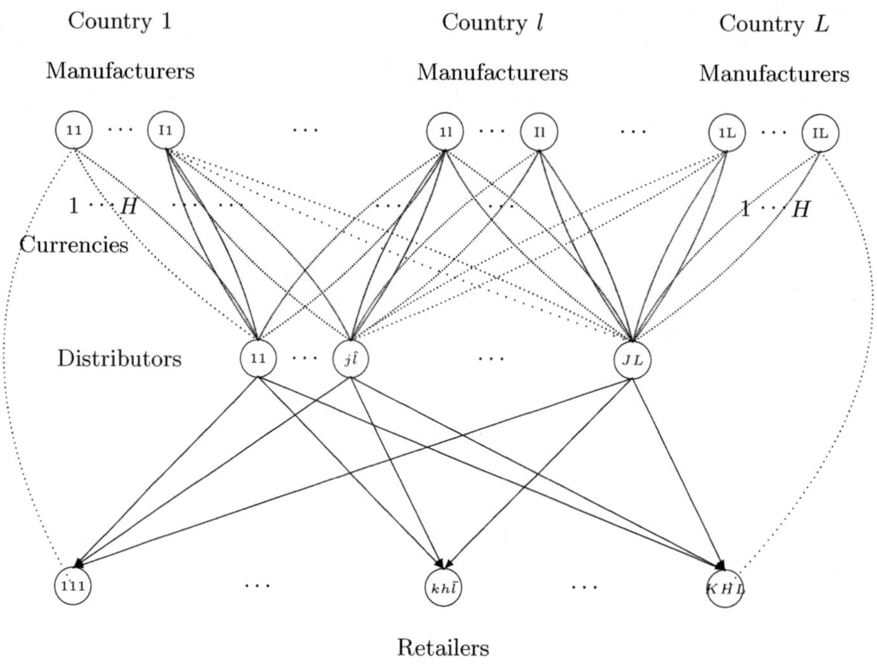

Fig. 1. The Structure of the Global Supply Chain Network

2.1 The Behavior of the Manufacturers and their Optimality Conditions

Let q^{il} denote the nonnegative production output of manufacturer il. Group the production outputs of all manufacturers into the column vector $q \in R_+^{IL}$. Here it is assumed that each manufacturer il has a production cost function f^{il}, which can depend, in general, on the entire vector of production outputs, that is,

$$f^{il} = f^{il}(q), \quad \forall i, l. \tag{1}$$

Hence, the production cost of a particular manufacturer can depend not only on his production output but also on the production outputs of the other manufacturers. This allows one to model competition.

Note that in Figure 1, there are H distinct links between a manufacturer and distributor pair. Each of these links represents the possibility of a transaction between manufacturer il and distributor $j\hat{l}$ in a certain currency h. Let $c^{il}_{jh\hat{l}}$ denote the transaction cost (which we assume includes the cost of transportation and other expenses) that manufacturer il is faced with transacting with distributor $j\hat{l}$ in currency h. Let $c^{il}_{kh\bar{l}}$, in turn, denote the transaction cost (which also includes the cost of transportation and other expenses) that manufacturer il is faced with transacting directly with retailer k in currency h and country $\bar{l}$. These transaction costs may depend upon the volume of transactions between each such pair in a particular currency, and their form depends on the type of transaction. They are given, respectively, by:

$$c^{il}_{jh\hat{l}} = c^{il}_{jh\hat{l}}(q^{il}_{jh\hat{l}}), \quad \forall i, l, j, h, \hat{l} \tag{2a}$$

and

$$c^{il}_{kh\bar{l}} = c^{il}_{kh\bar{l}}(q^{il}_{kh\bar{l}}), \quad \forall i, l, k, h, \bar{l}. \tag{2b}$$

Obviously, not every product can be purchased and shipped over electronic distribution channels. For example, in some cases, the purchase may occur through the Internet but the delivery requires a physical (and not virtual) means of transport. The generality of the above transaction cost functions (2a) and (2b) can represent such combined (or aggregated) activities, as well.

The following conservation of flow equation must hold:

$$q^{il} = \sum_{j=1}^{J}\sum_{h=1}^{H}\sum_{\hat{l}=1}^{L} q^{il}_{jh\hat{l}} + \sum_{k=1}^{K}\sum_{h=1}^{H}\sum_{\bar{l}=1}^{L} q^{il}_{kh\bar{l}}, \tag{3}$$

which states that the quantity of the product transacted by manufacturer il is equal to the amount produced by the manufacturer. For easy reference in the subsequent sections the product transactions between all pairs of manufacturers and distributors are grouped into the column vector denoted by $Q^1 \in R^{ILJHL}_{+}$. In addition, the product transactions between all pairs of manufacturers and retailers are grouped into the column vector denoted by $Q^2 \in R^{ILKHL}_{+}$. With this notation, one can express the production cost function of manufacturer il (cf. (1)) as a function of the vectors Q^1 and Q^2: $f^{il}(Q^1, Q^2)$.

It is assumed that each manufacturer seeks to maximize his profit which is the difference between his revenue and the total costs incurred. The revenue is equal to the product of the price of the product and the total quantity sold to all the distributors and all the retailers. Since we allow the transactions to take place in different currencies, the prices of the product in different currencies may be distinct. Let $\rho^{il*}_{1jh\hat{l}}$ denote the price associated with the product transacted between manufacturer il and distributor $j\hat{l}$ in currency h, and let $\rho^{il*}_{1kh\bar{l}}$ denote the price of the product associated with a transaction between manufacturer il and retailer k in currency h and country $\bar{l}$.

We now introduce the currency appreciation rate e_h, which is the appreciation(exchange) rate of currency h relative to the base currency (see [7] and [21] for further details). This is necessary since the revenue of a given manufacturer needs to be expressed in a base currency. Hence, the total revenue of manufacturer il is given by:

$$\sum_{j=1}^{J}\sum_{h=1}^{H}\sum_{\hat{l}=1}^{L}(\rho_{1jh\hat{l}}^{il*} \times e_h)q_{jh\hat{l}}^{il} + \sum_{k=1}^{K}\sum_{h=1}^{H}\sum_{\bar{l}=1}^{L}(\rho_{1kh\bar{l}}^{il*} \times e_h)q_{kh\bar{l}}^{il}.$$

The total costs incurred by the manufacturer il, in turn, are equal to the sum of the manufacturer's production costs and the total transaction costs. We assume that all the cost functions are in the base currency.

Hence, using the conservation of flow equation (3), the production cost functions, and the transaction cost functions (2a) and (2b), one can express the profit maximization criterion for manufacturer il as:

$$\text{Maximize} \quad \sum_{j=1}^{J}\sum_{h=1}^{H}\sum_{\hat{l}=1}^{L}(\rho_{1jh\hat{l}}^{il*} \times e_h)q_{jh\hat{l}}^{il} + \sum_{k=1}^{K}\sum_{h=1}^{H}\sum_{\bar{l}=1}^{L}(\rho_{1kh\bar{l}}^{il*} \times e_h)q_{kh\bar{l}}^{il}$$

$$-f^{il}(Q^1,Q^2) - \sum_{j=1}^{J}\sum_{h=1}^{H}\sum_{\hat{l}=1}^{L}c_{jh\hat{l}}^{il}(q_{jh\hat{l}}^{il}) - \sum_{k=1}^{K}\sum_{h=1}^{H}\sum_{\bar{l}=1}^{L}c_{kh\bar{l}}^{il}(q_{kh\bar{l}}^{il}), \tag{4}$$

subject to: $q_{jh\hat{l}}^{il} \geq 0$, for all $j, h, \hat{l}$ and $q_{kh\bar{l}}^{il} \geq 0$, for all $k, h, \bar{l}$.

In addition to the criterion of profit maximization, we also assume that each manufacturer is concerned with risk minimization. Here, for the sake of generality, we assume, as given, a risk function r^{il}, for manufacturer i in country l, which is assumed to be continuous and convex and a function of not only the product transactions associated with the particular manufacturer but also of those of other manufacturers. Hence, we assume that

$$r^{il} = r^{il}(Q^1,Q^2), \quad \forall i,l. \tag{5}$$

Note that according to (5) the risk as perceived by a manufacturer is dependent not only upon his product transactions but also on those of other manufacturers. Hence, the second criterion of manufacturer il can be expressed as:

$$\text{Minimize} \quad r^{il}(Q^1,Q^2), \tag{6}$$

subject to: $q_{jh\hat{l}}^{il} \geq 0$, for all $j, h, \hat{l}$ and $q_{kh\bar{l}}^{il} \geq 0$, for all $k, h, \bar{l}$. The risk function may be distinct for each manufacturer/country combination and can assume whatever form is necessary.

The Multicriteria Decision-Making Problem for a Manufacturer in a Particular Country

Each manufacturer il associates a nonnegative weight α^{il} with the risk minimization criterion (6), with the weight associated with the profit maximization criterion (4) serving as the numeraire and being set equal to 1. Hence, we can construct a *value function* for each manufacturer (cf. [22], [23], [24], [25]) using a constant additive weight value function. Consequently, the multicriteria decision-making problem for manufacturer il is transformed into:

$$\text{Maximize} \quad \sum_{j=1}^{J}\sum_{h=1}^{H}\sum_{\hat{l}=1}^{L}(\rho_{1jh\hat{l}}^{il*} \times e_h)q_{jh\hat{l}}^{il} + \sum_{k=1}^{K}\sum_{h=1}^{H}\sum_{\bar{l}=1}^{L}(\rho_{1kh\bar{l}}^{il*} \times e_h)q_{kh\bar{l}}^{il}$$

$$-f^{il}(Q^1,Q^2)-\sum_{j=1}^{J}\sum_{h=1}^{H}\sum_{\hat{l}=1}^{L}c_{jh\hat{l}}^{il}(q_{jh\hat{l}}^{il})-\sum_{k=1}^{K}\sum_{h=1}^{H}\sum_{\bar{l}=1}^{L}c_{kh\bar{l}}^{il}(q_{kh\bar{l}}^{il})-\alpha^{il}r^{il}(Q^1,Q^2), \quad (7)$$

subject to: $q_{jh\hat{l}}^{il} \geq 0$, for all $j, h, \hat{l}$ and $q_{kh\bar{l}}^{il} \geq 0$, for all $k, h, \bar{l}$.

The manufacturers are assumed to compete in a noncooperative fashion. Also, it is assumed that the production cost functions and the transaction cost functions for each manufacturer are continuous and convex. The governing optimization/equilibrium concept underlying noncooperative behavior is that of Nash [26, 27], which states, in this context, that each manufacturer will determine his optimal production quantity and transactions, given the optimal ones of the competitors. Hence, the optimality conditions for all manufacturers *simultaneously* can be expressed as the following inequality (see also [20], [28], [29], [30]): determine the solution $(Q^{1*}, Q^{2*}) \in R_+^{IL(JHL+KHL)}$, which satisfies:

$$\sum_{i=1}^{I}\sum_{l=1}^{L}\sum_{j=1}^{J}\sum_{h=1}^{H}\sum_{\hat{l}=1}^{L}\left[\frac{\partial f^{il}(Q^{1*},Q^{2*})}{\partial q_{jh\hat{l}}^{il}} + \frac{\partial c_{jh\hat{l}}^{il}(q_{jh\hat{l}}^{il*})}{\partial q_{jh\hat{l}}^{il}} + \alpha^{il}\frac{\partial r^{il}(Q^{1*},Q^{2*})}{\partial q_{jh\hat{l}}^{il}}\right.$$

$$\left. -\rho_{1jh\hat{l}}^{il*} \times e_h\right] \times \left[q_{jh\hat{l}}^{il} - q_{jh\hat{l}}^{il*}\right]$$

$$+\sum_{i=1}^{I}\sum_{l=1}^{L}\sum_{k=1}^{K}\sum_{h=1}^{H}\sum_{\bar{l}=1}^{L}\left[\frac{\partial f^{il}(Q^{1*},Q^{2*})}{\partial q_{kh\bar{l}}^{il}} + \frac{\partial c_{kh\bar{l}}^{il}(q_{kh\bar{l}}^{il*})}{\partial q_{kh\bar{l}}^{il}} + \alpha^{il}\frac{\partial r^{il}(Q^{1*},Q^{2*})}{\partial q_{kh\bar{l}}^{il}}\right.$$

$$\left. -\rho_{1kh\bar{l}}^{il*} \times e_h\right] \times \left[q_{kh\bar{l}}^{il} - q_{kh\bar{l}}^{il*}\right] \geq 0, \quad \forall (Q^1, Q^2) \in R_+^{IL(JHL+KHL)}. \quad (8)$$

The inequality (8), which is a *variational inequality* (cf. [30]) (for fixed prices and appreciation rates) has a nice economic interpretation. In particular, from the first term one can infer that, if there is a positive amount of the product transacted between a manufacturer and a distributor, then the marginal cost of production plus the marginal cost of transacting plus the weighted marginal risk associated with that transaction must be equal to the price (converted to the base currency) that the distributor is willing to pay for the product. If the marginal cost of production plus the marginal cost and the weighted marginal risk of transacting exceeds that price, then there will be zero volume of flow of the product between the two.

The second term in (8) has a similar interpretation; in particular, there will be a positive volume of transaction of the product from a manufacturer to a retailer if the marginal cost of production of the manufacturer plus the marginal cost of transacting with the retailer via the Internet and the weighted marginal risk is equal to the price (in the base currency) that the retailer is willing to pay for the product.

Note that, in the above framework, we explicitly allow each of the manufacturers to not only have distinct risk functions but also distinct weights associated with their respective risk functions. We emphasize that the study of risk has had a long and prominent history in the field of finance, dating to the work of Markowitz [31, 32], whereas the incorporation of risk into supply chain management has only been addressed fairly recently. Furthermore, we emphasize that by allowing the risk function to be general one can then construct an appropriate function pertaining to the specific situation and decision-maker. For example, in the field of finance, measurement of risk has included the use of variance-covariance matrices, yielding quadratic expressions for the risk (see also, e.g.,[21]). In addition, in finance, the bicriterion optimization problem of net revenue maximization and risk minimization is fairly standard (see also, e.g., [33]).

2.2 The Behavior of the Distributors and their Optimality Conditions

As was mentioned earlier, the distributors transact with both the manufacturers since they need to obtain the product for distribution, and with the retailers, who sell the product to the consumers. Similar to the manufacturers, and for the sake of modeling generality and flexibility, we assume that the distributors can transact in any of the H currencies.

Let $q^{j\hat{l}}_{kh\bar{l}}$ denote the amount of the product transacted between retailer $kh\bar{l}$ and distributor $j\hat{l}$ in currency h. We group these transaction quantities into the column vector $Q^3 \in R^{JLKHL}_+$.

A distributor $j\hat{l}$ is faced with certain expenses, which may include, for example, loading/unloading costs, storage costs, etc., associated with the product. We refer collectively to such costs as a *handling* cost and denote it by $c_{j\hat{l}}$. In a simple situation, one might have that

$$c_{j\hat{l}} = c_{j\hat{l}}\Big(\sum_{i=1}^{I}\sum_{l=1}^{L}\sum_{h=1}^{H} q^{il}_{jh\hat{l}}, \sum_{k=1}^{K}\sum_{h=1}^{H}\sum_{\bar{l}=1}^{L} q^{j\hat{l}}_{kh\bar{l}}\Big), \tag{9a}$$

that is, the *handling* cost of a distributor is a function of how much of the product he has obtained and how much of the product he has transacted with the various retailers. However, for the sake of generality, and to enhance the modeling of competition, we follow Dong, Zhang, and Nagurney [18] and allow the function to depend also on the amount of the product acquired and transacted by other distributors. Hence, we assume that, for all $j\hat{l}$:

$$c_{j\hat{l}} = c_{j\hat{l}}(Q^1, Q^3). \tag{9b}$$

Let $\rho_{2kh\bar{l}}^{j\hat{l}*}$ denote the price in currency h associated with the transaction between distributor $j\hat{l}$ and retailer $kh\bar{l}$. This price, as will be shown, will be endogenously determined in the model and will be, in the case of a positive volume of flow between a distributor-retailer pair, equal to a clearing-type price. The total amount of revenue the distributor obtains from his transactions is equal to the sum of the price (transformed into the base currency) and the amount of the product transacted with the various retailers in the distinct countries and currencies. Indeed, since transactions can be made in distinct currencies and with different retailers, who, in fact, may even be virtual, the total revenue of distributor $j\hat{l}$ can be expressed in the base currency as follows:

$$\sum_{k=1}^{K}\sum_{h=1}^{H}\sum_{\bar{l}=1}^{L}(\rho_{2kh\bar{l}}^{j\hat{l}*} \times e_h)q_{kh\bar{l}}^{j\hat{l}}. \tag{10}$$

Assuming that the distributors are profit-maximizers, the profit maximization problem for distributor $j\hat{l}$ can be expressed as:

$$\text{Maximize} \quad \sum_{k=1}^{K}\sum_{h=1}^{H}\sum_{\bar{l}=1}^{L}(\rho_{2kh\bar{l}}^{j\hat{l}*} \times e_h)q_{kh\bar{l}}^{j\hat{l}} - c_{j\hat{l}}(Q^1, Q^3) - \sum_{i=1}^{I}\sum_{l=1}^{L}\sum_{h=1}^{H}(\rho_{1jh\hat{l}}^{il*} \times e_h)q_{jh\hat{l}}^{il} \tag{11}$$

subject to:

$$\sum_{k=1}^{K}\sum_{h=1}^{H}\sum_{\bar{l}=1}^{L} q_{kh\bar{l}}^{j\hat{l}} \leq \sum_{i=1}^{I}\sum_{l=1}^{L}\sum_{h=1}^{H} q_{jh\hat{l}}^{il}, \tag{12}$$

and the nonnegativity assumptions: $q_{jh\hat{l}}^{il} \geq 0$, for all $i, \bar{l}, h$ and $q_{kh\bar{l}}^{j\hat{l}} \geq 0$, for all $k, h, \bar{l}$.

Constraint (12) states that a distributor in a country cannot sell more of the product than he has obtained from the various manufacturers.

In addition, each distributor seeks to also minimize his risk associated with obtaining and shipping the product to the various retailers. Each distributor $j\hat{l}$ is faced with his own individual risk denoted by $r^{j\hat{l}}$ with the function being assumed to be continuous and convex and dependent on the transactions to and from all the distributors, that is,

$$r^{j\hat{l}} = r^{j\hat{l}}(Q^1, Q^3), \quad \forall j, \hat{l}. \tag{13}$$

The Multicriteria Decision-Making Problem for a Distributor in a Particular Country

We assume that each distributor associates a weight of 1 with the profit criterion (11) and a weight of $\beta^{j\hat{l}}$ with his risk level. Therefore, the multicriteria decision-making problem for distributor $j\hat{l}$; $j = 1, \ldots, J$; $\hat{l} = 1, \ldots, L$, can be transformed directly into the optimization problem:

$$\text{Maximize} \quad \sum_{k=1}^{K}\sum_{h=1}^{H}\sum_{\bar{l}=1}^{L}(\rho_{2kh\bar{l}}^{j\hat{l}*} \times e_h)q_{kh\bar{l}}^{j\hat{l}} - c_{j\hat{l}}(Q^1, Q^3) - \sum_{i=1}^{I}\sum_{l=1}^{L}\sum_{h=1}^{H}(\rho_{1jh\hat{l}}^{il*} \times e_h)q_{jh\hat{l}}^{il}$$

$$-\beta^{j\hat{l}} r^{j\hat{l}}(Q^1, Q^3) \tag{14}$$

subject to:

$$\sum_{k=1}^{K} \sum_{h=1}^{H} \sum_{\bar{l}=1}^{L} q_{kh\bar{l}}^{j\hat{l}} \leq \sum_{i=1}^{I} \sum_{l=1}^{L} \sum_{h=1}^{H} q_{jh\hat{l}}^{il}, \tag{15}$$

and the nonnegativity constraints: $q_{jh\hat{l}}^{il} \geq 0$, and $q_{kh\bar{l}}^{j\hat{l}} \geq 0$, for all i, h, k, l, and $\bar{l}$.

Objective function (14) represents a value function for distributor $j\hat{l}$ with $\beta^{j\hat{l}}$ having the interpretation as a conversion rate in dollar value.

Here it is assumed that the distributors compete in a noncooperative manner, given the actions of the other distributors. Note that, at this point, we consider that the distributors seek to determine not only the optimal amounts purchased by the retailers, but, also, the amount that they wish to obtain from the manufacturers. In equilibrium, all the transactions between the tiers of global supply chain network will have to coincide.

Assuming that the handling cost $c_{j\hat{l}}$ for each distributor is a continuous and convex function, the optimality conditions of the distributors *simultaneously* can be stated as the following variational inequality: determine the solution $(Q^{1*}, Q^{3*}, \gamma^*) \in R_+^{JL(ILH+KHL+1)}$, which satisfies:

$$\sum_{j=1}^{J} \sum_{\hat{l}=1}^{L} \sum_{i=1}^{I} \sum_{l=1}^{L} \sum_{h=1}^{H} \left[\frac{\partial c_{j\hat{l}}(Q^{1*}, Q^{3*})}{\partial q_{jh\hat{l}}^{il}} + \beta^{j\hat{l}} \frac{\partial r^{j\hat{l}}(Q^{1*}, Q^{3*})}{\partial q_{jh\hat{l}}^{il}} + \rho_{1jh\hat{l}}^{il*} \times e_h - \gamma_{j\hat{l}}^* \right] \times \left[q_{jh\hat{l}}^{il} - q_{jh\hat{l}}^{il*} \right]$$

$$+ \sum_{j=1}^{J} \sum_{\hat{l}=1}^{L} \sum_{k=1}^{K} \sum_{h=1}^{H} \sum_{\bar{l}=1}^{L} \left[-\rho_{2kh\bar{l}}^{j\hat{l}*} \times e_h + \frac{\partial c_{j\hat{l}}(Q^{1*}, Q^{3*})}{\partial q_{kh\bar{l}}^{j\hat{l}}} + \beta^{j\hat{l}} \frac{\partial r^{j\hat{l}}(Q^{1*}, Q^{3*})}{\partial q_{kh\bar{l}}^{j\hat{l}}} + \gamma_{j\hat{l}}^* \right] \times \left[q_{kh\bar{l}}^{j\hat{l}} - q_{kh\bar{l}}^{j\hat{l}*} \right]$$

$$+ \sum_{j=1}^{J} \sum_{\hat{l}=1}^{L} \left[\sum_{i=1}^{I} \sum_{l=1}^{L} \sum_{h=1}^{H} q_{jh\hat{l}}^{il*} - \sum_{k=1}^{K} \sum_{h=1}^{H} \sum_{\bar{l}=1}^{L} q_{kh\bar{l}}^{j\hat{l}*} \right] \times \left[\gamma_{j\hat{l}} - \gamma_{j\hat{l}}^* \right] \geq 0,$$

$$\forall (Q^1, Q^3, \gamma) \in R_+^{JL(ILH+KHL+1)}, \tag{16}$$

where $\gamma_{j\hat{l}}$ is the Lagrange multiplier associated with constraint (12) for distributor $j\hat{l}$ and γ is the column vector of all the distributors' multipliers. In inequality (16), as in inequality (8), the prices charged are not variables.

The economic interpretation of the distributors' optimality conditions is now highlighted. From the second term in inequality (16), one has that, if retailer $kh\bar{l}$ purchases the product from a distributor $j\hat{l}$, that is, if the $q_{kh\bar{l}}^{j\hat{l}*}$ is positive, then the price charged by retailer $j\hat{l}$, converted into the base currency, $\rho_{2kh\bar{l}}^{j\hat{l}*} \times e_h$, is equal to the marginal handling cost plus the weighted marginal risk, plus the shadow price term $\gamma^*_{j\hat{l}}$, which, from the third term in the inequality, serves as the price to clear the market from distributor $j\hat{l}$. Furthermore, from the first term in inequality (16), one can infer that, if a manufacturer transacts with a distributor resulting in a positive flow of the product between the two, then the price $\gamma^*_{j\hat{l}}$ is precisely equal to distributor $j\hat{l}$'s payment to the manufacturer in a certain currency h transformed into the base currency, $\rho^{il*}_{1jh\hat{l}} \times e_h$, plus his marginal cost of handling the product associated with transacting with the particular manufacturer plus the weighted marginal risk associated with the transaction.

In the above derivations, we have considered supply-side risk from the perspective of the manufacturers as well as from the distributors. We now turn to the retailers and consider demand-side risk, which is modeled in a stochastic manner to represent uncertainty. Clearly, if the demands were known with certainty there would be no risk associated with them and the retailers and manufacturers could make their production and distribution decisions accordingly.

2.3 The Retailers and their Optimality Conditions

The retailers, in turn, must decide how much to order from the distributors and from the manufacturers in order to cope with the random demand while still seeking to maximize their profits. A retailer $kh\bar{l}$ is also faced with what we term a *handling* cost, which may include, for example, the display and storage cost associated with the product. We denote this cost by $c_{kh\bar{l}}$ and, in the simplest case, we would have that $c_{kh\bar{l}}$ is a function of $s_{kh\bar{l}} = \sum_{i=1}^{I}\sum_{l=1}^{L} q^{il}_{kh\bar{l}} + \sum_{j=1}^{J}\sum_{\hat{l}=1}^{L} q^{j\hat{l}}_{kh\bar{l}}$, that is, the holding cost of a retailer is a function of how much of the product he has obtained from transactions with the various manufacturers directly through orders via the Internet and from the various distributors. However, for the sake of generality, and to enhance the modeling of competition, we allow this function to depend, in general, also on the amounts of the product held by other retailers and, therefore, we may write:

$$c_{kh\bar{l}} = c_{kh\bar{l}}(Q^2, Q^3), \quad \forall k, h, \bar{l}. \tag{17}$$

Let $\rho_{3kh\bar{l}}$ denote the demand price of the product associated with retailer $kh\bar{l}$. We assume that $\hat{d}_{kh\bar{l}}(\rho_{3kh\bar{l}})$ is the demand for the product at the demand price of $\rho_{3kh\bar{l}}$ at retail outlet $kh\bar{l}$, where $\hat{d}_{kh\bar{l}}(\rho_{3kh\bar{l}})$ is a random variable with a density function of $F_{kh\bar{l}}(x, \rho_{3kh\bar{l}})$, with $\rho_{3kh\bar{l}}$ serving as a parameter. Hence, we assume that the density function may vary with the demand price. Let $P_{kh\bar{l}}$ be the probability distribution function of $\hat{d}_{kh\bar{l}}(\rho_{3kh\bar{l}})$, that is, $P_{kh\bar{l}}(x, \rho_{3kh\bar{l}}) = P_{kh\bar{l}}(\hat{d}_{kh\bar{l}} \leq x) = \int_0^x F_{kh\bar{l}}(x, \rho_{3kh\bar{l}})dx$.

Retailer $kh\bar{l}$ can sell to the consumers no more than the minimum of his supply or his demand, that is, the actual sale of the product at retailer $kh\bar{l}$ cannot exceed $\min\{s_{kh\bar{l}}, \hat{d}_{kh\bar{l}}\}$. Let

$$\Delta^+_{kh\bar{l}} \equiv \max\{0, s_{kh\bar{l}} - \hat{d}_{kh\bar{l}}\} \tag{18}$$

and

$$\Delta^-_{kh\bar{l}} \equiv \max\{0, \hat{d}_{kh\bar{l}} - s_{kh\bar{l}}\}, \tag{19}$$

where $\Delta^+_{kh\bar{l}}$ is a random variable representing the excess supply (inventory), whereas $\Delta^-_{kh\bar{l}}$ is a random variable representing the excess excess demand (shortage).

Note that the expected values of excess supply and excess demand of retailer $kh\bar{l}$ are scalar functions of $s_{kh\bar{l}}$ and $\rho_{3kh\bar{l}}$. In particular, let $\pi^+_{kh\bar{l}}$ and $\pi^-_{kh\bar{l}}$ denote, respectively, the expected values $E(\Delta^+_{kh\bar{l}})$ and $E(\Delta^-_{kh\bar{l}})$, that is,

$$\pi^+_{kh\bar{l}}(s_{kh\bar{l}}, \rho_{3kh\bar{l}}) \equiv E(\Delta^+_{kh\bar{l}}) = \int_0^{s_{kh\bar{l}}} (s_{kh\bar{l}} - x)F_{kh\bar{l}}(x, \rho_{3kh\bar{l}})dx, \tag{20}$$

and

$$\pi^-_{kh\bar{l}}(s_{kh\bar{l}}, \rho_{3kh\bar{l}}) \equiv E(\Delta^-_{kh\bar{l}}) = \int_{s_{kh\bar{l}}}^{\infty} (x - s_{kh\bar{l}})F_{kh\bar{l}}(x, \rho_{3kh\bar{l}})dx. \tag{21}$$

Assume that retailer $kh\bar{l}$ is faced with certain penalties for having an excess or shortage in regards to the supply. Let $\lambda^+_{kh\bar{l}} \geq 0$ denote the unit penalty of having excess supply at retail outlet $kh\bar{l}$, and let $\lambda^-_{kh\bar{l}} \geq 0$ denote the unit penalty of having excess demand at outlet $kh\bar{l}$. Then the expected total penalty of retailer $kh\bar{l}$ can be expressed as:

$$E(\lambda^+_{kh\bar{l}}\Delta^+_{kh\bar{l}} + \lambda^-_{kh\bar{l}}\Delta^-_{kh\bar{l}}) = \lambda^+_{kh\bar{l}}\pi^+_{kh\bar{l}}(s_{kh\bar{l}}, \rho_{3kh\bar{l}}) + \lambda^-_{kh\bar{l}}\pi^-_{kh\bar{l}}(s_{kh\bar{l}}, \rho_{3kh\bar{l}}).$$

Assuming profit-maximizing behavior of the retailers, one can state the following optimization problem for retailer $kh\bar{l}$:

$$\text{Maximize} \quad E((\rho^*_{3kh\bar{l}} \times e_h) \cdot \min\{s_{kh\bar{l}}, \hat{d}_{kh\bar{l}}\}) - E(\lambda^-_{kh\bar{l}}\Delta^+_{kh\bar{l}} + \lambda^-_{kh\bar{l}}\Delta^-_{kh\bar{l}})$$

$$+c_{kh\bar{l}}(Q^2, Q^3) - \sum_{i=1}^{I}\sum_{l=1}^{L}(\rho^{il*}_{1kh\bar{l}} \times e_h)q^{il}_{kh\bar{l}} - \sum_{j=1}^{J}\sum_{\hat{l}=1}^{L}(\rho^{j\hat{l}*}_{2kh\bar{l}} \times e_h)q^{j\hat{l}}_{kh\bar{l}}, \tag{22}$$

subject to: $q^{il}_{kh\bar{l}} \geq 0$, $q^{j\hat{l}}_{kh\bar{l}} \geq 0$, for all $i, l, j, \hat{l}$.

Objective function (22) expresses that the expected profit of retailer $kh\bar{l}$, which is the difference between the expected revenues and the sum of the expected penalty, the handling cost, and the payouts to the manufacturers and to the distributors, should be maximized.

Applying now the definitions of $\Delta^+_{kh\bar{l}}$, and $\Delta^-_{kh\bar{l}}$, we know that $\min\{s_{kh\bar{l}}, \hat{d}_{kh\bar{l}}\}$ $= \hat{d}_{kh\bar{l}} - \Delta^-_{kh\bar{l}}$. Therefore, the objective function (22) can be expressed as

$$\text{Maximize} \quad (\rho^*_{3kh\bar{l}} \times e_h)d_{kh\bar{l}}(\rho_{3kh\bar{l}}) - (\rho^*_{3kh\bar{l}} \times e_h + \lambda^-_{kh\bar{l}})\pi^-_{kh\bar{l}}(s_{kh\bar{l}}, \rho^*_{3kh\bar{l}})$$

$$-\lambda^+_{kh\bar{l}}\pi^+_{kh\bar{l}}(s_{kh\bar{l}}, \rho^*_{3kh\bar{l}}) - c_{kh\bar{l}}(Q^2, Q^3) - \sum_{i=1}^{I}\sum_{l=1}^{L}(\rho^{il*}_{1kh\bar{l}} \times e_h)q^{il}_{kh\bar{l}}$$

$$-\sum_{j=1}^{J}\sum_{\hat{l}=1}^{L}(\rho_{2kh\bar{l}}^{j\hat{l}*}\times e_h)q_{kh\bar{l}}^{j\hat{l}}, \tag{23}$$

where $d_{kh\bar{l}}(\rho_{3kh\bar{l}}) \equiv E(\hat{d}_{kh\bar{l}})$ is a scalar function of $\rho_{3kh\bar{l}}$.

We now consider the optimality conditions of the retailers assuming that each retailer is faced with the optimization problem (23), subject to the nonnegativity assumption on the variables. Here, we also assume that the retailers compete in a noncooperative manner so that each maximizes his profits, given the actions of the other retailers. Note that, at this point, we consider that retailers seek to determine the amount that they wish to obtain from the manufacturers and from the distributors. First, however, we make the following derivation and introduce the necessary notation:

$$\frac{\partial\pi_{kh\bar{l}}^{+}(s_{kh\bar{l}},\rho_{3kh\bar{l}})}{\partial q_{kh\bar{l}}^{il}} = \frac{\partial\pi_{kh\bar{l}}^{+}(s_{kh\bar{l}},\rho_{3kh\bar{l}})}{\partial q_{kh\bar{l}}^{j\hat{l}}} = P_{kh\bar{l}}(s_{kh\bar{l}},\rho_{3kh\bar{l}})$$

$$= P_{kh\bar{l}}(\sum_{i=1}^{I}\sum_{l=1}^{L}q_{kh\bar{l}}^{il}+\sum_{j=1}^{J}\sum_{\hat{l}=1}^{L}q_{kh\bar{l}}^{j\hat{l}},\rho_{3kh\bar{l}}) \tag{24}$$

$$\frac{\partial\pi_{kh\bar{l}}^{-}(s_{kh\bar{l}},\rho_{3kh\bar{l}})}{\partial q_{kh\bar{l}}^{il}} = \frac{\partial\pi_{kh\bar{l}}^{-}(s_{kh\bar{l}},\rho_{3kh\bar{l}})}{\partial q_{kh\bar{l}}^{j\hat{l}}} = P_{kh\bar{l}}(s_{kh\bar{l}},\rho_{3kh\bar{l}})-1$$

$$= P_{kh\bar{l}}(\sum_{i=1}^{I}\sum_{l=1}^{L}q_{kh\bar{l}}^{il}+\sum_{j=1}^{J}\sum_{\hat{l}=1}^{L}q_{kh\bar{l}}^{j\hat{l}},\rho_{3kh\bar{l}})-1. \tag{25}$$

Assuming that the handling cost for each retailer is continuous and convex, then the optimality conditions for all the retailers satisfy the variational inequality: determine $(Q^{2*},Q^{3*})\in R_{+}^{(IL+JL)KHL}$, satisfying:

$$\sum_{k=1}^{K}\sum_{h=1}^{H}\sum_{\bar{l}=1}^{L}\Bigg\{\sum_{i=1}^{I}\sum_{l=1}^{L}\Big[\lambda_{kh\bar{l}}^{+}P_{kh\bar{l}}(s_{kh\bar{l}}^{*},\rho_{3kh\bar{l}}^{*})-(\lambda_{kh\bar{l}}^{-}+\rho_{3kh\bar{l}}^{*}\times e_h)$$

$$\times(1-P_{kh\bar{l}}(s_{kh\bar{l}}^{*},\rho_{3kh\bar{l}}^{*}))+\frac{\partial c_{kh\bar{l}}(Q^{2*},Q^{3*})}{\partial q_{kh\bar{l}}^{il}}+\rho_{1kh\bar{l}}^{il*}\times e_h\Big]\times\left[q_{kh\bar{l}}^{il}-q_{kh\bar{l}}^{il*}\right]$$

$$+\sum_{j=1}^{J}\sum_{\hat{l}=1}^{L}\Big[\lambda_{kh\bar{l}}^{+}P_{kh\bar{l}}(s_{kh\bar{l}}^{*},\rho_{3kh\bar{l}}^{*})-(\lambda_{kh\bar{l}}^{-}+\rho_{3kh\bar{l}}^{*}\times e_h)(1-P_{kh\bar{l}}(s_{kh\bar{l}}^{*},\rho_{3kh\bar{l}}^{*}))$$

$$+\frac{\partial c_{kh\bar{l}}(Q^{2*},Q^{3*})}{\partial q_{kh\bar{l}}^{j\hat{l}}}+\rho_{2kh\bar{l}}^{j\hat{l}*}\times e_h\Big]\times\left[q_{kh\bar{l}}^{j\hat{l}}-q_{kh\bar{l}}^{j\hat{l}*}\right]\Bigg\}\geq 0,$$

$$\forall(Q^2,Q^3)\in R_{+}^{(IL+JL)KHL}. \tag{26}$$

In this derivation, as in the derivation of inequalities (8) and (16), we have not had the prices charged be variables. They become endogenous variables in the integrated global supply chain network equilibrium model. A similar derivation but in the absence of electronic commerce (and in the case of only a two-tiered rather than a three-tiered supply chain network) was obtained in Dong, Zhang, and Nagurney [19]. See Dong, Zhang, and Nagurney [18] for a three-tiered (single-country and currency) supply chain network model with random demands.

We now highlight the economic interpretation of the retailers' optimality conditions. In inequality (26), we can infer that, if a manufacturer il transacts with a retailer $kh\bar{l}$ resulting in a positive flow of the product between the two, then the price at retail outlet $kh\bar{l}$, $\rho^*_{3kh\bar{l}}$, with the probability of $(1-P_{kh\bar{l}}(\sum_{i=1}^{I}\sum_{l=1}^{L} q^{il*}_{kh\bar{l}}+\sum_{j=1}^{J}\sum_{\hat{l}=1}^{L} q^{j\hat{l}*}_{kh\bar{l}}, \rho^*_{3kh\bar{l}}))$, that is, when the demand is not less then the total order quantity, is precisely equal to the retailer $kh\bar{l}$'s payment to the manufacturer, $\rho^{il*}_{1kh\bar{l}} \times e_h$, plus his marginal cost of handling the product and the penalty of having excess demand with probability of $P_{kh\bar{l}}(\sum_{i=1}^{I}\sum_{l=1}^{L} q^{il*}_{kh\bar{l}} + \sum_{j=1}^{J}\sum_{\hat{l}=1}^{L} q^{j\hat{l}*}_{kh\bar{l}}, \rho^*_{3kh\bar{l}})$, (which is the probability when actual demand is less than the order quantity), subtracted by the penalty of having shortage with probability of $(1-P_{kh\bar{l}}(\sum_{i=1}^{I}\sum_{l=1}^{L} q^{il*}_{kh\bar{l}}+\sum_{j=1}^{J}\sum_{\hat{l}=1}^{L} q^{j\hat{l}*}_{kh\bar{l}}, \rho^*_{3kh\bar{l}}))$ (when the actual demand is greater than the order quantity).

Similarly, if a distributor $j\hat{l}$ transacts with a retailer $kh\bar{l}$ resulting in a positive flow of the product between the two, then the selling price at retail outlet $kh\bar{l}$, $\rho^*_{3kh\bar{l}}$, with the probability of $(1-P_{kh\bar{l}}(\sum_{i=1}^{I}\sum_{l=1}^{L} q^{il*}_{kh\bar{l}}+\sum_{j=1}^{J}\sum_{\hat{l}=1}^{L} q^{j\hat{l}*}_{kh\bar{l}}, \rho^*_{3kh\bar{l}}))$, that is, when the demand is not less then the total order quantity, is precisely equal to the retailer k's payment to the manufacturer, $\rho^{j\hat{l}*}_{2kh\bar{l}} \times e_h$, plus his marginal cost of handling the product and the penalty of having excess demand with probability of $P_{kh\bar{l}}(\sum_{i=1}^{I}\sum_{l=1}^{L} q^{il*}_{kh\bar{l}} + \sum_{j=1}^{J}\sum_{\hat{l}=1}^{L} q^{j\hat{l}*}_{kh\bar{l}}, \rho^*_{3kh\bar{l}})$, (which is the probability when actual demand is less than the order quantity), subtracted by the penalty of having shortage with probability of $(1-P_{kh\bar{l}}(\sum_{i=1}^{I}\sum_{l=1}^{L} q^{il*}_{kh\bar{l}} + \sum_{j=1}^{J}\sum_{\hat{l}=1}^{L} q^{j\hat{l}*}_{kh\bar{l}}, \rho^*_{3kh\bar{l}}))$ (when the actual demand is greater than the order quantity).

2.4 The Equilibrium Conditions

We now turn to a discussion of the market equilibrium conditions. Subsequently, we construct the equilibrium conditions for the entire global supply chain network.

The equilibrium conditions associated with the transactions that take place between the retailers and the consumers are the stochastic economic equilibrium conditions, which, mathematically, take on the following form: for $k, h, \bar{l}$; $k =$

$1, ..., K;\ h = 1, ..., H;\ \bar{l} = 1, ..., L$:

$$\hat{d}_{kh\bar{l}}(\rho^*_{3kh\bar{l}}) \begin{cases} = \sum_{i=1}^{I}\sum_{l=1}^{L} q^{il*}_{kh\bar{l}} + \sum_{j=1}^{J}\sum_{\hat{l}=1}^{L} q^{j\hat{l}*}_{kh\bar{l}} & \textbf{a.e.} \quad \text{if } \rho^*_{3kh\bar{l}} > 0 \\ \leq \sum_{i=1}^{I}\sum_{l=1}^{L} q^{il*}_{kh\bar{l}} + \sum_{j=1}^{J}\sum_{\hat{l}=1}^{L} q^{j\hat{l}*}_{kh\bar{l}} & \textbf{a.e.} \quad \text{if } \rho^*_{3kh\bar{l}} = 0, \end{cases} \tag{27}$$

where **a.e.** means that the corresponding equality or inequality holds almost everywhere.

Conditions (27) state that, if the demand price at outlet $kh\bar{l}$ is positive, then the quantities purchased by the retailer from the manufacturers and from the distributors in the aggregate are equal to the demand, with exceptions of zero probability. These conditions correspond to the well-known economic equilibrium conditions (cf. [30]). Related equilibrium conditions, but without electronic transactions allowed, and in a single country context, were first proposed in Dong, Zhang, and Nagurney [18].

Equilibrium conditions (27) are equivalent to the following variational inequality problem, after taking the expected value of the demand and summing over all retailers $kh\bar{l}$: determine $\rho_3^* \in R_+^{KHL}$ satisfying:

$$\sum_{k=1}^{K}\sum_{h=1}^{H}\sum_{\bar{l}=1}^{L}\left[\sum_{i=1}^{I}\sum_{l=1}^{L} q^{il*}_{kh\bar{l}} + \sum_{j=1}^{J}\sum_{\hat{l}=1}^{L} q^{j\hat{l}}_{kh\bar{l}} - d_{kh\bar{l}}(\rho^*_{3kh\bar{l}})\right] \times \left[\rho_{3kh\bar{l}} - \rho^*_{3kh\bar{l}}\right] \geq 0,$$

$$\forall \rho_3 \in R_+^{KHL}, \tag{28}$$

where ρ_3 is the KHL-dimensional column vector with components: $\rho_{3111}, ..., \rho_{3kh\bar{l}}$.

2.5 The Equilibrium Conditions of the Global Supply Chain

In equilibrium, we must have that the sum of the optimality conditions for all manufacturers, as expressed by inequality (8), the optimality conditions of the distributors, as expressed by condition (16), the optimality conditions for all retailers, as expressed by inequality (26), and the market equilibrium conditions, as expressed by inequality (28) must be satisfied. Hence, the product transactions from the manufacturers to the retailers must be equal to the product transactions that the retailers accept from the manufacturers. In addition, the product transactions from the manufacturers to the distributors, must be equal to those accepted by the distributors, and, finally, the product transactions from the distributors to the retailers must coincide with those accepted by the retailers. We state this explicitly in the following definition:

Definition 1: Global Supply Chain Network Equilibrium with Supply-Side and Demand-Side Risk
The equilibrium state of the global supply chain network with supply and demand-side risk is one where the product transactions between the tiers of the decision-makers coincide and the product transactions and prices satisfy the sum of the optimality conditions (8), (16), and (26), and the conditions (28).

The summation of inequalities (8), (16), (26), and (28) (with the prices at the manufacturers, the distributors, and at the retailers denoted, respectively, by their values at the equilibrium, after algebraic simplification, yields the following result:

Theorem 1: Variational Inequality Formulation
A product transaction and price pattern $(Q^{1*}, Q^{2*}, Q^{3*}, \gamma^*, \rho_3^*) \in \mathcal{K}$ *is an equilibrium pattern of the global supply chain network model according to Definition 1 if and only if it satisfies the variational inequality problem:*

$$\sum_{i=1}^{I}\sum_{l=1}^{L}\sum_{j=1}^{J}\sum_{h=1}^{H}\sum_{\hat{l}=1}^{L}\left[\frac{\partial f^{il}(Q^{1*},Q^{2*})}{\partial q^{il}_{jh\hat{l}}}+\frac{\partial c^{il}_{jh\hat{l}}(q^{il*}_{jh\hat{l}})}{\partial q^{il}_{jh\hat{l}}}+\frac{\partial c_{j\hat{l}}(Q^{1*},Q^{3*})}{\partial q^{il}_{jh\hat{l}}}\right.$$

$$\left.+\alpha^{il}\frac{\partial r^{il}(Q^{1*},Q^{2*})}{\partial q^{il}_{jh\hat{l}}}+\beta^{j\hat{l}}\frac{\partial r^{j\hat{l}}(Q^{1*},Q^{3*})}{\partial q^{il}_{jh\hat{l}}}-\gamma^*_{j\hat{l}}\right]\times\left[q^{il}_{jh\hat{l}}-q^{il*}_{jh\hat{l}}\right]$$

$$+\sum_{i=1}^{I}\sum_{l=1}^{L}\sum_{k=1}^{K}\sum_{h=1}^{H}\sum_{\bar{l}=1}^{L}\left[\frac{\partial f^{il}(Q^{1*},Q^{2*})}{\partial q^{il}_{kh\bar{l}}}+\frac{\partial c^{il}_{kh\bar{l}}(q^{il*}_{kh\bar{l}})}{\partial q^{il}_{kh\bar{l}}}+\frac{\partial c_{kh\bar{l}}(Q^{2*},Q^{3*})}{\partial q^{il}_{kh\bar{l}}}\right.$$

$$\left.+\alpha^{il}\frac{\partial r^{il}(Q^{1*},Q^{2*})}{\partial q^{il}_{kh\bar{l}}}+\lambda^+_{kh\bar{l}}P_{kh\bar{l}}(s^*_{kh\bar{l}},\rho^*_{3kh\bar{l}})-(\lambda^-_{kh\bar{l}}+\rho^*_{3kh\bar{l}}\times e_h)(1-P_{kh\bar{l}}(s^*_{kh\bar{l}},\rho^*_{3kh\bar{l}}))\right]$$

$$\times\left[q^{il}_{kh\bar{l}}-q^{il*}_{kh\bar{l}}\right]+\sum_{j=1}^{J}\sum_{\hat{l}=1}^{L}\sum_{k=1}^{K}\sum_{h=1}^{H}\sum_{\bar{l}=1}^{L}\left[\lambda^+_{kh\bar{l}}P_{kh\bar{l}}(s^*_{kh\bar{l}},\rho^*_{3kh\bar{l}})\right)$$

$$-(\lambda^-_{kh\bar{l}}+\rho^*_{3kh\bar{l}}\times e_h)(1-P_{kh\bar{l}}(s^*_{kh\bar{l}},\rho^*_{3kh\bar{l}}))$$

$$\left.+\frac{\partial c_{j\hat{l}}(Q^{1*},Q^{3*})}{\partial q^{j\hat{l}}_{kh\bar{l}}}+\frac{\partial c_{kh\bar{l}}(Q^{2*},Q^{3*})}{\partial q^{j\hat{l}}_{kh\bar{l}}}+\beta^{j\hat{l}}\frac{\partial r^{j\hat{l}}(Q^{1*},Q^{3*})}{\partial q^{j\hat{l}}_{kh\bar{l}}}+\gamma^*_{j\hat{l}}\right]\times\left[q^{j\hat{l}}_{kh\bar{l}}-q^{j\hat{l}*}_{kh\bar{l}}\right]$$

$$+\sum_{j=1}^{J}\sum_{h=1}^{H}\sum_{\hat{l}=1}^{L}\left[\sum_{i=1}^{I}\sum_{l=1}^{L}q^{il*}_{jh\hat{l}}-\sum_{k=1}^{K}\sum_{\bar{l}=1}^{L}q^{j\hat{l}*}_{kh\bar{l}}\right]\times\left[\gamma_{j\hat{l}}-\gamma^*_{j\hat{l}}\right]$$

$$+\sum_{k=1}^{K}\sum_{h=1}^{H}\sum_{\bar{l}=1}^{L}\left[\sum_{i=1}^{I}\sum_{l=1}^{L}q^{il*}_{kh\bar{l}}+\sum_{j=1}^{J}\sum_{\hat{l}=1}^{L}q^{j\hat{l}}_{kh\bar{l}}-d_{kh\bar{l}}(\rho^*_{3kh\bar{l}})\right]\times\left[\rho^*_{3kh\bar{l}}-\rho^*_{3kh\bar{l}}\right]\geq 0, \tag{29}$$

$$\forall(Q^1,Q^2,Q^3,\gamma,\rho_3)\in\mathcal{K},$$

where
$\mathcal{K} \equiv \{(Q^1, Q^2, Q^3, \gamma, \rho_3)|(Q^1, Q^2, Q^3, \gamma, \rho_3) \in R_+^{IL(JHL+KHL)+JL(KHL+1)+KHL}\}$.

For easy reference in the subsequent sections, variational inequality problem (29) can be rewritten in standard variational inequality form (cf. [30]) as follows: determine $X^* \in \mathcal{K}$ satisfying:

$$\langle F(X^*), X - X^* \rangle \geq 0, \quad \forall X \in \mathcal{K} \equiv R_+^{IL(JHL+KHL)+JL(KHL+1)+KHL}, \tag{30}$$

where $X \equiv (Q^1, Q^2, Q^3, \gamma, \rho_3)$, and

$$F(X) \equiv (F^{il}_{jh\hat{l}}, F^{il}_{kh\bar{l}}, F^{j\hat{l}}_{kh\bar{l}}, F_{j\hat{l}}, F_{kh\bar{l}})_{i=1,\ldots,I;l=\hat{l}=\bar{l}=1,\ldots,L;j=1,\ldots,J;h=1,\ldots,H;k=1,\ldots,K},$$

with the specific components of F being given by the functional terms preceding the multiplication signs in (29). The term $\langle \cdot, \cdot \rangle$ denotes the inner product in N-dimensional Euclidean space.

Note that the variables in the model (and which can be determined from the solution of either variational inequality (29) or (30)) are: the equilibrium product transactions between manufacturers and the distributors given by Q^{1*}, the equilibrium product transactions transacted electronically between the manufacturers and the retailers denoted by Q^{2*}, and the equilibrium product transactions between the distributors and the retailers given by Q^{3*}, as well as the equilibrium demand prices ρ_3^* and the equilibrium shadow prices γ^*. We now discuss how to recover the prices ρ_1^* associated with the top tier of nodes of the global supply chain network and the prices ρ_2^* associated with the middle tier.

First, note that from (8), we have that (as already discussed briefly) that if $q^{il*}_{jh\hat{l}} > 0$, then the price $\rho^{il*}_{1jh\hat{l}} \times e_h = \frac{\partial f^{il}(Q^{1*},Q^{2*})}{\partial q^{il}_{jh\hat{l}}} + \frac{\partial c^{il}_{jh\hat{l}}(q^{il*}_{jh\hat{l}})}{\partial q^{il}_{jh\hat{l}}} + \alpha^{il} \frac{\partial r^{il}(Q^{1*},Q^{2*})}{\partial q^{il}_{jh\hat{l}}}$. Also, from (8) it follows that if $q^{il*}_{kh\bar{l}} > 0$, then the price $\rho^{il*}_{1kh\bar{l}} \times e_h = \frac{\partial f^{il}(Q^{1*},Q^{2*})}{\partial q^{il}_{kh\bar{l}}} + \frac{\partial c^{il}_{kh\bar{l}}(q^{il*}_{kh\bar{l}})}{\partial q^{il}_{kh\bar{l}}} + \alpha^{il} \frac{\partial r^{il}(Q^{1*},Q^{2*})}{\partial q^{il}_{kh\bar{l}}}$. Hence, the product is priced at the manufacturer's level according to whether it has been transacted physically or electronically; and also according to the distributor or retailer with whom the transaction has taken place. On the other hand, from (16) it follows that if $q^{j\hat{l}*}_{kh\bar{l}} > 0$, then $\rho^{j\hat{l}*}_{2kh\bar{l}} \times e_h = \frac{\partial c_{j\hat{l}}(Q^{1*},Q^{3*})}{\partial q^{j\hat{l}}_{kh\bar{l}}} + \gamma^*_{j\hat{l}} + \beta^{j\hat{l}} \frac{\partial r^{j\hat{l}}(Q^{1*},Q^{3*})}{\partial q^{j\hat{l}}_{kh\bar{l}}}$.

3 Qualitative Properties

In this Section, we provide some qualitative properties of the solution to variational inequality (29) (equivalently, variational inequality (30)). In particular, we derive existence and uniqueness results. We also investigate properties of the function F (cf. (30)) that enters the variational inequality of interest here.

Since the feasible set is not compact we cannot derive existence simply from the assumption of continuity of the functions. Nevertheless, we can impose a rather weak condition to guarantee existence of a solution pattern.

Let

$$\mathcal{K}_b = \{(Q^1, Q^2, Q^3, \gamma, \rho_3) | 0 \leq Q^m \leq b_m, m = 1, 2, 3; 0 \leq \gamma \leq b_4; 0 \leq \rho_3 \leq b_5\}, \tag{31}$$

where $b = (b_1, ..., b_5) \geq 0$ and $Q^m \leq b_m$; $\gamma \leq b_4$; $\rho_3 \leq b_5$ means that $q^{il}_{jh\hat{l}} \leq b_1$, $q^{il}_{kh\bar{l}} \leq b_2$, $q^{j\hat{l}}_{kh\bar{l}} \leq b_3$, and $\gamma_{j\hat{l}} \leq b_4$, $\rho_{3kh\bar{l}} \leq b_5$ for all i, l, $\hat{l}$, $\bar{l}$, j, k, h. Then $\mathcal{K}_b$ is a bounded closed convex subset of $R_+^{IL(JHL+KHL)+JL(KHL+1)+KHL}$. Thus the following variational inequality

$$\langle F(X^b), X - X^b \rangle \geq 0, \quad \forall X^b \in \mathcal{K}_b, \tag{32}$$

admits at least one solution $X^b \in \mathcal{K}_b$, since $\mathcal{K}_b$ is compact and F is continuous. Following Kinderlehrer and Stampacchia [34](see also Theorem 1.5 in [30]),we then have:

Theorem 2
Variational inequality (29) admits a solution if and only if there exists a $b > 0$, such that variational inequality (32) admits a solution in $\mathcal{K}_b$ with

$$Q^{1b} < b_1; \quad Q^{2b} < b_2; \quad Q^{3b} < b_3; \quad \gamma^b < b_4; \quad \rho_3^b < b_5 \tag{33}$$

Theorem 3: Existence
Suppose that there exist positive constants M, N, R with $R > 0$, such that:

$$\frac{\partial f^{il}(Q^1, Q^2)}{\partial q^{il}_{jh\hat{l}}} + \frac{\partial c^{il}_{jh\hat{l}}(q^{il}_{jh\hat{l}})}{\partial q^{il}_{jh\hat{l}}} + \frac{\partial c_{j\hat{l}}(Q^1, Q^3)}{\partial q^{il}_{jh\hat{l}}} + \alpha^{il}\frac{\partial r^{il}(Q^1, Q^2)}{\partial q^{il}_{jh\hat{l}}}$$

$$+\beta^{j\hat{l}}\frac{\partial r^{j\hat{l}}(Q^1, Q^3)}{\partial q^{il}_{jh\hat{l}}} \geq M, \quad \forall Q^1 \quad with \quad q^{il}_{jh\hat{l}} \geq N \quad \forall i, l, j, h, \hat{l}; \tag{34a}$$

$$\frac{\partial f^{il}(Q^1, Q^2)}{\partial q^{il}_{kh\bar{l}}} + \frac{\partial c^{il}_{kh\bar{l}}(q^{il}_{kh\bar{l}})}{\partial q^{il}_{kh\bar{l}}} + \alpha^{il}\frac{\partial r^{il}(Q^1, Q^2)}{\partial q^{il}_{kh\bar{l}}} + \lambda^+_{kh\bar{l}} P_{kh\bar{l}}(s_{kh\bar{l}}, \rho_{3kh\bar{l}})$$

$$-(\lambda^-_{kh\bar{l}} + \rho_{3kh\bar{l}} \times e_h)(1 - P_{kh\bar{l}}(s_{kh\bar{l}}, \rho_{3kh\bar{l}})) + \frac{\partial c_{kh\bar{l}}(Q^2, Q^3)}{\partial q_{ikh}}\beta^{j\hat{l}}\frac{\partial r^{j\hat{l}}(Q^1, Q^3)}{\partial q^{j\hat{l}}_{kh\bar{l}}} \geq M,$$

$$\forall Q^2 \quad with \quad q^{il}_{kh\bar{l}} \geq N \quad \forall i, l, k, h, \bar{l}; \tag{34b}$$

$$\lambda^+_{kh\bar{l}} P_{kh\bar{l}}(s_{kh\bar{l}}, \rho_{3kh\bar{l}}) - (\lambda^-_{kh\bar{l}} + \rho_{3kh\bar{l}} \times e_h)(1 - P_{kh\bar{l}}(s_{kh\bar{l}}, \rho_{3kh\bar{l}})) + \frac{\partial c_{j\hat{l}}(Q^1, Q^3)}{\partial q^{j\hat{l}}_{kh\bar{l}}}$$

$$+\frac{\partial c_{kh\bar{l}}(Q^2, Q^3)}{\partial q^{j\hat{l}}_{kh\bar{l}}} \geq M, \quad \forall Q^3 \quad with \quad q^{j\hat{l}}_{kh\bar{l}} \geq N \quad \forall j, \hat{l}, k, h, \bar{l}; \tag{34c}$$

and

$$d_{kh\bar{l}}(\rho_{3kh\bar{l}}) \leq N, \quad \forall \rho_3 \quad with \quad \rho_{3kh\bar{l}} \geq R, \quad \forall k, h, \bar{l}. \tag{35}$$

Then, variational inequality (30) admits at least one solution.

Proof: Follows using analogous arguments as the proof of existence for Proposition 1 in Nagurney and Zhao [35] (see also existence proof in [20]). □

Assumptions (34a), (34b), (34c) and (35) can be economically justified as follows. In particular, when the volume of the product transacted, $q^{il}_{jh\hat{l}}$, between manufacturer il and distributor j in currency h, and that transacted between manufacturer il and retailer k in country $\bar{l}$, $q^{il}_{kh\bar{l}}$, are large, one can expect the corresponding sum of the marginal costs associated with the production, transaction, and holding plus the weighted marginal risk to exceed a positive lower bound, say, M. At the same time, the large $q^{il}_{jh\hat{l}}$ and $q^{il}_{kh\bar{l}}$ causes a greater $s_{kh\bar{l}}$, which, in turn, causes the probability distribution $P_{kh\bar{l}}(s_{kh\bar{l}}, \rho_{3kh\bar{l}})$ to be close to 1. Consequently, the sum of the middle two terms on the left-hand side of (36b), $\lambda^+_{kh\bar{l}} P_{kh\bar{l}}(s_{kh\bar{l}}, \rho_{3kh\bar{l}}) - (\lambda^-_{kh\bar{l}} + \rho_{3kh\bar{l}} \times e_h)(1 - P_{kh\bar{l}}(s_{kh\bar{l}}, \rho_{3kh\bar{l}}))$ is seen to be positive. Therefore, the left-hand sides of (34b) and (364c), respectively, are greater than or equal to the lower bound M. On the other hand, a high price $\rho_{3kh\bar{l}}$ at retailer k and country $\bar{l}$, will drive the demand at that retailer down, in line with the decreasing nature of any demand function, which ensures (35).

We now recall the concept of an *additive production cost*, which was introduced by Zhang and Nagurney [36] in the stability analysis of dynamic spatial oligopolies, and has also been employed in the qualitative analysis of supply chains by Nagurney, Dong, and Zhang [20].

Definition 2: Additive Production Cost
Suppose that for each manufacturer il, the production cost f^{il} is additive, that is

$$f^{il}(q) = f^{il^1}(q^{il}) + f^{il^2}(\bar{q}^{il}), \tag{36}$$

where $f^{il^1}(q^{il})$ is the internal production cost that depends solely on the manufacturer's own output level q^{il}, which may include the production operation and the facility maintenance, etc., and $f^{il^2}(\bar{q}^{il})$ is the interdependent part of the production cost that is a function of all the other manufacturers' output levels $\bar{q}^{il} = (q^{11}, \cdots, q^{il-1}, q^{il+1}, \cdots, q^{il})$ and reflects the impact of the other manufacturers' production patterns on manufacturer il's cost. This interdependent part of the production cost may describe the competition for the resources, consumption of the homogeneous raw materials, etc.

We now explore additional qualitative properties of the vector function F that enters the variational inequality problem. Specifically, we show that F is monotone as well as Lipschitz continuous. These properties are fundamental in establishing the convergence of the algorithmic scheme in the subsequent section.

Lemma 1
Let $g_{kh\bar{l}}(s_{kh\bar{l}}, \rho_{3kh\bar{l}})^T = (P_{kh\bar{l}}(s_{kh\bar{l}}, \rho_{3kh\bar{l}}) - \rho_{3kh\bar{l}}(1 - P_{kh\bar{l}}(s_{kh\bar{l}}, \rho_{3kh\bar{l}})), s_{kh\bar{l}} - \rho_{3kh\bar{l}})$, where $P_{kh\bar{l}}$ is a probability distribution with the density function of $F_{kh\bar{l}}(x, \rho_{3kh\bar{l}})$. Then $g_{kh\bar{l}}(s_{kh\bar{l}}, \rho_{3kh\bar{l}})$ is monotone, that is,

$$[-\rho'_{3kh\bar{l}}(1 - P_{kh\bar{l}}(s'_{kh\bar{l}}, \rho'_{3kh\bar{l}})) + \rho''_{3kh\bar{l}}(1 - P_{kh\bar{l}}(s''_{kh\bar{l}}, \rho''_{3kh\bar{l}}))] \times [q^{j\bar{l}'}_{kh\bar{l}} - q^{j\bar{l}''}_{kh\bar{l}}]$$

$$+[s'_{kh\bar{l}} - d_{kh\bar{l}}(\rho'_{3kh\bar{l}}) - s''_{kh\bar{l}} + d_{kh\bar{l}}(\rho''_{3kh\bar{l}})] \times [\rho'_{3kh\bar{l}} - \rho''_{3k}] \geq 0,$$
$$\forall (s'_{kh\bar{l}}, \rho'_{3kh\bar{l}}), (s''_{kh\bar{l}}, \rho''_{3kh\bar{l}}) \in R^2_+ \tag{37}$$

if and only if $d'_{kh\bar{l}}(\rho_{3kh\bar{l}}) \leq -(4\rho_{3kh\bar{l}}F_{kh\bar{l}})^{-1}(P_{kh\bar{l}} + \rho_{3kh\bar{l}}\frac{\partial P_{kh\bar{l}}}{\partial \rho_{3kh\bar{l}}})^2$.

Proof: In order to prove that $g_{kh\bar{l}}(s_{kh\bar{l}}, \rho_{3kh\bar{l}})$ is monotone with respect to $s_{kh\bar{l}}$ and $\rho_{3kh\bar{l}}$, we only need to show that its Jacobian matrix is positive semidefinite, which will be the case if all all eigenvalues of the symmetric part of the Jacobian matrix are nonnegative real numbers.

The Jacobian matrix of $g_{kh\bar{l}}$ is

$$\nabla g_{kh\bar{l}}(s_{kh\bar{l}}, \rho_{3kh\bar{l}}) =$$
$$\begin{bmatrix} \rho_{3kh\bar{l}}F_{kh\bar{l}}(s_{kh\bar{l}}, \rho_{3kh\bar{l}}) & -1 + P_{kh\bar{l}}(s_{kh\bar{l}}, \rho_{3kh\bar{l}}) + \rho_{3kh\bar{l}}\frac{\partial P_{kh\bar{l}}(s_{kh\bar{l}}, \rho_{3kh\bar{l}})}{\partial \rho_{3kh\bar{l}}} \\ 1 & -d'_{kh\bar{l}}(\rho_{3kh\bar{l}}) \end{bmatrix}, \tag{38}$$

and its symmetric part is

$$\frac{1}{2}[\nabla g_{kh\bar{l}}(s_{kh\bar{l}}, \rho_{3kh\bar{l}}) + \nabla^T g_{kh\bar{l}}(s_{kh\bar{l}}, \rho_{3kh\bar{l}})]$$
$$= \begin{bmatrix} \rho_{3kh\bar{l}}F_{kh\bar{l}}(s_{kh\bar{l}}, \rho_{3kh\bar{l}}), & \frac{1}{2}\left(\rho_{3kh\bar{l}}\frac{\partial P_{kh\bar{l}}}{\partial \rho_{3kh\bar{l}}} + P_{kh\bar{l}}(s_{kh\bar{l}}, \rho_{3kh\bar{l}})\right) \\ \frac{1}{2}\left(\rho_{3kh\bar{l}}\frac{\partial P_{kh\bar{l}}}{\partial \rho_{3kh\bar{l}}} + P_{kh\bar{l}}(s_{kh\bar{l}}, \rho_{3kh\bar{l}})\right), & -d'_{kh\bar{l}}(\rho_{3kh\bar{l}}) \end{bmatrix}. \tag{39}$$

The two eigenvalues of (39) are

$$\gamma_{min}(s_{kh\bar{l}}, \rho_{3kh\bar{l}}) = \frac{1}{2}\left[(\rho_{3kh\bar{l}}F_{kh\bar{l}} - d'_{kh\bar{l}})\right.$$
$$\left. -\sqrt{(\rho_{3kh\bar{l}}F_{kh\bar{l}} - d'_{kh\bar{l}})^2 + (\rho_{3kh\bar{l}}\frac{\partial P_{kh\bar{l}}}{\partial \rho_{3kh\bar{l}}} + P_{kh\bar{l}})^2 + 4\rho_{3kh\bar{l}}F_{kh\bar{l}}d'_{kh\bar{l}}}\right], \tag{40}$$

$$\gamma_{max}(s_{kh\bar{l}}, \rho_{3kh\bar{l}}) = \frac{1}{2}\left[(\rho_{3kh\bar{l}}F_{kh\bar{l}} - d'_{kh\bar{l}})\right.$$
$$\left. +\sqrt{(\rho_{3kh\bar{l}}F_{kh\bar{l}} - d'_{kh\bar{l}})^2 + (\rho_{3kh\bar{l}}\frac{\partial P_{kh\bar{l}}}{\partial \rho_{3kh\bar{l}}} + P_{kh\bar{l}})^2 + 4\rho_{3kh\bar{l}}F_{kh\bar{l}}d'_{kh\bar{l}}}\right]. \tag{41}$$

Moreover, since what is inside the square root in both (40) and (41) can be rewritten as

$$\left(\rho_{3kh\bar{l}}F_{kh\bar{l}} + d'_{kh\bar{l}}\right)^2 + \left(\rho_{3kh\bar{l}}\frac{\partial P_{kh\bar{l}}}{\partial \rho_{3kh\bar{l}}} + P_{kh\bar{l}}\right)^2$$

and can be seen as being nonnegative, both eigenvalues are real. Furthermore, under the condition of the lemma, $d'_{kh\bar{l}}$ is non-positive, so the first item in (40) and in (41) is nonnegative. The condition further implies that the second item in (40) and in (41), the square root part, is not greater than the first item, which guarantees that both eigenvalues are nonnegative real numbers. □

The condition of Lemma 1 states that the expected demand function of a retailer is a nonincreasing function with respect to the demand price and its first order derivative has an upper bound.

Theorem 4: Monotonicity
The function F that enters the variational inequality problem (29) is monotone, if the condition assumed in Lemma 1 is satisfied for each k; $k = 1, \cdots, K$, and if the following conditions are also satisfied. Suppose that the production cost functions $f^{il}; i = 1, ..., I;\ l = 1, ..., L$, are additive, as defined in Definition 2, and that the $f^{il^1}; i = 1, ..., I;\ l = 1, ..., L$, are convex functions. If the $c^{il}_{jh\hat{l}}$, $c^{il}_{kh\bar{l}}$, $c_{kh\bar{l}}$, $c_{j\hat{l}}$, r^{il}, and $r^{j\hat{l}}$ functions are convex, for all i, l, $\hat{l}$, $\bar{l}$, j, k, h, then the vector function F that enters the variational inequality (30) is monotone, that is,

$$\langle F(X') - F(X''), X' - X'' \rangle \geq 0, \quad \forall X', X'' \in \mathcal{K}. \tag{42}$$

Proof: Let $X' = (Q^{1'}, Q^{2'}, Q^{3'}, \gamma', \rho_{3'})$, $X'' = (Q^{1''}, Q^{2''}, Q^{3''}, \gamma'', \rho_3'')$. Then, inequality (42) can been seen from the following:

$$\langle F(X') - F(X''), X' - X'' \rangle$$

$$= \sum_{i=1}^{I}\sum_{l=1}^{L}\sum_{j=1}^{J}\sum_{h=1}^{H}\sum_{\hat{l}=1}^{L}\left[\frac{\partial f^{il}(Q^{1'}, Q^{2'})}{\partial q^{il}_{jh\hat{l}}} - \frac{\partial f^{il}(Q^{1''}, Q^{2''})}{\partial q^{il}_{jh\hat{l}}}\right] \times \left[q^{il'}_{jh\hat{l}} - q^{il''}_{jh\hat{l}}\right]$$

$$+ \sum_{i=1}^{I}\sum_{l=1}^{L}\sum_{j=1}^{J}\sum_{h=1}^{H}\sum_{\hat{l}=1}^{L}\left[\frac{\partial c_{j\hat{l}}(Q^{1'}, Q^{3'})}{\partial q^{il}_{jh\hat{l}}} - \frac{\partial c_{j\hat{l}}(Q^{1''}, Q^{3''})}{\partial q^{il}_{jh\hat{l}}}\right] \times \left[q^{il'}_{jh\hat{l}} - q^{il''}_{jh\hat{l}}\right]$$

$$+ \sum_{i=1}^{I}\sum_{l=1}^{L}\sum_{j=1}^{J}\sum_{h=1}^{H}\sum_{\hat{l}=1}^{L}\left[\frac{\partial c^{il}_{jh\hat{l}}(q^{il'}_{jh\hat{l}})}{\partial q^{il}_{jh\hat{l}}} - \frac{\partial c^{il}_{jh\hat{l}}(q^{il''}_{jh\hat{l}})}{\partial q^{il}_{jh\hat{l}}}\right] \times \left[q^{il'}_{jh\hat{l}} - q^{il''}_{jh\hat{l}}\right]$$

$$+ \sum_{i=1}^{I}\sum_{l=1}^{L}\sum_{j=1}^{J}\sum_{h=1}^{H}\sum_{\hat{l}=1}^{L}\alpha^{il}\left[\frac{\partial r^{il}(Q^{1'}, Q^{2'})}{\partial q^{il}_{jh\hat{l}}} - \frac{\partial r^{il}(Q^{1''}, Q^{2''})}{\partial q^{il}_{jh\hat{l}}}\right] \times \left[q^{il'}_{jh\hat{l}} - q^{il''}_{jh\hat{l}}\right]$$

$$+ \sum_{i=1}^{I}\sum_{l=1}^{L}\sum_{j=1}^{J}\sum_{h=1}^{H}\sum_{\hat{l}=1}^{L}\beta^{j\hat{l}}\left[\frac{\partial r^{j\hat{l}}(Q^{1'}, Q^{3'})}{\partial q^{il}_{jh\hat{l}}} - \frac{\partial r^{j\hat{l}}(Q^{1''}, Q^{3''})}{\partial q^{il}_{jh\hat{l}}}\right] \times \left[q^{il'}_{jh\hat{l}} - q^{il''}_{jh\hat{l}}\right]$$

$$+ \sum_{i=1}^{I}\sum_{l=1}^{L}\sum_{k=1}^{K}\sum_{h=1}^{H}\sum_{\bar{l}=1}^{L}\left[\frac{\partial f^{il}(Q^{1'}, Q^{2'})}{\partial q^{il}_{kh\bar{l}}} - \frac{\partial f_i(Q^{1''}, Q^{2''})}{\partial q^{il}_{kh\bar{l}}}\right] \times \left[q^{il'}_{kh\bar{l}} - q^{il''}_{kh\bar{l}}\right]$$

$$+ \sum_{i=1}^{I}\sum_{l=1}^{L}\sum_{k=1}^{K}\sum_{h=1}^{H}\sum_{\bar{l}=1}^{L}\left[\frac{\partial c_{kh\bar{l}}(Q^{2'}, Q^{3'})}{\partial q^{il}_{kh\bar{l}}} - \frac{\partial c_{kh\bar{l}}(Q^{2''}, Q^{3''})}{\partial q^{il}_{kh\bar{l}}}\right] \times \left[q^{il'}_{kh\bar{l}} - q^{il''}_{kh\bar{l}}\right]$$

$$+ \sum_{i=1}^{I}\sum_{l=1}^{L}\sum_{k=1}^{K}\sum_{h=1}^{H}\sum_{\bar{l}=1}^{L}\left[\frac{\partial c^{il}_{kh\bar{l}}(q^{il'}_{kh\bar{l}})}{\partial q_{ik}} - \frac{\partial c^{il}_{kh\bar{l}}(q^{il''}_{kh\bar{l}})}{\partial q^{il}_{kh\bar{l}}}\right] \times \left[q^{il'}_{kh\bar{l}} - q^{il''}_{kh\bar{l}}\right]$$

$$+\sum_{i=1}^{I}\sum_{l=1}^{L}\sum_{k=1}^{K}\sum_{h=1}^{H}\sum_{\bar{l}=1}^{L}\alpha^{il}\left[\frac{\partial r^{il}(Q^{1'},Q^{2'})}{\partial q^{il}_{kh\bar{l}}}-\frac{\partial r^{il}(Q^{1''},Q^{2''})}{\partial q^{il}_{kh\bar{l}}}\right]$$

$$+\sum_{i=1}^{I}\sum_{l=1}^{L}\sum_{k=1}^{K}\sum_{h=1}^{H}\sum_{\bar{l}=1}^{L}[\lambda^{+}_{kh\bar{l}}P_{kh\bar{l}}(s'_{kh\bar{l}},\rho'_{3kh\bar{l}})-\lambda^{+}_{kh\bar{l}}P_{kh\bar{l}}(s''_{kh\bar{l}},\rho''_{3kh\bar{l}})]\times[q^{il'}_{kh\bar{l}}-q^{il''}_{kh\bar{l}}]$$

$$+\sum_{i=1}^{I}\sum_{l=1}^{L}\sum_{k=1}^{K}\sum_{h=1}^{H}\sum_{\bar{l}=1}^{L}[-\lambda^{-}_{kh\bar{l}}(1-P_{kh\bar{l}}(s'_{kh\bar{l}},\rho'_{3kh\bar{l}}))+\lambda^{-}_{kh\bar{l}}(1-P_{kh\bar{l}}(s''_{kh\bar{l}},\rho''_{3kh\bar{l}}))]$$
$$\times[q^{il'}_{kh\bar{l}}-q^{il''}_{kh\bar{l}}]$$

$$+\sum_{i=1}^{I}\sum_{l=1}^{L}\sum_{k=1}^{K}\sum_{h=1}^{H}\sum_{\bar{l}=1}^{L}[-\rho'_{3kh\bar{l}}\times e_h(1-P_{kh\bar{l}}(s'_{kh\bar{l}},\rho'_{3kh\bar{l}}))+\rho''_{3kh\bar{l}}$$
$$\times e_h(1-P_{kh\bar{l}}(s''_{kh\bar{l}},\rho''_{3kh\bar{l}}))]\times[q^{il'}_{kh\bar{l}}-q^{il''}_{kh\bar{l}}]$$

$$+\sum_{j=1}^{J}\sum_{\hat{l}=1}^{L}\sum_{k=1}^{K}\sum_{h=1}^{H}\sum_{\bar{l}=1}^{L}\left[\frac{\partial c_{j\hat{l}}(Q^{1'},Q^{3'})}{\partial q^{j\hat{l}}_{kh\bar{l}}}-\frac{\partial c_{j\hat{l}}(Q^{1''},Q^{3''})}{\partial q^{j\hat{l}}_{kh\bar{l}}}\right]\times\left[q^{j\hat{l}'}_{kh\bar{l}}-q^{j\hat{l}''}_{kh\bar{l}}\right]$$

$$+\sum_{j=1}^{J}\sum_{\hat{l}=1}^{L}\sum_{k=1}^{K}\sum_{h=1}^{H}\sum_{\bar{l}=1}^{L}\left[\frac{\partial c_{kh\bar{l}}(Q^{2'},Q^{3'})}{\partial q^{j\hat{l}}_{kh\bar{l}}}-\frac{\partial c_{kh\bar{l}}(Q^{2''},Q^{3''})}{\partial q^{j\hat{l}}_{kh\bar{l}}}\right]\times\left[q^{j\hat{l}'}_{kh\bar{l}}-q^{j\hat{l}''}_{kh\bar{l}}\right]$$

$$+\sum_{j=1}^{J}\sum_{\hat{l}=1}^{L}\sum_{k=1}^{K}\sum_{h=1}^{H}\sum_{\bar{l}=1}^{L}\beta^{j\hat{l}}\left[\frac{\partial r^{j\hat{l}}(Q^{1'},Q^{3'})}{\partial q_{jk}}-\frac{\partial r^{j\hat{l}}(Q^{1''},Q^{3''})}{\partial q^{j\hat{l}}_{kh\bar{l}}}\right]\times\left[q^{j\hat{l}'}_{kh\bar{l}}-q^{j\hat{l}''}_{kh\bar{l}}\right]$$

$$+\sum_{j=1}^{J}\sum_{\hat{l}=1}^{L}\sum_{k=1}^{K}\sum_{h=1}^{H}\sum_{\bar{l}=1}^{L}[\lambda^{+}_{kh\bar{l}}P_{kh\bar{l}}(s'_{kh\bar{l}},\rho'_{3kh\bar{l}})-\lambda^{+}_{kh\bar{l}}P_{kh\bar{l}}(s''_{kh\bar{l}},\rho''_{3kh\bar{l}})]\times[q^{j\hat{l}'}_{kh\bar{l}}-q^{j\hat{l}''}_{kh\bar{l}}]$$

$$+\sum_{j=1}^{J}\sum_{\hat{l}=1}^{L}\sum_{k=1}^{K}\sum_{h=1}^{H}\sum_{\bar{l}=1}^{L}[-\lambda^{-}_{kh\bar{l}}(1-P_{kh\bar{l}}(s'_{kh\bar{l}},\rho'_{3kh\bar{l}}))+\lambda^{-}_{kh\bar{l}}(1-P_{kh\bar{l}}(s''_{kh\bar{l}},\rho''_{3kh\bar{l}}))]$$
$$\times[q^{j\hat{l}'}_{kh\bar{l}}-q^{j\hat{l}''}_{kh\bar{l}}]$$

$$+\sum_{j=1}^{J}\sum_{\hat{l}=1}^{L}\sum_{k=1}^{K}\sum_{h=1}^{H}\sum_{\bar{l}=1}^{L}[-\rho'_{3kh\bar{l}}\times e_h(1-P_{kh\bar{l}}(s'_{kh\bar{l}},\rho'_{3kh\bar{l}}))+\rho''_{3kh\bar{l}}$$
$$\times e_h(1-P_{kh\bar{l}}(s''_{kh\bar{l}},\rho''_{3kh\bar{l}}))]\times[q^{j\hat{l}'}_{kh\bar{l}}-q^{j\hat{l}''}_{kh\bar{l}}]$$

$$+\sum_{k=1}^{K}\sum_{h=1}^{H}\sum_{\bar{l}=1}^{L}[s'_{kh\bar{l}}-d_{kh\bar{l}}(\rho'_{3kh\bar{l}})-s''_{kh\bar{l}}+d_{kh\bar{l}}(\rho''_{3kh\bar{l}})]\times[\rho'_{3kh\bar{l}}-\rho''_{3kh\bar{l}}]$$
$$=(I)+(II)+(III)+\cdots+(XV)+\cdots+(XIX). \tag{43}$$

Since the f^{il}; $i=1,...,I$; $l=1,...,L$, are additive, and the f^{il^1}; $i=1,...,I$; $l=1,...,L$ are convex functions, one has

$$(I)+(VI)=\sum_{i=1}^{I}\sum_{l=1}^{L}\left\{\sum_{j=1}^{J}\sum_{h=1}^{H}\sum_{\hat{l}=1}^{L}\left[\frac{\partial f^{il^1}(Q^{1'},Q^{2'})}{\partial q^{il}_{jh\hat{l}}}-\frac{\partial f^{il^1}(Q^{1''},Q^{2''})}{\partial q^{il}_{jh\hat{l}}}\right]\right.$$

$$\times\left[q^{il'}_{jh\hat{l}}-q^{il''}_{jh\hat{l}}\right]$$

$$\left.+\sum_{k=1}^{k}\sum_{h=1}^{H}\sum_{\bar{l}=1}^{L}\left[\frac{\partial f^{il}(Q^{1'},Q^{2'})}{\partial q^{il}_{kh\bar{l}}}-\frac{\partial f^{il}(Q^{1''},Q^{2''})}{\partial q^{il}_{kh\bar{l}}}\right]\times\left[q^{il'}_{kh\bar{l}}-q^{il''}_{kh\bar{l}}\right]\right\}\geq 0. \quad (44)$$

The convexity of $c_{j\hat{l}}$, for all $j,\hat{l}$; $c^{il}_{jh\hat{l}}$, for all $i,l,j,h,\hat{l}$; $c_{kh\bar{l}}$, for all $k,h,\bar{l}$, $c^{il}_{kh\bar{l}}$, for all i,l,k,h, r^{il}, for all i,l; and $r^{j\hat{l}}$, for all $j,\hat{l}$, gives, respectively,

$$(II)+(XIII)=\sum_{i=1}^{I}\sum_{l=1}^{L}\left\{\sum_{j=1}^{J}\sum_{h=1}^{H}\sum_{\hat{l}=1}^{L}\left[\frac{\partial c_{j\hat{l}}(Q^{1'},Q^{3'})}{\partial q^{il}_{jh\hat{l}}}-\frac{\partial c^{j\hat{l}}(Q^{1''},Q^{3''})}{\partial q^{il}_{jh\hat{l}}}\right]\right.$$

$$\times\left[q^{il'}_{jh\hat{l}}-q^{il''}_{jh\hat{l}}\right]$$

$$\left.+\sum_{k=1}^{K}\sum_{h=1}^{H}\sum_{\bar{l}=1}^{L}\left[\frac{\partial c_{j\hat{l}}(Q^{1'},Q^{3'})}{\partial q^{j\hat{l}}_{kh\bar{l}}}-\frac{\partial c_{j\hat{l}}(Q^{1''},Q^{3''})}{\partial q^{j\hat{l}}_{kh\bar{l}}}\right]\times\left[q^{j\hat{l}}_{kh\bar{l}'}-q^{j\hat{l}''}_{kh\bar{l}}\right]\right\}\geq 0, \quad (45)$$

$$(III)=\sum_{i=1}^{I}\sum_{l=1}^{L}\sum_{j=1}^{J}\sum_{h=1}^{H}\sum_{\hat{l}=1}^{L}\left[\frac{\partial c^{il}_{jh\hat{l}}(q^{il'}_{jh\hat{l}})}{\partial q^{il}_{jh\hat{l}}}-\frac{\partial c^{il}_{jh\hat{l}}(q^{il''}_{jh\hat{l}})}{\partial q^{il}_{jh\hat{l}}}\right]\times\left[q^{il'}_{jh\hat{l}}-q^{il''}_{jh\hat{l}}\right]\geq 0, \quad (46)$$

$$(VII)+(XIV)=\sum_{k=1}^{K}\sum_{h=1}^{H}\sum_{\bar{l}=1}^{L}\left\{\sum_{i=1}^{I}\sum_{l=1}^{L}\left[\frac{\partial c_{kh\bar{l}}(Q^{2'},Q^{3'})}{\partial q^{il}_{kh\bar{l}}}-\frac{\partial c_{kh\bar{l}}(Q^{2''},Q^{3''})}{\partial q^{il}_{kh\bar{l}}}\right]\right.$$

$$\times\left[q^{il'}_{kh\bar{l}}-q^{il''}_{kh\bar{l}}\right]$$

$$\left.+\sum_{j=1}^{J}\sum_{\hat{l}=1}^{L}\left[\frac{\partial c_{kh\bar{l}}(Q^{2'},Q^{3'})}{\partial q^{j\hat{l}}_{kh\bar{l}}}-\frac{\partial c_{kh\bar{l}}(Q^{2''},Q^{3''})}{\partial q^{j\hat{l}}_{kh\bar{l}}}\right]\times\left[q^{j\hat{l}'}_{kh\bar{l}}-q^{j\hat{l}''}_{kh\bar{l}}\right]\right\}\geq 0, \quad (47)$$

$$(VIII)=\sum_{i=1}^{I}\sum_{l=1}^{L}\sum_{k=1}^{K}\sum_{h=1}^{H}\sum_{\bar{l}=1}^{L}\left[\frac{\partial c^{il}_{kh\bar{l}}(q^{il'}_{kh\bar{l}})}{\partial q^{il}_{kh\bar{l}}}-\frac{\partial c^{il}_{kh\bar{l}}(q^{il''}_{kh\bar{l}})}{\partial q^{il}_{kh\bar{l}}}\right]\times\left[q^{il'}_{kh\bar{l}}-q^{il''}_{kh\bar{l}}\right]\geq 0, \quad (48)$$

$$(IV)+(IX)=\sum_{i=1}^{I}\sum_{l=1}^{L}\alpha^{il}\left\{\sum_{j=1}^{J}\sum_{h=1}^{H}\sum_{\hat{l}=1}^{L}\left[\frac{\partial r^{il}(Q^{1'},Q^{2'})}{\partial q^{il}_{jh\hat{l}}}-\frac{\partial r^{il}(Q^{1''},Q^{2''})}{\partial q^{il}_{jh\hat{l}}}\right]\right.$$

$$\times\left[q^{il'}_{jh\hat{l}}-q^{il''}_{jh\hat{l}}\right]+\sum_{k=1}^{K}\sum_{h=1}^{H}\sum_{\bar{l}=1}^{L}\left[\frac{\partial r^{il}(Q^{1'},Q^{2'})}{\partial q^{il}_{kh\bar{l}}}-\frac{\partial r^{il}(Q^{1''},Q^{2''})}{\partial q^{il}_{kh\bar{l}}}\right]\Bigg\}\geq 0, \quad (49)$$

$$(V)+(XV)=\sum_{j=1}^{J}\sum_{\hat{l}=1}^{L}\beta^{j\hat{l}}\Bigg\{\sum_{i=1}^{I}\sum_{l=1}^{L}\sum_{h=1}^{H}\left[\frac{\partial r^{j\hat{l}}(Q^{1'},Q^{3'})}{\partial q^{il}_{jh\hat{l}}}-\frac{\partial r_j(Q^{1''},Q^{3''})}{\partial q^{il}_{jh\hat{l}}}\right]$$

$$\times\left[q^{il'}_{jh\hat{l}}-q^{il''}_{jh\hat{l}}\right]$$

$$+\sum_{k=1}^{K}\sum_{h=1}^{H}\sum_{\bar{l}=1}^{L}\left[\frac{\partial r^{j\hat{l}}(Q^{1'},Q^{3'})}{\partial q^{j\hat{l}}_{kh\bar{l}}}-\frac{\partial r^{j\hat{l}}(Q^{1''},Q^{3''})}{\partial q^{j\hat{l}}_{kh\bar{l}}}\right]\times\left[q^{j\hat{l}'}_{kh\bar{l}}-q^{j\hat{l}''}_{kh\bar{l}}\right]\Bigg\}\geq 0. \quad (50)$$

Since the probability function $P_{kh\bar{l}}$ is an increasing function w.r.t. $s_{kh\bar{l}}$, for all k, and $s_{kh\bar{l}}=\sum_{i=1}^{I}\sum_{l=1}^{L}q^{il}_{kh\bar{l}}+\sum_{j=1}^{J}\sum_{\hat{l}=1}^{L}q^{j\hat{l}}_{kh\bar{l}}$, hence, we have the following:

$$(X)+(XVI)=\sum_{k=1}^{K}\sum_{h=1}^{H}\sum_{\bar{l}=1}^{L}[\lambda^{+}_{kh\bar{l}}P_{kh\bar{l}}(s'_{kh\bar{l}},\rho'_{3kh\bar{l}})-\lambda^{+}_{kh\bar{l}}P_{kh\bar{l}}(s''_{kh\bar{l}},\rho''_{3kh\bar{l}})]$$

$$\times[s'_{kh\bar{l}}-s''_{kh\bar{l}}]\geq 0, \quad (51)$$

$$(XI)+(XVII)=\sum_{k=1}^{K}\sum_{h=1}^{H}\sum_{\bar{l}=1}^{L}[-\lambda^{-}_{kh\bar{l}}(1-P_{kh\bar{l}}(s'_{kh\bar{l}},\rho'_{3kh\bar{l}}))$$

$$+\lambda^{-}_{kh\bar{l}}(1-P_{kh\bar{l}}(s''_{kh\bar{l}},\rho''_{3kh\bar{l}}))]\times[s'_{kh\bar{l}}-s''_{kh\bar{l}}]\geq 0. \quad (52)$$

Since for each k, applying Lemma 1, we can see that $g_{kh\bar{l}}(s_{kh\bar{l}},\rho_{3kh\bar{l}})$ is monotone, hence, we have:

$$(XII)+(XVIII)+(XIX)=\sum_{k=1}^{K}\sum_{h=1}^{H}\sum_{\bar{l}=1}^{L}[-\rho'_{3kh\bar{l}}\times e_h(1-P_{kh\bar{l}}(s'_{kh\bar{l}},\rho'_{3kh\bar{l}}))$$

$$+\rho''_{3kh\bar{l}}\times e_h(1-P_{kh\bar{l}}(s''_{kh\bar{l}},\rho''_{3kh\bar{l}}))]\times[s'_{kh\bar{l}}-s''_{kh\bar{l}}]$$

$$+\sum_{k=1}^{K}\sum_{h=1}^{H}\sum_{\bar{l}=1}^{L}[s'_{kh\bar{l}}-d_{kh\bar{l}}(\rho'_{3kh\bar{l}})-s''_{kh\bar{l}}+d_{kh\bar{l}}(\rho''_{3kh\bar{l}})]\times[\rho'_{3kh\bar{l}}-\rho''_{3kh\bar{l}}]\geq 0. \quad (53)$$

Therefore, we conclude that (44) is nonnegative in $\mathcal{K}$. The proof is complete. □

Theorem 5: Strict Monotonicity

The function F that enters the variational inequality problem (30) is strictly monotone, if the conditions mentioned in Lemma 1 for $g_{kh\bar{l}}(s_{kh\bar{l}},\rho_{3kh\bar{l}})$ are satisfied strictly for all k and if the following conditions are also satisfied. Suppose that the production cost functions f^{il}; $i=1,...,I$; $l=1...L$, are additive, as defined in Definition 2, and that the ${f^{il}}^1$; $i=1,...,I$; $l=1,...,L$, are strictly convex functions. If the $c^{il}_{jh\hat{l}}$, $c^{il}_{kh\bar{l}}$, $c_{kh\bar{l}}$, $c_{j\hat{l}}$, r^{il}, and $r^{j\hat{l}}$ functions are strictly convex, *for all i,l, $\hat{l}$, $\bar{l}$, j, k, h, then the vector function F that enters the variational inequality (30) is strictly monotone, that is,*

$$\langle F(X') - F(X''), X' - X'' \rangle > 0, \quad \forall X', X'' \in \mathcal{K}. \tag{54}$$

Theorem 6: Uniqueness
Under the conditions indicated in Theorem 5, the function F that enters the variational inequality (30) has a unique solution in $\mathcal{K}$.

From Theorem 6 it follows that, under the above conditions, the equilibrium product shipment pattern between the manufacturers and the retailers, as well as the equilibrium price pattern at the retailers, is unique.

Theorem 7: Lipschitz Continuity
The function F that enters the variational inequality problem (30) is Lipschitz continuous, that is,

$$\|F(X') - F(X'')\| \leq L\|X' - X''\|, \quad \forall X', X'' \in \mathcal{K}, \text{ with } L > 0, \tag{55}$$

under the following conditions:
(i). Each function f^{il}; $i = 1, ..., I$; $l = 1, ..., L$, is additive and has a bounded second order derivative;
(ii). The $c^{il}_{jh\hat{l}}$, $c^{il}_{kh\bar{l}}$, $c_{kh\bar{l}}$, $c_{j\hat{l}}$, r^{il}, and $r^{j\hat{l}}$ functions have bounded second order derivatives, for all i, j, k, h, l, $\hat{l}$, $\bar{l}$.

Proof: Since the probability function $P_{kh\bar{l}}$ is always less than or equal to 1, for each retailer $kh\bar{l}$, the result is direct by applying a mid-value theorem from calculus to the vector function F that enters the variational inequality problem (30). □

4 The Algorithm

In this Section, an algorithm is presented which can be applied to solve any variational inequality problem in standard form (see (30)), that is:

Determine $X^* \in \mathcal{K}$, satisfying:

$$\langle F(X^*), X - X^* \rangle \geq 0, \quad \forall X \in \mathcal{K}. \tag{56}$$

The algorithm is guaranteed to converge provided that the function F that enters the variational inequality is monotone and Lipschitz continuous (and that a solution exists). The algorithm is the modified projection method of Korpelevich [37].

The statement of the modified projection method is as follows, where $\mathcal{T}$ denotes an iteration counter:

Modified Projection Method

Step 0: Initialization

Set $X^0 \in \mathcal{K}$. Let $\mathcal{T} = 1$ and let a be a scalar such that $0 < a \leq \frac{1}{L}$, where L is the Lipschitz continuity constant (cf. [37]) (see (55)).

Step 1: Computation
Compute $\bar{X}^{\mathcal{T}}$ by solving the variational inequality subproblem:

$$\langle \bar{X}^{\mathcal{T}} + aF(X^{\mathcal{T}-1}) - X^{\mathcal{T}-1}, X - \bar{X}^{\mathcal{T}} \rangle \geq 0, \quad \forall X \in \mathcal{K}. \tag{57}$$

Step 2: Adaptation
Compute $X^{\mathcal{T}}$ by solving the variational inequality subproblem:

$$\langle X^{\mathcal{T}} + aF(\bar{X}^{\mathcal{T}}) - X^{\mathcal{T}-1}, X - X^{\mathcal{T}} \rangle \geq 0, \quad \forall X \in \mathcal{K}. \tag{58}$$

Step 3: Convergence Verification
If max $|X_l^{\mathcal{T}} - X_l^{\mathcal{T}-1}| \leq \epsilon$, for all l, with $\epsilon > 0$, a prespecified tolerance, then stop; else, set $\mathcal{T} =: \mathcal{T} + 1$, and go to Step 1.

We now state the convergence result for the modified projection method for this model.

Theorem 8: Convergence
Assume that the function that enters the variational inequality (29) (or (30)) has at least one solution and satisfies the conditions in Theorem 4 and in Theorem 7. Then the modified projection method described above converges to the solution of the variational inequality (29) or (30).

Proof: According to Korpelevich [37], the modified projection method converges to the solution of the variational inequality problem of the form (30), provided that the function F that enters the variational inequality is monotone and Lipschitz continuous and that a solution exists. Existence of a solution follows from Theorem 3. Monotonicity follows Theorem 5. Lipschitz continuity, in turn, follows from Theorem 7. □

We emphasize that, in view of the fact that the feasible set $\mathcal{K}$ underlying the global supply chain network model with supply and demand side risk is the nonnegative orthant, the projection operation encountered in (57) and (58) takes on a very simple form for computational purposes. Indeed, the product transactions as well as the product prices at a given iteration in both (57) and in (58) can be exactly and computed in closed form,

5 Numerical Examples

In this Section,we apply the modified projection method to several numerical examples. The algorithm was coded in FORTRAN and the computer used was a SUN system at the University of Massachusetts at Amherst. The convergence criterion utilized was that the absolute value of the product transactions and prices between two successive iterations differed by no more than 10^{-4}. The parameter a in the modified projection method (cf. (57) and (58)) was set to .01 for all the examples.

The structure of the global supply chain network for the examples is given in Figure 2. Specifically, we assumed that there were two countries, with two manufacturers in each country, and two distributors in each country. In addition, we assumed a single currency (for example, the euro) and two retailers in each country. Note that electronic transactions were permitted between the manufacturers and the retailers. Hence, we had that $I = 2$, $L = 2$, $J = 2$, $K = 2$, and $H = 1$ and, hence $e_h = e_1 = 1$.

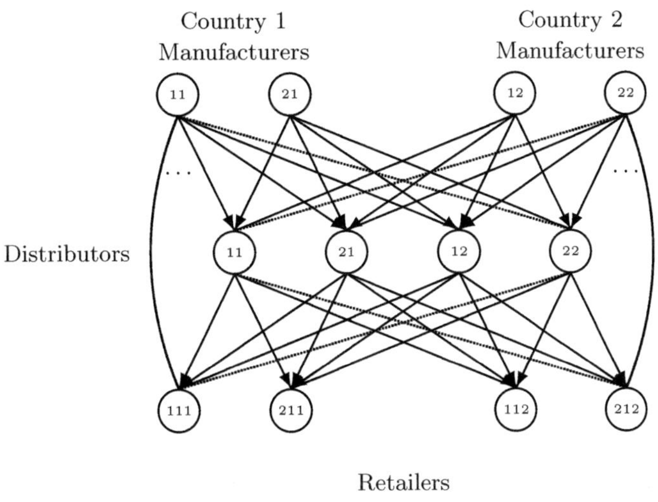

Fig. 2. Global supply chain network for the numerical examples

In all the examples, we assumed that the demands associated with the retail outlets followed a uniform distribution. In particular, we assumed that the random demand $\hat{d}_{kh\bar{l}}(\rho_{3kh\bar{l}})$, of retailer $kh\bar{l}$, is uniformly distributed in $\left[0, \frac{b_{kh\bar{l}}}{\rho_{3kh\bar{l}}}\right]$, with $b_{kh\bar{l}} > 0$; $k = 1, 2$; $h = 1$, and $\bar{l} = 1, 2$. Therefore, we have that

$$P_{kh\bar{l}}(x, \rho_{3kh\bar{l}}) = \frac{x\rho_{3kh\bar{l}}}{b_{kh\bar{l}}}; \quad k = 1, 2; h = 1; \bar{l} = 1, 2, \tag{59}$$

$$F_{kh\bar{l}}(x, \rho_{3kh\bar{l}}) = \frac{\rho_{3kh\bar{l}}}{b_{kh\bar{l}}}; \quad k = 1, 2; h = 1; \bar{l} = 1, 2; \tag{60}$$

$$d_{kh\bar{l}}(\rho_{3kh\bar{l}}) = E(\hat{d}_{kh\bar{l}}) = \frac{1}{2}\frac{b_{kh\bar{l}}}{\rho_{3kh\bar{l}}}; \quad k = 1, 2; h = 1; \bar{l} = 1, 2. \tag{61}$$

It is straightforward to verify that the expected demand function $d_{kh\bar{l}}(\rho_{3kh\bar{l}})$ associated with retailer $kh\bar{l}$ is a decreasing function of the price at the demand market in the particular country.

The modified projection method was initialized as follows: all variables were set equal to zero, except for the initial retail prices $\rho_{3kh\bar{l}}$, which were set to 1 for all $k, h, \bar{l}$.

Example 1

The data for this example were constructed for easy interpretation purposes. The production cost functions of the manufacturers in the two countries (cf. (1)) were given, respectively, by:

$$f^{11}(q) = 2.5(q^{11})^2 + q^{11}q^{21} + 2q^{11}, \quad f^{21}(q) = 2.5(q^{21})^2 + q^{11}q^{21} + 2q^{21},$$

$$f^{12}(q) = 2.5(q^{12})^2 + q^{12}q^{22} + 2q^{12}, \quad f^{22}(q) = 2.5(q^{12})^2 + q^{12}q^{22} + 2q^{22}.$$

The transaction costs faced by the manufacturers and associated with transacting with the distributors (cf. (2a)) were given by:

$$c^{il}_{jh\hat{l}} = .5(q^{il}_{jh\hat{l}})^2 + 3.5q^{il}_{jh\hat{l}}, \quad \text{for } i = 1,2; l = 1,2; j = 1,2; h = 1; \hat{l} = 1,2.$$

The transaction costs faced by the manufacturers but associated with transacting with the retailers electronically (cf. (2b)) were given by:

$$c^{il}_{kh\bar{l}} = .5(q^{il}_{kh\bar{l}})^2 + 5q^{il}_{kh\bar{l}}, \quad \text{for } i = 1,2; l = 1,2; k = 1,2; h = 1; \bar{l} = 1,2.$$

The handling costs of the distributors in the two countries, in turn, (cf.(9b)), were given by:

$$c_{j\hat{l}} = .5(\sum_{i=1}^{2}\sum_{l=1}^{2} q^{il}_{jh\hat{l}})^2, \quad \text{for } j = 1,2; \hat{l} = 1,2.$$

The handling costs of the retailers (cf. (17)) were:

$$c_{kh\bar{l}} = .5(\sum_{j=1}^{2}\sum_{\hat{l}=1}^{2} q^{j\hat{l}}_{kh\bar{l}})^2, \quad \text{for } k = 1,2; h = 1; \bar{l} = 1,2.$$

The $b_{kh\bar{l}}$s (cf. (59) – (61)) were set to 100 for all $k, h, \bar{l}$. The weights associated with excess supply and with excess demand at the retailers were (see following (21)): $\lambda^+_{kh\bar{l}} = \lambda^-_{kh\bar{l}} = 1$ for $k = 1,2$; $h = 1$, and $\bar{l} = 1,2$. Thus, we assigned equal weights for each retailer in each country for excess supply and excess demand.

In Example 1, we set all the weights associated with risk minimization to zero, that is, we had that $\alpha^{il} = 0$ for $i = 1,2$ and $l = 1,2$ and $\beta^{j\hat{l}} = 0$ for $j = 1,2$ and $\hat{l} = 1,2$. This means that in the first example all the manufacturers and all the distributors were concerned with profit maximization exclusively.

The modified projection method converged and yielded the following equilibrium pattern. All physical transactions were equal to .186, that is, we had that $q^{il*}_{jh\hat{l}} = q^{j\hat{l}*}_{kh\bar{l}} = .186$ for all $i, l, j, h, \hat{l}$, and $k, \bar{l}$. All product transactions conducted electronically via the Internet, in turn, were equal to .177, that is, we had that $q^{il*}_{kh\bar{l}} = .177$ for all $i, l, k, h, \bar{l}$. Note that there was a larger volume of product transacted physically than electronically in this example.

The computed equilibrium prices, in turn, were as follows. The equilibrium prices at the distributors were: $\gamma^*_{j\hat{l}} = 15.091$ for $j = 1,2$ and $\hat{l} = 1,2$, whereas the demand market equilibrium prices were: $\rho^*_{3kh\bar{l}} = 32.320$ for $k = 1, 2$, $h = 1$, and $\bar{l} = 1, 2$. Note that, as expected, the demand market prices exceed the prices for the product at the distributor level. This is due to the fact that the prices increase as the product propagates down through the supply chain since costs accumulate.

Example 2

Example 2 was constructed from Example 1 as follows. All the data were as in Example 1 except that now we set the $b_{kh\bar{l}}$s =1000. This means (cf. (59) – (61)) that, in effect, the demand has increased for the product at all retailers in all countries.

The modified projection method converged and yielded the following new equilibrium pattern: the product transactions between the manufacturers and the distributors were: $q^{il*}_{jh\hat{l}} = .286$ for all i, l, $j, h, \hat{l}$; whereas the volumes of the product transacted electronically between the manufacturers and the retailers were: $q^{il*}_{kh\bar{l}} = 1.071$ for all i, l and $k, h, \bar{l}$. Hence, the volumes of electronic transactions exceeded the physical ones. Finally, the computed equilibrium product transactions between the distributors and the retailers were: $q^{j\hat{l}*}_{kh\bar{l}} = .286$ for all $j, \hat{l}$ and $k, h, \bar{l}$.

The computed equilibrium prices associated with the distributors, in turn, were: $\gamma^*_{j\hat{l}} = 39.487$ for $j = 1, 2$ and $\hat{l} = 1, 2$, whereas the equilibrium demand market prices were: $\rho^*_{3kh\bar{l}} = 90.395$ for $k = 1, 2$; $h = 1$, and $\bar{l} = 1, 2$.

Note that since the demand increased, the product transactions also increased. In this example, there were more transactions conducted electronically than physically. Also, observe that since demand increased, the demand prices also increased as did the prices at the distributors.

Example 3

Example 3 was constructed from Example 2 as follows. We kept the data as in Example 2 but we assumed now that the first manufacturer in the first country was a multicriteria decision-maker and concerned with risk minimization with his risk function being given by:

$$r^{11} = (\sum_{kh\bar{l}} q^{11}_{kh\bar{l}} - 2)^2,$$

that is, the manufacturer sought to achieve, in a sense, a certain goal target associated with his electronic transactions. The weight associated with his risk measure was $\alpha^{11} = 2$.

The modified projection method yielded the following new equilibrium pattern. The computed equilibrium product transactions between the first manufacturer in the first country and the distributors were now: $q^{11*}_{jh\hat{l}} = .569$, for $j = 1, 2$, $h = 1$, and $\hat{l} = 1, 2$. The analogous transactions, but from the second manufacturer in the first country were; $q^{21*}_{jh\hat{l}} = .216$. All other product transactions between the manufacturers and the distributors were equal to .215.

The equilibrium product transactions associated with electronic transactions were as follows. For the first manufacturer in the first country (who is now concerned with risk associated with electronic transactions) the product transactions were: $q_{111}^{11*} = .300$, $q_{211}^{11*} = .303$, $q_{112}^{11*} = .307$, and $q_{212}^{11*} = .307$. The analogous transactions but from the second manufacturer in the first country, were all 1.145, with the remainder of the electronic transactions at equilibrium equal to 1.143. Hence, the first manufacturer in the first country reduced the volume of his transactions conducted electronically since there was increased risk associated with such transactions.

The product transactions at equilibrium between the distributors and the retailers, in turn, were: $q_{111}^{j\hat{l}*} = .300$, for $j = 1,2$ and $\hat{l} = 1,2$; $q_{211}^{j\hat{l}*} = .303$ for $j = 1,2$ and $\hat{l} = 1,2$; $q_{112}^{j\hat{l}*} = .307$ and $q_{212}^{j\hat{l}*} = .307$ for $j = 1,2$ and $\hat{l} = 1,2$.

The equilibrium prices were now: $\gamma_{j\hat{l}}^* = 39.514$ for $j = 1,2$; $\hat{l} = 1,2$ whereas $\rho_{3kh\bar{l}}^* = 90.565$ for $k = 1,2$; $h = 1$, and $\bar{l} = 1,2$.

Obviously, the above examples are highly stylized but they, nevertheless, demonstrate the efficacy of the model and the computational procedure. One can now conduct numerous simulations by altering the data as well as adding decision-makers with their associated functions and weights.

6 Summary and Conclusions

This paper has developed a three-tiered global supply chain network equilibrium model consisting of manufacturers, distributors, and retailers with electronic commerce and with risk management. In particular, the manufacturers as well as the distributors are assumed to be multicriteria decision-makers and concerned not only with profit maximization but also with risk minimization. The demands for the product, in turn, are random.

The developed multi-agent framework permits for the handling of as many countries, as many manufacturers in each country, as many currencies in which the products can be obtained, as many retailers, and as many demand markets, as mandated by the specific application. Moreover, the generality of the multi-agent framework allows for the demand to have almost any distribution as long as it satisfies certain technical conditions. In addition, the retailers need not be country-specific and can transact either virtually or physically with both the manufacturers and the consumers.

Finite-dimensional variational inequality theory was used to formulate the derived equilibrium conditions, to study the model qualitatively, and also to obtain convergence results for the proposed algorithmic scheme. Finally, numerical examples were presented to illustrate the model and computational procedure.

This multi-agent framework generalizes the recent work of Nagurney, Cruz, and Matsypura [7] to include supply-side risk modeled as a multicriteria decision-making problem from the perspective of both the manufacturers and the distributors, and demand-side risk, handled through random demands. The results herein provide a contribution to the burgeoning topic of the development of quantitative tools for multi-agent modeling, analysis, and computations that bridge finance and operations in the setting of supply chain modeling and analysis with a specific focus on the global arena.

7 Acknowledgments

The research of the authors was supported, in part, by NSF Grant No.: IIS-0002647, and that of the first two by a 2002 AT&T Industrial Ecology Faculty Fellowship. This support is gratefully acknowledged.

The authors thanks Dr. Brahim Chaib-draa for editing this volume and for support in the preparation of this paper.

References

1. Engardio, P., Shari, M., Weintraub, A., Arnst, C.: Deadly Virus. BusinessWeek Online (2003) see: http://www.businessweek.com/magazine/content/03_15/b3828002_mx 046.htm
2. Sheffi, Y.: Supply Chain Management under the Threat of International Terrorism. The International Journal of Logistics Management **12** (2001) 1–11
3. Nagurney, A., Dong, J.: Supernetworks: Decision-Making for the Information Age. Edward Elgar Publishers, Cheltenham, England (2002)
4. Mullany, T. J., Green, H., Arndt, M., Hof, R. D., Himelstein, L.: The E-Biz Surprise. Business Week Online (2003) see: http://www.businessweek.com/magazine/content/03_19/b3832601.htm
5. Kogut, B., Kulatilaka, N.: Options Thinking and Platform Investments: Investing in Opportunity: California Management Review **36** (1994) No. 4
6. Cohen, M. A., Mallik, S.: Global Supply Chains: Research and Applications. Production and Operations Management **6** (1997) 193–210
7. Nagurney, A., Cruz, J., Matsypura D.: Dynamics of Global Supply Chain Supernetworks. Mathematical and Computer Modelling **37** (2003) 963–983
8. Erenguc, S. S., Simpson, N. C., Vakharia, A. J.: Integrated Production/Distribution Planning in Supply Chains: An Invited Review. European Journal of Operational Research **115** (1999) 219–236
9. Cohen, M. A., Huchzermeier, A.: Global Supply Chain Management: A Survey of Research and Applications. In: Tayur, S., Magazine, M., and Ganeshan, R. (eds.): Quantitative Models for Supply Chain Management. Kluwer Academic Publishers, New York (1998)
10. Smeltzer, L. R., Siferd S. P.: Proactive Supply Management: The Management of Risk. International Journal of Purchasing and Materials Management, Winter (1998) 38–45
11. Agrawal, V., Seshadri, S.: Risk Intermediation in Supply Chains. IIE Transactions **32** (2000) 819 831
12. Johnson, M. E.: Learning from Toys: Lessons in Managing Supply Chain Risk from the Toy Industry. California Management Review **43** (2001) 106–124
13. Van Mieghem, J. A.: Capacity Portfolio Investment and Hedging: Review and New Directions. To appear in Manufacturing & Service Operations Management (2002)
14. Zsidisin, G. A.: Managerial Perceptions of Supply Risk. The Journal of Supply Chain Management: A Global Review of Purchasing and Supply Winter (2003) 14–25
15. Huchzermeier, A., Cohen, M. A.: Valuing Operational Flexibility Under Exchange Rate Uncertainty. Operations Research **44** (1996) 100–113

16. Nagurney, A., Loo, J., Dong, J., Zhang , D.: Supply Chain Networks and Electronic Commerce: A Theoretical Perspective. Netnomics **4** (2002) 187–220
17. Dong, J., Zhang, D., Nagurney, A.: Supply Chain Networks with Multicriteria Decision-Makers. In: M. A. P. (ed.): Transportation and Traffic Theory in the 21st Century. Pergamon Press, Amsterdam, The Netherlands (2002)
18. Dong, J., Zhang, D., Nagurney, A.: Supply Chain Supernetworks with Random Demands. In: Lee, D. -H. (ed.): Urban and Regional Transportation Modeling: Essays in Honor of David E. Boyce. Edward Elgar Publishers, Cheltenham, England (2004a) 289–313
19. Dong, J., Zhang, D., Nagurney, A.: A Supply Chain Network Equilibrium Model with Random Demands. European Journal of Operational Research **156** (2004b) 194–212
20. Nagurney, A., Dong, J., Zhang, D.: A Supply Chain Network Equilibrium Model. Transportation Research E **38** (2002) 281–303
21. Nagurney, A., Siokos, S.: Financial Networks: Statics and Dynamics. Springer-Verlag, Heidelberg, Germany (1997)
22. Fishburn, P. C.: Utility Theory for Decision Making. John Wiley & Sons, New York (1970)
23. Chankong, V., Haimes, Y. Y.: Multiobjective Decision Making: Theory and Methodology. North-Holland, New York (1997)
24. Yu, P. L.: Multiple Criteria Decision Making – Concepts, Techniques, and Extensions. Plenum Press, New York (1985)
25. Keeney, R. L., Raiffa, H.: Decisions with Multiple Objectives: Preferences and Value Tradeoffs. Cambridge University Press, Cambridge, England (1993)
26. Nash, J. F.: Equilibrium Points in N-Person Games. In: Proceedings of the National Academy of Sciences, USA **36** (1950) 48–49
27. Nash, J. F.: Noncooperative Games. Annals of Mathematics **54** (1951) 286–298
28. Gabay, D., Moulin, H.: On the Uniqueness and Stability of Nash Equilibria in Noncooperative Games. In: Bensoussan, A., Kleindorfer , P., Tapiero, C. S. (eds.): Applied Stochastic Control of Econometrics and Management Science. North-Holland, Amsterdam, The Netherlands (1980)
29. Bazaraa, M. S., Sherali, H. D., Shetty, C. M.: Nonlinear Programming: Theory and Algorithms. John Wiley & Sons, New York (1993)
30. Nagurney, A.: Network Economics: A Variational Inequality Approach. Second and revised edition, Kluwer Academic Publishers, Dordrecht, The Netherlands (1999)
31. Markowitz, H. M.: Portfolio Selection. The Journal of Finance **7** (1952) 77–91
32. Markowitz, H. M.: Portfolio Selection: Efficient Diversification of Investments. John Wiley & Sons, New York (1959)
33. Dong, J., Nagurney A.: Bicriteria Decision Making and Financial Equilibrium: A Variational Inequality Perspective. Computational Economics **17** (2001) 19–42.
34. Kinderlehrer, D., Stampacchia, G.: An Introduction to Variational Inequalities and Their Application. Academic Press, New York (1980)
35. Nagurney, A., Zhao, L.: Networks and Variational Inequalities in the Formulation and Computation of Market Disequilibria: The Case of Direct Demand Functions. Transportation Science **27** (1993) 4–15
36. Zhang, D. , Nagurney, A.: Stability Analysis of an Adjustment Process for Oligopolistic Market Equilibrium Modeled as a Projected Dynamical System. Optimization **36** (1996) 263–285
37. Korpelevich, G. M.: The Extragradient Method for Finding Saddle Points and Other Problems. Matekon **13** (1977) 35–49

RedAgent: An Autonomous, Market-based Supply-Chain Management Agent for the Trading Agents Competition

Doina Precup, Philipp W. Keller, and Felix-Olivier Duguay

McGill University

Abstract. The Supply Chain Management track of the international Trading Agents Competition (TAC SCM) was introduced in 2003 as a test-bed for researchers interested in building autonomous agents that act in dynamic supply chains. TAC SCM provides a challenging scenario for existing AI decision-making algorithms, due to the high dimensionality and the non-determinism of the environment, as well as the combinatorial nature of the problem. In this paper we present RedAgent, the winner of the first TAC SCM competition in 2003. RedAgent is based on a multi-agent design, in which many simple heuristic agents manage tasks such as fulfilling customer orders or procuring particular resources. The key idea is to use internal markets as the main decision mechanism, in order to determine what products to focus on and how to allocate the existing resources. The internal markets ensure the coordination of the individual agents, and at the same time provide price estimates for the goods that RedAgent has to sell and purchase, a key feature in this domain. We describe the architecture of the agent, analyze its behavior based on data from the competition, and discuss modifications that we experimented with in 2004.

1 Introduction

Supply-chain management is a key economic task, as it impacts directly the ability of an enterprise to adapt to changing market demands. The Trading Agents Competition Supply Chain Management scenario (TAC SCM), designed by researchers at SICS and CMU [1], provides a challenging test-bed for automated trading agents acting in dynamic supply chains. In TAC SCM, agents representing personal computer (PC) manufacturers compete in markets for components (supplies), as buyers, as well as in markets for finished goods (as sellers), with the purpose of maximizing profits over a simulated year. TAC SCM was designed with the goal of providing a realistic simulation of a supply chain environment. As a result, the agents have to solve a very challenging combinatorial optimization problem, regarding which supplies to purchase and how to allocate the existing resources optimally. Moreover, they have to act in the face of tremendous uncertainty regarding the behavior of suppliers, clients, and competitor agents. For instance, agents have to decide which components to buy before they know how much demand there will be in the market. And, of course, TAC is a very competitive setting.

In this chapter, we present RedAgent, the winner of the first TAC SCM competition, held in 2003. RedAgent has a highly distributed architecture, in which simple, heuristic-based agents are assigned to deal with individual aspects of the game, such as component procurement and production of customer orders. These agents communicate

D. Precup et al.: RedAgent: An Autonomous, Maket-based Supply-Chain Management Agent for the Trading Agents Competition, Studies in Computational Intelligence (SCI) 28, 135–154 (2006)
www.springerlink.com

through a market mechanism in order to determine, collectively, which components to purchase, which types of PCs to produce, how to allocate the available components and production cycles, and what offers to send to customers.

Market-based mechanisms have been used successfully for complex resource allocation problems in grid computing (e.g., [2]), computer networks (e.g., [3]), and scheduling (e.g., [4]). Markets have several advantages over more traditional optimization approaches. First, they offer the possibility of a modular design, in which different agents, responsible for different resources, can communicate through the market. Second, markets provide a more efficient search mechanism than traditional approaches, since alternative courses of action are not considered explicitly. Third, in addition to an allocation, the market also estimates a value attached to each resource. This is perhaps the most important feature from our point of view, as we will describe below.

In this chapter, we first provide an overview of the TAC SCM scenario and the possible high-level strategies. Then, we present RedAgent's architecture and the heuristics used by the different components of its multi-agent system. We analyze various aspects of RedAgent's performance based on data from the semi-final and final rounds of the TAC SCM competition from 2003. Finally, we discuss some variations that we tried, with less success, in the 2004 competition, and present avenues for future work.

2 Overview of TAC SCM

In TAC SCM, six agents representing PC assemblers operate in a common market environment, simulated over a period of one year (220 days). The agents are involved in a supply chain, in the sense that they compete both in markets for components (CPUs, motherboards, memory and hard drives) and in the market for finished PCs. These agents can assemble 16 different types of PCs, each using different combinations of components and different numbers of production cycles. Trades with the suppliers, as well as with the customers, are negotiated through a *request-for-quotes (RFQ)* mechanism. In this mechanism, the buyer issues RFQs to one or more sellers. The sellers respond with *offers*, which are then accepted or rejected by the buyer. An accepted offer becomes an *order*, which the seller has to deliver. The suppliers and customers in the TAC SCM market follow fixed strategies, as described in the game specification [5] (note that new rules were in effect in 2004 and 2005). The goal of the PC manufacturers is to obtain the highest profit (i.e., have the highest bank balance at the end of the game).

During each simulated day (which lasts for 15 seconds), the PC manufacturer has to decide which RFQs to issue for components, which of the offers from suppliers to accept, which PCs to manufacture with the available resources, which of the customer orders to ship, and what offers to send to customers in response to their RFQs. This complex decision making process can be simplified by considering two possible high-level strategies: buy-to-build and build-to-order. In the *buy-to-build* strategy, an agent stocks up on components, and starts producing PCs without necessarily having orders for all the production. In the *build-to-order* strategy, the first concern of a PC maker is to secure orders from customers; then, PCs are mostly built with the purpose of delivering these existing orders. The buy-to-build strategy has the advantage of ensuring a large stock, which can then be "dumped" on the market at any time. If the other competing

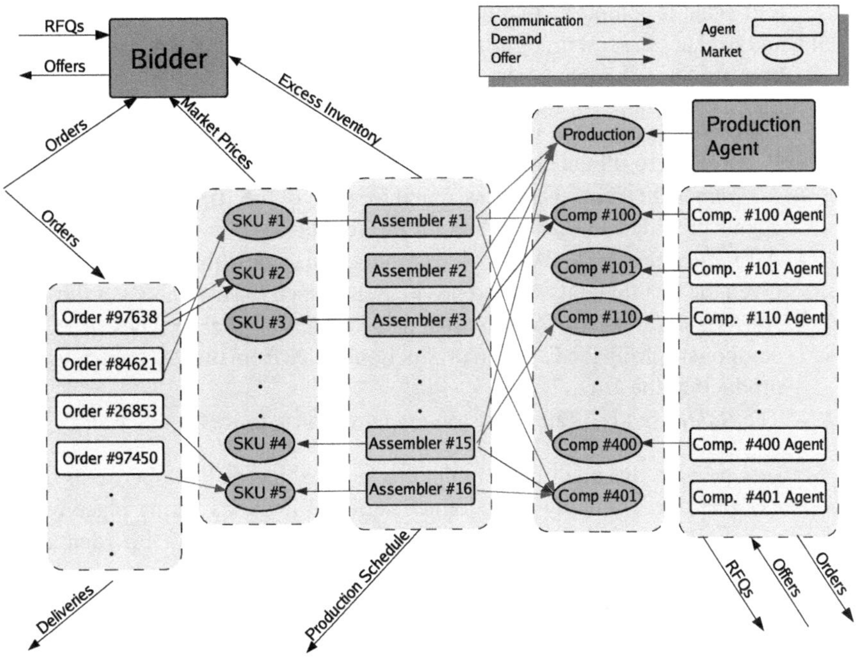

Fig. 1. Overview of RedAgent's components

manufacturers have low stocks, this has the added benefit of being able to obtain a lot of customer orders at a high profit margin. On the other hand, in a low-demand market, this strategy can be detrimental, given that sales would be low, and profits may not be high enough to cover the cost of the unsold PCs. Even given this potential pitfall, we adopted the buy-to-build strategy, because the absence of costs for maintaining inventory in the TAC SCM 2003 scenario made it appealing.

3 Architecture of RedAgent

The TAC SCM problem can be decomposed, at a high level, into three complex, interacting tasks: the procurement of supplies, the allocation of these supplies and (most importantly) of the production cycles to different goods, and the competition for market share. Hence, it seems natural to tackle these problems separately. However, the strong interdependence between these tasks means that the different modules need to exchange a lot of information. RedAgent proposes a unified solution to all these tasks: the use of internal auctions. We initially explored the idea of using auctions to solve the resource allocation problem. However, from a different perspective, auctions provide a relatively simple protocol for the agents to exchange information about their needs and valuations. Most importantly, auctions are a great way of assigning meaningful values to components, as well as finished goods.

RedAgent is composed of a number of simple, heuristic-based agents communicating and exchanging resources almost exclusively through auctions, as shown in Figure 1. There are five types of agents:

- An *Order Agent* (OA) is created for each received order; its goal is to obtain the PCs needed to fill the order and ship them to the customer.
- A *Component Agent* (CA) is assigned for each of the 10 types of components; these agents provide RFQs and orders for the suppliers.
- A *Production Agent* (PA) provides production cycles; this is a bottleneck resource, given the fact that only a fixed production capacity is available per day.
- An *Assembler Agent* (AA) is assigned for each of the 16 types of PC; it obtains components from the CAs and production cycles from the PA, then delivers finished products to the OAs.
- The *Bidder* sends offers to customers in response to RFQs.

Markets provide the main communication and exchange links between the different agents. A *market* is simply a predefined series of auctions taking place on each simulated day. RedAgent has one internal market for each type of PC (denoted SKU_i in Figure 1), for each component type, and for the production cycles

4 Internal Markets and Auctions

All the markets use sequential, sealed-bid double auctions. In each auction, buyers and sellers submit secret offer and demand bids for the type of resource (components, PCs or production cycles) being traded. The *offer* and *demand profiles* are the lists of offer and demand bids, in ascending and descending order of bid price respectively. These two profiles are matched, and a single exchange price is determined, which will be used for all trades. We used the midpoint between the highest satisfied offer and the lowest satisfied demand as the exchange price. An example of such an auction closing is given in Figure 2. In this case, the top three demands are satisfied, and the exchange prices are all set at $900, given that both the lowest demand partly satisfied and the highest accepted offer are of this value. This mechanism is designed to maximize the overall utility gain, assuming that the bids are indicative of the true valuations of the buyers and sellers. This assumption is justified in our case because this is an internal market in which the participants work towards a common goal. Note that in a general setting with self-interested agents the strategy of bidding true valuations may not be dominant.

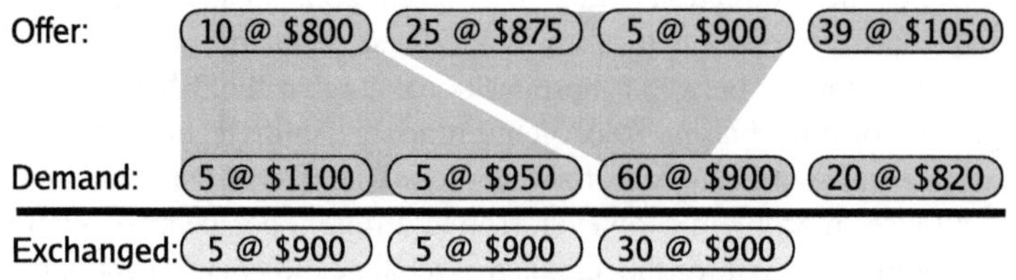

Fig. 2. Example of an offer and demand profile, and the resulting exchanges

After an auction closes, the offer and demand profiles are made available to all the participating agents, in order to allow them to adjust their future bids. In our implementation, only the sellers make use of the demand profiles to adjust their offer prices, although this information could be used more extensively.

The auctions take place in a fixed order on each simulated game day. First, the auctions for each type of PC close. These auctions can be considered orthogonal: no agent participates in more than one PC auction, since each order agent only bids for the PCs in its assigned order, and each assembler only offers a single type of PC. The PC auctions provide fresh demand information, which the assembler agents can use to determine how to bid in the resource markets.

The markets for resources hold 5 rounds of auctions each day, with the auctions taking place in predefined order: production cycles (usually the most contentious resource), CPUs (the most expensive resource), motherboards, memory, hard drives. This order reflects the importance of each type of resource for the manufacturing process. In each auction, the buyers are the assembler agents (which participate only in the markets trading the resources they need), and the main seller is the corresponding component agent or the production agent. However, assembler agents also have the option of selling excess resources. For instance, an assembler agent may purchase a lot of production cycles but then realize that it cannot acquire enough CPUs to sustain its production. In this case, it will attempt to sell the excess production cycles in the next round of auctions, to an assembler agent which can use them more productively. The bidding strategy used by the assemblers is described in detail in section 6.

Since we have a setting where auction participants need to obtain a bundle of goods, it would seem natural to consider using combinatorial auctions. However, resolving combinatorial auctions is quite complicated, and they do not necessarily assign a meaningful value to each individual good. Sequential auctions on the other hand can be resolved quickly, are much simpler to implement, and allow the participating agents to gain more information about the market by observing the successive closing prices and bid profiles. The idea of using sequential auctions for allocating complementary resources has been explored before in [6], but in the context of bids computed by dynamic programming. Instead, we will use simple heuristic bidding, as described below.

5 Order Agents

The order agents are created for each customer order, and their goal is to deliver the requested PCs, by acquiring them from the assembler agents. By participating in the PC market, these agents collectively determine how to allocate the available PCs to existing orders. A crucial effect of their bidding, however, is to provide an estimate of the true valuations of different types of PCs. The valuations are used both in the bidding process, as explained in Section **??**, as well as in the decision-making performed by the assembler agents, regarding which types of PCs to produce. Because of the need to produce accurate estimates, good bidding on the part of the order agents is very important for RedAgent's performance.

Suppose that an order agent is bidding for one PC unit, ordered for a price p, and let d be the index of the day *relative* to the due date of the order (e.g., $d = -1$ one day

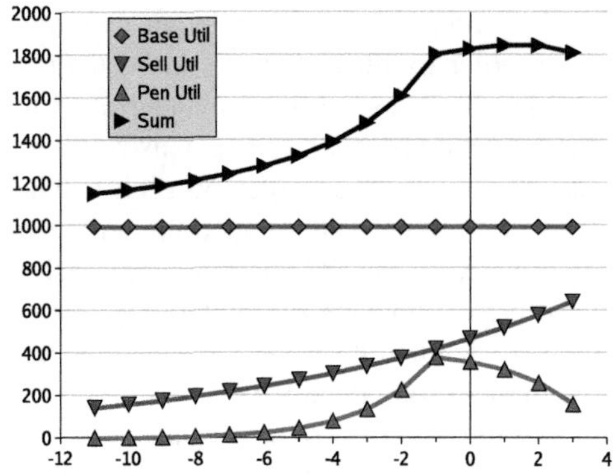

Fig. 3. Typical bidding profile for an order agent as a function of number of days after the due date. The profile (top curve) is obtained by adding the baseline, selling and penalty-avoidance utilities

before the order is due). The bid is computed as a sum of 3 terms: the *base price* of the PC, the *estimated discounted profit* for this unit, and the *discounted penalties* that would be avoided by shipping the PC now. The base price accounts for the components necessary to produce the PC. For each such component i, let c_i be the average price that was paid to the external suppliers for the units that are still in RedAgent's stock. Then $c = \sum_i c_i$ is the estimated material cost of the PC, and the base price is computed as $b(p,c) = \min(p,c)$. Note that if the material cost is lower than the order price, the agent is bidding lower, in order to push down the price for PCs of this type in the internal market.

The estimated discounted profit for the PC is computed by subtracting the base price from the order price, and discounting it in a standard fashion, based on the number of days left until the order "expires":

$$s(p,c,d) = (p - b(p,c))D_{sale}^{3-d}.$$

In our implementation, $D_{sale} = 0.9$. Finally, the penalty-avoidance term represents the money saved by avoiding late deliveries. In the TAC SCM rules, a daily penalty q is charged for 5 days after the due date if the order has not been delivered, after which the order is canceled. Because of their fairly large amount (between 5% and 15% of the order price), penalties play a significant role in the final profits, so they must be carefully considered when assigning available PCs to orders. The penalty avoidance term is computed as:

$$r(q,d) = \sum_{i=max(d,-1)}^{3} D_{pen}^{i-d} q,$$

where D_{pen} is a discount factor used to model the uncertainty in the possibility of shipping at a later date. In our implementation, $D_{pen} = 0.7$.

The bid placed by the order agent is computed as $b(c,p) + s(p,c,d) + r(q,d)$. A typical bidding profile, as a function of the day around the due date, is shown in Figure 3. Figure 4 illustrates two interesting kinds of bidding curves. The black line corre-

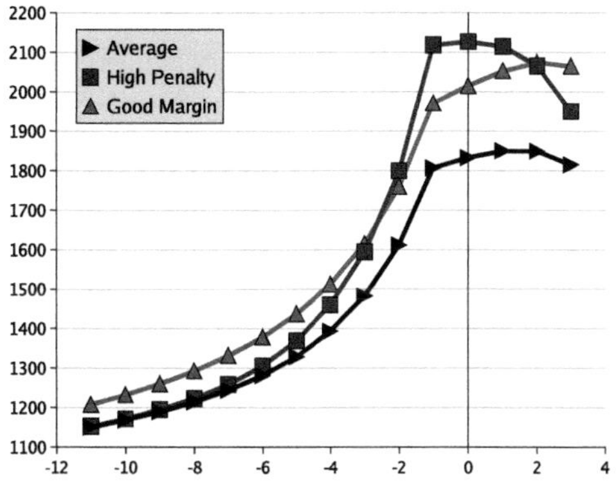

Fig. 4. Bidding profiles for order agents with higher penalties, or larger profit margins

sponds to an "average" profile. The blue curve corresponds to a high-penalty order, and it dominates the other curves around the due date. The sharp drop at the end is due to the fact that all penalty has now been incurred. The red curve corresponds to a high margin order, which should also be assigned high priority. We have to note here that our discount factors have been chosen such that orders get "prioritized" in this way, but RedAgent is very robust to the actual values of these factors. All that matters is that D_{sale} is close to 1 and D_{pen} is smaller.

Order agents generally submit a demand bid in the corresponding PC market on each day (according to their profile) until they obtain the needed PCs, but they do not ship as soon as possible. Instead, after they have acquired the PCs, they offer to sell them at their utility value, thus allowing other agents who need them more to obtain their inventory. This is reasonable from a cooperative perspective, but also if the order agents were self-interested, since they are compensated. The larger number of bids generated in this way is also helpful in ensuring that the closing prices of the auctions will be representative of the cost of taking on new orders. Note that if a PC type is in plentiful supply, the order agents obtain it long before the due dates at a lower price, whereas a shortage will force them to wait until their profile shows maximum valuation.

6 Assembler Agents

The 16 assembler agents, one for each PC model, deal with the task of scheduling production. They compete in the component and production cycle auctions in order to acquire the resources necessary for assembling PCs, which they sell in the PC auctions.

The assemblers construct their bids with the goal of maintaining a target inventory, which can be viewed as a buffer to counter spikes in PC demand or temporary shortages of components/production capacity. The target inventory is computed as the number of PCs expected to be needed for 10 days of operation, truncated between static lower and upper bounds (set to 40 and 400 respectively). The expected demand is computed as a running average (over 20 days) of the number of profitable PCs requested by the order agents. The target inventory is linearly increased from 0 in the beginning of the game (over the first 40 days), to avoid "panicked" purchasing. A similar linear decrease is performed at the end of the game (during the last 60 days) to eliminate excess inventory.

Assemblers place offers in the PC market to sell their inventory. In order to form an offer, the base price, b_a, for a given PC is the sum of the component and production costs, estimated using the latest closing prices of the corresponding resource auctions. The whole PC inventory is divided into three batches. The first two batches are equal to the target buffer. The computers in the first batch are priced linearly between b_a and $1.3b_a$, in order to increase the likelihood that this stock is kept. The second batch is priced linearly between $0.7b_a$ and b_a. All other computers are offered at $0.7b_a$. Hence, if an assembler starts selling computers from its safety buffer, the price it demands for the corresponding type of PC will increase, and vice versa. This information reaches the bidder agent through the PC market.

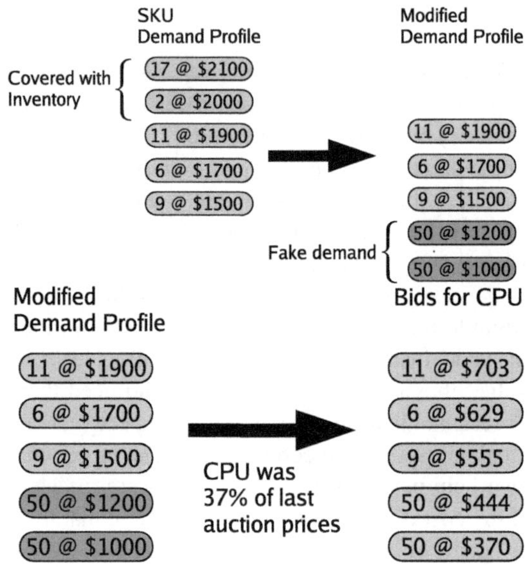

Fig. 5. Example of constructing resource bids

The auctions on the supply-side are used as a search method for finding a good allocation of resources (components and production capacity) among the 16 assembler agents. To determine what bids to place, the assembler takes the demand profile from the most recent PC auction. This is a sorted list of demand bids, and each bid is a price-quantity pair. An example is shown in Figure 5. Based on the current resource market prices, the assembler eliminates unprofitable bids. Then it adds a series of "fake" bids, so as to maintain its target inventory. The prices for these bids are regularly spaced in the range of $[c, 1.3c]$, where c is the same true average material cost used by the order agents. Another set of bids is added at prices regularly spaced between $0.7c$ and c. This is to indicate the desire to build extra PCs if some of the components are available at a discount. The demands that can be covered by the existing inventory are taken off the list, leaving only PCs that the assembler wants to produce. This process of determining the desired production level is depicted in the top part of Figure 5.

Now the demand curve must be transformed into one for component types (or production cycles). The bottom part of Figure 5 presents such an example. In this case, the assembler wants to determine how much to bid in the CPU auction. To do this, the assembler agent computes the ratio ρ between the latest closing price in the CPU market, and the sum of closing prices from all the resource auctions that are needed. The demand profile for CPUs is obtained simply by multiplying the prices from the PC demand profile by ρ ($= 0.37$ in our example). The assembler then eliminates components that it has in stock, and thus obtains a profile for the upcoming CPU auction.

Since the auctions are repeated for 5 daily rounds, the assembler can modify the number of bids it actually submits in accordance with the resources it has already acquired. For instance, if an assembler's demand profile calls for 100 PCs to be produced, but it only secures enough motherboards to build 50, it will avoid buying more than 50 memory units, and it will attempt to sell any excess CPUs and production cycles already acquired.

Because production is the most constrained resource, the assembler first attempts to buy cycles for the total number of PCs it wants to manufacture, denoted Q_{total}. Suppose it only obtains cycles to manufacture $Q_{obtained}$ PCs. The number of components that it will attempt to buy in the next auction is adjusted to:

$$Q_{cap} = Q_{obtained} + \frac{\text{RoundsLeft}}{\text{TotalRounds}}(Q_{total} - Q_{obtained}) \tag{1}$$

After each component auction, the maximum number of bids in the next auction is decreased (if necessary) by a similar adjustment. When a new production auction happens, Q_{cap} may increase or decrease, according to (1). The process is repeated for all auction rounds.

In addition to buying resources, the assemblers also sell resources that exceed Q_{cap}, at a price lower or equal to the lowest previous offer price. This reduces the possibility of resources being held by an agent that is unable to make use of them.

At the end of each day, after all the auctions close, each assembler builds all the PCs for which it has the necessary resources. They are available for sale to the order agents on the following day.

7 Component and Production Agents

The 10 component agents also aim to maintain their inventory in a desired target region. Hence, they need to estimated the rate at which their inventory is getting depleted. We use for this estimate a running average of the daily sales to the assemblers, $sale_c$. The lower bound L of the target inventory region is determined so as to maintain enough inventory to operate for 10 game days, assuming that the daily sales will all be equal to $sale_c$. This buffer is increased from 0 over 40 days in the beginning of the game, and decreased back to 0 at the end of the game. The upper bound U for the target inventory region on a given day is simply L plus the expected sales over the rest of the game, based on $sale_c$.

The bids of a component agent are directly determined as a function of these upper and lower bounds, in a very similar fashion to the bids of the assembler agents. The base price b_c for the sale of components is the *true average cost*, the average of the price paid for all the components ordered (but not necessarily received) to date. Then, the entire inventory is offered for sale as follows:

Inventory portion	Offer price
$[0, L]$	$[1.0b_c, 2.5b_c]$
$[L, U]$	b_c
$[U, U+L]$	$[0.5b_c, 1.0b_c]$
$[U, \infty$	$0.5b_c$

The part of the inventory acting as a buffer is priced at a very high premium in order to quickly push up the price of a resource if it becomes scarce. We allow for a similar buffer over the upper bound to avoid abruptly discounting the price if we have a slight excess in inventory.

The main role of a component agent is sending RFQs to suppliers and responding to returned offers. Determining which RFQs to send is done by projecting the inventory assuming that sales in d days are given by $(0.98)^d sale_c$. Whenever the projected inventory falls below L, an RFQ is submitted to all potential suppliers, with the due date on that day d. The quantity requested is the quantity expected to be used from d until the end of the game, capped by a constant (3000 components per order) and by a function of the number of days before the order is due (no more than 200 for every day until the due date). If multiple suppliers can fill the order, half the amount is requested from each instead of requesting the full order twice. A single component agent can send multiple RFQs on a day (with some caveats, described below). If several component agents send RFQs to the same supplier, their order in the RFQ bundle is chosen randomly.

When offers are received in response to these RFQs, they are sorted by price, and the inventory projection is repeated. For every day when components are required, the lowest priced satisfying offer is accepted. It should be noted that if a supplier cannot deliver the requested quantity on the date requested, it sends an earliest-complete offer as well as a partial offer with the requested delivery date. Hence there is no risk in sending too large RFQs, except that it may push up the price of subsequent RFQs for the same component if the supplier allocated production capacity to the first RFQ.

The production agent's role is somewhat similar to that of the component agents in that it sells resources to the assemblers, but since factory capacity is fixed in the TAC SCM scenario, it offers all the production cycles for a fixed base price. The price paid in the production auction is entirely determined by the bids of the assembler agents.

As the preliminary and seeding rounds of TAC SCM 2003 progressed, most agents started ordering their components on the first day of the game. A detailed discussion of this phenomenon is given in [7]. Essentially, due to the game design, all RFQs sent to suppliers on the first day generated an offer with the lowest possible price, so there was a strong incentive to order large amounts of components on the first day. However, the more manufacturers sent large early orders, the more the suppliers' capacity tended to be saturated, and the delays for obtaining components became long. We modified the component agent especially for this peculiarity of the game scenario.

On the first day of the game, each component agent sends 5 RFQs to every supplier, requesting the total amount of components needed for an "average" game (as assessed from prior games). The proportions of the RFQs were $\langle 0.55, 0.30, 0.10, 0.035, 0.015 \rangle$. These RFQs are sent with a due date of 0, in order to received earliest-complete offers. We expected that most participants would adopt similar first-day strategies and that RedAgent could only hope to receive the first, and possibly the second, of these orders within a reasonable delay. A simple mechanism was added to avoid accepting any of these large first-day offers for which more than 15% was expected to go unused. The smaller RFQs were left in the eventuality that it was possible to receive some cheap components near the end of the game, even if the larger RFQs were unusable. This strategy also gave RedAgent the capability to adapt to changes in the other agents. In particular, DeepMaize, the entry from the University of Michigan, changed tactics during the semi-final and final round, adopting a pre-emptive strategy to stop other agents from ordering on the first day[7]. As a result, many agents suffered from lack of components. RedAgent just started to order in smaller RFQs throughout the game.

8 Bidder Agent

The bidder's task is to respond to customer RFQs with offers. Unlike the other internal agents, the bidder does not directly participate in auctions. However, it relies heavily on the closing prices in the PC auctions in order to compute the offer prices.

Initially we attempted to use machine learning techniques in order to estimate the winning bid distributions for the RFQs; we were unsuccessful due to the highly stochastic nature of the bidding process and the changing competitors. Therefore, starting in the seeding rounds, we used an *adaptive-margin bidder*. The basic idea is to compute the offer price for a PC as the closing price of the corresponding internal market, incremented by an absolute margin *per production cycle* needed to produce the PC.

To compensate for the noise in the daily market closing prices, we used a running average over four days. A shorter averaging period would cause price oscillations in the customer bidding process because of overcorrections, which would be detrimental to all competitiors. On the other hand, the longer the averaging period is, the slower RedAgent reacts to internal changes (shortages, overstocks, etc.).

The per-cycle margin is meant to ensure that RedAgent does not receive more orders than it can produce PCs for, given its limited production capacity. At the same time, it aims to maximize the profit per cycle, since production capacity tends to be the bottleneck resource. The margin increases if RedAgent is receiving too many orders, the total number of RFQs increases, or the orders already do saturate production capacity around the due date of the RFQ being considered. It decreases in the opposite situations.

The resulting strategy is essentially to bid on all RFQs by a certain margin above the internal market value of the PCs being sold, and to adjust this margin so as to use all production capacity after observing the success rate of our offers.

9 Internal Behavior

The complex setting of TAC SCM makes it difficult to analyze quantitatively the impact of the different components of RedAgent on performance. In this section, we focus just on markets, and illustrate their merit both as resource allocation tools, as well as for the purpose of communication and price estimation.

Round	Producion	CPUs	All other components
1	26.1 (22%)	78.2 (12%)	17.1(7.5%)
2	13.8 (16%)	39.7 (5.2%)	12.4 (5.8%)
3	12.2 (14%)	27.0 (3.6%)	11.1 (5.7%)
4	11.0 (11%)	15.0 (2.1%)	8.1 (4.0%)
5	9.0 (8.1%)	8,9 (1.2%)	6.2 (2.9%)

Table 1. Average change in daily prices after the five auctions

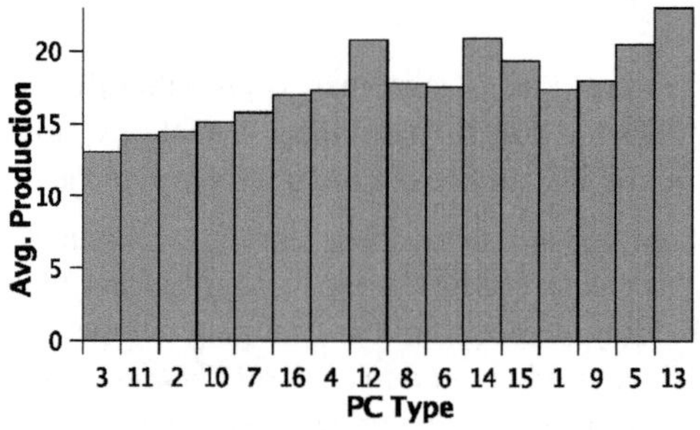

Fig. 6. Average production for different typs of PCs. The PCs are sorted in increasing order of material cost / number of production cycles

In terms of resource allocation, we hoped that markets would provide a good answer in a fairly timely manner. Table 1 shows the absolute and relative change in closing prices after each of the 5 daily auctions, averaged over all days except the first and last 10 days (which are subject to large inventory variations). The first auction has much higher variations on average than all the other auctions, because the bids take into account new data from the PC markets and suppliers. The changes decrease in further auctions. The auctions for production cycles are the most volatile because cycles are a bottleneck resource, and the assembler agents often divert a lot of funds to purchase them. The CPU auctions show high absolute changes because CPUs are expensive, but they converge the most rapidly. All other components behave very similarly. It is also important to note that in almost all games, there are no price variations over the last 2-3 auctions of the day. Hence, markets converge very quickly to a valuation.

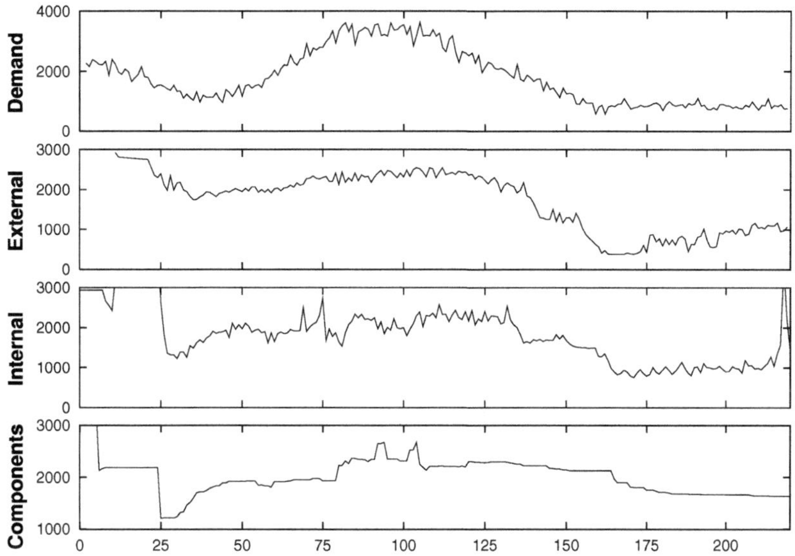

Fig. 7. Illustration of market communication

In order to examine the quality of the allocation produced by the markets, we looked at what RedAgent chooses to produce. Figure 6 shows the average number of PCs produced daily, during the 31 final games, as a function of the ratio of material cost vs. productions cycles needed for each PC type. As shown, RedAgent has a strong preference for PC types with a large ratio. These are either PCs that can be built quickly, or PCs with high component costs, for which it can apply a large margin.

Figure 7 illustrates the role of the internal markets in ensuring communication between the different agents during a typical game with variable demand. The top graph shows the total demand during the game, computed as the total number of customer RFQs, to all players. The second panel shows the external market price for PCs of a given type, computed as the average of the winning bids on the given day. The third panel shows the internal market closing price for PCs of this type. The bottom panel shows the sum of the internal market prices for all resources needed to build PCs of this type. The external market prices are always high in the beginning of a game because few agents have the necessary components, then decreases later. The spikes in the internal component price represent shortages, while the dips are temporary surplusses. The internal market price is clearly influenced both by the external market and by the component price. It approximates quite successfully the external market price, but spikes when there are component shortages, as well as at the end, when we have depleted our stock. Note that internal market price is effectively a lower bound on any offer sent to customers.

Figure 8 shows the evolution of the margin, production utilization and stock. The margin mostly follows the demand curve, ensuring steady sales. However, when the production is not utilized, the margin also drops in order to ensure that the production

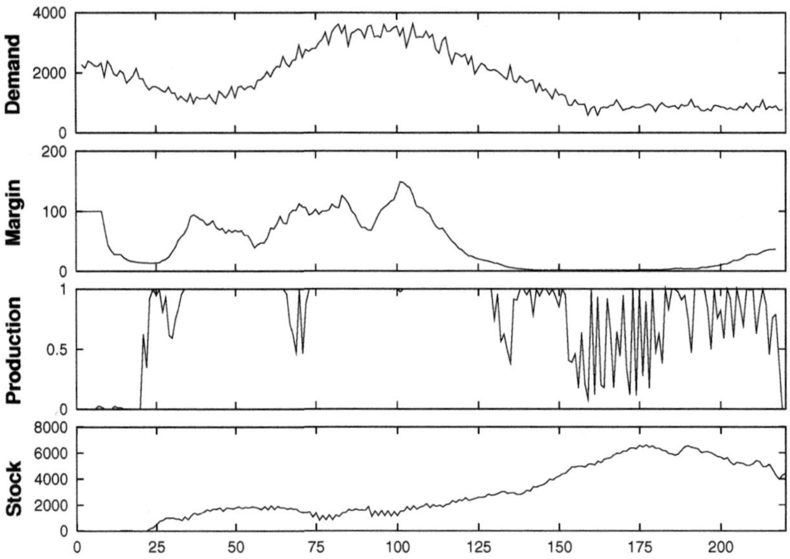

Fig. 8. Illustration of production control

cycles are filled up. Note that the margin increases at the end of the game because the internal prices are dropping, but also because other competitors typically run out of stock and stop bidding. Our stock is also decreasing at this time, as RedAgent dumps PCs on the market. In games ending with low demand, the margin decreases to near 0 to sell off the stock.

10 Competition Performance

In this section, we try to provide some insight into the strategy of RedAgent, as well as that of other competitors in the 2003 TAC SCM competition. A broader analysis of all the agents is presented by the competition organizers in [8], The data we present here comes from analyzing the semi-final group in which RedAgent played, and finals. In all the following graphs, the agents are ordered from left to right, in the order in which they ranked during the competition.

RedAgent's buy-to-build strategy is quite apparent from the PC inventory graphs, presented in Figure 9. RedAgent has by far the highest stock of PCs in the final round. During the semi-finals, only one other team, HarTAC, had similar inventory levels for some systems. We note that HarTAC came last during the semi-final round due to technical networking problems, which caused one catastrophic game. Otherwise, their performance would have been much better.

The buy-to-build strategy has two important effects. On one hand, RedAgent is able to respond very well to customer RFQs, given that there are always PCs in stock. Figure 10 shows the ratio of offers sent to customers, divided by the number of RFQs

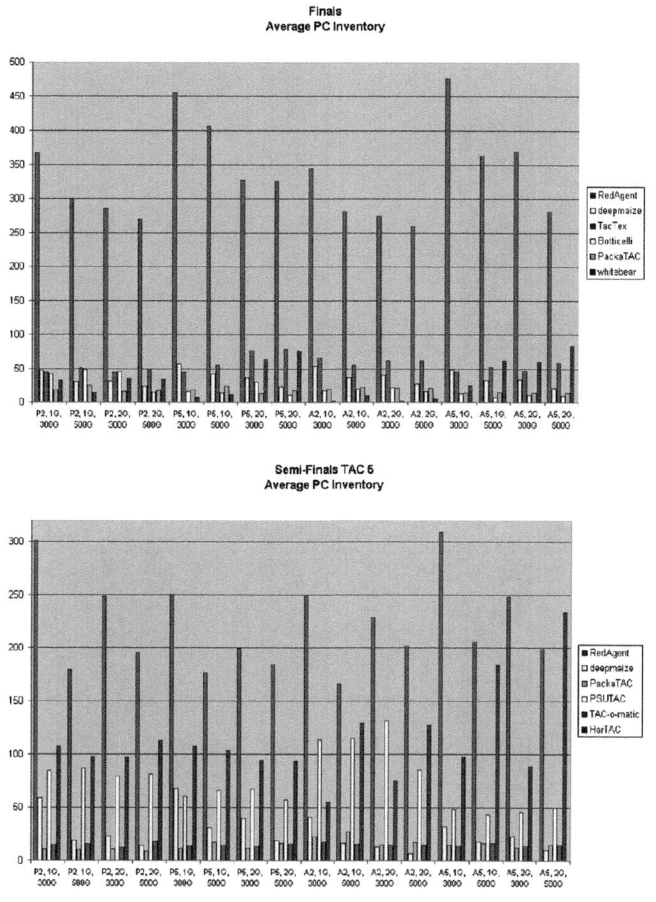

Fig. 9. Average PC inventory during the finals (top) and semi-finals (bottom)

received. As can be seen, RedAgent has a significantly response rate than the other teams. We also note that both of the top two teams, RedAgent and deepmaize, have higher response rates than the rest of the competitors. A very similar profile was seen in the semi-final round, where only HarTAC had similar RFQ response rates.

A second important effect is that, because RedAgent has large stocks, it can sell when other manufacturers are out of stock, thus obtaining very good prices. Figure 11 shows the average price obtained per PC. RedAgent manages to obtain the best prices in *all PC categories*, both in the semi-finals and in the finals. This is due mostly to the fact that RedAgent sells PCs at a very good price especially in the beginning and in the middle part of the game (until day 180 or so). In the end of the game, RedAgent lowers its prices, in order to get rid of its stock, but at the same time, it dominates sales during this period, so it is still able to ensure high revenues.

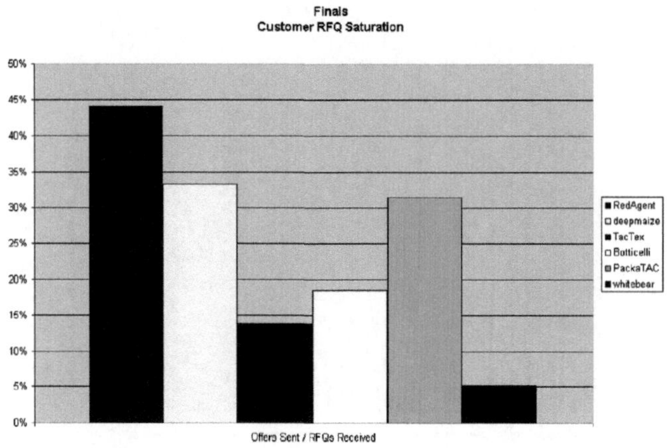

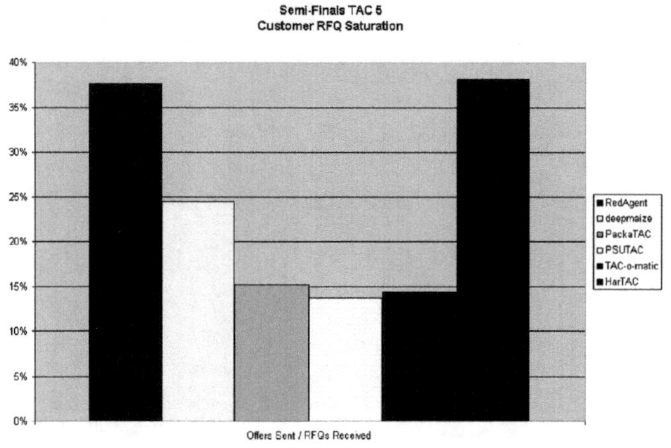

Fig. 10. Percentage of customer RFQs for which offers were sent

A dual perspective on performance is offered by the inventory and price paid for parts, shown in Figure 12 and Figure 13 respectively. As can be seen, RedAgent keeps a fairly large inventory (in order to support its production), but not too large, compared to other competitors. It is also able to secure components at prices that are lower or very similar to those of the competition, thus ensuring higher profit margins than those of other agents. It is worth mentioning here that the design of the game favored agents that buy their components at the beginning of the game. Some agents (e.g., whitebear) relied exclusively on first-day supplies. RedAgent orders most of the supplies in the beginning, but continues to order supplies later on, as needed. As described in [7], this aspect of the game had strategic importance in the 2003 competition. In order to avoid this behavior, in 2004 storage costs for components were introduced. However, this still did not prevent agents from placing large component orders in the beginning of the game. A game-theoretic analysis presented in [9] shows that the costs imposed were

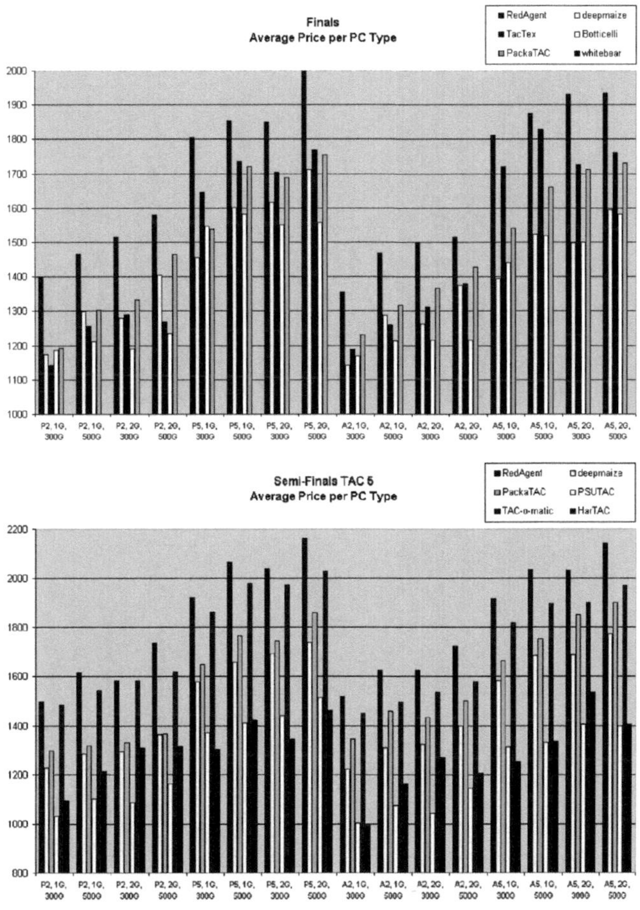

Fig. 11. Average price per PC type

actually too low, and the rational behavior was still to order as much as possible on the first day despite the resulting large inventories. The 2005 competition introduced significantly more changes in the game rules [10], in an attempt to make the behavior of the suppliers more realistic.

11 Discussion

The use of internal markets has provided RedAgent with a very efficient search mechanism for resource allocation, while at the same time ensuring the communication between different agents in our distributed architecture. It is worth noting that RedAgent 2003 used significantly fewer computational resources than other competitors [11, 12]. However, the reliance on markets has the disadvantage of losing a lot of information that

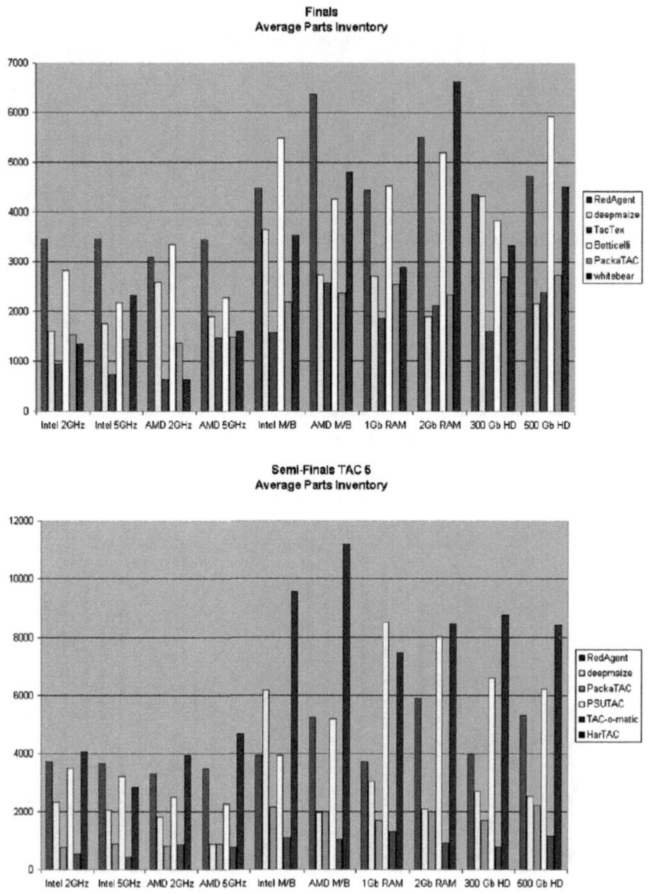

Fig. 12. Average inventory maintained for the different components during the finals (top) and semi-finals (bottom)

could be captured in a richer communication protocol. In 2004, we replaced the market mechanism with a protocol for negotiation between the different internal agents. In this case, instead of just participating in an auction, the agents could exchange information such as the due date of an order, or expected penalties, in an explicit way. We envisioned a negotiation between the internal agents much like the negotiations that happen with customers and suppliers in the TAC game. Unfortunately, this design did not work nearly as well as the less structured auctions. Good heuristics that would take into account the information being exchanged proved much more difficult to find than the simple heuristics needed in our first design.

Another major change that we tried in 2004 was to use machine learning and probabilistic reasoning techniques to improve our prediction of the future behavior of the markets. One technique that worked fairly well was to use Kalman filtering to predict

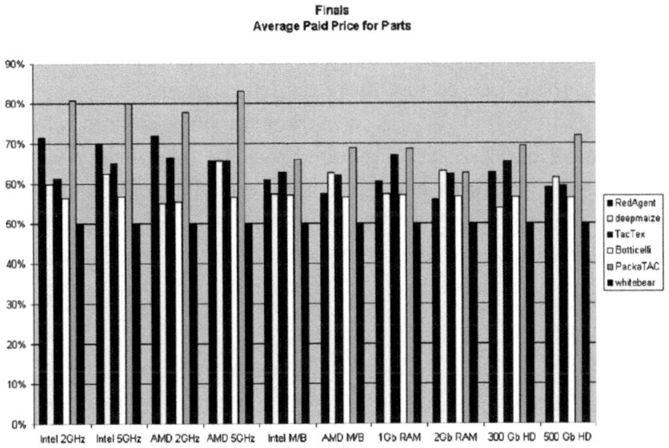

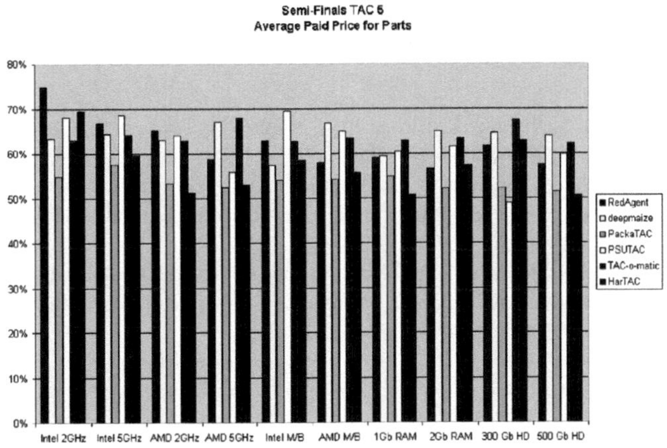

Fig. 13. Average price paid for the different components during the finals (top) and semi-finals (bottom). The price is expressed as a percentage of the base price of each part

the state of the suppliers based on the component offers received. Unfortunately this information was of little use, since most components continued to be ordered at the beginning of the game, before any negotiations with suppliers took place.

We also attempted to use an instance-based learning mechanism to estimate the variation that we will see in the price for different kinds of PCs in the future. The choice of instance-based learning was due to the fact that it makes it easy to incorporate new data. We trained off-line using data from past games, and then used this component as part of the bidding mechanism and for guiding production decisions. Unfortunately, there is too much variance in pricing behavior in the TAC SCM games, both during one game, and especially between different games, due to the big differences in the strategies of the participating agents. As a result, the generalization from one game to the next was very poor. The learned predictions were sufficiently good only towards the end of a game, after most purchasing and production decisions have already been made. The most likely reason is that agent strategies are more constrained in the end of the game. This technique may perform better if the strategies of the competing agents were less erratic, as can be expected in a more realistic scenario.

We strongly believe that the use of markets in supply-chain management problems is very promising. RedAgent demonstrated that the use of very simple, but numerous agents in cooperative markets can provide an excellent mechanism for resource allocation, decision making as well as utility computation, in a complex scenario. The use of markets in supply-chain management, as well as in other domains in which good valuations for resources are needed should be explored more in the future.

Acknowledgments

We gratefully acknowledge the help of Matthew Wahab and Francois Plamondon, participants in the 2003 and 2004 teams respectively. This research was supported in part by funding from NSERC, FQRNT and CFI.

References

1. Sadeh, N., Arunachalam, R., Eriksson, J., Finne, N., Janson, S.: TAC-03–a supply-chain trading competition. AI Magazine **24**(1) (2003) 92–94
2. Wolski, R., Plank, J.S., Brevik, J., Bryan, T.: Analyzing market-based resource allocation strategies for the computational Grid. The International Journal of High Performance Computing Applications **15**(3) (2001) 258–281
3. Kuwabara, K., Ishida, T.: Equilibratory approach to distributed resource allocation: Toward coordinated balancing. In Castelfranchi, C., Werner, E., eds.: Artificial Social Systems (LNAI Volume 830), Springer (1994) 133–146
4. Walsh, W.E., Wellman, M.P., Wurman, P.R., MacKie-Mason, J.K.: Some economics of market-based distributed scheduling. In: International Conference on Distributed Computing Systems. (1998) 612–621
5. Arunachalam, R., Eriksson, J., Finne, N., Jansson, S., Sadeh, N.: The TAC supply chain management game. Technical Report Draft version 0.62, SICS (2003)
6. Boutilier, C., Goldszmidt, M., Sabata, B.: Sequential auctions for the allocation of resources with complementarities. In: IJCAI. (1999) 527–523
7. Wellman, M.P., Estelle, J., Singh, S., Vorobeychik, Y., Kiekintveld, C., Soni, V.: Strategic interactions in a supply-chain game. Computational intelligence **21**(1) (2005) 1–26
8. Arunachalam, R., Sadeh, N.: The supply chain trading agent competition. Electronic Commerce Research and Applications **4** (2005) 63–81
9. Kiekintveld, C., Vorobeychik, Y., Wellman, M.P.: An analysis of the 2004 tac supply chain management tournament. In: IJCAI-05 Workshop on Trading Agent Design and Analysis. (2005)
10. Collins, J., Arunachalam, R., Sadeh, N., Eriksson, J., Finne, N., Sverker: The supply chain management game for the 2005 trading agent competition. Technical Report CMU-ISRI-04-139, Carnegie Mellon University (2004)
11. Kiekintveld, C., Wellman, M.P., Singh, S., Estelle, J., Vorobeychik, Y., Soni, V., Rudary, M.: Distributed feedback control for decision making on supply chains. In: ICAPS. (2004)
12. Benisch, M., Greenwald, A., Grypari, I., Lederman, R., Naroditskiy, V., Tschantz, M.: Botticelli: A supply chain management agent designed to optimize under uncertainty. SIGecom Exchanges **4**(3) (2004) 2937

A Framework of Optimization Agent for Supply Chain Management

Jae Kyu Lee[1] and Yong Sik Chang[2]

[1] School of Information Systems, Singapore Management University
(and Korea Advanced Institute of Science and Technology)
80 Stamford Road, Singapore 178902
jklee@smu.edu.sg
[2] Department of e-Business, Hanshin University, 411 Yangsan-dong, Osan, Kyunggi
447-791 South Korea
yschang@hs.ac.kr

Abstract. Effective supply chain management requires formulating and solving optimization models embedded in the multiple software agents along the supply chain. The stakeholders like manufacturers and third party delivers have their objectives, and are restricted by the constraints of delivery requirements and resource limitations. Since a variety of situations require many combinations of models, it is not easy to prepare all the necessary models in advance. To resolve this problem, this research proposes a framework of automatically formulating and modifying optimization models. For this purpose a base model template is identified, and then it is modified to the target model. To implement the idea, we design an architecture of optimization agents AGENT-OPT2 specifically for supply chain management. AGENT-OPT2 is equipped with rule-based interoperable and flexible model modification capability. This framework is validated by applying a prototype to a cooperative delivery scheduling problem.

1 Introduction

Many optimization models have been used to improve supply chain performance within and between enterprises [32]. Thus in an effective agent-based supply chain management (SCM) system, it is necessary to embed the optimization function in the distributed software agents. Since the combination of situations that involve many stakeholders along the supply chain is very diverse, and many corresponding models are necessary, it is too expensive to prepare all the combination of models in advance. To overcome this problem, we need to develop a framework that can dynamically formulate and modify the optimization models specifically tailored to the supply chain domain.

In this line of research, Chang and Lee [7] contrasted three approaches of optimization model formulation and modification – the *primitive model approach, the*

J.K. Lee and Y.S. Chang: *A Framework of Optimization Agent for Supply Chain Management*, Studies in Computational Intelligence (SCI) **28,** 155–178 (2006)
www.springerlink.com

most similar case approach, and the *full model approach* depending upon how a ***base model*** is selected to start with the modifications.

In this study, we adopt the ***primitive model approach*** as a foundation of optimization agent design. The primitive model implies a *minimal scale base model whose modification requires only the addition of terms and constraints*, thus the modification operator for the target model is simple: ***INSERT***.

In the Web Services environment, a third party Supply Chain Optimization Service Provider can provide the dynamic formulation and solution to many users. To handle the dynamic formulation based on the users' request and resource constraints, we need to focus the following issues:

1) Designing the architecture of the SCM optimization modeling agent system,
2) Identifying a base model from the candidate models in SCM domain,
3) Designing of SCM optimization modification operators and rules,
4) Designing the procedure of modifying the base models,
5) Designing a canonical and semantic representation of target optimization model,
6) Transforming the canonical model to commercial solver.

A prototype AGENT-OPT2 is developed with this framework. To illustrate the validity of this approach, we applied the framework to the delivery scheduling problem which is a typical model for efficient logistics management in the supply chain. As an illustrational scenario, we adopted the Multi-depot Vehicle Routing Problem with Time Windows (M-VRPTW) as the base model for the delivery scheduling category. In this problem, manufacturers produce their products at factories, store them in multiple depots distributed over the regions, and deliver products to multiple buyers. Manufacturers request that the third party logistics company delivers products to buyers after picking them up from one of the depots. The modified models from the base model M-VRPTW are suitable for solving a variation of problems in this category.

When multiple base models exist, we need to select a suitable base model. For the selection of a base model, we adopt the rule-based backward chain reasoning approach. On the other hand, for the rule-based modification of base model, we adopt the forward chain reasoning approach. For the canonical representation of the optimization model, we adopt the Document Type Definition (DTD) form that is used for the structural definition of XML statements. As a solver of integer programming models, we adopt the package LINGO.

The following sections describe the above issues one by one. Section 2 reviews the literature on the approaches to optimization model management. Section 3 describes the structure of optimization models in supply chain. Section 4 describes the architecture of AGENT-OPT2 and the procedure of automatic formulation of optimization models. Section 5 illustrates the procedure of SCM modeling with an example case. The study concludes with a summary of its contributions and limitations.

2 Review of Model Management for Optimization Agent

Model management research aims at investigating the establishment of tools and methodologies that can support effective model formulation, modification, and maintenance in the context of real world scale and multiple models in the dynamically changing business environment

Earlier research on the model management system (MMS) has focused on the representation of models and the reasoning methods in order to map specific problems with the model structure. In this line of research, a number of model management studies were conducted from the perspective of data and object management, network data modeling [16], relational data modeling [3], entity-relationship data modeling [4], and object-oriented data modeling [13, 19, 24, 30]. Liang proposed a framework that included both relational and network concepts [26].

On the other hand, the researchers in the Artificial Intelligence community explored the knowledge representation techniques to represent the models and focused on automating the model formulation processes. These representations are based on predicate calculus [6], semantic information net [10], knowledge abstraction [8], first order logic [9, 17], structured modeling [11, 12, 31, 32], rule-base formulation [20, 21], and frame based representation which is a precursor of objects in the object oriented program paradigm [2, 22].

They recognized that it is very difficult to formulate a model from scratch. So they attempted to modify the existing models by analogy [3, 14, 15, 27, 28, 34] and case-based reasoning approaches [7, 33]. These approaches require complex knowledge for model modification. In case of modeling by analogy, it takes a huge computational effort because feature mapping is theoretically NP-complete [25]. Modeling by analogy attempted to compose a model using the model components without modifying its structure. Although the analogy approach was introduced for this purpose [1], it was not entirely satisfactory because it handles only the conceptual level.

Model management specifically for the Operations Research models are investigated by Lee and Kim [22], Yeom and Lee [35], and Chang and Lee [7]. Since optimization models are the primary solution tools for the agent's decision in supply chains, management of optimization models is the foundation of optimizing agents for the supply chain. The time-bounded negotiation framework is studied from the multiple agents context [23].

Since the previous studies do not provide dynamic model formulation capability to cope with diverse situations happening in SCM, automatic formulation of optimization models for supply chains is necessary. So this study aims at creating a methodology and tool AGENT-OPT2 that generates an automated modeling procedure within the software agents along the supply chain.

3 Structure of Optimization Models in the Supply Chain

To organize the potential primitive models in the supply chain, we need to classify the optimization models for the supply chain. Typical models adopted in the supply chain domains are *transportation*, *order selection*, *production*, and *supplier selection* as depicted in Fig. 1. Each model domain has several optimization model types. Fig.1 shows an illustrative example of SCM modeling hierarchy. It consists of four layers: ***SCM model domains, optimization model types, primitive models***, and ***target model components***. Each optimization model types can have a corresponding primitive model which consists of a primitive objective and primitive constraints. The upper three layers are generic, while the bottom layer of the target model components is very case specific.

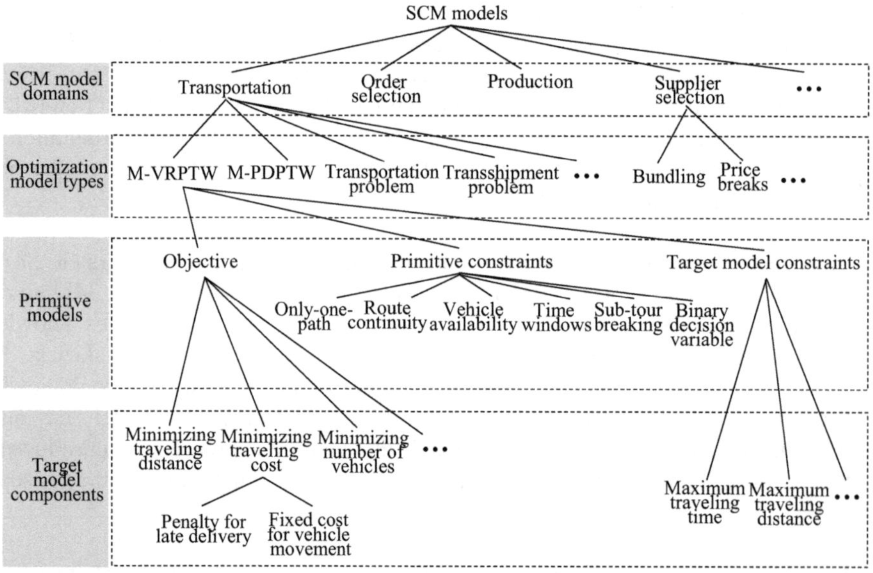

Fig. 1. An illustration of SCM Modeling Hierarchy

For instance, in the transportation domain, there are optimization model types like M-VRPTW, M-PDPTW (Multi-Depot Pickup and Delivery Problem with Time Windows), the transportation problem, and transshipment problem. In the supplier selection domain, there are models for bundling [31] or price break [7]. From the selected optimization model types, a primitive model is selected as a base model which it will be modified to synthesize a target model.

Let us illustrate the structure with the optimization model type M-VRPTW in the transportation domain.

3.1 Primitive Model of M-VRPTW

The primitive model of M-VRPTW is based on the vehicle routing problem [18]. The constraints of the primitive model are stated in Table 1 without objective function. The primitive constraints are the only-one-path, route continuity, vehicle capacity, vehicle availability, time window, sub-tour breaking, and binary decision variable constraints.

Table 1. Constraints for the M-VRPTW Primitive Model

Constraint Name	Constraints in Mathematical Notation	
Only-one- path	$\sum_{i=1}^{n+m} \sum_{k=1}^{v} X_{ijv} = 1 \quad \text{for } j = 1,2, \ldots, n$	(1)
	$\sum_{j=1}^{n+m} \sum_{k=1}^{v} X_{ijv} = 1 \quad \text{for } i = 1,2, \ldots, n$	(2)
Route continuity	$\sum_{i=1}^{n+m} X_{ihk} - \sum_{j=1}^{n+m} X_{hjk} = 1 \quad \text{for } h = 1,2, \ldots, n+m, \; k = 1,2, \ldots, v$	(3)
Vehicle capacity	$\sum_{i=1}^{n+m} \sum_{j=1}^{n+m} q_i X_{ijk} \leq p_k \quad \text{for } k = 1,2, \ldots, v$	(4)
Vehicle availability	$\sum_{i=n+1}^{n+m} \sum_{j=1}^{n} X_{ijk} \leq 1 \quad \text{for } k = 1,2, \ldots, v$	(5)
	$\sum_{j=n+1}^{n+m} \sum_{i=1}^{n} X_{ijk} \leq 1 \quad \text{for } k = 1,2, \ldots, v$	(6)
Time windows	if $X_{ijk} \geq 1$ then $T_i + s_i + t_{ij} \leq T_j$ for i=1,2, …,n, j=1,2, …,n, k=1,2,…,v	(7)
	$et_i \leq T_i \leq lt_i$ for i=1,2, …,n	(8)
Sub-tour breaking	$Y_i - Y_j + (m + n) X_{ijk} \leq n + m - 1$ for $1 \leq i \neq j \leq n$, $1 \leq k \leq v$	(9)
Binary decision variable	X_{ijk} = binary for all i, j, k	(10)

<u>Notations</u>

i, j, h: the indices *i*, *j*, and *h* imply the locations of buyer delivery points and depots {*1, 2, …, n+m*}, where

n: the number n implies the number of delivery points,
m: the number m implies the number of depots.
k: the index k implies the vehicles $\{1, 2, \ldots, v\}$, where
v: the number of vehicles.

q_i: demand to delivery point i.
p_k: capacity of vehicle k.
s_i: service time at delivery point i.
t_{ij}: traveling time between delivery point i and j.
et_i: earliest delivery time at delivery point i.
lt_i: latest delivery time at delivery point i.
T_i: arrival time at delivery point i.
f_k: maximum traveling time for vehicle k.
g_k: maximum traveling distance for vehicle k.

X_{ijk}= 1 if a pair, starting node i and ending node j, is in the route of vehicle k,
X_{ijk} = 0 otherwise.

Y_i: the real number that breaks sub-tours.

The constraints (1) – (10) implies the following:

Constraints (1) and (2) ensure that each delivery point is served by one and only one vehicle.
Constraint (3) means route continuity.
Constraint (4) describes the vehicle capacity.
Constraints (5) and (6) verify vehicle availability.
Constraints (7) and (8) specify the time window constraints in which vehicles have to visit the delivery points.
Constraint (9) prohibits sub-tours.
Constraint (10) limits the decision variable to binary number of 0 or 1.

3.2 Options of the Objective Function

In the M-VRPTW case, the objectives of M-VRPTW are dependent on deliverer's policy. For example, a deliver may require minimizing traveling distance, cost, or number of vehicles. The cost related objectives may be reclassified into three categories: minimizing travel cost, penalty cost for late delivery, and fixed cost for vehicle movement. The notational expression of these constraints is shown in Table 2. Multiple objectives may be combined with each other.

Table 2. Illustrative Options of Objectives for M-VRPTW

Type	Objective factors	Objectives

Distance	Minimizing traveling distance (*O1*)	$\sum_{i=1}^{n+m} \sum_{j=1}^{n+m} \sum_{k=1}^{v} d_{ij} X_{ijk}$	(11)
Cost	Minimizing traveling cost (*O2*)	$\sum_{i=1}^{n+m} \sum_{j=1}^{n+m} \sum_{k=1}^{v} a\, d_{ij} X_{ijk}$	(12)
	Penalty cost for late delivery (*O3*)	$\sum_{i=1}^{n+m} b\, (lt_i - T_i)\, Z_i$	(13)
	Fixed cost for vehicle movement (*O4*)	$\sum_{k=1}^{v} c_k W_k$	(14)
Vehicle	Minimizing number of vehicles (*O5*)	$\sum_{k=1}^{v} U_k$	(15)

Here, the three constants a, b, and c_k imply the following:

a: a proportional constant that transforms the distance unit into a cost unit.
b: a proportional constant that transforms the time unit into a cost unit.
c_k: a constant for fixed cost when a vehicle k moves.

3.3 Objective-Driven Constraints

When the deliverer decides the objectives, some constraints should be generated to make the constraints consistent with the objective function. We illustrate three constraints of this kind as summarized in Table 3.

(i) The objectives of minimizing traveling cost, penalty cost for late delivery, and fixed cost require searching for delivery routes minimizing the sum of their corresponding costs. For instance, the objective in (13) requires the constraints (16) and (17), and the objective (14) requires the constraints (18) and (19) to make the base model valid.

(ii) The objective of minimizing the number of vehicles requires searching for delivery routes minimizing the sum of the total number of traveling vehicles. For instance, the objective (15) requires the constraints (20) and (21).

Table 3. Illustrative Objective-Driven Constraints for M-VRPTW

Type	Objective factors	Constraints	
Cost	Penalty cost for late delivery (*O3*)	if $lt_i - T_i > 0$, then $Z_i = 1$, for $i = 1, 2, \ldots, n$, Z_i = binary, for $i = 1, 2, \ldots, n$.	(16) (17)

	Fixed cost for vehicle movement (*O4*)	if $\sum_{i=n+1}^{n+m} \sum_{j=1}^{n} X_{ijk} = 1$ then $W_k = 1$ for $k = 1,2, \ldots, v$,	(18)
		W_k = binary, for $k = 1, 2, \ldots, v$.	(19)
Vehicle	Minimizing Number of vehicles (*O5*)	if $\sum_{i=n+1}^{n+m} \sum_{j=1}^{n} X_{ijk} = 1$ then $U_k = 1$ for $k = 1,2, \ldots, v$,	(20)
		U_k = binary, for $k = 1, 2, \ldots, v$.	(21)

3.4 Optional Constraints for the Target Problem

To define a target model, additional constraints may be added to the base model. For example, the required restrictions in the M-VRPTW case may be the maximum traveling time and maximum traveling distance. Table 4 demonstrates such constraints.

The maximum traveling time in (22) means that each vehicle has to take time less than or equal to the required time limit to complete the work on a route. Similarly, the maximum traveling distance in (23) means that each vehicle has to travel less than or equal to the required distance. The notations of these constraints are shown in Table 4.

Table 4. Illustrative Constraints for the Target Model of M-VRPTW

Type	Constraint factors	Constraints	
Time	Maximum traveling time (*C1*)	If $X_{hjk} = 1$ AND $X_{ihk} = 1$ THEN $T_i + t_{ih} - (T_j - t_{hj}) \leq f_k$ for i=1,2,...,n, j=1,2,...n, h=n+1,n+2,...,n+m, k=1,2,…,v.	(22)
Distance	Maximum traveling distance (*C2*)	$\sum_{i=1}^{n+m} \sum_{j=1}^{n+m} d_{ij} X_{ijk} \leq g_k$ for k=1,2,…,v.	(23)

In this manner, knowledge of the primitive models, optional objective functions, objective-driven constraints, and optional constraints for the target problem needs to be specified to support the dynamic formulation of various target problems. So the basis of the optimization agent for any application domain is the analysis of this kind of knowledge for the particular domain. In this research, we demonstrate the analysis for the supply chain domain.

4 Architecture of Optimization Agents

Based on the knowledge we have analyzed in section 3, we build the optimization agent AGENT-OPT2 tailored to supply chain modeling. Fig. 2 illustrates the architecture of AGENT-OPT2 for the transportation domain. The optimization agent takes five steps to perform its mission.

Step 1: Identify a base model
The optimization agent identifies a base model using the Base Model Identification Rules in the SCM model domains. Backward chain reasoning can be applied considering the delivery requirements received from the requesting manufacturers.

Step 2: Identify the requirements of the target model
The optimization agent receives the message on resource constraints from the deliverers, and identifies a target model. The target is specified by the base model and objectives and constraints. The modification can be automatically executed by forward chain reasoning using the Target Model Identification Rules.

Step 3: Generate a canonical target model
The target model formulated is generated in a canonical document. The canonical representation adopted in this study is the DTD form.

Step 4: Transform to a commercial solver's formulation
The formulated model is transformed to the format of a commercial IP Solver. In this study, LINGO is adopted for this purpose.

Step 5: Solve and report
The model is solved by the solver and the answer is reported to relevant stakeholders such as manufacturers and deliverers.

The steps are elaborated in the next sub-sections.

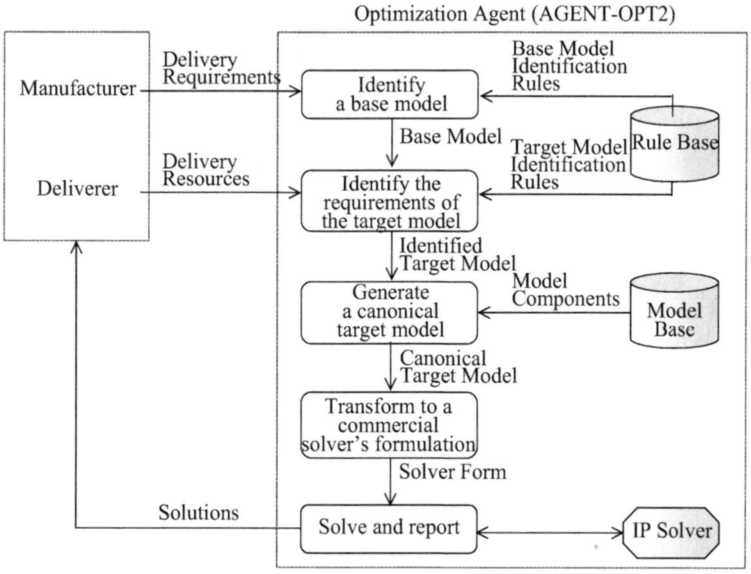

Fig. 2. Architecture of AGNET-OPT2 for Transportation Problems

4.1 Identification of a Base Model

In Step 1, the optimization agent recognizes the SCM model domain and identifies a base model after having received the delivery requirements from multiple manufacturers. Fig. 3 illustrates an AND/OR graph for base model identification in the transportation domain. M-VRPTW will be selected when the *SCM model domain is transportation*, *items of each depot are delivered to multiple delivery locations*, and *a route is composed of a depot and several delivery locations*. The backward chain rules that identify the base models - Base Model Identification Rules - are stated in Table 5.

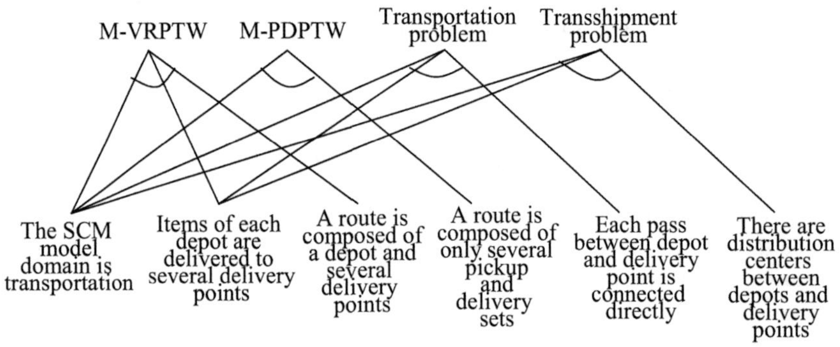

Fig. 3. AND/OR Graph for Base Model Identification

Table 5. Base Model Identification Rules in the Transportation Domain

Rule name	Rule Statements
RULE_DOMAIN_ M_VRPTW	IF the *SCM model domain is transportation* AND *Items of each depot are delivered to several delivery points* AND *A route is composed of a depot and several delivery points* THEN *Base model is vehicle routing problem*
RULE_DOMAIN_ M_PDPTW	IF the *SCM model domain is transportation* AND *A route is composed of only several pickup and delivery sets* THEN *Base model is pickup and delivery problem*
RULE_DOMAIN_ TRANSPORTATION	IF the *SCM model domain is transportation* AND *Items of each depot are delivered to several delivery points* AND *Each pass between depot and delivery point is connected directly*

	THEN *Base model is transportation problem*
RULE_DOMAIN_ TRANSSHIPMENT	IF the *SCM model domain is transportation* AND *Items of each depot are delivered to several delivery points* AND *There are distribution centers between depots and delivery points* THEN *Base model is transshipment problem*

Base models need data for their coefficients. Some of them are given from manufacturer delivery requirements and deliverer resources. The others are retrieved by the optimization agent from the common database. Table 6 shows three data groups.

In the M-VRPTW case, depots and delivery points, demands of delivery points, and time windows of delivery points are given by the manufacturer. Capacities of vehicles are given by the deliverer. The optimization agent may retrieve the service time at a delivery location and the traveling time between points based on historical data.

Table 6. Modeling Data for the M-VRPTW Base Model

Data Source	Coefficient Data
Delivery requirements from manufacturer	Distance between depots and delivery points (i, j) Demand to the delivery points (q_i) Time windows at the delivery points (et_i, lt_i)
Delivery resources of deliverer	Capacity of vehicles (p_k)
Basic data in the optimization agent	Service time at delivery points (s_i) Traveling time between delivery points (t_{ij})

4.2 Identification of a Target Model

In Step 2, AGENT-OPT2 with a selected base model makes a request for information about the delivery resources. In this example, the M-VRPTW is selected as the base model in Step 1. To identify the target model, AGNET-OPT2 needs to specify the base model, additional objective terms, objective-driven constraints, and target model constraints.

The identification can be conducted by the rules described in the Target Model Identification Rules in Tables 7 to 9. Since we have adopted the primitive model approach, the necessary operators in model modification is simply the ***INSERT*** statement to insert terms to objectives and constraints, and to insert new constraints as shown in Fig. 4. If we have adopted other approaches such as Most Similar Case Approach, we would need the *DELETE* operator as well.

INSERT <Expression> INTO OBJECTIVE
INSERT <Expression > INTO CONSTRAINT

Fig. 4. Operators for Optimization Model Modification in the Primitive Model Approach

The five rules that identify the additional objective terms to be added are illustrated in Table 7. The additional term is selected by the definition of the identified objectives, and more than one rule may be fired.

Table 7. Objective Identification Rules for M-VRPTW

Rule Name	Rule Statements
RULE_M_VRPTW_OBJ_ TRAVELING_DISTANCE	**IF OBJECTIVE IS** *minimizing_traveling_distance* **THEN** **INSERT** *Term (11)* **INTO OBJECTIVE**
RULE_M_VRPTW_OBJ_ TRAVELING_COST	**IF OBJECTIVE IS** *minimizing_traveling_cost* **THEN** **INSERT** *Term(12)* **INTO OBJECTIVE**
RULE_M_VRPTW_OBJ_ PENALTY_COST	**IF OBJECTIVE IS** *minimizing_traveling_cost* **AND** *penalty_cost_for_late_delivery* **THEN** **INSERT** *Term (13)* **INTO OBJECTIVE**
RULE_M_VRPTW_OBJ_ FIXED_COST	**IF OBJECTIVE IS** *minimizing_traveling_cost* **AND** *fixed_cost_for_vehicle_movement* **THEN** **INSERT** *Term (14)* **INTO OBJECTIVE**
RULE_M_VRPTW_OBJ_ VEHICLE_NO	**IF OBJECTIVE IS** *minimizing_number_of_vehicles* **THEN** **INSERT** *Term (15)* **INTO OBJECTIVE**

Three rules in Table 8 derive constraints from the identified objectives. These constraints are necessary to make the added objective semantically meaningful. Two rules identified by the target of traveling time and/or distance can derive the constraints described in Table 9.

Table 8. Objective-Driven Constraints for M-VRPTW

Rule Name	Rule Statements
RULE_M_VRPTW_CON_ PENALTY_COST	**IF OBJECTIVE IS** *minimizing_traveling_cost*

	AND *penalty_cost_for_late_delivery* **THEN** **INSERT** *Eq(16)* **INTO CONSTRAINT** **INSERT** *Eq(17)* **INTO CONSTRAINT**
RULE_M_VRPTW_CON_ FIXED_COST	**IF OBJECTIVE IS** *minimizing_traveling_cost* *AND* *fixed_cost_for_vehicle_movement* **THEN** **INSERT** *Eq(18)* **INTO CONSTRAINT** **INSERT** *Eq(19)* **INTO CONSTRAINT**
RULE_M_VRPTW_CON_ VEHICLE_NO	**IF OBJECTIVE IS** *minimizing_number_of_vehicles* **THEN** **INSERT** *Eq(20)* **INTO CONSTRAINT** **INSERT** *Eq(21)* **INTO CONSTRAINT**

Table 9. Target-Driven Constraints for M-VRPTW

Rule Name	Rule Statements
RULE_M_VRPTW_CON_ TOTAL_TRAVELING_TIME	**IF CONSTRAINT IS** *maximum_total_traveling_time* **THEN** **INSERT** *Eq(22)* **INTO CONSTRAINT**
RULE_M_VRPTW_CON_ TOTAL_TRAVELING_ DISTANCE	**IF CONSTRAINT IS** *maximum_total_traveling_distance* **THEN** **INSERT** *Eq(23)* **INTO CONSTRAINT**

As such, the delivery requirements, delivery resources and policy identify the target model derived by these rules. The derived model is canonically represented in DTD form.

4.3 Canonical Representation of the Target Model

The identified target model is represented in a canonical DTD form to equip the semantics and ability of transforming to any commercial solver. In this study, we have adopted LINGO as the solver of integer programming. To define the complete model, we need to identify the base model, added objective terms, and target constraints along with the relevant data set. This structure is depicted in Fig. 5. In this model, the necessary data set includes the delivery points, vehicles, and constants. Fig. 6 expresses the semantic structure in DTD form, and an illustrative formulation is illustrated in Fig. 15 in the next section.

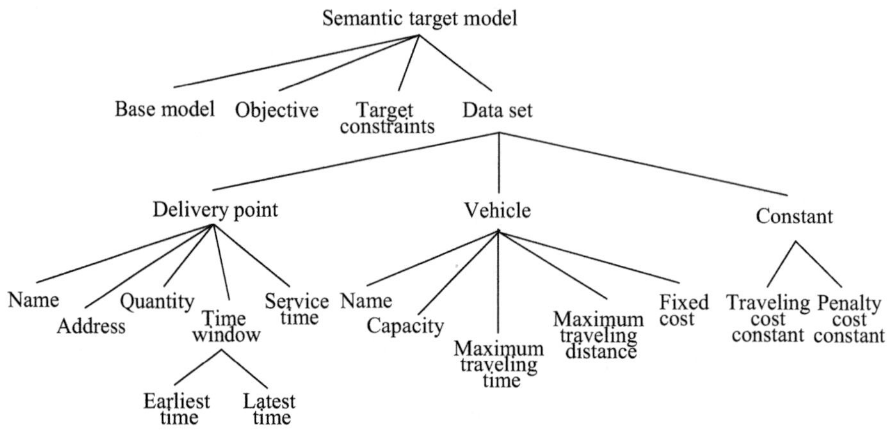

Fig. 5. Semantic Structure of the Target Optimization Model in M-VRPTW

```
<?xml version="1.0" encoding="euc-kr" ?>
<!ELEMENT canonical_target_model (base_model, objective, target_constraints, data_set)>
<!ELEMENT base_model (#PCDATA)>
<!ELEMENT objective (#PCDATA)>
<!ELEMENT target_constraints (constraint+)>
<!ELEMENT constraint (#PCDATA)>
<!ELEMENT data_set (delivery_points, vehicles, constants?)>
<!ELEMENT delivery_points (delivery_point+)>
<!ELEMENT delivery_point (l_name, l_addr, quantity?, unit_weight?,
          time_window?, service_time?)>
<!ATTLIST delivery_point type CDATA #REQUIRED
          seq CDATA #REQUIRED>
<!ELEMENT l_name (#PCDATA)>
<!ELEMENT l_addr (#PCDATA)>
<!ELEMENT quantity (#PCDATA)>
<!ELEMENT unit_weight (#PCDATA)>
<!ATTLIST unit_weight unit CDATA #REQUIRED>
<!ELEMENT time_window (earliest_time, latest_time)>
<!ELEMENT earliest_time (#PCDATA)>
<!ELEMENT latest_time (#PCDATA)>
<!ELEMENT service_time (#PCDATA)>
<!ELEMENT vehicles (vehicle+)>
<!ELEMENT vehicle (v_name, capacity, maximum_traveling_time?,
          maximum_traveling_distance?, fixed_cost?)>
<!ATTLIST vehicle seq CDATA #REQUIRED>
<!ELEMENT v_time (#PCDATA)>
<!ELEMENT capacity (#PCDATA)>
<!ATTLIST capacity unit CDATA #REQUIRED>
<!ELEMENT maximum_traveling_time (#PCDATA)>
<!ATTLIST maximum_traveling_time unit CDATA #REQUIRED>
<!ELEMENT maximum_traveling_distance (#PCDATA)>
<!ATTLIST maximum_traveling_distance unit CDATA #REQUIRED>
```

```
<!ELEMENT fixed_cost (#PCDATA)>
<!ATTLIST fixed_cost unit CDATA #REQUIRED>
<!ELEMENT constants (traveling_cost_constant, penalty_cost_constant)>
<!ELEMENT traveling_cost_constant (#PCDATA)>
<!ELEMENT penalty_cost_constant (#PCDATA)>
```

Fig. 6. DTD Form of the Canonical Optimization Model for M-VRPTW

4.4 Formulation of the Target Model for the IP Solver

The modeling factors described in Tables 1 to 4 corresponds to the terms in the objective function and equations in the constraints. This representation is effective if the optimization models can be managed as a unit of objective terms and constraint equations. Fortunately, a popular IP solver LINGO effectively handles the optimization models in this manner, and data can be managed separately. So we adopt LINGO as executing solver.

Table 10 illustrates the objective function of (11) and constraints of (23) in the form of LINGO [29]. The formulation in DTD form can be transformed to the LINGO form, and solved by the solver. An illustrative formation is illustrated in the Fig. 16 and 17 in the next section.

Table 10. Illustrative Model Components in the Form of LINGO

Model Component	LINGO Formats
Minimizing traveling distance in (11)	MIN = @SUM(LINKS(i,j,k): D(i,j) * X(i,j,k))
Maximum traveling distance in (23)	@FOR(VEHICLE(k): @SUM(LINK2(i, j): D(i,j) * X(i,j,k)) <= G(k))

5 Illustrative Automatic Modeling Procedure with AGENT-OPT2

Let us illustrate the modeling procedure with AGENT-OPT2 for a supply. In this example, a manufacturer needs to deliver its products from one depot to six delivery points. Let us see the procedure step by step.

5.1 Identification of a Base Model

First, the manufacturer sees a question through AGENT-OPT2 so that he selects an answer in the SCM model domain. Suppose the manufacturer chooses '*Transportation*' in Fig 7. In the transportation domain, there are several optimization model types. AGENT-OPT2 asks several questions to the manufacturer to find which optimization model type is suitable as a base model. Suppose the manufacturer answers the next two questions in Fig 8 and 9 with 'yes'. These questions are generated by backward chain reasoning against Base Model Identification Rules as specified in Table 5. From this reasoning, the M-VRPTW model is selected as the base model as depicted in Fig 10.

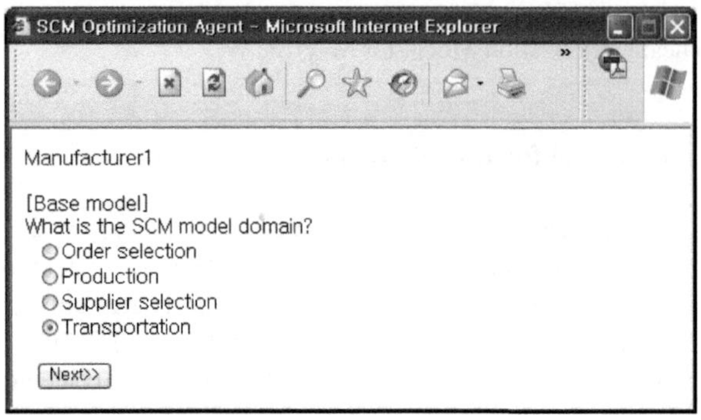

Fig. 7. Query 1 to Identify a Base Model

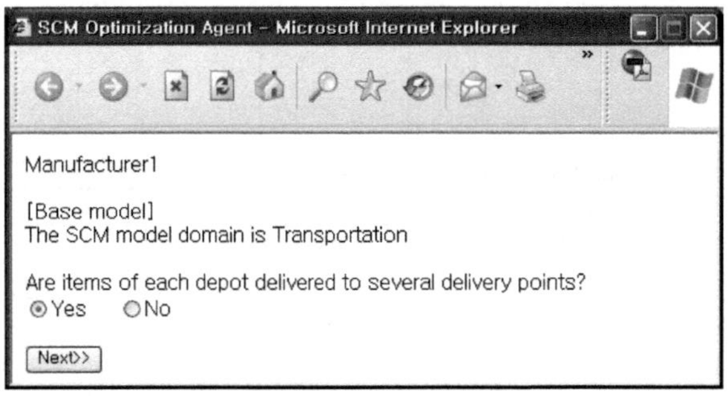

Fig. 8. Query 2 to Identify a Base Model

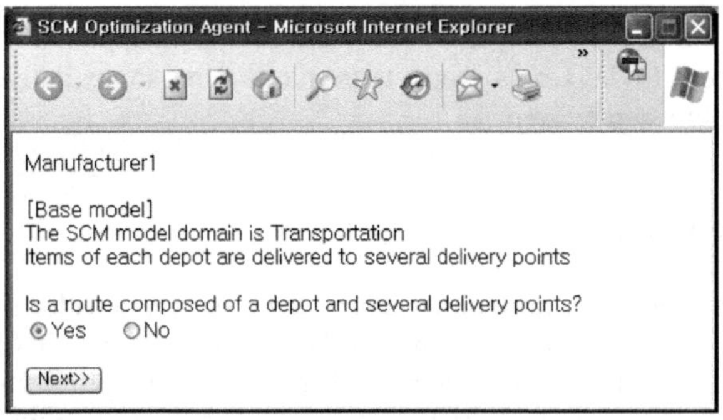

Fig. 9. Query 3 to Identify a Base Model

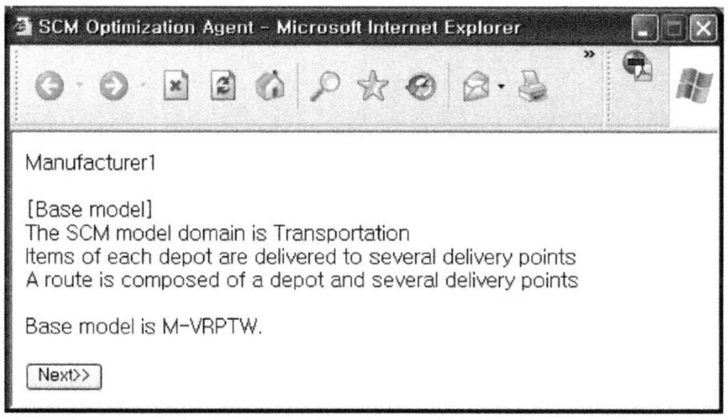

Fig. 10. An Identified Base Model

5.2 Identification of a Target Model

The formulation of M-VRPTW requires information about depot and delivery points. So the manufacturer sees the inquiry through AGENT-OPT2 as illustrated in Fig. 11. The depot includes the information about its name and location, and the delivery point includes the information about the name of the recipient, location, quantity, weight, and required delivery time window. Based on the input, AGENT-OPT2 identifies the delivery requirement summary as in Fig 12.

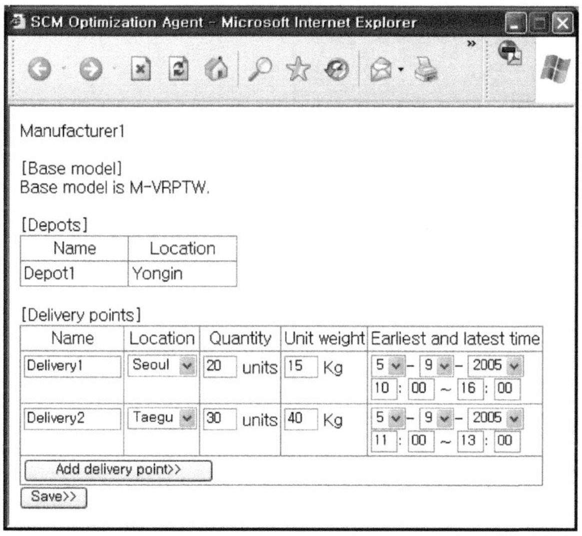

Fig. 11. A Manufacturer's Delivery Request on Depot and Delivery Points

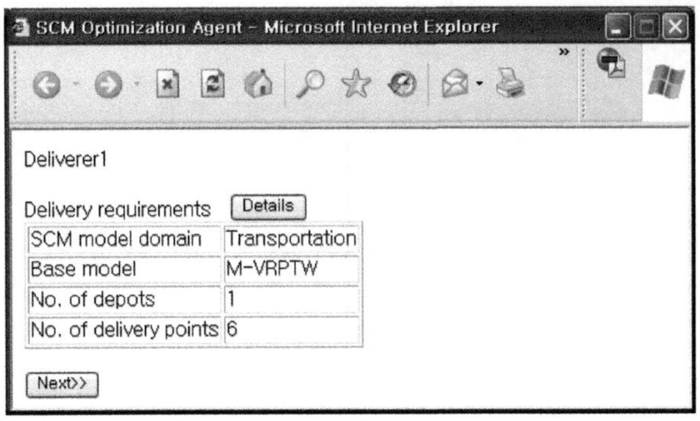

Fig. 12. Delivery Requirement Summary for a Manufacturer

AGENT-OPT2 continues to show the three candidate objective terms in the M-VRPTW model as in Fig. 13. In this example, the third party deliverer selects the objective of minimizing traveling distance, and the target model constraints of maximum traveling distance for the vehicle to be 600 km.

Fig. 14 shows the necessary data on the vehicles. It describes the number of vehicles and vehicle capacity. The other data such as fixed cost, traveling cost, and penalty cost are not included because the model components related to them were not chosen.

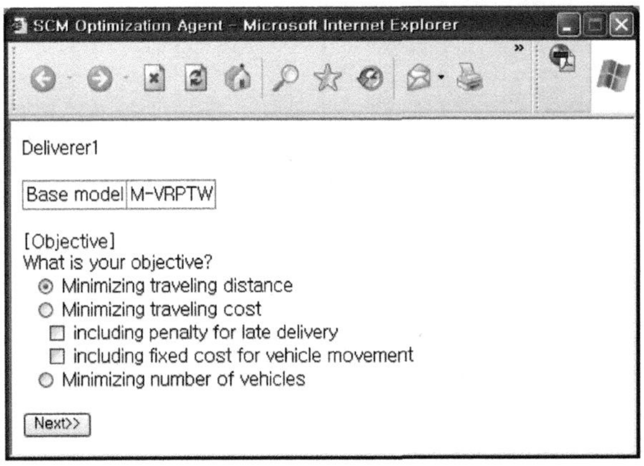

Fig. 13. Query on the Objectives of the Target Model from the Deliverer

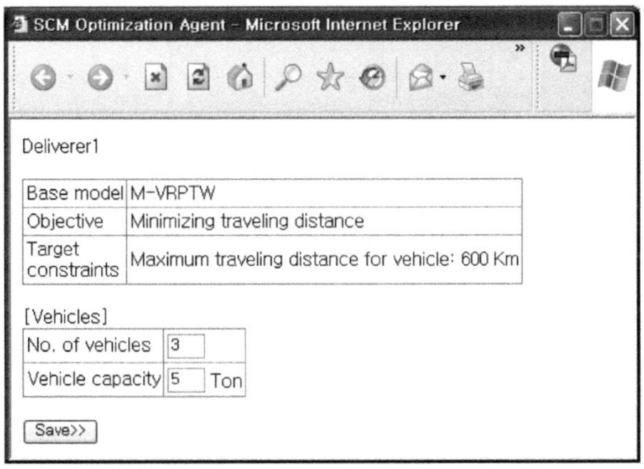

Fig. 14. Illustrative Data Set on Vehicles

5.3 Canonical Representation of the Target Model

After receiving delivery resources from the deliverer, the optimization agent identifies a target model in the canonical representation according to the DTD schema in Fig. 6. Fig. 15 shows an example target model for M-VRPTW that has a depot and six delivery points for a manufacturer with three vehicles of a deliverer.

According to the semantic representation in XML format, the example model consists of the base model M-VRPTW, an objective of *Minimizing_traveling_distance,* and a constraint of *Maximum_traveling_distance_for_vehicle* along with the coefficient data on delivery points and vehicles.

```
- <canonical_target_model>
    <base_model>M-VRPTW</base_model>
    <objective>Minimizing_traveling_diatance</objective>
  - <target_constraints>
      <constraint>Maximum_traveling_distance_for_vehicle</constraint>
    </target_constraints>
  - <data_set>
    - <delivery_points>
      - <delivery_point type="delivery_point" seq="1">
          <l_name>Delivery1</l_name>
          <l_addr>Seoul</l_addr>
          <quantity>20</quantity>
          <unit_weight unit="Kg">15</unit_weight>
        - <time_window>
            <earliest_time>10:00</earliest_time>
            <latest_time>16:00</latest_time>
          </time_window>
          <service_time>0.5</service_time>
        </delivery_point>
      + <delivery_point type="delivery_point" seq="2">
      + <delivery_point type="delivery_point" seq="3">
      + <delivery_point type="delivery_point" seq="4">
      + <delivery_point type="delivery_point" seq="5">
      + <delivery_point type="delivery_point" seq="6">
      - <delivery_point type="depot" seq="7">
          <l_name>Depot1</l_name>
          <l_addr>Yongin</l_addr>
        </delivery_point>
      </delivery_points>
    - <vehicles>
      - <vehicle seq="1">
          <v_name>Truck-01</v_name>
          <capacity unit="ton">5</capacity>
          <maximum_traveling_distance
            unit="Km">600</maximum_traveling_distance>
        </vehicle>
      </vehicles>
    + <vehicle seq="2">
    + <vehicle seq="3">
    </data_set>
  </canonical_target_model>
```

Fig. 15. A Canonical Target Model of M-VRPTW in Semantic XML Tags

5.4 Formulating a Target Model for Solver LINGO

The canonical target model M-VRPTW in Fig. 15 needs to be transformed to a specific IP solver. In this study, we use LINGO version 8.0, which is one of the most popular IP solvers. In LINGO, the model structure and data can be separately specified. The model structure in Fig. 16 describes a base model template, objective, and target model constraints. In this example, the base model template of M-VRPTW defines a set of variables and constraints, and a set of data. As specified in the canonical model, the model specifies that the objective of M-VRPTW is the *minimization of traveling distance*, and the target model constraint is the *maximum traveling distance.*

Fig. 17 shows the data to use in the model. It gives the information on indices of delivery points and vehicles, the number of depots, the number of delivery points, maximum traveling distance, service time, demand, time windows for earliest and latest time, capacity of the vehicle, distance, and traveling time.

Finally, the LINGO solves the target model, and returns to AGENT-OPT2. Then AGENT-OPT2 constructs the messages necessary for the relevant manufacturer and deliverer in XML format. The solver component may be designed to be either as a part of the optimization agent or as an external unit.

Fig. 16. Target Model in LINGO Format

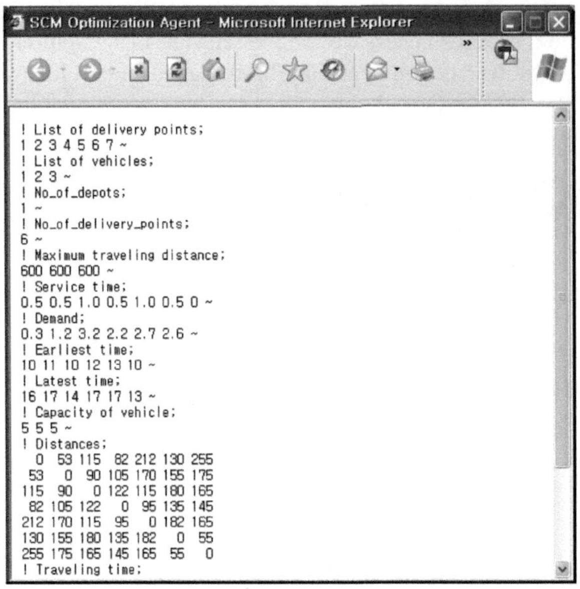

Fig. 17. An Illustrative Data Set in LINGO Format

6 Conclusion: Toward Ontology for Supply Chain Model Warehouse Services

We have seen how the primitive model approach along with modified objectives and constraints can easily formulate a set of optimization models for SCM problems. This study proposed a rule-based model modification scheme based on backward chain and forward chain reasoning that can formulate optimization models automatically. We showed the viability of this approach through the prototype AGENT-OPT2, and illustrated the framework using delivery scheduling problems.

We have confirmed that the proposed framework can be used for the management of multiple models in this manner because the modeling components can be shared by multiple instance models. In this sense, this architecture can be adopted for use in public model warehouses [5] which are equipped with external solvers. In the Web Services environment, optimization model management service can be provided to multiple and diverse customers, such as the third party deliverers and virtual manufacturing schedulers. This implies that the framework can be effectively scaled up to diverse supply chain modeling services.

The application of AGENT-OPT2 developed in this spirit is not limited to SCM alone, but the rules of adding objectives and constraints need an intensive analysis in terms of the application domain. In this sense, the supply chain is the most typical application with such analyzable modeling knowledge. Let us call such knowledge

Modeling Ontology. In order to more comprehensively support SCM automatic modeling, the modeling ontology study needs detailed scrutiny. This implies huge research potential for supply chain modeling experts and information system designers.

If a customer repetitively uses a formulated model with minor modifications, the formulated model can be regarded as the base model of next step. Then the adoption of the *primitive model approach* can evolve into the *most similar case approach* reducing the effort of modification.

References

1. Binbasioglu, M.: Process-based Reconstructive Approach to Model Building, Decision Support Systems 12 (2) (1994) 97-113
2. Binbasioglu, M., Jarke, M.: Domain Specific DSS Tools for Knowledge-based Model Building, Decision Support Systems 2 (3) (1986) 213-223
3. Blanning, R.W.: A Relational Framework for Model Management in Decision Support Systems, DSS-82 Transactions (1982) 16-28
4. Blanning, R.W.: An Entity-relationship Approach to Model Management, Decision Support Systems 2 (1) (1986) 65-72
5. Bolloju, N., Khalifa, M., Turban, E.: Integrating Knowledge Management into Enterprise Environments for the Next Generation Decision Support, Decision Support Systems 33 (2) (2002) 163-176.
6. Bonczek, R., Holsapple, C., Whinston, A.: Foundations of Decision Support Systems, Academic Press, New York, (1981)
7. Chang, Y.S., Lee, J.K.: Case-based Modification for Optimization Agents: AGENT-OPT, Decision Support Systems 36 (2004) 355-370
8. Dolk, D.R., Konsynski, B.R.: Knowledge Representation for Model Management Systems, IEEE Transactions on Software Engineering SE-10 (6) (1984) 619-628
9. Dutta, A., Basu, A.: An Artificial Intelligence Approach to Model Management in Decision Support Systems, IEEE Computer 17 (9) (1984) 89-97
10. Elam, J.J., Henderson, J.C., Miller, L.W.: Model Management Systems: an Approach to Decision Support in Complex Organizations, Proceedings of the First International Conference on Information System (1980) 98-110
11. Geoffrion, A.M.: Introduction to Structured Modeling, Management Science 33 (5) (1987) 547-588
12. Geoffrion, A.M.: The Formal Aspects of Structured Modeling, Operations Research 37 (1) (1989) 30-51
13. Huh, S.Y.: Model-base Construction with Object-oriented Constructs, Decision Sciences 24 (2) (1993) 409-434
14. Ishikawa, T., Terano, T.: Analogy by Abstraction: Case Retrieval and Adaptation for Inventive Design Expert Systems, Expert Systems with Applications 10 (3/4) (1996) 351-356
15. Kedar-Cabelli, S.T.: Toward a Computational Model of Purpose-directed Analogy, Kaufmann, Analogica, CA, (1988)
16. Konsynski, B., Dolk, D.R.: Knowledge Abstractions in Model Management, DSS-82 Transactions (1982) 187-202
17. Krishnan, R.: A Logic Modeling Language for Automated Model Construction, Decision

Support Systems 6 (3) (1990) 123-152
18. Kulkarni, R.V., Bhave, P.R.: Integer Programming Formulations of Vehicle Routing Problems, European Journal of Operational Research 20 (1985) 58-67
19. Le Claire, B., Sharda, R.: An Object-oriented Architecture for Decision Support Systems, 1990 ISDSS Conference Proceedings (1990) 567-586
20. Lee, J.S.: A Model base for Identifying Mathematical Programming Structures, Decision Support Systems 7 (2) (1991) 99-105
21. Lee, J.S., Jones, C.V., Guignard, M.: MAPNOS: Mathematical Programming Formulation Normalization System, Expert Systems with Applications 1 (1990) 367-381
22. Lee, J.K., Kim, M.Y.: Knowledge-assisted Optimization Model Formulation: UNIK-OPT, Decision Support Systems 13 (2) (1995) 111-132
23. Lee, K.J., Chang, Y.S., Lee, J.K.: Time-bound Negotiation Framework for Electronic Commerce Agents, Decision Support Systems 28 (4) (2000) 319 - 331
24. Lenard, M.L.: An Object-oriented Approach to Model Management, Decision Support Systems 9 (1) (1993) 67-73
25. Liang, T.P.: Analogical Reasoning and Case-based Learning in Model Management Systems, Decision Support Systems 10 (2) (1993) 137-160
26. Liang, T.P.: Integrating Model Management with Data Management in Decision Support Systems, Decision Support Systems 1 (1) (1985) 221-232
27. Liang, T.P.: Modeling by Analogy: A Case-based Approach to Automated Linear Program Formulation, IEEE (1991) 276-283
28. Liang, T.P., Konsynski, B.R.: Modeling by Analogy: Use of Analogical Reasoning in Model Management Systems, Decision Support Systems 9 (1) 1993 113-125
29. Lingo, Lindo Systems Inc. Chicago, IL, (2003)
30. Muhanna, W.A.: An Object-oriented Framework for Model Management and DSS Environment, 1990 ISDSS Conference Proceedings (1990) 553-565
31. Park, S.J., Kim, H.D.: Constraint-based Metaview Approach for Modeling Environment Generation, Decision Support Systems 9 (4) (1993) 325-348
32. Tsai, Y.: Model Integration using SML, Decision Support Systems 22 (4) (1998) 355-377
33. Vellore, R.C., Sen, A., Vinze, A.S.: A Case–based Planning Approach to Model Formulation, 1990 ISDSS Conference Proceedings (1990) 353-382
34. Winston, W.L.: Operations Research, 3rd ed., Duxbury Press, Belmont, CA, (1994)
35. Yeom, K., Lee, J.K.: Logical Representation of Integer Programming Models, Decision Support Systems 18 (3-4) (1996) 227-251

Multi-Agent Modeling and Fuzzy Task Assignment for Real-Time Operation in a Supply Chain

Umesh Deshpande [1], Arobinda Gupta [2], and Anupam Basu [3]

[1] Dept. of Electronics and Computer Science,
Visvesvaraya National Institute of Technology, Nagpur, India
uad@vnitnagpur.ac.in

[2] Dept. of Computer Science and Engineering
and School of Information Technology,
Indian Institute of Technology, Kharagpur, India
agupta@cse.iitkgp.ernet.in

[3] Dept. of Computer Science and Engineering,
Indian Institute of Technology, Kharagpur, India
<anupam>@cse.iitkgp.ernet.in

Abstract. In many situations, real-time operation is required in a supply chain, where it is desirable to re-optimize the schedules each time a new order is submitted. Usually, multiple objectives (or criteria) and different preference ratings for the criteria are associated with a task submitted to a supply chain. In a practical scenario, it is more convenient to specify the preference ratings using ordinal values or linguistic terms. The assignment of the tasks from the manufacturers to the suppliers for real-time operation in a supply chain is an important multiobjective decision making problem. In this chapter, a supply chain is first modeled using a multiagent framework and an architecture for the agents located at every organization in the chain is presented. We propose a real-time scheduler that can schedule new orders with soft real-time deadlines. An algorithm, based on the fuzzy set approach, for task assignment is then presented. We have implemented a simulated environment for a supply chain. In that, the algorithm is compared with other commonly used heuristic methods for task assignment. Experimental results show significant performance gains of our scheme over other schemes.

1 Introduction

A supply chain can be defined as a network of autonomous or semi-autonomous business entities collectively responsible for procurement, manufacturing, and distribution activities associated with one or more families of related products [15]. Coordinating the activities of the various entities in a supply chain is important for improving performance in terms of multiple objectives such as on-time delivery, quality assurance, and cost minimization. A typical supply

U. Deshpande et al.: Multi-Agent Modeling and Fuzzy Task Assignment for Real-Time Operation in a Supply Chain, Studies in Computational Intelligence (SCI) 28, 179–202 (2006)
www.springerlink.com

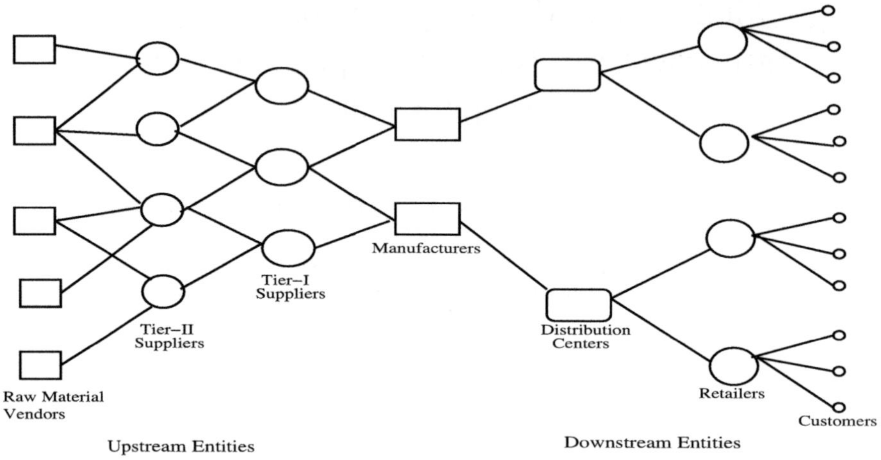

Fig. 1. A Supply Chain Network

chain is shown in Figure 1. It shows the manufacturers as the focal organizations, the upstream entities as the suppliers, and the downstream entities as the consumers. As is evident, a supply chain is a distributed system.

Usually, there are three phases of decision making in supply chain management depending on the frequency of each decision and the time-frame over which a decision phase has an impact [4]. These are the supply chain strategy phase, the planning phase, and the operation phase. In supply chain operation, the time horizon is weekly or daily, and companies make decisions regarding individual customer orders. Here, firms allocate individual orders to inventory or production, set a date by which an order has to be filled, allocate an order to a particular shipping mode, set delivery schedules of trucks etc. For many situations, it is desirable to re-optimize the schedules each time a new order is being considered [13]. This would enable details for scheduling the new order to be efficiently integrated with all other unscheduled orders. A new order may have a deadline associated with it and hence the whole operation has to be executed with soft real-time constraints. This is referred to as real-time operation in a supply chain.

Multiagent systems consist of multiple autonomous or semi-autonomous agents where knowledge and control is distributed and agents communicate through messages [3]. Since supply chain operation is concerned with coordination among multiple decision makers, a multiagent framework for modeling it is a natural choice. In the current scenario, the consumer demand is ever changing and there are fluctuations in resource costs and availability. Hence companies must respond rapidly to maintain consumer satisfaction and profits. As these changes are occurring at increasing speeds, scales, and complexity unmanageable by humans, the need for automated techniques becomes acute [17].

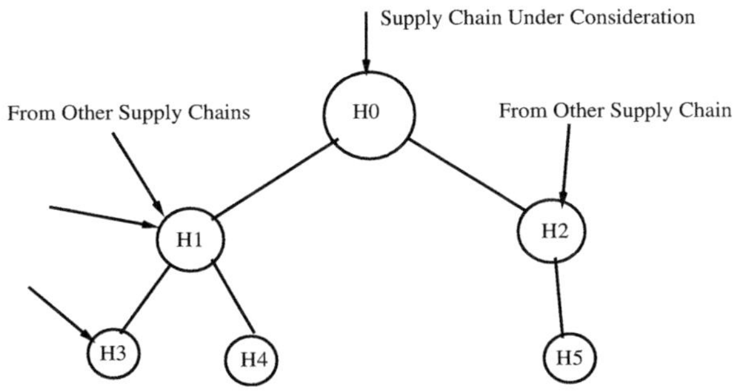

Fig. 2. A Part of a Supply Chain

Let us consider a part of a supply chain with a focal organization H_0 and its upstream entities (the suppliers) as shown in Figure 2. H_0 may not have all the required resources to completely perform all the subtasks of a task submitted to it. Those subtasks, which cannot be performed at H_0, have to be assigned to its suppliers quickly in real-time operational environments. Since the suppliers are autonomous entities, they may belong to more than one supply chain. Hence H_0 does not have a complete control of the scheduling activities of its suppliers. At the same time, more than one supplier may have the resources to perform a subtask. H_0 has to make a decision to choose one of the suppliers to which a subtask needs to be delegated. A similar task assignment decision has to be done by all the suppliers, in turn, for the tasks delegated to them.

Usually there are multiple objectives (or criteria) like minimizing cost and time, maximizing quality, etc. associated with a task. Different preference ratings may be assigned to each of the criteria. These preference ratings, which are the relative importances for each of the criteria, may be assigned by the focal organization, downstream entities, or the end-users themselves. The problem of the assignment of a task to a *node* (we will henceforth use the common term *node* for an organization, a company, or a firm in a supply chain) is important since a good assignment would make it feasible to satisfy the criteria. Task assignment in a supply chain is a multiobjective decision making problem.

In practice, it is very convenient for the user to express the preference rating of one criterion as a comparison with the rating for another criterion using a *linguistic term* or a *subjective ordinal value.* For example, for a patient in a hospital, the quality may have a *strong* importance over cost but the timely response may have *absolute* importance over the other two. Such a procedure of comparing two elements at a time using ordinal values is considerably simpler than trying to work with all the objectives at once, specially if the number of objectives is quite large.

Since we are considering a dynamic distributed system, a node does not have a complete view of the entire chain and hence does not have the global system state. Hence the global state has to be collected before the task assignment.

Since the utilization of the resources and hence the schedules at various nodes change continuously, it is not possible to get a precise system state. The task assignment decision should be done allowing for tolerance for the imprecision both in the global state collection and the preference rating specification.

In this chapter, a model for the supply chain as a multiagent system is presented first with agents located at each of the entities in the supply chain. We design a real-time scheduler for the agents that would integrate schedules of new orders having soft real-time deadlines with the already existing schedules. To deal with the imprecision mentioned, a fuzzy set approach [19] is used in the context of multiobjective decision making for task assignment. We adapt the multiobjective decision process using fuzzy sets for the task assignment problem in a supply chain management system. The proposed technique is evaluated against other commonly used heuristic approaches for task assignment. It is seen that for a large range of task arrival rates the fuzzy technique has superior performance. However, its performance deteriorates for very high arrival rates.

To overcome this limitation, we next design a hybrid algorithm which augments the multiobjective decision process using fuzzy sets. This approach does not suffer from performance degradation even at very high task arrival rates. The hybrid algorithm is compared with other techniques and it is found that it outperforms all of them for all arrival rates.

The chapter is organized as follows. In Section 2, we review the related works. The supply chain is modeled using a multiagent framework in Section 3. The real-time scheduler used in the agents is explained in Section 4. In Section 5, the task assignment in a supply chain is formulated as a multiobjective decision making problem. Simulation studies are presented in Section 6. In Section 7 the hybrid algorithm is proposed. Section 8 concludes the chapter.

2 Related Works

Modeling of the supply chain using a multiagent framework has been done in the past. In [11], a reconfigurable, multi-level, agent based planning and scheduling architecture called MASCOT (Multi-Agent Supply Chain cOordination Tool) has been designed. It is aimed at supporting dynamic, effective, and efficient coordination among supply chain partners for various activities like production, transportations, and other strategic decisions. It discusses new coordination protocols for better exploiting the power of finite capacity scheduling functionalities across the supply chain. Fox et.al. [9] explain the construction of a supply chain using an agent-oriented software architecture. The approach uses an agent building shell which provides generic components for communication, coordination, role-based organization etc. Using these components, the technique explores the support of complex cooperative work and management of perturbation caused by stochastic events. In [15], a modeling framework using multiagent systems is proposed for simulating supply chain environments. Using this approach, supply chain models are composed from software components that represent types of supply chain agents (like retailers, manufacturers etc.), their constituent control elements (like inventory policy), and their interaction protocols (like message types). It provides a reusable base of domain specific primitives that enables rapid development of customized decision support tools. In [17], the problem of automating supply chain formation is addressed within an economic framework. Price systems are used for guiding the decisions of the self-interested agents in a supply chain. Different market protocols are proposed for agents to negotiate the allocation of resources to form supply chains.

In this chapter, we handle the scheduling and the task assignment issues for real-time operation while taking into account multiple objectives and preference ratings associated with tasks. We model the supply chain as a multiagent system where each node has a hierarchical structure. The agents coordinate within and across organizations using commitments [16]. The agents at the different nodes cooperate with one another by sharing some of their local state.

3 Agent based System Model of the Supply Chain

A supply chain can be represented as a directed acyclic graph. Let us define a set $UpS(H)$ that includes all those nodes which are in the immediate upstream (i.e. immediate suppliers) of a node H. Note that the set $UpS(H)$ may be empty and that the sets $UpS(H_x)$ and $UpS(H_y)$ for two nodes H_x and H_y need not be disjoint. Let H_0 be the focal node under consideration. Let $UpS(H_0) = \{H_1, H_2, \ldots, H_n\}$. The nodes communicate only through messages and it is assumed that the delay for exchanging messages between H_0 and H_i is a constant Ψ_i^0. For a node $H_{ij} \in UpS(H_i)$, the communication delay between H_i and H_{ij} is Ψ_j^i. The supply chain can be modeled using a multiagent framework in which each of the nodes in the supply chain is itself a multiagent system. The architecture of a node is explained below. If a subtask T' of a task T, submitted at H_0, requires a resource r_k that is not present locally, then T' has to be migrated to one of the nodes in $UpS(H_0)$ where r_k is present.

3.1 The Architecture of a Node

When a node H does not have all possible resources, it has to coordinate with the nodes in $UpS(H)$. We identify a Liaison Unit (LU) present in every node that is responsible for this coordination. Figure 3 shows the architecture of a typical node. It is a hierarchical organization with the managerial units (MU) at the top, the departments (D) and sub-departments (SD) at the middle, and the functional units (FU) at the bottom.

In the architecture, a task enters any of the managerial units. We assume that every MU, D, SD, FU, and the LU has an agent present in it. A functional unit FU_i controls the resource R_i. We assume that the duration of usage of a resource R, which is the amount of time a task needs to use R, is constant.

A task T can be described using a task graph as shown in Figure 4. The task graph specification is similar to that of the TAEMS (Task Analysis Environment Modeling and Simulation) language [6]. In this figure, a task T consists of three

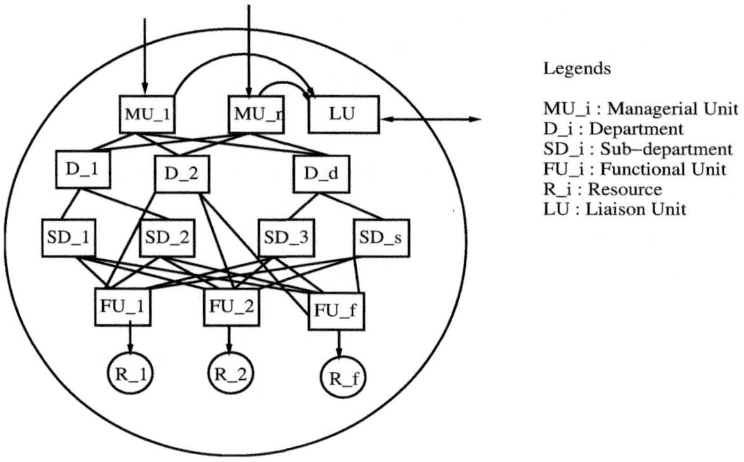

Fig. 3. The Architecture of a Node

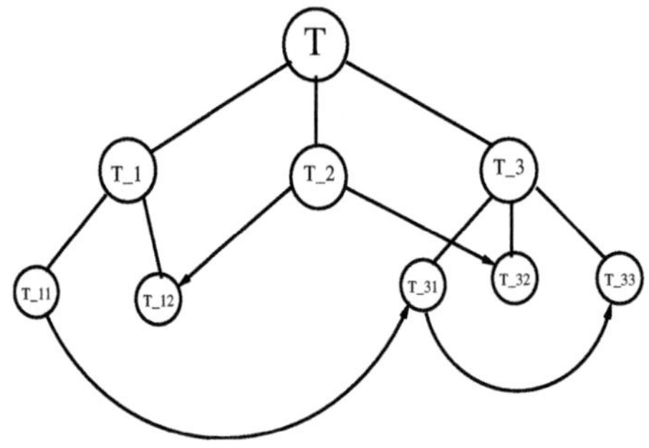

Fig. 4. The Task Graph of a task T

subtasks T_1, T_2, and T_3. An arrow between a subtask T_{11} and T_{31} indicates the *enables* relationship with T_{11} enabling T_{31}. Hence the execution of T_{31} should be started only after T_{11} has finished execution. An appropriate deadline may be specified along with the task T. We assume that each subtask requires a distinct resource and it is the responsibility of the managerial units to assign it to the correct functional unit. If a MU recognizes that a subtask T_i of T needs a resource that is not present locally, it sends the request to the Liaison unit. The Liaison unit decides the appropriate remote node where T_i should be migrated.

The modeling of the individual agents at the MU, D, and SD are similar to those defined in [7]. In the next subsection, we describe the model of the functional unit (FU) followed by the description of the data structures of the LU in subsection 3.3. These data structures constitute the state of a node.

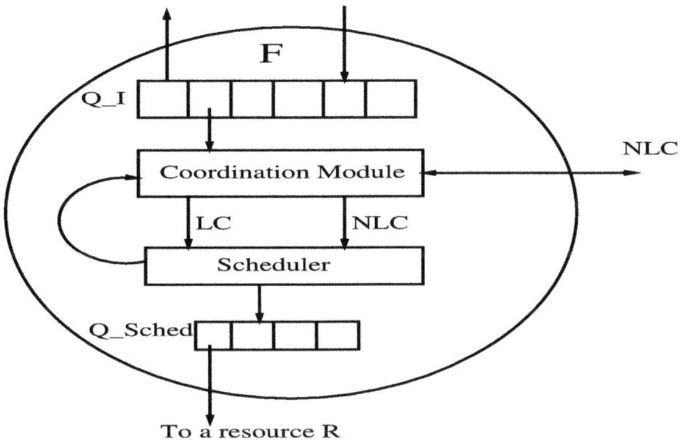

Fig. 5. The Model of an Agent F at a FU

3.2 Modeling of the Functional Unit (FU)

An agent F at a FU controlling a resource R present at a node H is modeled as shown in Figure 5. It consists of an input queue Q_I where the tasks input to F are kept. The coordination module is responsible for the coordination of the activities of F. The coordination module and the scheduler work together to prepare schedules for the input tasks. The schedule queue Q_{Sched} contains the dispatch times of the tasks. The agent F and the agents at other FUs coordinate with one another using commitments. The local commitments (LC) for F are those which the coordination module of F generates to constrain the local scheduler of F. The nonlocal commitments (NLC) for F are those which the coordination module of F exchanges with that of some other agent. Assume that a subtask T submitted at F is at the source end of an *enables* relationship with another subtask T' submitted at the agent F'. F gives an *earliest start time* commitment to the agent F' for the task T' after F has computed the schedule for T. This is the enabling time for T' at F'. The details of the commitments and the coordination module are explained in [8]. The scheduler is explained in Section 4.

3.3 The Local State

As mentioned earlier, the Liaison Unit at a node H is responsible for coordination with other nodes in $UpS(H)$ if a resource is not available at H. The LU at a node H maintains the following two tables which constitute the local state at every node.

1. **Local Resource Table (LRT)** - Each entry of the table is discussed below.

r_l	Duration	Surplus time

   - r_l is the name of the resource present locally.
   - *Duration* is the time required for the usage of r_l.
   - *Surplus time* is the ratio of the time r_l was not used in a past window of length WL_l. The window length is chosen appropriately depending on the duration for the usage of r_l. Surplus of 0 indicates that r_l was fully occupied in the past window and surplus of 1 indicates that r_l was always free in the past window. Surplus time is periodically updated and the period is directly proportional to the duration.
2. **Remote Resource Table (RRT)**: This table keeps the information about the resources present at the nodes in $UpS(H)$. The fields of the table are described below.

r_k	Id	Duration	Surplus time

   - *Id* is the identifier of the node where a resource with name r_k is present.
   - *Duration* is the time for the usage of r_k at that node.
   - *Surplus time* is the one that was most recently obtained from the upstream node. The surplus values are exchanged between a node and its upstream entities periodically and also are piggybacked over messages during message exchanges.
   For each resource, the entries are ordered in the decreasing order of the surplus. If the surplus is the same for two resources, then the ordering is done according to the duration (the less duration one comes earlier).

4 The Real-Time Scheduler at each FU

This section presents the scheduler at an agent F of a FU which controls a resource R having duration Δ. The scheduler is invoked periodically with the period $\delta * \Delta$ (δ is a constant). At each invocation, the scheduler first finds out all the enabled tasks from the input queue Q_I. The enabled tasks are those for which all incoming enabling commitments are known. The enabling time for a task is the maximum of all the incoming enabling commitments. The *Estimated Latest Start Time (ELST)* (explained in section 4.1) of all the enabled tasks is computed. The tasks are rated in proportion to the earliness of their $ELST$, i.e. a task having the least $ELST$ value is given the highest rating and a task with the highest $ELST$ value is given the lowest rating. The scheduler prepares a set of tasks $Q_{Enabled}$ in which tasks are kept in the decreasing order of their ratings. The following is the scheduler algorithm.

```
For each task T in Q_Enabled do
begin
      sched-time = enabling time of T;
      best-time = get-bestime(sched-time, ELST(T));
      Schedule T from best-time. Make an entry in Q_Sched.
      Give the earliest start time of best-time + Δ to all
      the agents at the other end of the outgoing enables
      relationships.
end.

get-bestime(sched-time, ELST(T))
begin
      The interval of scheduling for T is [sched-time, ELST(T)].

      Let Overlap = {T_1, T_2, ..., T_n} be a set of tasks from Q_Sched
      that overlap with T. The tasks in Overlap are ordered
      in the non-decreasing order of their dispatch times.
      Let t_k be the dispatch time of the task T_k ∀k : 1..n.
      A task T_k belongs to Overlap
        if (((t_k ≥ sched-time) and (t_k ≤ ELST(T)))
        or ((t_k + Δ ≥ sched-time) and (t_k + Δ ≤ ELST(T)))).

      If Overlap = φ then best-time = sched-time.
      else if t_1 > sched-time + Δ then best-time = sched-time.
      else find a task T_i(i > 0) such that (t_{i+1} - t_i ≥ 2 * Δ)
      and (t_i + Δ < ELST(T)) and (t_i + Δ > sched-time)
      and T_i is the first such task in Overlap.
      The best-time for T is t_i + Δ.

      If no such task is found, then the above step is repeated
      for finding the best-time among all those tasks in Q_Sched
      whose dispatch time is greater than t_n (t_n is the dispatch
      time of the last task T_n of Overlap).

      If no such task is found even then, best-time is set to
      t_last + Δ where t_last is the dispatch time of the last task
      T_last in Q_Sched.
end.
```

Example

Let Δ be 20 and at a point of time $Q_{Sched} = \{(T_1, 20), (T_2, 70), (T_3, 100), (T_4, 140), (T_5, 220)\}$. Assume that a subtask T_6 arrives with an ELST of 160, enabled time of 90, and deadline of 200. The interval of T_6 is $[90, 160]$ and hence the set $Overlap = \{T_2, T_3, T_4\}$. There is not enough time for T_6 to execute after T_3 finishes and before T_4 starts but there is just enough time to place T_6 after T_4. Hence the *best-time* $= 120$. The new $Q_{Sched} = \{(T_1, 20), (T_2, 70), (T_3, 100),$

$(T_6, 120), (T_4, 140), (T_5, 220)\}$.

Now let us assume that a subtask T_7 arrives with an ELST of 150, enabled time of 100, and deadline of 240. The interval of T_7 is $[100, 200]$ and hence the set $Overlap = \{T_3, T_6, T_4\}$. T_7 can be placed only after T_4, hence the new interval for T_7 is $[160, 240]$, $Overlap = \{T_5\}$, and $best\text{-}time = 160$.

4.1 Estimated Latest Start Time (ELST) Computation

$ELST$ of a subtask T' of a task T is an approximate measure of the time by which T' should start execution at the agent F so that T does not miss its deadline. For an exact computation of the $ELST$, all the possible paths of the *enables* relationships in the task graph of T have to be explored. Hence, the exact computation is intractable. An approximation of the measure is computed as explained below.

Identify a set of subtasks $E = \{T_{e_1}, T_{e_2}, \ldots, T_{e_e}\}$ present in T such that T_{e_1} is enabled by T' and has the maximum duration among those tasks that T' enables. Similarly each subsequent task $T_{e_{k+1}}$ is enabled by T_{e_k} and $T_{e_{k+1}}$ has the maximum duration among those tasks that T_{e_k} enables. Let $\Delta_e = \Sigma_{i=1}^{e} duration(T_{e_i})$. The Estimated Latest Start Time ($ELST$) of T' is approximated as $D - \Delta_e - \Delta$ where D is the deadline of T.

5 Task Assignment under Imprecision

In this section, we describe how the task assignment in a supply chain, which is a distributed system, can be formulated as a multiobjective decision making problem. In a distributed system, task assignment has to be done when the global system state is not available at any single node. The node where the decision is taken has to collect the global state first. There is imprecision in the state collection itself since we are considering a dynamic system. Since the resource utilization changes continuously, the system state might have changed in real time from the one that was collected.

In the next subsection we explain how multiobjective decision making can be done using fuzzy logic. In Section 5.2, a procedure for obtaining the preference ratings of the objectives is described. In Section 5.3, we adapt this approach for the supply chain.

5.1 Multiobjective Decision Making using Fuzzy Logic

The typical multiobjective decision making problem involves the selection of one alternative, a_i, from a universe of alternatives containing n elements, $A = \{a_1, a_2, \ldots, a_n\}$ given a set of p elements, $O = \{O_1, O_2, \ldots, O_p\}$, of criteria or objectives that are important to the user. We want to evaluate how well each alternative, or choice, satisfies each objective, and we wish to combine the weighted objectives into an overall decision function.

In [2], Bellman and Zadeh suggest that each objective can be represented as a fuzzy set over the set of alternatives $\{A\}$. Thus if O_i indicates the ith objective, then the grade of membership of an alternative a in O_i, denoted by $\mu_{O_i}(a)$, is the degree to which the alternative a satisfies the criteria specified for this objective. It is then pointed out that when one has a collection of objectives to be satisfied, the overall objective is $D = O_1$ *and* $O_2 \ldots$ *and* O_p, i.e., one is seeking solutions that satisfy O_1 *and* O_2 *and* O_3, etc. They then suggest that the appropriate mathematical form for the *and* operation is the *min* operator that was originally described for the intersection of fuzzy sets by Zadeh [19]. Thus the decision function is

$$D = O_1 \cap O_2 \cap \ldots \cap O_p$$

Hence, the grade of membership that the decision function, D, has for each alternative a is given by

$$\mu_D(a) = min[\mu_{O_1}(a), \mu_{O_2}(a), \ldots, \mu_{O_p}(a)]$$

The optimum decision, a^*, will then be the alternative that satisfies

$$\mu_D(a^*) = Max_{a \in A}(\mu_D(a))$$

In this approach, if $\{S\}$ is a set of elements that has the structure of an ordinal scale, then the grades of the fuzzy sets can be chosen from this set and we can work out the optimal. This approach does not demand precise values from the user. However, it does not permit one to distinguish between the importance of the objectives.

In [18], Yager extends this approach to allow the introduction of preference ratings with the objectives. With each objective O_i, a non-negative number, α_i, is associated that indicates the power or importance of the objective in the decision. A higher number indicates that the corresponding objective is more important. Yager suggests that each fuzzy objective should be raised to its power, with the exponent being the preference rating, before combining to form the decision function D. Thus, the sets $O_1^{\alpha_1}, O_2^{\alpha_2}, \ldots, O_p^{\alpha_p}$ are formed where

$$\boldsymbol{\mu}_{O_i^{\alpha_i}}(a) = [\boldsymbol{\mu}_{O_i}(a)]^{\alpha_i}$$

The new decision function, D, is

$$D = O_1^{\alpha_1} \cap O_2^{\alpha_2} \cap \ldots \cap O_p^{\alpha_p}$$

Therefore, the grade of membership that the new decision function, D, has for each alternative a is given by

$$\boldsymbol{\mu}_D(a) = min[\{\boldsymbol{\mu}_{O_1}(a)\}^{\alpha_1}, \{\boldsymbol{\mu}_{O_2}(a)\}^{\alpha_2}, \ldots, \{\boldsymbol{\mu}_{O_p}(a)\}^{\alpha_p}] \tag{1}$$

The optimum decision, a^*, will be the alternative that satisfies

$$\boldsymbol{\mu}_D(a^*) = Max_{a \in A}(\boldsymbol{\mu}_D(a)) \tag{2}$$

Table 1. Pairwise Importance Ratings

Value of r_{ij}	Definition
1	equal importance
3	weak importance of one over other
5	strong importance of one over other
7	demonstrated importance of one over other
9	absolute importance of one over other
2,4,6,8	intermediate values between two adjacent judgments

5.2 A Procedure for Determining the Importance of the Objectives

Saaty has developed a procedure for obtaining a ratio scale of importance for a group of p elements based upon paired comparisons [12]. In this section, this method is described.

There are p objectives and a scale should be constructed rating these objectives as to their importances with respect to the decision. The decision maker (the user) is asked to compare the objectives in paired comparisons. If an objective O_i is being compared with another objective O_j, the values r_{ij} and r_{ji} are assigned as follows.

1. If O_i is more important than O_j, then r_{ij} gets assigned a number according to Table 1. (If O_j is more important than O_i, then r_{ji} is assigned the value according to Table 1.)
2. $r_{ij} = \frac{1}{r_{ji}}$.

Having obtained the above judgments from the user, a $p \times p$ matrix, $\mathcal{R}$, is constructed as follows.

1. $r_{ii} = 1$.
2. For $r_{ij}(i \neq j)$, either r_{ij} or r_{ji} is determined by Table 1 and the other by the relationship $r_{ij} = \frac{1}{r_{ji}}$.

Saaty [12] has shown that an eigenvector corresponding to the maximum eigenvalue, λ_{max}, is a cardinal ratio scale for the elements compared. In this case, the eigenvector whose total is p should be found out, since this makes the average importance 1. Thus

$$W = \begin{bmatrix} \alpha_1 \\ \alpha_2 \\ \vdots \\ \alpha_p \end{bmatrix}$$

where

$$\mathcal{R}W = \lambda_{max} \tag{3}$$

$$\sum \alpha_i = p \tag{4}$$

W is the vector corresponding to the decision maker's subjective assessment of the importance of each objective in the decision.

It is to be noted that the procedure of comparing two elements at a time is considerably simpler than trying to work with all the objectives at once. Furthermore, Saaty has shown that the procedure does not require consistency nor transitivity in the formulation of the matrix $\mathcal{R}$.

5.3 The Computation of the Objectives for the Supply Chain Operation

Suppose a task T is input to the focal node H_0 and one of the subtasks of T, say T', cannot be accomplished at H_0. Assume that for the task assignment, the alternatives, $A = \{H_1, H_2, \ldots, H_n\}$, where $H_1, H_2, \ldots, H_n$ are nodes in $UpS(H_0)$. We have defined four objectives - *Execution Time (ET), Quality (Q), Cost (C),* and *Migration Overhead (MO)* in the supply chain setup. The *Execution Time, Quality,* and *Cost* objectives have the obvious meanings but the *Migration Overhead* is not so obvious. Let us assume that T' has been assigned to H_i. It may not be possible to execute some of the subtasks of T' at H_i due to the lack of resources at H_i. Hence those subtasks have to be migrated to the nodes in $UpS(H_i)$. The *Migration Overhead,* both in terms of time and cost, is the overhead incurred for migration of these subtasks. One can give a preference rating for the overhead depending on the number of migrations that can be tolerated, the cost of the migrations, and the available time.

The objectives $O = \{ET, Q, C, MO\}$ are defined for every node H_i present in $UpS(H_0)$. Along with T', the Estimated Latest Start Time of T', i.e. $ELST(T')$ and the pairwise preference matrix $\mathcal{R}$ is also input. From $\mathcal{R}$, the preference ratings for the individual objectives as specified by the vector $W = \{\alpha_1, \alpha_2, \alpha_3, \alpha_4\}$ are computed. We define the four objectives, $O = \{O_1, O_2, O_3, O_4\} =$ $\{ET, Q, C, MO\}$ using fuzzy sets as follows.

$$ET = \{\frac{h_{11}}{H_1}, \frac{h_{21}}{H_2}, \ldots, \frac{h_{n1}}{H_n}\} \qquad Q = \{\frac{h_{12}}{H_1}, \frac{h_{22}}{H_2}, \ldots, \frac{h_{n2}}{H_n}\}$$

$$C = \{\frac{h_{13}}{H_1}, \frac{h_{23}}{H_2}, \ldots, \frac{h_{n3}}{H_n}\} \qquad MO = \{\frac{h_{14}}{H_1}, \frac{h_{24}}{H_2}, \ldots, \frac{h_{n4}}{H_n}\}$$

In our case, all the values of h_{ij} $\forall i : 1..n$, $\forall j : 1..4$ and b_i $\forall i : 1..4$ are from the unit interval $[0, 1]$. The multiobjective decision process discussed in Section 5.1 is executed and T' is assigned to the node identified by the process. The fuzzy sets of the objectives are computed taking into account the current available state of the system. The value h_{ij} denotes the goodness of a node H_i with respect to the objective O_j for the task T'.

The basic idea of the computation of the fuzzy sets for the objectives is illustrated with an example of the fuzzy set for *Execution Time(ET)*. First, the time required, TR_i, for T' at every node H_i is estimated. We then normalize the estimated values. That is, we find the minimum of the TR_i values, TR_{min}, and divide TR_{min} by the individual TR_i values. With this normalization, the node with the least estimated time required gets a value 1 and all others will get values between $[0, 1]$. A node having a lower estimated time required value will have a higher normalized value.

Second, a representative fuzzy set called *Time Criticality(TC)* is constructed whose universe of discourse is the interval $[0, 1]$. The values of membership function of TC, μ_{TC}, are also in the range $[0, 1]$. This fuzzy set maps the normalized TR values to membership values such that a higher normalized value gets a higher membership value in TC. Then, h_{i1}, which is the membership value of a node H_i in the fuzzy set ET, is computed as follows.

$$h_{i1} = \mu_{TC}(TR_{min}/TR_i)$$

The fuzzy sets for the other objectives are also computed in a similar way. These are explained in the next subsection. Representative fuzzy sets called *Good Quality (GQ)*, *Less Cost (LC)*, and *Less Migration (LM)* are constructed similar to the set *Time Criticality (TC)*. Default fuzzy sets for these are provided at the system initialization time, or, if the user so desires, the representative fuzzy sets may be provided by the user.

It is to be noted that the computation of the h_{ij} values cannot be precise because of the dynamic nature of the system. For example, the time required for T' at a node H_i can only be known after T' has been scheduled completely. We do not know this at the point when the assignment decision has to be made. Also, those subtasks of T' that require a resource not present at H_i would have to be migrated. The destination node of the migration has to be found out by another invocation of the task assignment procedure at H_i. It is also not known when the assignment decision of T' at H_0 has to be made. Hence in all the following computations, the destination node is always assumed to be the one which has the maximum surplus for the required resource.

Computation of the Execution Time Objective The values h_{i1} for all i between 1 and n are computed as follows. First, an estimate of the time required for the execution of T' at H_i, TR_i, is computed. To obtain this, first consider a resource r present at H_i and required by T. The duration of usage of r at H_i is divided by the surplus of r at H_i (found from the Local Resource Table at H_i). This gives an approximate measure of the time required for using r. When the surplus is very low, a bound on this measure is made. Summation Sum_1 of this computation for all such resources is done.

Next, consider a resource required by T' but not present at H_i. In this case, the duration of usage of r at the best node $H_{ib} \in UpS(H_i)$ is divided by the surplus of r at H_{ib}. The node H_{ib} is the one that has the maximum surplus for r among all the nodes in $UpS(H_i)$ (found from the Remote Resource Table at H_i). In addition to this, we also have to consider the time required for migrating subtasks of T' to H_{ib}. Hence the term $2 * \Psi_b^i$ is added. The multiplication factor 2 is used since two messages between H_i and H_{ib} would be exchanged - one for sending the subtask by H_i and the other for sending the finished product by H_{ij}. Summation Sum_2 of this computation for all such resources is done. The estimated time required by T' at H_i, TR_i, is the addition of the two summations Sum_1 and Sum_2.

In the following expression for TR_i, we use the term $\forall r \in T' \wedge r \in H_i$ to denote a resource r that is required by T' and is present at H_i and the term $\forall r \in T' \wedge r \notin H_i$ to denote a resource r that is required by T' but is not present at H_i. We calculate TR_i for all values of i between 1 and n as follows.

$$TR_i = \sum_{\forall r \in T' \wedge r \in H_i} \frac{\Delta_i^r}{S_i^r} + \sum_{\forall r \in T' \wedge r \notin H_i} \frac{\Delta_b^r}{S_b^r} + 2 * \Psi_b^i \tag{5}$$

where Δ_i^r is the duration of usage of r at H_i and Δ_b^r is the duration of usage of r at the best node H_{ib}. Similarly S_i^r is the surplus of r at the node H_i and S_b^r is the surplus of r at the best node H_{ib}. Let $TR_{min} = MIN_i(TR_i)$. Then $h_{i1} = \boldsymbol{\mu}_{TC}(TR_{min}/TR_i)$. If $ELST(T') + TR_q > D$ (where D is the deadline of the task T) for some node H_q, then h_{q1} is assigned the value zero.

Computation of the Quality Objective The values h_{i2} for all i between 1 and n are computed as follows. We first find Q_i for all values of i using the following expression.

$$Q_i = \sum_{\forall r \in T' \wedge r \in H_i} Q_i^r + \sum_{\forall r \in T' \wedge r \notin H_i} Q_b^r \tag{6}$$

where Q_i^r is the quality of r at H_i and Q_b^r is the quality of r at the best node H_{ib}. For resources that are absent we choose the quality of the best node. Let $Q_{max} = MAX_i(Q_i)$. Then $h_{i2} = \boldsymbol{\mu}_{GQ}(Q_i/Q_{max})$. Recall that GQ is the fuzzy set *Good Quality.*

Computation of the Cost Objective The computation of h_{i3} is similar to that of h_{i2}. We first find C_i for all values of i between 1 and n using the following expression.

$$C_i = \sum_{\forall r \in T' \wedge r \in H_i} C_i^r + \sum_{\forall r \in T' \wedge r \notin H_i} C_b^r \tag{7}$$

where C_i^r is the cost of r at H_i and C_b^r is the cost of r at the best node H_{ib}. Let $C_{min} = MIN_i(C_i)$. Then $h_{i3} = \boldsymbol{\mu}_{LC}(C_{min}/C_i)$.

Computation of the Migration Overhead Objective The values h_{i4} for all i between 1 and n are computed as follows. First the estimation of the migration overhead, MO_i, for T' at H_i is carried out. The time required for the actual migration and the cost due to the migration are included in the calculation of MO_i as shown below.

$$MO_i = \sum_{r \in T' \wedge r \notin H_i} (2 * \Psi_b^i + MAX_{\forall H_{ij} \in UpS(H_i) \wedge r \in H_{ij}} (C_b^r - C_j^r)) \tag{8}$$

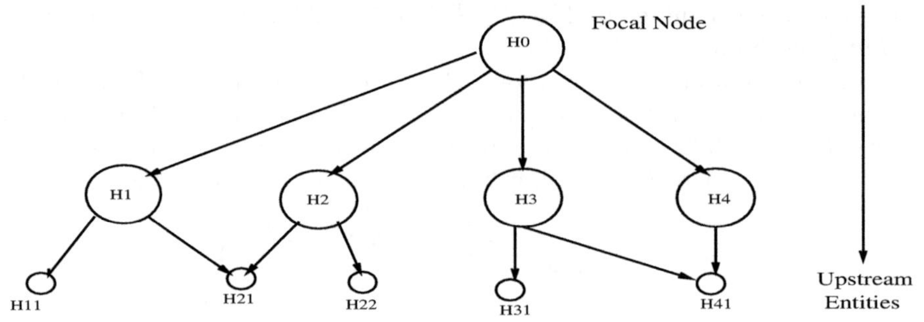

Fig. 6. The Supply Chain Used in Experiments

The first term in the summation, $2 * \Psi_b^i$, represents the actual migration time of the task and getting back the finished products. The second term indicates the cost overhead. C_j^r is the cost of the resource at a node $H_{ij} \in UpS(H_i)$ and C_b^r is the cost of the resource at H_{ib}. The maximum of the difference is chosen as a pessimistic estimate. MO_i is computed for all values of i between 1 and n. Then $h_{i4} = \boldsymbol{\mu}_{LM}(MO_{min}/MO_i)$.

Using the technique of Section 5.1, we formulate the task assignment as a multiobjective decision making problem and solve it. The solution returned is the node where T' should be assigned.

6 Performance Evaluation

In the experiments carried out below, we have simulated a single focal node H_0 and nine nodes in its upstream entities. The supply chain used for simulation is shown in Figure 6. The resources are distributed randomly across all nodes in the network. The tasks and the pairwise preferences of the objectives in the matrix $\mathcal{R}$ associated with every task are randomly generated. The task arrival at H_0 is modeled as a Poisson process with the arrival rate X. When a task execution is over, we calculate a performance measure which is the weighted average of the time required, the quality accumulated, the cost, and the migration overhead incurred for the task with the weights being the preference ratings for the respective objectives for the task. All the objectives need to be minimized except the quality objective. Hence we multiply the value of quality accumulated by -1 in calculating the performance measure. We sum up the performance measures for all tasks and calculate the *average performance measure P = (sum of performance measures for all tasks / total number of tasks)*. For the simulation study, we use a single set to represent all the following fuzzy sets - *Time Criticality(TC), Good Quality(GQ), Less Cost (LC),* and *Less Migration (LM)*. The membership function of the fuzzy set is generated using a table of points. Its graph is shown in Figure 7.

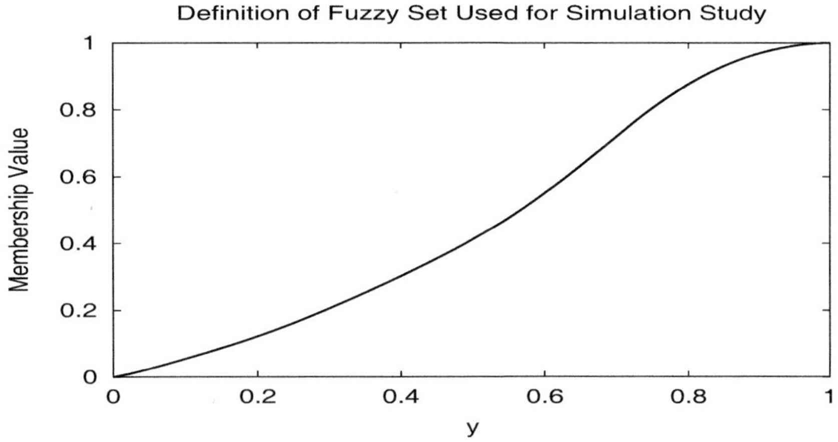

Fig. 7. The Membership Function for the Representative Fuzzy Sets

In this section, we evaluate the performance of the *fuzzy* technique and compare it against commonly used techniques for task assignment. We first present a comparison study with the *random* task assignment algorithm. In this, a task is assigned randomly to any node of the system without taking into account any system state information. In Section 6.1, we compare the *fuzzy* technique with the *Preference-Based Heuristic (PBH)* algorithm. The *PBH* algorithm uses the system state information and performs task assignment based only on that objective for which the user has given the highest preference rating.

Let P_{fuz} and P_{ran} be the performance measures using the *fuzzy* and the *random* approaches respectively. We plot $P_1 = P_{fuz}/P_{ran}$ (which is the ratio of the performance measure of the *fuzzy* technique against the *random* technique) for X varying from 50 tasks/day to 500 tasks/day. For getting every point on the plots, we run the simulation for a total simulated time of 30 days. Each such experiment is performed for three different values of R, the number of resources required by a task; the values being $R = 5$, $R = 10$, and $R = 15$. Figure 8 shows the results. The graph is to be interpreted as follows. Any point on a curve is a ratio, comparing the fuzzy technique performance measure versus the random algorithm performance. For example, if the point is 1.2 then it means that the fuzzy technique performance is 20 percent better than the random algorithm performance. If the point is 0.9 then it means that the fuzzy technique performance is 10 percent worse than the random algorithm performance.

1. For a large range of the arrival rates (till 350 tasks/day for $R = 5$, till 300 for $R = 10$, and till 250 for $R = 15$), the fuzzy approach gives consistently high gains (between 25% and 10%) over the random technique. This is because of the fact that the random algorithm assigns tasks without consideration of the goodness of a node and the preference ratings of the tasks.

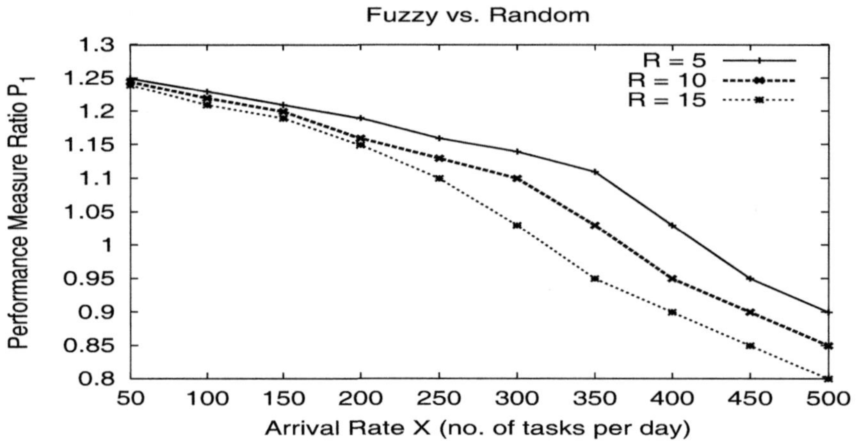

Fig. 8. Comparison of Fuzzy and Random Task Assignment Methods

2. When the arrival rate is high (between 350 and 400 for $R = 5$, between 300 and 350 for $R = 10$, and between 250 and 300 for $R = 15$), the fuzzy technique has small gains (less than 10%). The reason is that the system has reached the saturation point. Each node is highly loaded in this case and the fuzzy performance is only slightly better.
3. For very high arrival rates (above 400 for $R = 5$, above 350 for $R = 10$, and above 300 for $R = 15$), the fuzzy technique performance is worse than the random algorithm performance. At this point, the system is overloaded. For example, an arrival rate of 300 tasks/day for $R = 15$ means that a new task arrives every 5 minutes (approximately) and the average task duration time is 450 minutes (of simulated time). It is to be noted that we are considering a constantly evolving system where the surplus values of the nodes keep varying depending on the arrival rate (or the system load). The fuzzy technique relies heavily on the surplus values and in this overload condition the surplus values of the other nodes kept at a node are updated at a much lower rate than the rate at which they are actually changing. Hence the fuzzy technique cannot make a good decision with the stale surplus values even after spending a longer time and the performance drops down. It has also been shown in [8] that in such an overload case, a large number of tasks cannot finish within their deadlines.

   One possible solution to this problem is to increase the frequency of updating of the surplus values with the increase in the arrival rate. However, if this is done, the system throughput plunges. This is because, if the updating frequency is very high, there are a huge number of message exchanges. A lot of time is spent in processing those messages and the task execution gets hampered.

4. It is observed from the graphs that the performance with a lower R value is better than that with a higher R value. This can be explained as follows. In the computation of the fuzzy sets for the objectives, whenever a resource is not present at a node, we predict the target node for the subtask migration as the one which has the maximum amount of the surplus. This prediction could be erroneous for some of the tasks and hence the performance goes down. For lower R values, the probability of the absence of a resource at a node is lower, and hence the error would be introduced less number of times.

5. Based on the same argument as above, the *overload point*, i.e. the point from which the fuzzy technique performs worse than the random technique also comes later, i.e. with higher arrival rates, for lower values of R. This gives an indication that for larger tasks requiring a higher number of resources, the arrival rate of the supply chain operation should be chosen judiciously such that the system is not overloaded. The task mix and the complexity of the tasks should be used to make this choice.

6.1 Comparison with the Heuristic Algorithm

The fuzzy technique is subsequently compared with the *Preference-Based Heuristic (PBH)* task assignment algorithm. In the *PBH* task assignment algorithm, an input task is assigned to a node of the system that is optimum for the objective for which the user has given the highest preference rating. For example, if the user has the highest preference for the Execution Time objective, then equation (5) of Section 5.3 is used and the node giving the least value for that is chosen. All other objectives are ignored. The simulation is carried out with the same experimental setup as described earlier. Let P_p be the performance measure of the *PBH* algorithm. We plot $P_3 = P_{fuz}/P_p$ and $P_4 = P_p/P_{ran}$, which are the ratios of the performance measure of the *fuzzy* against the *PBH* technique and the *PBH* against the *random* technique respectively. Figures 9 and 10 show these plots. The following observations are made.

1. For a large range of the arrival rates (till 450 tasks/day for $R = 5$, till 400 for $R = 10$, and till 350 for $R = 15$), the fuzzy approach performs better than the PBH technique. The maximum gain is 24% for $R = 5$, 22% for $R = 10$, and 20% for $R = 15$. Thus the fuzzy technique that uses the combination of all the objectives using Yager's model does perform better than the PBH technique that considers only a single objective.
2. For rates higher than these, the performance of the fuzzy technique is worse than that of the PBH technique but this degradation is not as significant as that of the fuzzy technique against the random technique. The reason again is in the infrequent updating of the surplus values as compared to the very high arrival rates. The fuzzy technique uses the surplus values for computation of the fuzzy sets for all the objectives while the PBH technique uses it for only one objective. Hence the chances of introduction of error in the fuzzy technique are more since it is heavily dependent on the surplus

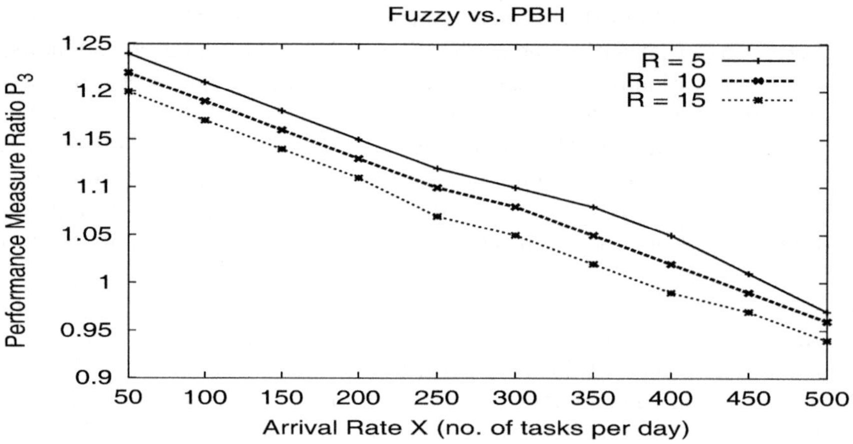

Fig. 9. Comparison of Fuzzy and PBH Task Assignment Methods

values. However, the worsening is less since there is some error introduced in the PBH technique as well because it also uses the surplus values, although for the computation of only one objective.

3. It is seen that the performance gain for the fuzzy technique drops down for higher values of R as compared to the PBH technique. This can be explained as follows. For a larger value of R, the probability of the absence of a resource for a task at a node is larger. In such a case we assume the *best node* (the node with the highest surplus value) as the target node for the subtask migration in the computation of the fuzzy sets. This assumption could be wrong for certain tasks. Since the computation is used for all the four objectives in the fuzzy technique and for only one in the case of the PBH technique, the chances of the introduction of error in the PBH technique is less. Hence the gain in the fuzzy technique over the PBH technique drops down for higher values of R. It is to be noted that this decrease is less than that of the fuzzy vs. random technique comparison.
4. The plot of the PBH versus the random technique is quite similar to that of the fuzzy versus the random technique except that the gain of the PBH technique is less than that of the fuzzy technique. The performance of the PBH technique drops down as compared to the random technique for very high arrival rates but this performance degradation is less than that of the fuzzy technique. The gain in the PBH technique with respect to the random technique decreases with increasing values of R but this reduction again is less than that of the fuzzy technique.

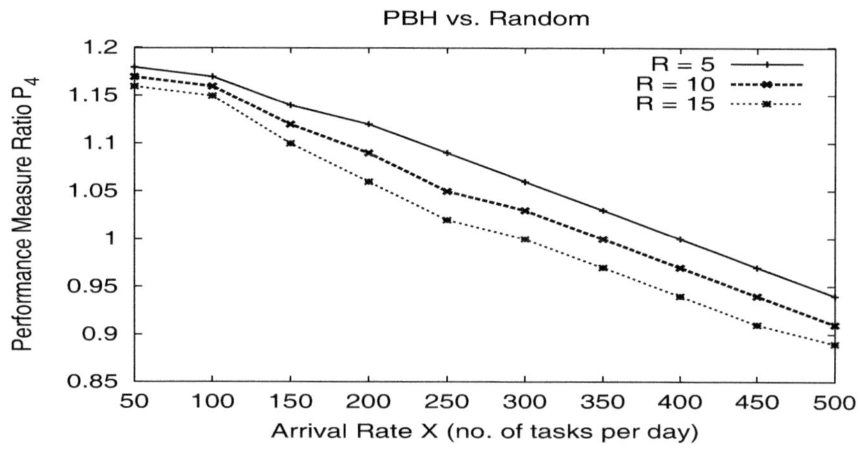

Fig. 10. Comparison of PBH and Random Task Assignment Methods

7 The Hybrid Algorithm

In order to avoid the performance degradation of the fuzzy technique at very high arrival rates, we design a *hybrid* algorithm for task assignment that reduces the dependency on the surplus values at very high arrival rates. It has been shown in [8] that the Guarantee Ratio - which is the ratio of the tasks that finish within their deadlines to the total number of input tasks - decreases drastically at very high arrival rates. We use this fact to design the *hybrid* algorithm which is described below.

1. The basic fuzzy technique remains the same but it is augmented by a continuous detection of the *overload point*. The identification of the *overload point* is as follows.
   - The *focal node* (recall that the *focal node* is the one where all the tasks first arrive in the supply chain management system) keeps a sliding task window of length S_W to monitor the finish time of the most recent S_W tasks.
   - At the end of every task, the finish time of the last S_W tasks in the sliding window is queried. If the number of tasks that could finish within their deadlines is less than a threshold Θ_S, we conclude that the *overload point* has been reached.
2. When the *overload point* is hit, we use the random algorithm for task assignment instead of the fuzzy technique.
3. When the system comes out of the overload state (again found out by the *overload point* identification), we begin reusing the fuzzy technique for task assignment.

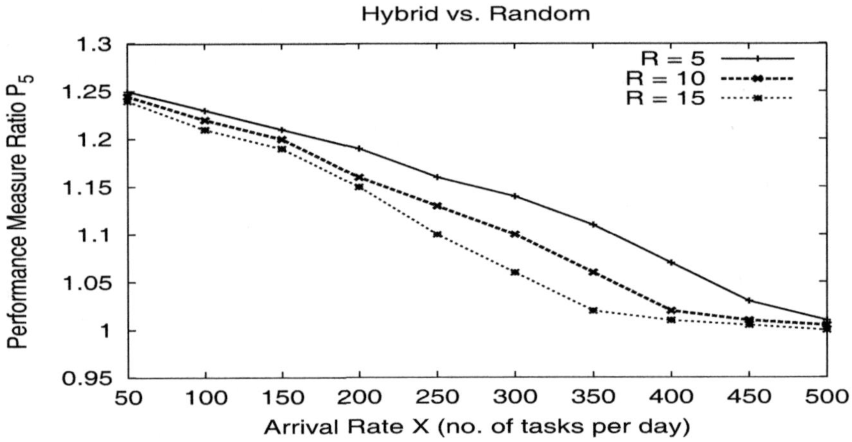

Fig. 11. Comparison of Hybrid and Random Task Assignment Methods

We introduce the *hybrid* algorithm in our simulated supply chain management system. The experimental setup is same as that in Section 6. The following parameters are used for simulation: $S_W = 10$ and $\Theta_S = 4$. The performance of the *hybrid* algorithm is evaluated against the *random* and the *PBH* task assignment techniques. Let P_h be the performance measure of the *hybrid* algorithm. We plot $P_5 = P_h/P_{ran}$ and $P_6 = P_h/P_p$ which are the ratios of the performance measure of the *hybrid* against the *random* and the *PBH* techniques respectively. Figures 11 and 12 show these plots. The observations are as follows.

1. From plot of Figure 11, it is clear that the *hybrid* algorithm has the advantages of the *fuzzy* technique for a large range of the arrival rates. The maximum performance gain is 25% which is the same as that of the *fuzzy* technique. For high arrival rates the performance of the *hybrid* algorithm is better than that of the *fuzzy* technique. It is also confirmed that the performance of the *hybrid* technique with respect to the random algorithm does not drop down with very high arrival rates (even beyond the *overload point*) and even for high values of R. Thus the technique of monitoring the *overload point* of the *hybrid* algorithm works well. We would like to point out that Θ_S has a significant influence on the performance at lower arrival rates. We also performed simulation experiments (not shown here) which demonstrated that for $\Theta_S = 6$, the performance gain for lower arrival rates is less (about 20% at the maximum). But the performance for high and very high arrival rates does not drop down even for $\Theta_S = 6$.
2. The plot of Figure 12 is similar to the plot of Figure 11 except that even for very high arrival rates the performance of the hybrid technique does not deteriorate with respect to the PBH technique. This is because at very high arrival rates, the hybrid technique detects the *overload point* and uses

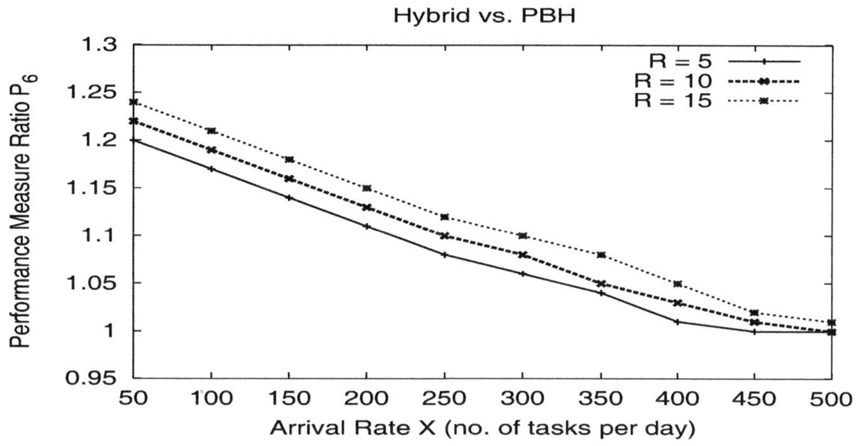

Fig. 12. Comparison of Hybrid and PBH Task Assignment Methods

the random task assignment algorithm. Thus stale values of the surplus due to the infrequent updating are not used and hence erroneous decisions are avoided.

In summary, we can conclude that the *hybrid* algorithm gives the advantages of the *fuzzy* task assignment technique for a large range of arrival rates. Its mechanism of continuous monitoring of the *overload point* ensures that the performance degradation for very high arrival rates is avoided. Hence, it is a good technique for task assignment for real-time operation in a supply chain.

8 Conclusions

In this chapter, a supply chain is modeled using a multiagent framework and an architecture for each of the agents is presented. We present a real-time scheduler that can accept new tasks with soft real-time deadlines. The schedules for the new tasks are integrated with the schedules of the already existing tasks in the system. The problem of task assignment for the real-time operation in a supply chain is then addressed. Multiple criteria like quality, cost, and duration may be associated with every task. Preference ratings may also be assigned to each of the criteria. The preference ratings may be described in an imprecise manner using only ordinal values or linguistic terms. An algorithm for the task assignment for real-time operation in a supply chain using the fuzzy set approach is proposed. It is felt that this approach is appropriate since there is imprecision in not only the specification of the preference ratings but also in the global system state collected for decision making since we are considering a dynamic distributed system. The simulation studies for evaluating the effectiveness of the task assignment in a simulated supply chain have been performed. The results suggest that the fuzzy technique presented in this chapter has significant performance gains for a large range of arrival rates over other commonly used heuristic techniques. However, its performance drops down for very high task arrival rates. In order to overcome this limitation, a hybrid algorithm is designed that uses the fuzzy approach but does not suffer from performance deterioration even at very high arrival rates.

References

1. Andersson, J.: A Survey of Multiobjective Optimization in Engineering Design. *Technical Report No. LiTH-IKP-R-1097*, Dept. of Mechanical Engineering, University of Linköping, Sweden, (2000)
2. Bellman, R., Zadeh, L.: Decision Making in a Fuzzy Environment. Management Science **17** (1970) 141–164
3. Bond, A. H., Gasser, Les: Readings in Distributed Artificial Intelligence. Morgan Kaufman, San Mateo, California, (1988)
4. Chopra, S., Meindl, P.: Supply Chain Management: Strategy, Planning, and Operation. Pearson Education Inc., (2001)
5. Deb, K.: Multiobjective Optimization using Evolutionary Algorithms. Wiley, Chichester, UK, (2001)
6. Decker, K. S.: Environment Centered Analysis and Design of Coordination Mechanisms. *Ph.D. thesis*, University of Massachusetts, Amherst, (1995)
7. Deshpande, U., Gupta, A., Basu, A.: Adaptive Fault Tolerant Hospital Resource Scheduling. in: Meersman, R., Tari, Z., et. al., eds., *Proceedings of Tenth International Conference on Cooperative Information Systems (CooPIS-2002), Irvine, USA, (Nov. 2002)*, LNCS - 2519, Springer Verlag, Berlin, 503–520
8. Deshpande, U., Gupta, A., Basu, A.: Coordinated Problem Solving through Resource Sharing in a Distributed Environment. IEEE Transactions on Systems, Man, and Cybernetics - Part B **34(2)** (April 2004) 1299–1304
9. Fox, M., Barbuceanu, M., Teigen, R.: Agent-Oriented Supply-Chain Management. The International Journal of Flexible Manufacturing Systems **12** (2000) 165–188
10. Goldberg, D.: Genetic Algorithms in Search, Optimization, and Machine Learning. Addison Wesley, Reading, Mass., (1989)
11. Sadeh, N. M., Hildum, D. W., Kjenstad, D., Tseng, A.: MASCOT: An Agent-Based Architecture for Dynamic Supply Chain Creation and Coordination in the Internet Economy. Production Planning and Control **12(3)** (2001)
12. Saaty, T. L.: The Analytic Hierarchy Process. McGraw Hill, (1980)
13. Shapiro, J. F.: Modeling the Supply Chain. Thomson Learning Inc., Duxbury, (2001)
14. Steuer, R.: Multiple Criteria Optimization: Theory, Computation and Application. John Wiley & Sons, Inc., New York, (1986)
15. Swaminathan, J. M., Smith, S. F., Sadeh, N. M.: Modeling Supply Chain Dynamics: A Multiagent Approach. Decision Sciences **29(3)** (April 1997)
16. Verdicchio, M., Colombetti, M.: Commitments for Agent-Based Supply Chain Management. ACM SIGecom Exchanges **3(1)** (2002) 13–23
17. Walsh, W. E.: Market Protocols for Decentralized Supply Chain Formation. *Ph.D. Thesis*, Dept. of Computer Science and Engineering, University of Michigan (2001)
18. Yager, R.: Fuzzy Decision Making including Unequal Objectives. Fuzzy Sets and Systems **1** (1978) 87–95
19. Zadeh, L.: Fuzzy Sets. Information and Control **8** (1965) 338–353

Multi-Agent based Supply Chain Modelling with Dynamic Environment

Toshiya Kaihara

Kobe University, Department of Computer and Systems Engineering,
1-1, Rokkodai, Nada, Kobe, 657-8501, Japan
kaihara@cs.kobe-u.ac.jp

Abstract. Nowadays, Supply Chain Management (SCM) is an important paradigm of business management in agile manufacturing environment. Clearly, there is a need for a mechanism through which these different functions can be integrated together. In this chapter, we focus on negotiation process in SCM formation under VE environment to clarify its effective management. Each enterprise in VE is defined as agent with multi-utilities, and a framework of multi-agent programming with marketing science and N-person game theoretic approach is newly proposed as negotiation algorithm amongst the agents. We classify business model into several types, and introduce Contract Net Protocol (CNP) with marketing science and game theory as a coordination and negotiation mechanism amongst the enterprises. We develop a computer simulation model to form SCM through multiple negotiations amongst several potential members in the negotiation domain, and finally clarify the formation dynamism with the negotiation process.

1 Introduction

Currently, SCM is a crucial paradigm of business management in agile manufacturing environment, such as Virtual Enterprise (VE) or e-manufacturing [1][2]. SC exists in both service and manufacturing organizations, although the complexity of the each enterprise may vary greatly from industry to industry. Realistic SC handles multiple end products with shared components, facilities and capacities. Since the flow of materials in SC is not always along an arborescent network, various modes of transportation may be considered, and the bill of materials for the end items may be both deep and large.

Traditionally, marketing, distribution, planning, manufacturing, and the purchasing organizations operated independently. These organizations have their own objectives and these are often conflicting. Marketing's objectives of high customer service and maximum sales conflict with manufacturing and distribution goals. Many manufacturing operations are designed to maximize throughput and lower costs with little consideration for the impact on inventory levels and distribution capabilities. Purchasing contracts are often negotiated with very little information beyond historical buying patterns. The result of these factors is that there is not a single, integrated plan for the organization - there were as many plans as businesses. Clearly, there is a need

T. Kaihara: Multi-Agent based Supply Chain Modelling with Dynamic Environment, Studies in Computational Intelligence (SCI) 28, 203–215 (2006)
www.springerlink.com

for a mechanism through which these different functions can be integrated together. Although cooperation is the fundamental characteristic of SCM concept, due to its distributed environment and the autonomous and heterogeneous nature of the SC members, cooperation can only be succeed if a proper management of dependencies between activities is in place.

In this chapter, we focus on negotiation process in SCM formation under VE environment to clarify its effective management. Each enterprise in VE is defined as agent, and a framework of multi-agent programming with marketing science and game theoretic approach is newly proposed as negotiation algorithm amongst the agents. Each unit is defined as agent in our SCM model, and their decision makings are formulated as marketing science models [3] and N-person game theoretic methodology [4].

We firstly classify business model into three types, such as vertically integrated business model, horizontally specialised business model and hybrid business model. Then we propose CNP based negotiation protocol amongst enterprises with marketing science models, such as lexicographic model and Maximum Likelihood Hierarchical (MLH) model, and N-person game theoretic approach. CNP models transfer of control in a distributed system with the metaphor of negotiation among autonomous intelligent beings [5]. CNP consists of a set of nodes that negotiate with one another through a set of message. Nodes generally represent the distributed computing resources to be managed, correspond to "enterprises" in this chapter. Marketing science based negotiation is applied into the vertically integrated business model considering a realistic enterprise management strategy. The horizontally specialised business model includes N-person game theoretic negotiations to realise the coordination amongst enterprises in the same business segment. We develop a computer simulation model to form SCM through multiple negotiations amongst several potential members in the negotiation domain, and finally clarify the formation dynamism with the negotiation process.

2 VE Business Model in Supply Chain Environment

2.1 VE Concept

Virtual enterprises are defined as "agile" enterprises, i.e. as enterprises with integration and reconfiguration capability in useful time, integrated from independent enterprises, primitive or complex, with the aim of taking profit from a specific market opportunity. After the conclusion of that opportunity, the virtual enterprise dissolves and a new virtual enterprise is integrated in order to achieve the necessary competitiveness to respond to another market opportunity. General definition of VE is as follows [2]:

> "A virtual enterprise is a temporary alliance of enterprises that come together to share skills or core competencies and resources in order to better respond to business opportunities and whose cooperation is supported by computer networks."

The knowledge and physical resources associated to the development and production of most of today's products often exceed what a single firm is able to accomplish. The new production enterprise is a network that shares experience, knowledge and capabilities – it is critical in this new environment for a manufacturing company to be able to efficiently tap these knowledge and information networks.

The organisational challenge of partitioning tasks amongst partners is of main concern, and can determine the success or failure of a project in the distributed manufacturing environment. They can fit and take advantage of the different competencies in VE, and that leads to their coordination and reconfigurability in order to keep alignment with the market requirements,

Faced to the requirements of competitiveness that the current environment is demanding, enterprises are expected to present at least the following characteristics:

- Fast reconfigurability or adaptability: the ability of fast change face to the unpredictable changes in the environment / market, implying substitution of resources (transition to a new A/V E instantiation)
- Evolutionary capability: the ability to learn with history.

A large number of diversified networked organisations of enterprises fall under the general definition of VE. We assumed our VE model in the possible simplest definition as a basic research, as follows:

i) Duration: Single business

An alliance of the enterprises is established towards a single business opportunity, and is dissolved at the end of such process.

ii) Topology: Fixed structure

There exist established supply chains with an almost fixed structure.

iii) Participation: Single alliance

All the enterprises are participating into only a single alliance at the same time.

iv) Coordination: Democratic alliance

A different organisation can be found in some supply chains without a dominant company. All the enterprises cooperate on an equal basis, preserving their autonomy except hybrid business model in this study.

v) Visibility scope: Single level

All the enterprises in VE communicate only to its direct neighbours in its architecture (figure 1). That is the case observed in most supply chains.

Needless to say, it is very important and difficult activity in forming a virtual enterprise to select appropriate business partners, i.e. partnering, because each enterprise considers not only pursuing its profit but also sharing the risk to join the virtual enterprise. The partnering is described as coordination activity amongst the enterprises, and some sophisticated coordination mechanism is required to realise efficient interactions.

The development of coordination mechanism in computer science can be found in the area of workflow management system, computer supported cooperative work (CSCW), and multi-agent systems. The area of multi-agent systems, especially when involving intelligent autonomous agents, has been discussing coordination issues and supporting mechanism [6][7][8]. The interaction capability, both amongst agents and between agents and their environment, is one of the basic characteristics of an agent. In this chapter we focus on the contract net protocol (CNP), that is one of the mechanism coming from the early works on multi-agent systems [5], as the coordination and negotiation mechanism amongst business units in VE.

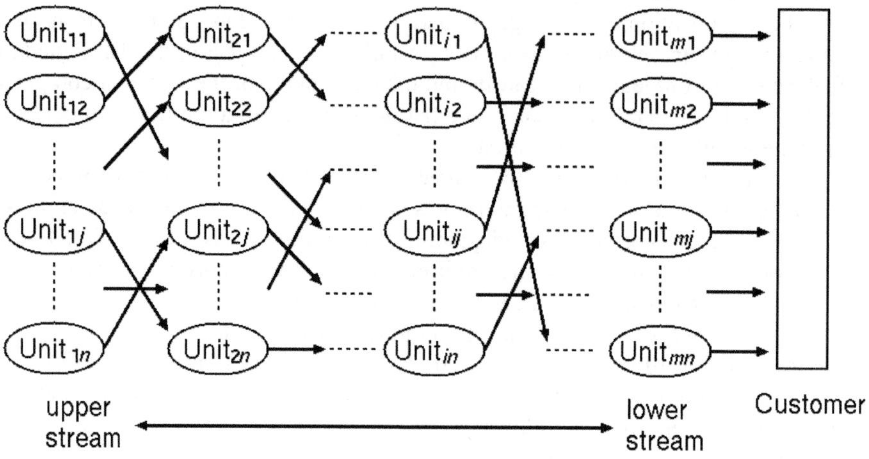

Fig. 1. VE model

Figure 1 shows the assumed VE model in this chapter. We call an enterprise as unit, and there exist m layers, which have mn units in the VE model. The lowest level corresponds to consumers who can create original task requests to the VE. As the layer number, m, increases, we describe it 'lower' based on the product flow order in this chapter.

At first, the customer dispatches new order to all the units in layer m, and then several units, which are satisfied with the order, responds and circulates the order toward upper units in the VE model. Finally a VE with single supply chain will be established for the order as a consequence of their negotiations through all the layers.

2.2 Business Model

As described in the previous section, there exist many business models in VE environment. We classify three types of business models in this study, such as vertically integrated business model, horizontally specialised business model and hybrid business model.

Generally we can observe vertical business models in traditional industries. In this business model the manufacturing business processes top-down as well as bottom-up from the process requirements, resulting in an integrated approach to overall business requirements. We will call this business model, which is based on end-to-end proprietary solutions that lock a customer to a manufacturer, a "vertically integrated business models". Each unit at the same layer in figure 1 never tries to cooperate in our vertically integrated business model, because each unit is keen to find its appropriate partner just in vertical directions. For example, Unitij tries to find an alliance unit at each neighbouring layer: layer (i-1) and (i+1) in figure 1. Consequently only one SC can be formed in this business model. We introduce consumer's behaviour in marketing science into unit behaviour, and that makes our VE model more practical.

On the other hand, enterprise relationships may represent a forerunning pattern of the learning alliance, whereby ongoing close interaction of horizontal alliance partners at single level or multiple hierarchical levels can be used to facilitate the mutual accumulation of superior organizational capabilities within the alliance firms. We will call this business model a "horizontally specialised business model". All the units at the same layer try to cooperate to maximise their profit in total in our horizontally specialised business model. They behave as if they are under joint management, and we can see a kind of this style of management in industrial cluster in Japan [9]. We apply N-persons game theory to formulate their behaviour in cooperation, because it describes a decision-making process to find their equilibrium solution in social manner.

Finally hybrid business model combines above-mentioned 2 business models. There exists a parent-child relationship amongst units in VE, and parent company has advantageous in their contract. Units in the child companies try to keep their harmony as a group sharing a common destiny in their management. We apply an alliance strategy based on cooperative game theory in this model.

3 Agent Definitions

3.1 Unit Structure

Each unit (enterprise) is defined as agent in our VE model, and its structure is described in figure 2. We adopt CNP as the coordination and negotiation mechanism amongst the units. CNP models transfer of control in a distributed system with the metaphor of negotiation among autonomous intelligent beings. CNP consists of a set of nodes that negotiate with one another through a set of message. Nodes generally represent the distributed computing resources to be managed, correspond to "units" in this chapter.

An agent (=unit) can act both as a manager and a contractor of a delivery sets. When a unit receives new order (= task announcement) *i*, it creates a contractor / manager set (Manager *i* / Contractor *i*) for the task inside. Manager *i* creates a new order towards the higher units to secure the contract with the lower layer.

3.2 Negotiation Algorithm

Negotiation steps according to agent roles are described as follows:

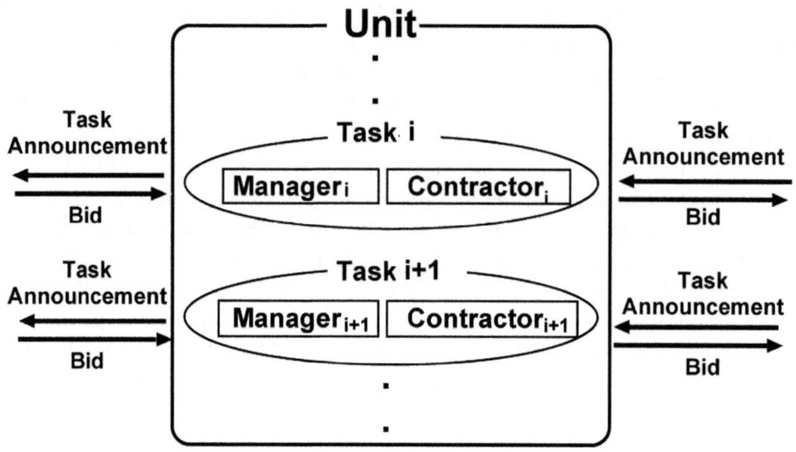

Fig.2. Unit structure

Manager (in layer x)

Step M1: Create a new task based on the received bid information.
Step M2: Task announcement (TA) to the lower units.
Step M3: After the bidding period expired, check all the acquired bids according to its standard. If there exists no bid to select, go to M4. Otherwise go to M5.
Step M4: Modify the task and go to M2.
Step M5: Select the task and send reward (Reward) to the corresponding unit.

Contractor (in layer x-1)

Step C1: Evaluate the received TA, and if satisfied, then create an estimated bid. Otherwise wait next TA.
Step C2: Send the bid to the Manager.
Step C3: If the contractor can get the contract, then request the manager in it to create TA to layer (x-2).

We prepare several parameters to define unit behaviours as follows:

P_{TA}^{ij} : TA price of Unit$_{ij}$ Q_{TA}^{ij} : TA quantity of Unit$_{ij}$
L_{TA}^{ij} : TA lead time of Unit$_{ij}$ $Q_{BID}^{(i+1)j}$: Bid quantity of Unit$_{(i+1)j}$
$L_{BID}^{(i+1)j}$: Bid lead time of Unit$_{(i+1)j}$ P_{BID}^{ij} : Bid price of Unit$_{ij}$
Q_{BID}^{ij} : Bid quantity of Unit$_{ij}$ L_{BID}^{ij} : Bid lead time of Unit$_{ij}$

$P_{TA}^{(i+1)j}$: TA price of Unit $_{(i+1)j}$ $Q_{TA}^{(i+1)j}$: TA quantity of Unit $_{(i+1)j}$
$L_{TA}^{(i+1)j}$: TA lead time of Unit $_{(i+1)j}$ cost_{ij} : process cost of Unit$_{ij}$ per product
$profit_{ij}$: profit rate of Unit$_{ij}$ $\text{processingtime}_{ij}$: process time of Unit$_{ij}$ per product
procuretime_{ij} : estimated procure time of Unit$_{ij}$ per product

Unit formulation in each business model is described in the following sections.

Each unit (enterprise) is defined as agent in our VE model, and its structure is described in figure 2. We adopt CNP as the coordination and negotiation mechanism

3.3 Vertically Integrated VE Model

CNP is only the skeleton in the negotiation algorithm, and it is necessarily to define how to select the appropriate bid (Step M5) to establish profitable contract. Marketing science based negotiation is applied into the managers' decision makings to realise sophisticated bid selection based on consumers' behaviours in the vertically integrated business model.

First of all, each attribute of manager$_{ij}$ in TA producing process (Step M1) is as follows:

$$P_{TA}^{ij} = \text{cost}_{ij} \times (1 - profit_{ij}) \tag{1}$$

$$Q_{TA}^{ij} = Q_{BID}^{(i+1)j} \tag{2}$$

$$L_{TA}^{ij} = L_{BID}^{(i+1)j} - Q_{BID}^{(i+1)j} \times \text{processing time}_{ij} \tag{3}$$

Then contractor$_{ij}$ tries to create its bid (Step C1) against TA from manager in layer (i+1) by the following equations:

$$P_{BID}^{ij} = \text{cost}_{ij} \times (1 + profit_{ij} / 2) \tag{4}$$

$$Q_{BID}^{ij} = Q_{TA}^{(i+1)j} \times (1 + plus_{ij}) \tag{5}$$

$$L_{BID}^{ij} = Q_{BID}^{ij} \times (\text{processing time}_{ij} + \text{procuretime}_{ij}) \tag{6}$$

Marketing science approach is applied in the bid selection mechanism (Step M5) in the vertically integrated VE model. We consider 2 types of marketing science model, named lexicographic model and MLH model [3]. Bid selection mechanism is described in each model as follows:

Lexicographic model:

Step L1: Set priorities on all the attributes.
Step L2: Select a bid with the highest value in the attribute. In case of a tie, move to the next attribute and check the highest value. This routine is continued until only one bid is selected.

MLH model:

Step MLH1: Set priorities on all the attributes.
Step MLH2: Standardise the highest attribute with the following equations:

$$\tilde{z}_{ij} = \frac{(highest \quad z_{ij}\ amongst \quad unselected \quad j \quad) - z_{ij}}{highest \quad z_{ij}\ amongst \quad unselected \quad j} \tag{7}$$

where z_{ij} : evaluation value of task *j* on *i* th attribute

$\tilde{z}_{ij}$: standardised z_{ij}

Step MLH3: If $\tilde{z}_{ij}$ is within the tolerable amount (i.e. tolerance error: τ_i %), then keep this attributes for the calculation in the next step.
Step MLH4: After all the attributes are evaluated, then the preference index V_j of task *j* is calculated:

$$V_j = (\tau_1 - \tilde{z}_{1j}) \times (\tau_2 - \tilde{z}_{2j}) \times \cdots \times (\tau_i - \tilde{z}_{ij}) \tag{8}$$

Finally selection probability of task$_j$ is calculated by the following equation:

$$P_j = V_j / \sum_{k=1}^{n} V_k \tag{9}$$

3.4 Horizontally Specialised VE Model

The horizontally specialised business model includes N-person game theoretic negotiations to realise the coordination amongst enterprises in the same business segment. In this model we consider there is a coordinator to manage the negotiation process amongst all the enterprises (i.e. contractors) in each layer. The coordinator in layer *i* manages all the units in the layer. It receives all TAs from lower layer (*i*+1) on behalf of coordinators in the layer *i*, and replies the manager. N-person cooperative game approach is applied to share its profit amongst all the units in the layer based on their attributes. So it is regarded as a kind of joint order organisation.

Firstly, TA producing process (Step M1) is in common to the vertically integrated business model. Then the Bidding process is automatically replied based on the received TA, unless it yields no profit in any cooperation of contractor units (Step C2). So the attributes in the bid are defined as follows:

$$P_{BID}^{ij} = P_{TA}^{(i+1)j} \tag{10}$$

$$Q^{ij}_{BID} = Q^{(i+1)j}_{TA} \tag{11}$$

$$L^{ij}_{BID} = L^{(i+1)j}_{TA} \tag{12}$$

Finally the order sharing process is formulated with axioms called Sharpley value. We applied the Shapley value because our VE cooperation model doesn't always satisfy super-additivity condition. The Shapley value describes one approach to the fair allocation of gains obtained by cooperation among several actors even in non super-additivity condition. The setup is as follows: a coalition of actors cooperates, and obtains a certain overall gain from that cooperation. Since some actors may contribute more to the coalition than others, the question arises how to fairly distribute the gains among the actors, in other words, how important is each actor to the overall operation, and what payoff can they reasonably expect? The Shapley value is one way to distribute the total gains to the actors, assuming that they all collaborate.
At first we define characteristic function ($V(S)$) as follows:

$$V(S) = U_{profit}(S) \times U_{leadtime}(S) \tag{13}$$

where $U_{profit}(S)$: profit in cooperation S,

$U_{leadtime}(S)$: Boolean of lead time constraint (1: OK, 0: NG)

Then the task is divided under the cooperation S', which maximises $V(S)$. The contribution of unit i in cooperation S is calculated as follows:

$$\phi_i = \sum_{S:i \in S \subseteq N} \frac{(|S|-1)!(n-|S|)!}{n!}(v(S) - v(S \setminus \{i\})) \tag{14}$$

where N: a set of all units, v: all the profit in cooperation S, $|S|$: the number of members in cooperation S, ($S\setminus\{i\}$): cooperation without unit i

Then naturally the following equation is acquired:

$$\sum_{i \in N} \phi_i(v) = v(N) \tag{15}$$

So the profit, which gained in the coordinator, is divided into each unit in the layer according to the Sharpley value in (14).

3.5 Hybrid VE Model

Hybrid business model combines above-mentioned 2 business models. There exists a parent-child relationship amongst units in VE, and parent company has advantageous in their contract, such as in automobile or space industries. Units in the child companies try to keep their harmony as a group in their management.

TA producing process (Step M1) has also no difference in above-mentioned 2 business models. Then contractor$_{ij}$ creates its bid (Step C1) against TA from manager in layer (i+1) by the following equations:

$$P_{BID}^{ij} = \cos t_{ij}(1 + profit_{ij}) \quad (16)$$

$$Q_{BID}^{ij} = Q_{TA}^{(i+1)j} / N_i \quad (17)$$

$$L_{BID}^{ij} = Q_{TA}^{ij} \times (\text{processing time}_{ij} + \text{procuretime}_{ij}) \quad (18)$$

where N: the number of cooperate units in layer i

Finally the bid selection process by manager (Step M5) is formulated by the cooperative game theory under super-additivity condition. We define characteristic function ($V(S)$) as follows:

$$V(S) = U'_{profit}(S) \times U_{leadtime}(S) \quad (19)$$

where $U'_{profit}(S)$: manager's profit in cooperation S

Then the characteristic function values in all the received bids are calculated and the manager finally selects the bid with the highest value amongst them.

4 Experimental Results

We have developed a computer simulation model to analyse VE formation dynamism through multiple negotiations amongst several potential members in the negotiation domain. Although many experiments have done in this study, only a part of our analytic results is explained in this chapter due to page limitation. The simulation parameters are shown in table 1.

m	n	τ_i	cost$_{0j}$	cost$_{1j}$	cost$_{1j}$	profit$_{ij}$	plus$_{ij}$	Processing time$_{ij}$	procuretime$_{ij}$
3	3,10	0.4	15-25	45-55	85-95	0.17-0.23	0.08-0.12	0.08-0.12	0.10-0.20

Table 1. Experimental parameters

The performances of finally acquired VE in three business models are compared in Table 2 (Model 1: vertically integrated business model), Table 3 (Model 2: horizontally specialised business model) and Table 4 (Model 3: hybrid model). We tried to analyse the VE robustness of each business model against "due date change" and "production volume change" in this experiment, and 500 trials are examined in each business model. The default due date and production volume are set to 30 and 100, respectively, and (0.5, 1.0), for example, means due date is shorten to 1/2 (0.5) and production volume is equivalent (1.0) to the default in these tables. Only the results of lexicographic model are described in Table 2.

The following points have been observed in those experiments:

- Any models don't satisfy lead time (15.00) in (0.5, 1.5), because the requirement change is too heavy to handle.
- Lead time is satisfied at Model 2 and 3 both in (0.5, 1.0) and (1.0, 1.5). These relatively slight changes are manageable in those business models except Model 1.
- Total profit is the highest at Model 1, because only the contributed units can take direct profit. Additionally stock is the least at this model.
- Model 3 attains the shortest lead time in most cases.

	(1.0, 1.0)		(0.5, 1.0)	
	Ave.	Std. Dev.	Ave.	Std. Dev.
Profit	8076.05	947.00	8076.05	947.00
Stock(WIP)	21.14	2.50	21.14	2.50
Lead time	24.10	2.82	24.10	2.82
	(1.0, 1.5)		(0.5, 1.5)	
	Ave.	Std. Dev.	Ave.	Std. Dev.
Profit	12098.30	1413.53	12098.30	1413.53
Stock(WIP)	31.64	3.79	31.64	3.79
Lead time	36.10	4.21	36.10	4.21

Table 2. Experimental results of vertically integrated business model (Model 1)

	(1.0, 1.0)		(0.5, 1.0)	
	Ave.	Std. Dev.	Ave.	Std. Dev.
Profit	6179.73	730.06	5919.06	708.20
Stock(WIP)	21.28	2.57	21.82	2.68
Lead time	16.44	2.14	13.47	1.81
	(1.0, 1.5)		(0.5, 1.5)	
	Ave.	Std. Dev.	Ave.	Std. Dev.
Profit	9263.05	1084.19	8889.12	1047.26
Stock(WIP)	32.12	3.82	33.04	3.81
Lead time	25.49	2.99	20.67	2.59

Table 3. Experimental results of horizontally specialised business model (Model 2)

	(1.0, 1.0)		(0.5, 1.0)	
	Ave.	Std. Dev.	Ave.	Std. Dev.
Profit	7928.55	930.86	7628.55	930.86
Stock(WIP)	22.32	2.70	22.32	2.70
Lead time	16.69	2.07	12.98	1.62
	(1.0, 1.5)		(0.5, 1.5)	
	Ave.	Std. Dev.	Ave.	Std. Dev.
Profit	11877.20	1390.56	11877.20	1390.56
Stock(WIP)	33.49	4.02	33.49	4.02
Lead time	24.95	2.99	19.41	2.31

Table 4. Experimental results of hybrid model (Model 3)

These results have been summarised as follows:

- It is difficult for vertically integrated business model to adapt due date change and production volume change, because it doesn't include any unit cooperation mechanisms. However, it performs best in profit, and this model is efficient in case there are enough margins in lead time (i.e. stable situations).
- Horizontally specialised business model is robust against due date change and production volume change, although the profit is less than vertically integrated business model. This business model is suitable to agile manufacturing situations with autonomous cooperation alliance.
- Hybrid model takes a middle position between vertically integrated business model and horizontally specialised business model. Although it takes advantages of both models, it has been observed that the shared profit is inclined into lower layer units in SC.

Each unit (enterprise) is defined as agent in our VE model, and its structure is described in figure 2. We adopt CNP as the coordination and negotiation mechanism

5 Conclusions

In this chapter, we focused on negotiation process in SCM formation under VE environment to clarify its effective management. We firstly classified business model into three types, such as vertically integrated business model, horizontally specialised business model and hybrid business model. Then we proposed CNP based negotiation protocol amongst enterprises with marketing science models, such as lexicographic model and MLH model, and N-person game theoretic approach. Marketing science based negotiation was applied into the vertically integrated business model considering a realistic enterprise management strategy. The horizontally specialised business model included N-person game theoretic negotiations to realise the coordination amongst enterprises in the same business segment. We developed a computer simulation model to form SCM through multiple negotiations amongst several potential members in the negotiation domain, and finally clarified the formation dynamism with the negotiation process. It has been confirmed that the vertically integrated business model is profit-oriented and it is the best in relatively stable business situations. On the contrary, horizontally specialised business model is robust against the order change, and it suits agile manufacturing situations. Hybrid business model is moderate characteristic between them, and it seems useful practically as often shown in real situations.

References

1. Fisher M. L., Making supply meet demand in uncertain world, Harvard Business Review, (1994) May/Jun
2. Camarinha-Matos, L. M. et al., The virtual enterprise concept, Infrastructures for virtual enterprises, Kluwer academic publishers, Boston (1999) 3-14
3. Katahira H., Marketing science (in Japanese), Tokyo university press, Tokyo (1987)
4. Von Neumann, J. et al., Theory of Games and Economic Behavior, Princeton University Press (1947)
5. Smith, R., The contract net protocol, IEEE Transaction on Computers, C-29 (1980) 1104-1113
6. Kaihara, T. and S. Fujii, A study on virtual enterprise coalition with multi-agent technology in agile manufacturing environment, International Journal of Advanced Manufacturing Systems, Vol.1, No.2 (2002) 125-139
7. Kaihara, T. and S. Fujii, IT based Virtual Enterprise Coalition Strategy for Agile Manufacturing Environment, Proc. of the 35th CIRP Int. Seminar on Manufacturing Systems (2002) 32-37
8. Durfee, E. et al., Coherent cooperation among communication problem solvers, IEEE Transaction on Computers, N 36 (1987) 1275-1291
9. Kansai Bureau of Economy, Trade and Industry, URL : http://www.kansai.meti.go.jp/3-2sanki/cluster-beam/english.html.

Agent-Based Technological Framework for Dynamic Configuration of a Cooperative Supply Chain

Alexander V. Smirnov[1], Leonid B. Sheremetov[2,1], Nikolai Chilov[1], Christian Sanchez-Sanchez[2]

[1]St.Petersburg Institute for Informatics and Automation of the Russian Academy of Sciences SPIIRAS, 39, 14th Line, St.Petersburg, 199178, Russia
e-mail: {smir, nick}@.iias.spb.su
[2] Mexican Petroleum Institute, Av. Lazaro Cardenas, 152, Col. San Bartolo Atepehuacan, México, D.F., CP 07730 , Mexico
e-mail: {sher, chsanche}@imp.mx

Abstract. In this chapter, an agent-based technological framework for dynamic configuration of a cooperative supply chain (CSC) as a flexible and agile system is proposed. Members of the CSC negotiate and compromise on the optimal configuration in a spirit of cooperation, in order to maximize individual benefits and overall outcome of the system. The proposed approach considers configuring as a sequence of interrelated tasks, from which, partner selection among available enterprises and resource allocation are described in details. These tasks are solved within the framework applying different techniques of computational intelligence: (i) the theory of games with fuzzy coalitions, (ii) genetic algorithms, and (iii) constraint satisfaction problem solving. Ontologies formalized as object oriented constraint networks are used for task description and decomposition. Implementation details of the multi-agent framework integrating these techniques are discussed and a case study illustrating the framework is described.

Key words: cooperative supply chain, multi-agent environment, fuzzy coalition games, genetic algorithm, constraint satisfaction

1 Introduction

Today's dynamic market trends require companies to provide low cost and highly responsible production and organizational solutions. On the other hand, current technologies assume appearance of complex and distributed organizational forms including virtual enterprises, supply chains, intelligent enterprises, etc. Cooperative supply-chain (CSC) is a philosophy for enterprise integration based on a society formed by autonomous agents, called members of the CSC, by bonding together to solve a common problem. In a CSC, enterprises forming it need to cooperate to fulfil product demands. The purpose of cooperative configuring of groups of agents is to maximise their benefits. Coordination is achieved within the framework of commitments made by CSC members to each other. Members negotiate and compromise in a spirit of cooperation, in order to meet these commitments [6], [23].

A.V. Smirnov et al.: *Agent-Based Technological Framework for Dynamic Configuration of a Cooperative Supply Chain*, Studies in Computational Intelligence (SCI) **28,** 217–246 (2006)
www.springerlink.com

A CSC can be designed in many different configurations. As a result an efficient and quick analysis of performance of such configurations and their optimization have become important. A number of approaches can be found in the literature. These approaches traditionally use optimization, statistical and simulation models. For instance, "cluster analysis" based techniques as a set of multivariate statistical procedures used to reorganize data sets containing information about a sample of entities into relatively homogenous groups of clusters, are applied to analysis of SC configurations in [22]. In other studies, supply chain configuration is considered from the inventory side. Some papers address optimizing safety stock placement across the supply chain [10], [12]. Various simulation modelling approaches have also been described in the literature. A SC simulation modelling and analysis technique based on operating data from General Motors and results of experimentation under various operating parameters and policies, such as Kanban and MRP for a fixed supply chain structure are presented in [4]. A method for modelling the dynamic behaviour for different configurations of food supply chains and evaluating alternative designs of the supply chain by applying discrete-event simulation is discussed in [32].

There are some general limitations related to those approaches. First, in the case of optimization methods, in many cases the optimal solution does not exist, or it is impossible to obtain this solution within the time limits because of high computational complexity of the task. Those methods also require central computational unit (usually with centralized information repository, which is difficult to achieve for the problem at hand). Next, in contrast to the crisp models, negotiations and bargaining associated with configuring generally take place before the application of coordinated strategies when agents have only a vague idea about the expected gains of the coalitions, so distribution of the gains can be vague, imprecise, and uncertain [23], [26]. Moreover, the relationships between the variables in the models being uncertain are not of the stochastic or erroneous nature. This means that the use of statistical methods to study those models can be non-adequate. Using the theory of the fuzzy sets it is possible to study the vagueness and to follow the way from the input of the vague idea about the gains via the bargaining process to the output corresponding to the possibilities of the vagueness of the distributions of the individual benefits. Heuristic solution techniques are expected to give us acceptable results close to optimal within less time.

In the chapter, an agent-based technological framework integrating different techniques of computational intelligence is proposed and studied to solve the problem of the CSC construction as a dynamic, flexible and agile system with dynamic configurations among groups of autonomous agents. The preliminary results of this study were published in [31]. In the chapter, an extended multi-agent framework is described. The demand is decomposed into a number of tasks. Each task is assigned to a group of agents (better to say, they try to form a coalition to perform this task) because a single agent (i) cannot perform them, (ii) cannot perform them efficiently or (iii) cannot meet system's constraints. Other important properties of the proposed techniques and algorithms solving the configuring problem are: (i) computational feasibility and (ii) increased both individual utility and the overall outcome of the system.

The proposed techniques are implemented as a multi-agent system (MAS) over the FIPA compliant CAPNET agent platform according to the general model of the information infrastructure for supply chain management support [6]. The implementation of the basic principle of cooperation in the CSC is based on distribution of procedures between different units / users (or different agents) concurrently in the common knowledge space described by shared application domain ontology [2], [5], [14], [15]. It is, therefore, natural to represent knowledge required for configuration management as a set of interacting autonomous agents, associated with the distributed knowledge sources with the integrated logic of knowledge processing and organizational rules of their integration in the multiagent environment. In the context of this chapter, agents are associated with each node of the CSC and exchange their information to configure a feasible structure for each demand using one of the developed techniques.

The chapter is organized as follows. In the following section, the intuitions behind the proposed problem statements are described, a configuration pattern of a CSC is identified. Configuration problem is defined as a collection of interrelated tasks, two of which, partner selection and resource allocation are described within the proposed framework in more details. Section 3 describes a generalized model of coalition games with fuzzy core applied to partner selection. Only effective fuzzy coalitions are considered, i.e. when redundant agents are offering the same services, the sub coalitions with smaller capacity of bargaining of payoffs or smaller possibility of being formed are eliminated. Also the algorithm to calculate the sum of the fuzzy utilities of the agents composing each coalition is presented. The imputations as a game solution are generated applying genetic algorithm (GA) that maximizes the payoffs of each coalition or that belongs to the core with higher possibility.

In section 4, the formalism of object-oriented constraint networks for the task representation in the task allocation problem is introduced. The following section illustrates the application of genetic algorithm and constraint satisfaction techniques for solving CSC configuration tasks. A GA is used as an heuristic approach to quasi-optimal solutions search for the resource allocation task. In section 6, the architecture of the developed multi-agent framework is described. Implementation of the framework integrates KSNet framework with a multi-agent system implemented over the CAPNET agent platform. A special type of adaptive agents capable of modifying themselves when solving a constraint-based problem solving is introduced.

Finally, a case study is presented in the context of the automotive CSC. A prototype that consists of seven companies is considered. The results obtained applying the developed methods are compared and discussed followed by the conclusions.

2 CSC Configuration Task

In this chapter, the supply chain configuration problem is considered as a system configuration task and a supply chain will be refereed to as a cooperative supply chain or CSC. When a production system (such as a supply chain) is considered, each new demand to be fulfilled assumes that there is some amount of work to be done and some facilities which can perform this work (CSC nodes with associated resources). The work consists of several operations (parallel and/or sequential). Supply chain system consists of production units capable to perform a number of operations (fig. 1). Every component (node) of CSC is described as a set of attributes/properties and a set of possible solutions/templates. Both products and units are described in the application domain ontology.

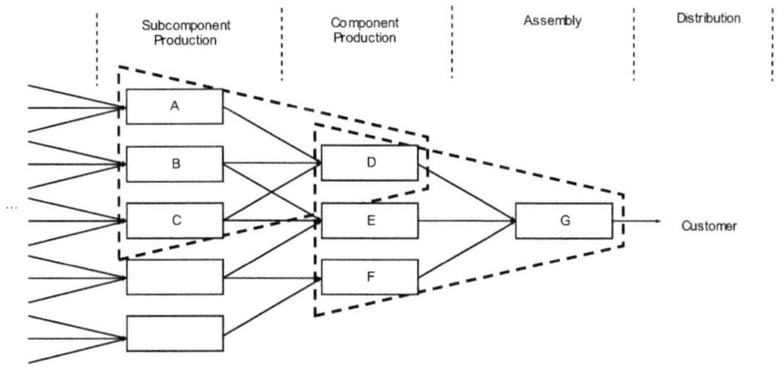

Fig. 1. Structure of a generic CSC pattern

Common CSC objective is to achieve "market-winning" fulfilment time with the least inventory risk for the product. Configuring deals with creating configuration solutions and selecting components and ways to configure these. In CSC each unit selects its direct suppliers ("first-level suppliers"). E.g. in fig. 1, unit D can choose its suppliers from units A, B and C, and unit G can choose its suppliers from units D, E and F. It is supposed that there is no central control unit that could influence upon a choice of other units. In this case, a generic CSC pattern can be defined (shown with dashed lines) as a unit with its first-level suppliers. This allows considering this generic CSC pattern as a configuration pattern.

Generic system configuration task can be described as follows. It is required to obtain a feasible composition of each configuration pattern of a system (fig. 1) meeting specified requirements, with system structure being known. System structure knowledge (described in the application domain ontology) includes system's components, their parameters, utility function and some constraints. Within the presented research the following important interrelated tasks required for supply chain configuration problem were identified: (1) product configuration in accordance with the customer requirements, (2) partner selection among available enterprises – potential supply chain members, (3) resource allocation, (4) transportation network configuration, (5) technological resource configuration. Conceptual architecture of the proposed framework is shown in fig. 2. The triad "input data – method – output data" is used for each task description.

Tasks 1, 4 and 5 are solved using constraint satisfaction/propagation technologies, based on existing algorithms. Tasks 2 and 3 do not have standard solutions. Though the proposed framework is aimed on solving the whole set of tasks, the latter tasks are described in more details. To solve these, the framework combines several techniques: games with fuzzy coalitions, genetic algorithm, and constraint satisfaction. The above tasks are integrated using the common integrated ontological model.

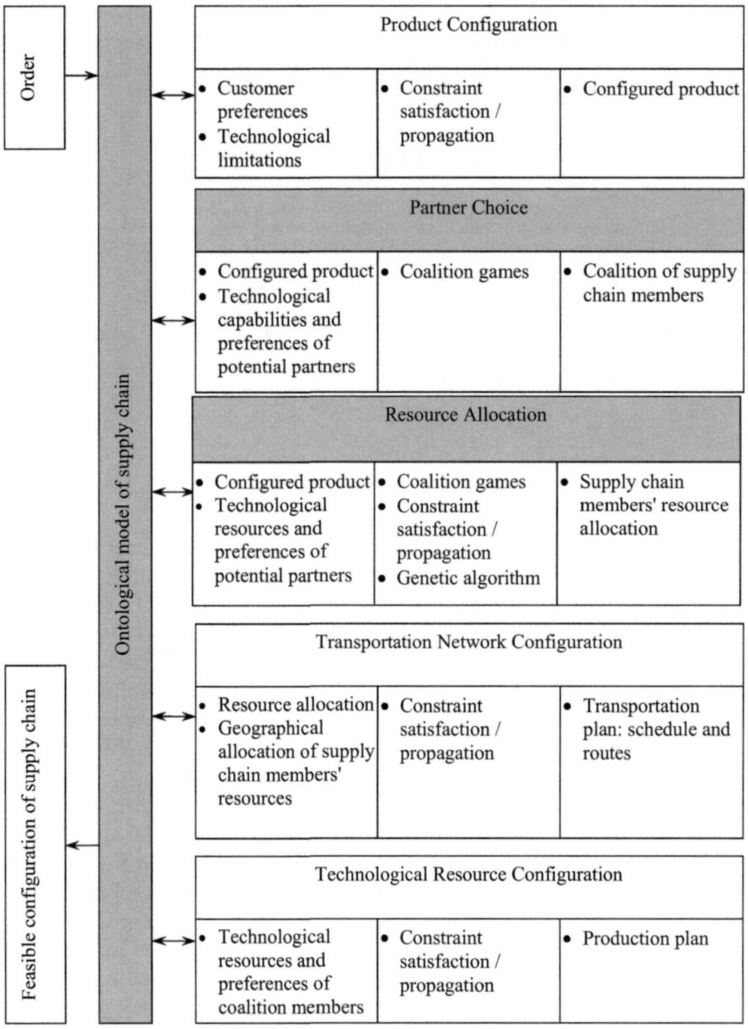

Fig. 2. Conceptual framework of the supply chain configuration problem

3 Partner Choice as a Coalition Game

Configuration patterns identified above are groups composed of rational self-interested agents. These groups are called coalitions. In the chapter, the case of formation of cooperative groups, which viability depends greatly upon the consensus between players as typical for the CSC, is considered. Game theory considers two types of coalitions. The first one corresponds to the situation when each agent can form part of only one group. This type of games is called games with coalitions with full involvement of the players. The second type of the game considers the possibility for agents to form part of different groups. This type of games is called games with a partial involvement of the players. Though for the CSC, the basic type of the involvement is full involvement, it is desirable to have a model capable of considering these both types.

The fundamental point of the agents' integration in coalitions is the economic attractiveness [2], [20]. This attractiveness is usually defined as a utility or membership function. The utilities of each effective coalition together with the distribution of the fuzzy payoff that will receive the grand coalition, conform the fuzzy core that represents the group of imputations (solutions) of the game of fuzzy coalitions. The solution of the game corresponds to the imputation with maximum payoff or with maximum possibility to belong to the fuzzy core of the game. In other words, it permits to maximize the benefits or viabilities (i) of each individual agent, (ii) of agents' coalition and even (iii) of a grand coalition that contains all the agents of the CSC. At the time of negotiation of the particular configuration, agents have only a vague idea about the expected gains of the coalitions, so distribution of the gains can be vague, imprecise, and uncertain. That is why we need to consider the type of games with fuzzy coalitions. The fuzziness is a fundamental component of realistic models of cooperation, which can be considered from the following two points: (i) when a simultaneous participation of players in several coalitions, similar to TU-games [1], takes place, and (ii) when fuzziness is in a preliminary shape or is a priori expressed by players and coalitions about their prospective earnings. In the first case, the fuzzy nature of the coalitions is based mainly on the partial participation of the players in the game. Considered approach is more oriented on the second case. In our case, a fuzzy coalition generates the membership function of the players' payments, which are fuzzy.

3.1 Cooperative Games with Fuzzy Coalitions

In [28], an approach to the problem of coalition forming in MAS based on the generalized model of fuzzy coalition games with associated core, first introduced by Milan Mareš [20], was introduced. The fuzzy core as a set of possible imputations of the game of the structure of effective coalitions is constructed, selecting via solution method based on genetic algorithms, the imputation that maximizes the payoffs for the coalitions or maximizes the membership degree for the fuzzy core. In other words, the possibility of viability of the considered structure of effective coalitions is maximized.

The approach developed in this chapter is based on fuzzy coalition game of integer players with full involvement [20], [21]. Let I be the players set with: $I = \{1,2,...,n\}$, n can be finite or infinite, where the set of possible coalitions is the power set:

$$2^I = \{\phi, \{1\}, \{2\}, ..\{n\}, \{1,2\}, ..., \{n-1, n\}, \{1,2,3\}, ..., \{n-2, n-1, n\}, ..., \{1,2,..n\}\}$$

A structure of n coalitions $K_1, ..., K_m$ is: $\kappa = \{K_1, ..., K_m\}$

Let (I,v) be a crisp coalition game where v is called a characteristic function of the game. The generalization of the game with fuzzy coalitions (I,w) as a fuzzy extension of (I,v) is proposed here according to the following definition of the core [28]. The fuzzy core for this game is defined as an extension of the crisp core. Let $w:2^I \rightarrow R^+$ be a mapping connecting every coalition $K \subset I$ with a fuzzy quantity $w(K) \in R^+$, with membership function $\mu_k : R \rightarrow [0,1]$, where $v(K)$ is the modal value of $w(K)$ with $w(\phi) = 0$.

Definition.- The core of the game with fuzzy coalitions (I,w) is defined as a fuzzy subset C_F of the R^I with the membership function $\gamma_{C_F} : R^I \rightarrow [0,1]$, such that for the imputation $X = (x_{ij})_{i \in I, j \in K} \in R^I$:

$$C_F = \left\{ X \in R^I : v \succ= (w(I), \left\langle \sum_{\substack{i \in I, \\ j \in K}} x_{ij} y_{ij} \right\rangle), \min_{K_i \subset K}(v \succ= (\left\langle \sum_{\substack{i \in I \\ j \in K_i}} x_{ii} y_{ij} \right\rangle, w(K_i)) \right\}. \quad (1)$$

where x_{ij} - is a fuzzy payoff of the agent *i* for the participation in the coalition *j*, $i = 1,2,\ldots,n \;\; j = 1,2,\ldots,K_i$,

$y_{ij} = 1$ if the agent *i* is in the coalition *j*,

$y_{ij} = 0$ if the agent *i* is not in the coalition *j*.

with associated memberships:

$$\gamma_{C_f}(x) = \min_{\substack{i \in K_i \\ j \in K}} \left\{ X \in R^I : v \succ= (w(I), \left\langle \sum_{\substack{i \in I, \\ j \in K}} x_{ij} y_{ij} \right\rangle), \min_{K_i \subset K}(v \succ= (\left\langle \sum_{\substack{i \in I \\ j \in K_i}} x_{ii} y_{ij} \right\rangle, w(K_i)) \right\} \quad (2)$$

where: $v \succ= (w(I), \left\langle \sum_{\substack{i \in I, \\ j \in K}} x_{ij} y_{ij} \right\rangle)$, with the preference $\succ=$ as a weak order fuzzy relation with membership function $v \succ= : R \times R \rightarrow [0,1]$.

To work with structures of effective coalitions in the sense that each coalition in the set cannot be eliminated by any sub coalition, the following definition is given [18].

Definition. - A coalition K is said to be effective if it cannot be eliminated from the structure of coalitions by a sub coalition $L \subset K$. The set of effective coalitions is called a structure of effective coalitions. The fact that K is effective happens with a possibility of:

$$\sup_{x \in R^l} (min(\mu_k(x), \mu^*_l(x) : L \subset K)) \tag{3}$$

The structure of effective coalitions is formed by incorporating of binary variables y_{ij} to the core (1). It means that agents can consider different strategies to join any coalition. Moreover, the proposed model is a game with non side off payments since it is possible to assume that fuzzy payoffs $\{(w(I), x_{ij}, w(K_i))\}$ have associated membership functions of any form, linear or non-linear ones, universal or not.

3.2 Construction of a Coalition Membership Function

One of the most important tasks is to construct a coalition membership function for effective coalition based on the expectations of each player. A model for construction of a coalition membership function consists in taking the individual aspirations of each agent of a coalition reflected in their prospective fuzzy utilities, and in generating the sum of the fuzzy quantities corresponding to these payments. It is based on the definition, a theorem and a lemma [8].

Definition.- A binary operation * in R is said to be incrementing if and only if:

If $x_1 > y_1$ *and* $x_2 > y_2$ *then* $x_1 * x_2 > y_1 * y_2$

Analogically, * is decrementing if and only if:

If $x_1 > y_1$ *and* $x_2 > y_2$ *then* $x_1 * x_2 < y_1 * y_2$.

Using the principle of the extension, * can be extended to Θ combining 2 fuzzy numbers (this is, convex fuzzy sets and normalized in R) M *and* N, with:

$$\mu_{M\Theta N}(z) = \sup_{z = x * y} \min(\mu_M(x), \mu_N(y)) \tag{4}$$

Lemma. - Let M *and* N be two fuzzy continuous numbers and * a binary continuous incrementing operation. Being $[\lambda_M, \rho_M]$ *and* $[\lambda_N, \rho_N]$ two intervals on growing parts of μ_m *and* μ_N respectively such that:

$$\forall x \in [\lambda_M, \rho_M],\ \forall y \in [\lambda_N, \rho_N],\ \mu_M(x) = \mu_N(y) = w$$
$$then:$$
$$\forall t \in [\lambda_M * \lambda_N, \rho_M * \rho_N],\ \mu_{M\Theta N}((t) = w \tag{5}$$

Theorem. - If M *and* N are continuous fuzzy numbers whose membership functions are onto and * is a binary continuous incrementing operation, then the extension Θ is a fuzzy continuous number whose membership function is onto. The effective construction of $M \Theta N$ can be developed separately increasing or decreasing parts of μ_M *and* μ_N, using the procedure pointed out by the previous lemma.

A general algorithm for an *n*-ary operation is presented which calculates a sum of *n* fuzzy quantities for our case:

1. Flattening: The n fuzzy sets are changed into fuzzy sets all having the same height.
2. Decomposition of each fuzzy set as described above into two sets, the set of no decreasing parts and the set of no increasing parts. The constant parts between two no decreasing parts (respectively, no increasing) belong to the no decreasing (respectively, no increasing) set. The constant parts, which are between parts of different kinds, belong to both.
3. In this chapter, the operation * is the sum (+). The operation * is performed for every n-tuple of parts (one part for each fuzzy set), all belonging to the same kind of sets (no decreasing or no increasing). The flattening effect may be used.
4. Union: For each n-tuple of parts a fuzzy set was built in step 3. The union of these fuzzy sets is the final result.

3.3 Solution of a Coalition Game with Genetic Algorithm

Considering solution methods, i.e. how to find the imputation x that satisfies (2), the fuzzy Pareto optimum gives the analytic solution, however it is very complicated and requires the definition of the fuzzy super optimum and fuzzy relation of domination. That is why genetic algorithm (GA) to find the optimum imputation is proposed as a solution method [28]. Some basic definitions of GA will be discussed in section 5. To find the solution of the cooperative game, the expression (1) is binary coded with the precision given by the numeric processor, taking the expression (2) as a fitness function. The influence of the parameters of the GA on the solution is beyond the scope of this chapter, so the crossover and mutation rates are considered fixed at the levels of 6 and 50% respectively, with a fixed population size. The algorithm forms sets of the dimension equivalent to that of the population, where each vector's coordinate is 0 or 1 representing chromosomes.

Starting the GA with the initial solution (feasible or not), the algorithm evolves in the region defined by the fuzzy core improving or looking for improvement, based on the fitness function in each iteration until the parameters of finishing the algorithm are completed, either for time, precision or number of iterations. The algorithm converges when the non empty core conditions are completed [21]. Once the algorithm fulfills the conditions settled by the assignment of the parameters, the fuzzy core, the membership function of the core and the optimum imputation are decoded from the binary form to the real one.

4. Ontological Model of the CSC Based on Object-Oriented Constraint Networks

The fact of existence of the relationships between resources and CSC components, allow considering the task of resource allocation as a constraint satisfaction problem [19]. In this section, we introduce the formalism of object-oriented constraint networks for the task representation in the task allocation problem. This abstract model unifies main concepts of languages, such as standard object-oriented languages with classes, and constraint programming languages. It supports the declarative representation, efficiency of dynamic constraint solving, and problem modeling capability, maintainability, reusability, and extensibility of the object-oriented technology. This formalism is used to define the application ontology (AO) describing the problem domain of automotive supply chain configuration.

4.1 Object-Oriented Constraint Network

The structure of a constraint network is defined as: $A = (O, Q, D, C)$ where

O – a set of *entities* or *object classes* ("*classes*"). Each of the entities in a class is considered as an *instance* of the class. The set $O = O^I \bigcup O^{II}$ consists of two subsets: $O^I = \{o : \exists\, instance(o)\}$ - a set of *non-primitive* classes i.e. classes which can have instances and $O^{II} = \{o : \neg\exists\, instance(o)\}$ - a set of *primitive* classes i.e. classes which cannot have instances;

Q – a set of class attributes ("*attributes*");

D – a set of attribute domains ("*domains*");

C – a set of *constraints*. Six types of constraints have been defined $C = C^I \bigcup C^{II} \bigcup C^{III} \bigcup C^{IV} \bigcup C^V \bigcup C^{VI}$:

$C^I = \{c^I\}$, $c^I = (o, q), o \in O, q \in Q$ - accessory of attributes to classes;

$C^{II} = \{c^{II}\}$, $c^{II} = (o, q, d), o \in O, q \in Q, d \in D$ - accessory of domains to attributes;

$C^{III} = \{c^{III}\}$, $c^{III} = (\{o\}, True \vee False), |\{o\}| \geq 2, o \in O$ - classes compatibility (compatibility structural constraints);

$C^{IV} = \{c^{IV}\}$, $c^{IV} = \langle o', o'', type\rangle, o' \in O, o'' \in O, o' \neq o''$ – hierarchical relationships (hierarchical structural constraints) "is a" defining class taxonomy (*type*=0), and "has part"/"part of" defining class hierarchy (*type*=1). The most abstract class is "Thing".

$C^V = \{c^V\}$, $c^V = (\{o\}), |\{o\}| \geq 2, o \in O$ - associative relationships ("one-level" structural constraints);

$C^{VI} = \{c^{VI}\}$, $c^{VI} = f(\{o\}, \{q\}) \to True \vee False$, $|\{o\}| \geq 0$, $|\{q\}| \geq 0$, $o \in O$, $q \in Q$ - functional constraints referring to the names of classes and attributes.

$|c|$ - is a number of parameters included into a constraint (constraint cardinality), $o \in O, q \in Q, d \in D, c \in C$ are considered as ontology elements.

A constraint network with the above structure can be described as $CNet = (A, I) = (O, Q, D, C, I)$ where I – is information content of the constraint network. Information content is a set of instances with fixed attribute values (known or not).

Fuzzy constraint network is an extension to the above constraint network and can be described as $(O, Q, D, C_\mu, W, T, I_p)$, where

O, Q, and D are the same as above;

C_μ is a set of constraints, where each constraint contains a function μ of membership to $[0:1]$ associated to weight ω_c representing its weight (importance) or priority;

W is a weight scheme, i.e. a function combining satisfaction degree of constraint $\mu(c)$ to ω, for estimation of weighted satisfaction degree of $\mu^\omega(c)$;

T is an aggregation function, which performs simple partial regulating on defined values, defining C_μ;

I_p is an information content of the constraint network, which has a nondeterministic or probabilistic nature.

Below, some example constraints are given:

- an attribute *costs* (q_1) belongs to a class *component* (o_1): $c^I{}_1 = (o_1, q_1)$;
- the attribute *costs* (q_1) belonging to the class *component* (o_1) may take positive values: $c^{II}{}_1 = (o_1, q_1, \mathrm{R}^+)$;
- a class *body* (o_2) is compatible with a class *body production* (o_3): $c^{III}{}_1 = (\{o_2, o_3\}, \mathit{True})$;
- an instance of the class *component* (o_1) can be a part of an instance of a class *facility* (o_4): $c^{IV}{}_1 = \langle o_2, o_3, 1\rangle$;
- the *body production* (o_3) is a *resource* (o_5): $c^{IV}{}_1 = \langle o_2, o_3, 0\rangle$;
- an instance of the class *body* (o_2) can be connected to an instance of the class *body production* (o_3): $c^V{}_1 = (o_2, o_3)$;
- the value of the attribute *cost* (q_1) of an instance of the class *facility* (o_4) depends on the values of the attribute *cost* (q_1) of instances of the class *component* (o_1) connected to that instance of the class *facility* and on the number of such instances: $c^{VI}{}_1 = f(\{o_1\}, \{(o_4, q_1), (o_1, q_1)\})$.

4.2 Construction of the Request Ontology

To solve the task of the CSC configuring within the multi-agent framework, the user's request should be translated into notations and terms of the AO. Let us consider a natural language specification of a typical configuration task:

"Configure supply chain (SC) in the part of car components production allocation to SC participants (SC participants are to be found). In accordance with the order the production costs must be minimal, car's engine volume must be 2.0 l, and cost is equal or less than $22000".

Request ontology describes structure of the user request. Instances contained in the request define an extensional content of the constraint network. This is done in two steps: preliminary request ontology (PRO) creation and request ontology (RO) creation. Preliminary RO describes request in system notation (object-oriented constraint networks described above): $\hat{A}(R) = (\hat{O}, \hat{Q}, \hat{D}, \hat{C})$, where

$\hat{O} \subset O, \hat{Q} \subset Q, \hat{D} \subset D, \hat{C} \subset C$ if the request is based on a template;

$\hat{O} \subset V_R$, $\hat{Q} \subset V_R$, $\hat{D} \subset D_R$, $\hat{C} \subset C_R$ in the other case, V_R, D_R, C_R - respectively vocabulary, domains, and constraints of the request.

To facilitate creation of PRO specially designed templates for user request input can be used. The process of request ontology creation includes two stages:

1. Preliminary request ontology $\hat{A}(R) = (\hat{O}, \hat{Q}, \hat{D}, \hat{C})$ in notation of the top-level ontology is created by a group of ontology engineers;
2. The request ontology in notation of the top-level ontology containing terms of AO and preliminary request ontology, and constraints describing correspondence between them (correspondence constraints): $A(R) = A'(R) \cup A''(R) \cup C^{\varphi}$, where:

$A'(R) = (O', Q', D,' C'), O' \subset O, Q' \subset Q, D' \subset D, C' \subset C$;

$A''(R) = (O'', Q'', D'', C''), O'' \subset \hat{O}, Q'' \subset \hat{Q}, D'' \subset \hat{D}, C'' \subset \hat{C}$;

C^{φ} – correspondence constraints: $\{c^{\varphi} = \langle e', e'', f, \mu \rangle\}$, where e', e'' are elements of the AO and preliminary request ontology respectively, f – logical condition of the correspondence validity, μ – reference to a method converting attributes values (optional);

φ denotes mapping between the preliminary request ontology and AO elements: $\varphi(E'') = E'$, where E', E'' – sets of corresponding elements of the AO and the preliminary request ontology respectively.

Object oriented constraint networks for user request (R_k) described in system terms and notation is defined as:

$$CNet(A''(R), R_k) = (O''(R_k), Q''(R_k), D''(R_k), C''(R_k), I''(R_k)),$$

$$O''(R_k) \subset O''(R),\ Q''(R_k) \subset Q''(R),\ D''(R_k) \subset D''(R),\ C''(R_k) \subset C''(R).$$

Fig. 3 shows a user request constituents example for a request corresponding to the case study (Section 7). This decomposition is required for problem representation to ILOG Configurator [13]. The Request Ontology obtained after association operation for the example request is shown in fig. 4. The request terms to be included in the structural constituent constituting the PRO are in italics; the terms to be included in the parametric constituent are underlined:

"*Configure supply chain (SC)* in the part of *car components* production allocation to *SC participants* (SC participants are to be found). In accordance with the *order* the *production costs* must be *minimal*, car's engine volume must be 2.0 l, and cost is equal or less than $22000".

This ontological model is used as an agglutinating centre of the framework.

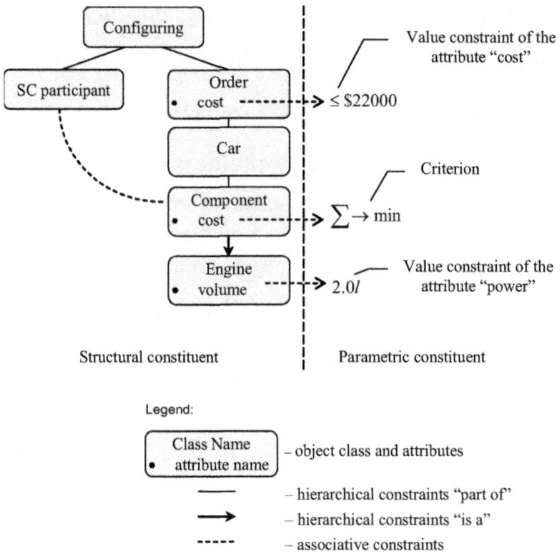

Fig. 3. Request constituents

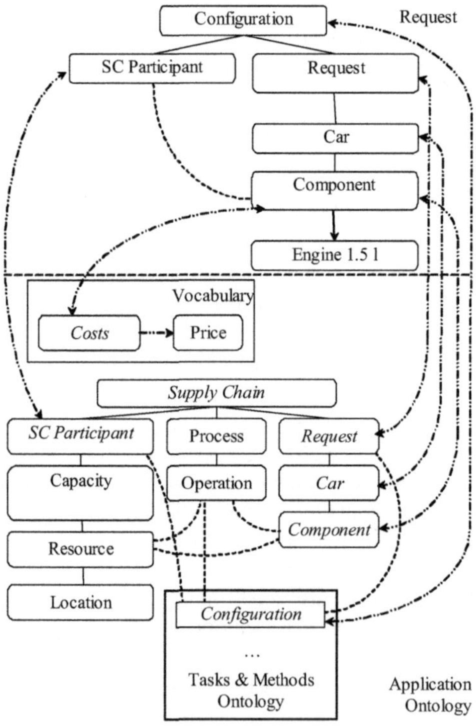

Fig. 4. A fragment from the Request Ontology

5 Methods for Resource Allocation Task Solution

The following illustrates the application of genetic algorithm and constraint satisfaction techniques for solving CSC resource allocation.

5.1 Genetic Algorithm

A typical GA is started with a set of solutions (represented by chromosomes) called population. Solutions from one population are taken and used to form a new population. This is motivated by a hope that the new population will be better than the old one. Solutions, which are selected to form new solutions (offspring) are selected according to their fitness - the more suitable they are the more chances they have to reproduce. This is repeated until some condition (e.g. number of populations or improvement of the best solution) is satisfied. The structure of basic GA is as follows:

1. Generate *random population* of n chromosomes (suitable solutions for the problem);
2. Evaluate the fitness $f(x)$ of each chromosome x in the population (*fitness*);
3. Create a *new population* by repeating following steps until the new population is complete:
   a) select two parent chromosomes from a population according to their fitness: the better fitness, the bigger chance to be selected (*selection*);
   b) with a crossover probability cross over the parents to form a new *offspring* or children. If no crossover was performed, offspring is an exact copy of parents;
   c) with a *mutation* probability mutate new offspring at each locus (position in chromosome);
   d) place new offspring in a new population;
4. Use new generated population for a further run of algorithm;
5. If the end condition is satisfied, stop, and return the best solution in current population;
6. Else go to step 2.

GA was used as an heuristic approach to quasi-optimal solutions search. Initially a random set of solutions is generated, then solutions are estimated; the set is sorted according to the chosen criteria, and mutation mechanism is applied to the best solution to generate new solutions. The newly generated solutions are sorted, and the new iteration is performed. The process stops after a predefined number of iterations.

Applied for the resource allocation task, a possible solution (a chromosome) can be represented as a tuple: $Dec = <Dec_1, Dec_2, ..., Dec_n>$, where n – is a quantity of components to be produced. Each element Dec_j (or a gene) is also a tuple: $S_j = <dec_{j1}, dec_{j2}, ..., dec_{jm}>$, where m – is a quantity of available capacities for the job, and each element dec_{ji} indicates capacity actually used for production. The fitness function indicates overall system costs which are to be minimized. It incorporates the same formula used in other techniques presented in this chapter. Several experiments have been performed with different mutation rates and no crossover mechanism applied. Results show that the GA's efficiency is highest when the mutation rate is about 20%. This value of mutation and population of 400 has been used for obtaining the results presented below.

5.2 Constraint Satisfaction Problem

A constraint satisfaction problem is defined as (X, D, C). It consists of n variables $X = \{x_1, x_2, ..., x_n\}$, whose values are taken from finite, discrete domains $D = \{D_1, D_2, ..., D_n\}$, and a set of constraints on their values $C = \{c_1, c_2, ... c_m\}$. A constraint is defined by a predicate. That is, the constraint $c_k(x_{k1}, ..., x_{kj})$ is a predicate that is defined on the Cartesian product $D_{k1} \times ... \times D_{kj}$. This predicate is true if the value assignment of these variables satisfies this constraint. Solving a constraint satisfaction problem assumes finding an assignment of values to all variables such that all constraints are satisfied. In other words, a constraint satisfaction problem is a task of finding a consistent assignment of values to variables [19], [33].

The components of the task have the following notations:

x_{ijt} – the quantity of the j component to be produced by unit i at a time interval $[t - 1, t]$;

D_{ijt} – the sets of integer positive numbers (I^+);

C – the constraints responsible for synchronization and consistency of the considered CSC system. The following groups of constraints were selected: capacity-based constraints, constraints on materials availability, delivery or synchronization constraints.

For the experiments the engine of ILOG Configurator was used. It represents constraint network to be processed in an object-oriented form (fig. 5). The system supports the following two types of relations: (i) "is a" between classes and objects, and (ii) "is connected to" between objects. In turn relations "is connected to" can be of the following two subtypes: (i) "has part" or "part of" and (ii) "uses".

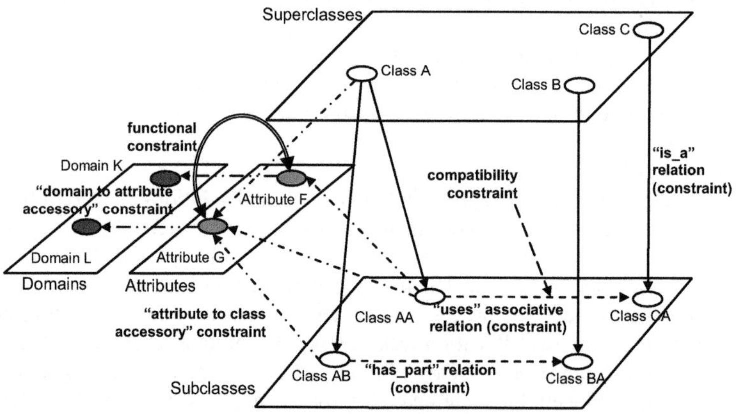

Fig. 5. ILOG Constraint Satisfaction Model adapted to the knowledge representation formalism

The objects own attributes whose values belong to defined domains and are controlled by constraints existing in the problem domain. This representation allows rapid and convenient model creation and maintenance. On the other hand an advanced ILOG Solver performs the task solving process in an efficient way.

6 Multi-agent Framework Description

The implementation of the framework illustrated in fig. 6, integrates Knowledge Source Network (KSNet) framework with a multi-agent system implemented over the CAPNET agent platform [31], [7]. KSNet was used for implementation of the constraint satisfaction technique, while CAPNET provides a generic infrastructure for the technological framework. Both infrastructures are based on the principles of the Service Oriented Architecture (SOA) environments. CAPNET agents can expose their functionality as Web Services (WS) and also to locate and consume web services (from KSNet) in a transparent fashion [29]. This way the whole framework can provides a unified interface for the user himself or for the user application hiding the implementation details.

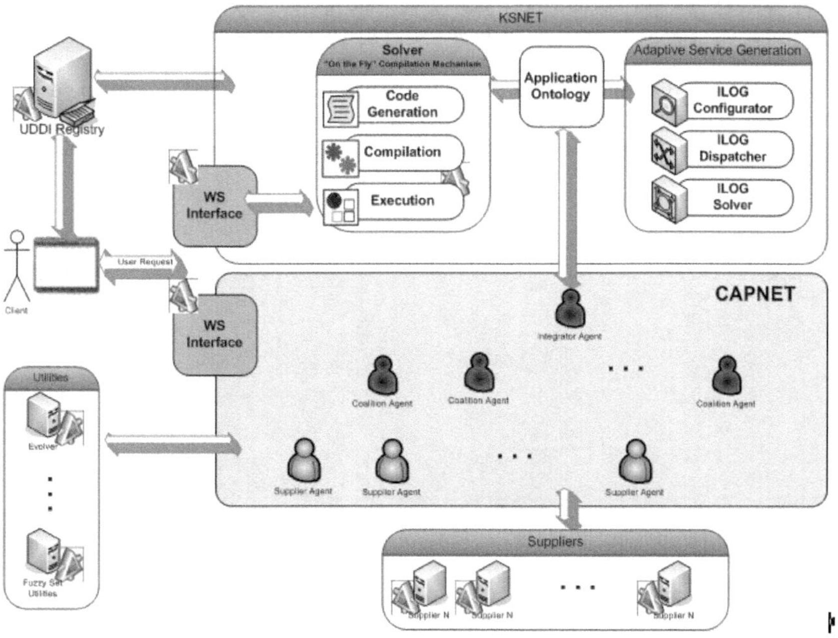

Fig 6. General Architecture of the Multi-agent Framework

Each node of the CSC was supplied with a domain Agent playing one of the following roles: Supplier Agent (for the suppliers of the raw materials and components), Coalition Agent (for each head of the configuration pattern like the suppliers of the car components, called Body, Motor and Transmission respectively in the case study) and Integrator Agent (called Assembly in the presented case study). All these agents are equipped with the contract net negotiation protocol (CNP) used to collect the biddings for the demand. Coalition Agents have been generated for each component every time a new demand occurred. The Integrator Agent (simulating the grand coalition) was the one initiating negotiation protocols and deciding the final configuration based on the computations performed by the game solvers. Utilities are implemented as external legacy software modules. Genetic algorithms are implemented using the Evolver software [11], accessed through the Excel Wrapper agent, while Fuzzy Set utilities are MATLAB components.

6.1 Coalition Game for Partner Selection

The phases of the algorithm implementing the model described in Section 4 are listed below:

1. *Demand Specification.* In this first phase, demand and parameters are specified and the product is identified by the application ontology available to the Integrator Agent. This operation can be made by the user via CSC Configurator or directly in the agent's API that gives the possibility to load the demand parameters specified by the request ontology and to run the simulation. CSC Configurator´s interface is also used to configure the multi-agent environment, visualize and document the results of the experiments (fig. 7). Once the demand is specified, the Integrator Agent chooses the components needed to satisfy it using the application ontology where it can find all the components classified by sub products.

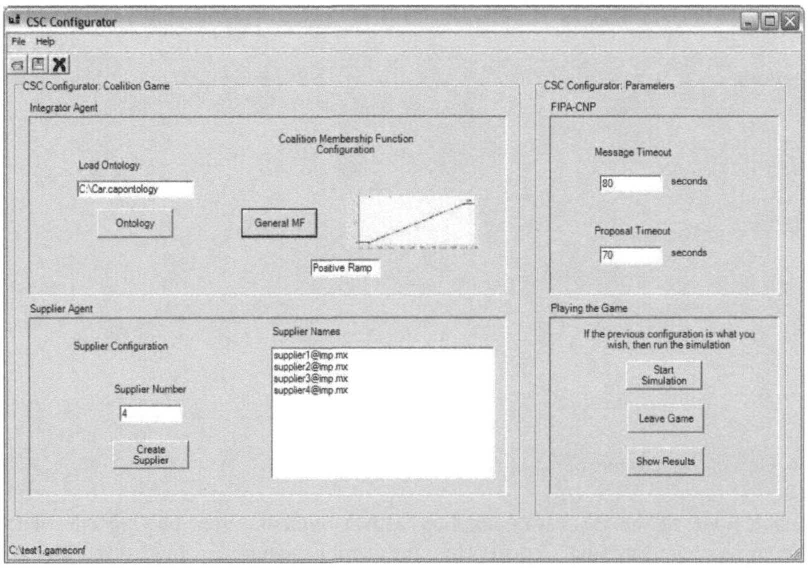

Fig 7. A screenshot of CSC Configurator

2. *Identifying possible suppliers.* Each component of the demand has an agent assigned to it for making the coalition that can produce this component. That is, the agent is responsible for getting all the sub products required to form a component. It is called the Coalition Agent. This agent gets from the Integrator

Agent the component requirement and consults the application ontology and demand specification. Once it has all the required information, it starts FIPA contract net protocol (FIPA-CNP) to get suppliers of each sub product. All the Supplier Agents with the capacity to produce a component will receive the CFP "call for proposals" message from the Coalition Agent, and if they are interested will return an offer with an associated membership function. Coalition Agents have a GUI to specify their parameters.

3. *Making Coalition Proposal.* Once the Coalition Agent received all the proposals, one for each sub product, it calculates the coalition membership function invoking the fuzzy set utilities component. Then the result is sent to the Integrator Agent.
4. *Game Resolution.* After the Integrator Agent has received the membership functions from each Coalition Agent, it creates the core of the game to be solved using the Evolver component that allows optimizing the distribution of payments and the possibility of the game.
5. *Results Distribution and Configuring of the SCN.* In the case a feasible solution in the previous step has been obtained, this is communicated to the Coalition Agents and further on to the Supplier Agents. Once all the suppliers have received the payment proposal, they can decide if it corresponds to their interests or not[1]. If they accept the payment then the CSC is formed, on the other hand the reject message is transmitted to all the agents in the system and the configuration attempt should be repeated by simply running the same game in order to analyze another feasible solution or by reconfiguring the game (e.g. changing membership functions).

Once the simulation is over the final results have to be visualized. To do this, each participant agent in the system when performing any action leaves an *agentprint* in the MAS log. This system log allows analyzing both the execution of each agent of the system (CAPNET functionality) as well as the results of the particular application. In order to implement this feature, the transducer automata was created which is responsible to validate the system execution, and at the same time to give a graphical output of the resulting configuration with the purpose of making an analysis of the results. This transducer verifies if during the execution of the system the agents register themselves, have the necessary communication and finally they are deregistered or stop operating when necessary. On the other hand, the analysis of the CNP permits to identify the coalition structures. The details of the transducer's implementation are out of the scope of this chapter; results visualization will be shown in the case study section.

6.2 Evolver Interface Integration

Step 4 of the described algorithm involves the solution of the game using GA (section 3.3). To obtain this, Evolver v. 4.0 for Microsoft Excel is used. To interact with Evolver, a particular Excel input spreadsheet with the parameters of the game should be configured. The whole operation is made by the Integrator Agent, when it receives

[1] It should be noted that the obtained solution belongs to their membership functions.

the information from the Coalition Agents, through executing an Excel macro that runs Evolver to get the results. The generated spreadsheet is shown in fig. 8 (for the game from the case study described in the following section).

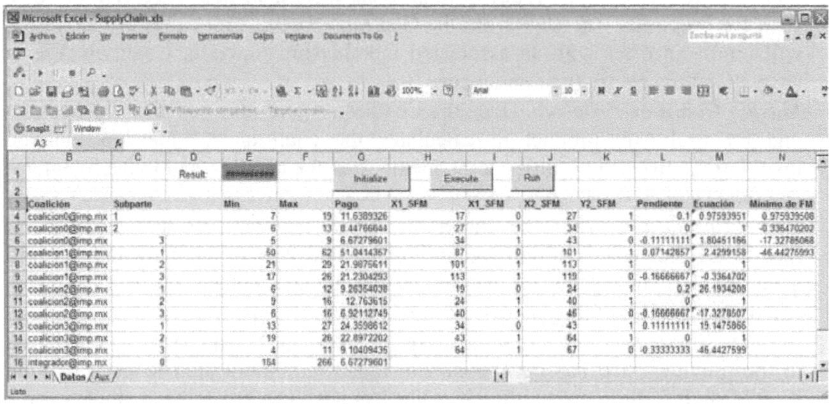

Fig. 8. A screenshot of the Excel spreadsheet

The meaning of the columns is the following. *Coalition*: contains the names of the formed coalitions. The last row of this column corresponds to the grand coalition. *Subpart*: each component of a coalition. *Min*: the minimum of the membership function's value of the selected proposal by supplier to offer the component service. Each supplier has a membership function from where the value is obtained. In the particular case of the grand coalition the minimum value is zero. *Max:* the maximum of the membership function value of the selected proposal by supplier to offer the component service. In the particular case of the grand coalition the minimum value is one. *Payment*: represents the game variables and the payments calculated by Evolver. For the study case, the number of considered variables equals to the number of suppliers participating in all coalitions. The cells F2-F13 contain these variables. In the F14 cell the formula is placed to obtain the minimum of the values between F2-F13 (the minimum of the calculated payments), that represents the game possibility.

The coalition membership function must be introduced to the Excel sheet. This is done breaking each function into rectilinear segments that form it. The number of rectilinear segments depends upon the type of a membership function: triangular one has 2 segments, trapezoidal - has 3 segments, the positive ramp has 1 (as in the case study), negative ramp has 1 and the step has 1. For each rectilinear segment we have to obtain two points that form it in the plane X, Y. By each segment the values or formulas that correspond to *X1_SFM, Y1_SFM, X2_SFM, Y2_SFM, Slope* and *Equation* that are mentioned below must be stored:

X1_SFM: the X value of the first segment point is placed.

Y1_SFM: the Y value of the first segment point is placed.

X2_SFM: the X value of the second segment point is placed.

Y2_SFM: the Y value of the second segment point is placed.

Slope: in each cell of this column is placed the formula to calculate the slope of each segment through the points defined in columns *X1_SFM, Y1_SFM, X2_SFM, Y2_SFM* respectively. The slope is calculated as follows:

(Y1_SFM - Y2_SFM)/(X1_SFM - X2_SFM).

Equation: the equation that corresponds to each segment is the sum of the payments of each coalition calculated as:

Y1_SFM - Pending * X1_SFM + Slope * SPC,

where *SPC* is the sum of payments of the coalition.

After introducing the segments data for each coalition, the same has to be made for those of the grand coalition. In this case, *SPC* is the sum of the payments of all the suppliers. *Minimum of FM:* contains the minimum of the equations of each membership function.

Finally, when all the data are stored in the Excel sheet, the Integrator Agent runs the macro to obtain the resultant payments. These payments are those that are sent like proposal to the Coalition Agents and to the respective suppliers.

6.3 Adaptive Agents for Resource Allocation

Additional innovative feature of the proposed approach is the use of adaptive agents. Since supply chains operate in a dynamic environment on-the-fly problem modification and solving is required. These agents may modify themselves when solving a particular problem. For example, within the KSNet-approach there is an agent responsible for constraint-based problem solving based on existing knowledge. The problem is described by application ontology. Upon receiving the request the application loads an appropriate ontology and generates an executable module for its solving on-the-fly. Roughly outlined this on-the-fly compilation mechanism within the KSNet is based on the following ideas (Fig. 6):

- a pre-processed user request defines (1) what ontologies are to be used for the problem domain description, and (2) what knowledge sources are to be used;
- C++ code is generated on the basis of information extracted from (1) the user request (goal, goal objects, etc.), (2) appropriate ontologies (classes, attributes, and constraints), and (3) suitable knowledge sources;
- the compilation is performed in an environment of the prepared in advance C++ project;
- failed compilations/executions do not interrupt the system work on the whole; an appropriate error message is generated.

Constraint solver is implemented using ILOG Configurator constraint programming engine. ILOG Configurator represents constraint network to be processed in an object-oriented form. The objects own attributes whose values belong to defined domains and are controlled by constraints existing in the problem domain. This representation allows rapid and convenient model creation and maintenance. On the other hand an advanced ILOG Solver performs the task solving process in an efficient way.

The essence of the proposed on-the-fly compilation mechanism is to write the ontology elements (classes, attributes, constraints) to C++ file directly so that it could be compiled into ILOG-based program. The agent responsible for the problem solving creates the C++ file based on these data and inserts the program source code to the program (the prepared in advance Microsoft Visual Studio project) prepared in advance. The program is compiled in order to create an executable file in the form of dynamic-link library (DLL). After that the agent calls a function from DLL to solve the problem. The experiments showed that for resource allocation problem the compilation time is significantly less than the time of the problem solving by the generated program.

7 Case Study Description

For the demonstration purposes a rather simple example of an automotive CSC configuration pattern was considered. The instantiation of the generic configuration patter for the case study is shown in fig. 9. The case study deals with production of a hypothetic vehicle (a *Car*). The price of the car is *$22,000*. The production process consists of the following two phases: *Component Production* and *Car Assembly*. *Component Production* consists of three parallel operations: *Body Production*, *Motor Production*, and *Transmission Production* [5].

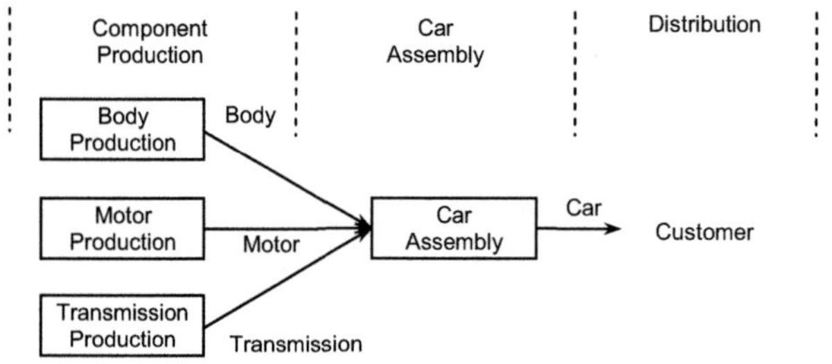

Fig. 9. Structure of the Production Process

Demand is represented by a uniform distribution around the linear trend:

$$d_t = a + b \cdot t + \mu \cdot \sigma, \text{ where} \tag{6}$$

t – time instant

d – demand (d_t corresponds to time interval [t -1, t])

a – basis value ($a = 100$)

b – trend component (*0* for demand without trend)

μ – random noise uniformly distributed within *[0:1]*

σ – distribution amplitude

For simplicity, an example with trend (b=5) and with low noise ($\sigma = 5$) is considered. The forecasting model for the demand is the following:

$$100 + 5t + 5\mu, \; for \; t = 1,...,5 \text{ and } \mu \text{ is uniform in } [0,1] \tag{7}$$

In this case the technique of Single Moving Average for the demand forecasting can be used:

$$f_{t+2} = f_{t+1} = \frac{\sum_{i=t-n+1}^{t} d_i}{n}, \text{ where} \quad (8)$$

f – forecasted data (forecast)

n – forecast base ($n = 5$);

Car Assembly can be performed by one unit (*car assembler* or Unit 7) with the following parameters:

- *Capacity = 150*. A case with fixed capacity (no changes) is considered.
- Stocks are *unlimited*
- Stocking costs are *$750* per car/week, *$300* per body/week, *$200* per motor/week, *$100* per transmission/week.
- Assembly costs are *$1500* per car

Component Production can be performed by 6 units, each with different facilities (**Table 1**). Payoffs for Component Production are *$7000* per body, *$5000* per motor, and *$4000* per transmission. Penalties for backorders are *$1500* per car/week (paid by the car assembler to the customer), *$400* per body/week, *$300* per motor/week, *$250* per transmission/week (paid by the component producers to the car assembler).

Table 1. Data for the production units (potential supply chain members)

Unit	Capacity	Facility	Production Costs	Stocking
Unit 1	*100*	Body Production	*$4500* per body	*$250* per body/week
Unit 2	*100*	Motor Production	*$3500* per motor	*$150* per motor/week
Unit 3	*100*	Transmission Production	*$2500* per transmission	*$50* per transmission/week
Unit 4	*303*	Body Production	*$4900* per body	*$300* per body/week
		Motor Production	*$3800* per motor	*$200* per motor/week
		Transmission Production	*$2700* per transmission	*$80* per transmission/week
Unit 5	*100*	Motor Production	*$3600* per motor	*$170* per motor /week
		Transmission Production	*$2600* per transmission	*$60* per transmission/week
Unit 6	*200*	Body Production	*$4700* per body	*$270* per body/week
		Motor Production	*$3600* per motor	*$170* per motor/week

7.1 Fuzzy Coalition Game Model of the Case Study

For the purposes of the chapter, the generalized model of the fuzzy core was used to include the capacities of each agent and the tasks to fulfill. These parameters are defined by the demand and are specified in the product bill and in the application ontology. The new core definition follows (without lacking of generality, for simplicity this model doesn't include the fluctuations caused by inventories):

$C = \{2500x_{11t} + 2100x_{41t} + 2300x_{61t} + 1500x_{22t} + 1200x_{42t} + 1400x_{52t} + 1400x_{62t} +$
$1500x_{33t} + 2500x_{74t} + 1300x_{43t} + 1400x_{53t} \geq (100 + 5t + \mu 5)w(I),$

$$
\begin{aligned}
&2500x_{11t} + 2100x_{41t} + 2300x_{61t} \le (100 + 5t + \mu 5)w(k_1) \\
&1500x_{22t} + 1200x_{42t} + 1400x_{52t} + 1400x_{62t} \le (100 + 5t + \mu 5)w(k_2) \\
&1500x_{33t} + 1300x_{43t} + 1400x_{53t} \le (100 + 5t + \mu 5)w(k_3) \\
&2500x_{74t} \le (100 + 5t + \mu 5)w(k_4) \\
&x_{11t} + x_{41t} + x_{61t} = 100 + 5t + 5\mu \\
&x_{22t} + x_{42t} + x_{52t} + x_{62t} = 100 + 5t + 5\mu \qquad (9) \\
&x_{33t} + x_{43t} + x_{53t} = 100 + 5t + 5\mu \\
&x_{74t} = 100 + 5t + 5\mu \\
&x_{11t} \le 100 \quad x_{41t} \le 302.5 \quad x_{61t} \le 200 \\
&x_{22t} \le 100 \quad x_{42t} \le 302.5 \quad x_{52t} \le 100 \quad x_{62t} \le 200 \\
&x_{33t} \le 100 \quad x_{43t} \le 302.5 \quad x_{53t} \le 100 \\
&x_{74t} \le 102.5
\end{aligned}
$$

$x_{ijt} \in R^+$, $i = 1,\ldots,7;\ j = 1,\ldots,4\ \ t = 1,\ldots,5\}$, where:

x_{ijt} = the quantity of the j component to be produced by agent i in time t.

$w(I)$ = fuzzy payoff per unit for car production.

$w(k_1)$ = fuzzy payoff per unit for Body Production.

$w(k_2)$ = fuzzy payoff per unit for Motor Production.

$w(k_3)$ = fuzzy payoff per unit for Transmission Production.

$w(k_4)$ = fuzzy payoff per unit for car assembly.

μ = uniform random variable in [0,1].

Results of game theoretic approach with fuzzy coalitions

For the case study, positive ramp membership functions (fig. 7) were selected: payoff per car - [7,100, 8,000]; gross payoffs for body - [6,500, 7,000]; motor - [4,500, 5,000]; transmission - [3,800, 4,000]; and assembly - [2,000, 4,000] respectively. Two solvers were used to find the solution. The first one, the Excel solver was able only to approximate the integer solution. Table 2 shows the imputations for each time interval, the last column shows the number of cars to be assembled according to the forecasting defined by (7). The final configuration of the SCS from the case study is shown in fig. 10.

Table 2. Production levels per unit for five time intervals

t	x_{11t}	x_{22t}	x_{33t}	x_{41t}	x_{42t}	x_{43t}	x_{52t}	x_{53t}	x_{61t}	x_{62t}	x_{74t}
1	100	100	100	0	0	0	5.299	5.3	5.3	.001	105.3
2	100	100	100	0	0	0	8.399	11.5	11.5	3.101	111.5
3	100	100	100	0	0	0	10.25	15.2	15.2	4.95	115.2
4	100	100	100	0	0	0	14.50	23.7	23.7	9.20	123.7
5	100	100	100	0	0	0	16.75	28.2	28.2	11.45	128.2

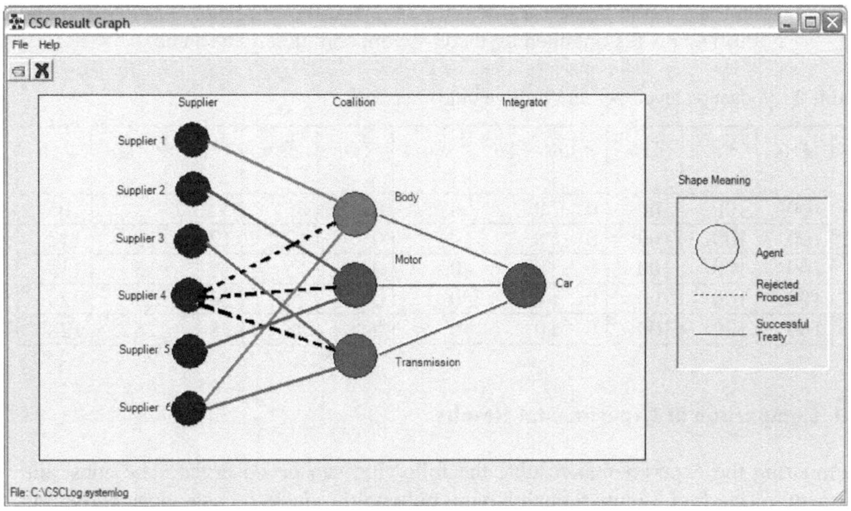

Fig. 10. Final configuration of the supply chain from the case study

The second solution method used GAs. The solution converged within several seconds (though communications between agents took considerable time). The rates of mutation and crossover in the typical values of 6 and 50% respectively, with a size of population of 50 organisms were fixed. Time convergence parameters and number of iterations were also fixed [28]. The solution of the game is obtained as a sequence of trials, each one of them corresponding to the players' movements. The solution evolves to the optimal value guided by the fitness function.

The common network payoffs per car obtained for each time interval are equal to 7,578.46, 7,578.22, 7,578.22, 7,577.84, 7,577.84 respectively. The payoffs (p) of the participating units per car/component are as follows: $p_{unit\ 1}$ = 2,500; $p_{unit\ 2}$ = 1,500; $p_{unit\ 3}$ = 1,500; $p_{unit\ 4}$ = 0; $p_{unit\ 5}$ = 1,400 (motor and transmission); $p_{unit\ 6}$ = 2,300 (body); $p_{unit\ 6}$ = 1,400 (motor); $p_{unit\ 7}$ = 2,500. The same gross payoffs per unit were obtained for each time interval for each component: $w(I)$ = 20,000; $w(k_1)$ = 7,000; $w(k_2)$ =5,000; $w(k_3)$ =4,000 $w(k_4)$ =4,000. The possibility of the fuzzy game $\gamma_c(I,w)$ = 1.00 (because of the simplicity of the case study), though the imputation obtained took into account the subjective estimations of the players defined by their fuzzy payments.

7.2 Results of the Constraint Satisfaction Approach

Obtained results applying constraint based model described in Section 5.2 for five time intervals are given in Table 3.

The common network payoffs per car obtained for each time interval are equal to 7,980.952; 7,957.143; 7,947.826; 7,922.581; 7,912.5 respectively. The payoffs (p) of the participating units per car/component are as follows: $p_{unit\ 1}$ = 2,500; $p_{unit\ 2}$ = 1,500;

$p_{unit\ 3} = 1{,}500$; $p_{unit\ 4} = 0$; $p_{unit\ 5} = 1{,}400$; $p_{unit\ 6} = 2{,}300$ (body); $p_{unit\ 6} = 1{,}400$ (motor); $p_{unit\ 7} = 2{,}500$. The results obtained by the GA were completely the same.

Table 3. Production levels per unit for five time intervals

t	x_{11t}	x_{22t}	x_{33t}	x_{41t}	x_{42t}	x_{43t}	x_{52t}	x_{53t}	x_{61t}	x_{62t}	x_{74t}
1	100	100	100	0	0	0	0	5	5	5	105
2	100	100	100	0	0	0	0	12	12	12	112
3	100	100	100	0	0	0	0	15	15	15	115
4	100	100	100	0	0	0	0	24	24	24	124
5	100	100	100	0	0	0	0	28	28	28	128

7.3 Comparison of Experimental Results

Comparing the experimental results, the following can be observed. The constraint capacity of the first 3 units though having minimal production costs, does not permit them to satisfy all the demand. That is why, while demand is increasing, other units are involved in the production process (in final configuration of the CSC). Let us compare the resultant configurations of the CSC. The results obtained by the two methods though being very similar in general, differs in some details. In the first case, the Unit 5 is not involved in the motor production. Instead of it, all the incrementing production of this component is assigned to the Unit 6 (Table 3). But if we compare the initial parameters of the units for the motor production, it can be seen that they are the same both for the production cost (*$3600* per motor) and for stocking (*$170* per motor /week). That is why, the both solutions can be considered valid. So, the difference is debt to the difference between the methods in solution strategies. In the case of a game, the more balanced final solution (see units 5 and 6 in the case of a motor component in Table 2) is preferred to the one, obtained by the constraint satisfaction solver. Some difference in common network payoffs is caused by the difference in the definition of the objective function.

8 Related Research and Discussion

In the chapter, CSC configuring tasks are considered as (i) fuzzy coalition formation and (ii) product and resource allocation in multi-agent environment. Coalition formation is a common approach to partners' selection in dynamic trading environments. In Chapter 2 of this book coalition deal negotiation approach was described. Examples of different approaches to coordination and negotiation inspired in game theory are also discussed in Chapters 8 and 10. In Chapter 7 the fuzzy set approach was applied to a task assignment problem.

The coalition formation approach permits to find optimal or sub-optimal solution for partner selection and resource allocation tasks applying the theory of games with fuzzy coalitions. Different models of the games to form fuzzy coalitions are described in the literature [1], [20], [25]. Traditional game theoretic approaches are typically computationally infeasible and usually only consider super-additive environments. Super-additive means that for any pair of coalitions, it is always beneficial for them to form one big coalition. While this assumption simplifies the analysis, it also limits the scope of the application.

Considering the models and solution algorithms for the formation of fuzzy coalitions in the context of the theory of games, the models of the fuzzy mathematical programming and the models of the fuzzy core basically exist. In the first case, the solution algorithms are well developed, however they are very restrictive in the sense that they are limited to incorporate the negotiations that are given among the agents. For this reason, it is more natural and efficient to use models that consider the fuzzy core (group of imputations of the game) of the game to form the alliances as a product of negotiations. In this chapter, a generalized class of the games when capacity constraints are added to the problem statement is presented. One of the key problems in this case is the task complexity, when solutions cannot be found with traditional methods. In this case, genetic algorithms are used to find solutions of the game.

The advantages of the proposed model are the following. First, it considers only effective coalitions; practically it means that it is necessary to consider all the possible coalitions and select those that accomplish with the effectivity criterion (3). This problem is NP-complete, i.e. solution algorithms are of exponential complexity and though the corresponding distributed algorithms with at least potential complexity can be developed, which probably will never be reduced to a linear one. Only for those cases when the number of agents (let's say about 20 players) and the number of coalitions are small, the centralized algorithms can be used, e.g. of combinatorial programming, of a graph theory or optimization models. Another serious limitation of the model discussed above is that the fuzzy quantities have the distributions of their values supposed to be linear, when in the case of CSC configuring they may have non-linear distributions. Moreover, though the calculation of the solution can be simplified greatly by considering the universal forms of fuzzy utilities for all agents, as for example, of a trapezoidal form to apply the method of Shapley fuzzy values, which make the calculations similar to the crisp case, this assumption is very restrictive for the dynamic MAS and converts the model into a game with side-off payoffs.

The theory of games with coalitions offers results showing the general form of possible cooperation and conditions under which this is achieved. In many cases, a broad class of attainable patterns of cooperation and of distributions of the final payments exists, and it is important to select the best one or the most unbiased. The ontology corresponding to the application domain is involved in such a way, that each coalition corresponds to a sub-system, with goals and resources well defined by the knowledge management.

The second approach to resource allocation uses (i) genetic algorithms directly and (ii) constraint satisfaction problem solving for configuring considered as a resource allocation task. Genetic Algorithms were invented by John Holland and developed by him and his students and colleagues [17]. They form a part of evolutionary computing, introduced in the 1960s by I. Rechenberg in his work "Evolution strategy" [24], which is now a rapidly growing area of artificial intelligence. In 1992 John Koza used GA to evolve programs to perform certain tasks

[18]. He called his method "genetic programming". LISP programs were used, because programs in this language can be expressed in the form of a "parse tree", which is the object the GA works on. GAs are used in the context of this chapter for two purposes: (i) as a solution technique for the game with fuzzy coalitions, and (ii) as a method for the task allocation problem solving.

The constraint-based models have been applied to a wide range of CSC problems in production logistics [3], product configuration [16], production program configuration [30], and planning [9]. These models require outlining constraint-based representation and constraint satisfaction/propagation mechanisms. In this chapter, object oriented constraint networks are used as a formal model of the ontological task description and decomposition. Resource allocation task is described by these ontologies and is solved by constraint-based techniques.

Using the combination of constraint-based technique allowing obtaining a global optimal solution, and ILOG Solver engine allowing utilizing both linear and non-linear constraints, makes it possible to solve large tasks with complex dependencies between their components.

One of the key issues of the developed approach is the scalability of the proposed techniques. As we show, the configuration scheme considered in the article though being rather simple (two tier) can be considered as a template to be integrated into the more complex multi-tier configuration schemes. Since the case study was rather simple (almost linear), there was no considerable computational cost for the methods applied. In all the cases, with four solvers used (ILOG for constraint satisfaction, GA, Excel solver and GA for the fuzzy game) the time was minimal (seconds). These results were obtained without the MAS test-bed to avoid the influence of communication costs.

Obviously, the problem complexity will influence the time required to find the solution. For example, the fuzzy game model defined in section 3.1 permits to consider non-linear membership functions, integer variables, multi-objective functions and non-linear constraints. Generating the imputations or elements of the fuzzy core (that is in turn a fuzzy set) is an arduous and complex task, analytically the optimum imputation can be obtained, for example, using the method of the fuzzy Pareto optimum, but paying a big mathematical and computational cost. It is for this reason that GAs are proposed not only to find the solution of the game, but also to find an effective coalition. They also allow to explore and to involve important elements such as, to consider non linear membership functions or to consider super additive, sub additive or additive games, that can be achieved only using the methods of soft-computing. For the same CSC structure but with non-linear membership functions, quite different performance results were obtained. For example, the Excel solver could not find a solution al all, while the GA found the solution in about five minutes, making more than 4000 trials (each trial corresponds to the movement of the players), about 1000 of which were viable. Generally speaking, problem complexity is one the key limitations of the game theoretic approach.

For the multi-agent framework, communication costs begin to play a considerable role (compared with the computational costs for this simple case study) for all the three approaches. Though in all the cases, the central solver was used, these communications were necessary for the implementation of the CNP for information interchange and result communication to all the participants. These costs depend on the implementation agent platform used for the test-bed, because the integrated communication mechanism is used (for example, remoting in the case of the CAPNET). Obviously, they also depend upon the physical distribution of the involved agents. As other experiments conducted up to date show, these costs can be from 5 seconds up to 10 minutes for each 1000 messages sent between agents deployed in the same and different network segments respectively. These costs can become a very important point to consider when highly distributed algorithms for resource allocation in MAS environment are used, as for example, described in [27].

Conclusions

Multi-agent framework integrating different solution techniques for configuring of cooperative supply chains have been discussed in this chapter. The presented methodology of the configuring implies implicit negotiations (vague, imprecise, open), which are guided to maximize the benefits for each node of the CSC. The specifics of the methods is that the best suiting solution is obtained according to the aspirations of the CSC units. The structure of a virtual company formed this way, has the distinctive point and the advantage in that it takes into account the convenience for the companies to participate in groups (virtual companies), it allows quick adaptations that respond to changes of the environmental dynamics.

The use of coalition formation as the method for the CSC configuring is explored through the game theoretic approach. These can be given in a collaborative or competitive (anti-coalitions) context. In the chapter, the case of formation of cooperative groups, this is that their viability depends greatly upon the consensus between players (CSC members), is considered. As discussed, only the use of heuristic and soft-computing techniques such as genetic algorithms permit to find the best converged imputation in a reasonable time for real world applications. The use of genetic algorithms has the following benefits: they allow to find a semi-optimal solution in the case when an optimal solution can't be found analytically because of the problem's complexity, they allow to find a local optimal solution in the cases when the optimal solution even doesn't exist (when the game is not convex), and also they can be used to find solutions for the mixed approaches similar to the discussed in the chapter when a game-theoretic approach is combined with the combinatorial one.

The constraint-based approach assumes *complete* information sharing while the game theoretic approach assumes *minimal* or *no* information sharing (actually what is negotiated implicitly while the solution of the game is being found, are the aspirations of the players defined by the corresponding membership functions). Since in real-world situations the both cases as well as some intermediate ones are possible, the application of presented here techniques in a combined way is preferable. Since this combination is not considered in this chapter, the subsequent research efforts should be devoted to investigation of combined techniques for tasks representing more realistic situations. Also some limitations of the presented research are also a subject to improvement. Among these limitations the following ones can be pointed out: wider range of parameters to optimise, more complicated strategies of the units (e.g., including safety stocks forming), more complicated demand patterns and forecasting techniques, and non-linear relationships in the model.

CAPNET FIPA compliant agent platform and a CASE tool for CSC development, modeling and simulation are used to conduct the experiments. The use of FIPA compliant agent platforms (like JADE in case of Chapter 2) enhances the interoperability of the solutions covering different aspects of the multi-agent support of supply chain management. The use of the CAPNET platform with enabled WS interface not only ensures interoperability among the agents and reusable components for the application of the proposed techniques in real enterprise-based applications, but also allows taking advantage of agent-based, as well as component-based services offered by the CSC units that can be used and customized as needed.

Acknowledgments

Partial support for this research work has been provided by the IMP within the project D.00006. The presented research was partially supported through the projects # 16.2.35 of the research program "Mathematical Modeling and Intelligent Systems", and # 1.9 of the research program “Fundamental Basics of Information Technologies and Computer Systems” of the Russian Academy of Sciences, and the project funded by grant # 05-01-00151 of the Russian Foundation for Basic Research. Some examples were developed using software granted by ILOG Inc.

References

1. J.P. Aubin: Cooperative fuzzy games, Mathematics of Operations Research, 6:1 (1981) 1-13.
2. R. Axelrod: The complexity of cooperation: agent based models of competition and collaboration, (Princeton University Press, 1997).
3. H. Baumgaertel: Distributed Constraint Processing for Production Logistics, IEEE Intelligent Systems, (2000) 40-48.
4. S. Bhaskaran: Simulation Analysis of a Manufacturing Supply Chain, Decision Sciences, 29:3 (1998) 633-657.
5. Bokma, CogNet: Integrated Information and Knowledge Management and its Use in Virtual Organisations, in: L.M. Camarinha-Matos, H. Afsarmanesh, R.J. Rabelo, eds., E-Business and Virtual Enterprises, Managing Business-to-Business Cooperation (Kluwer Academic Publishers, Boston, 2000) 361-370.
6. C., Chandra, A.V. Smirnov, and L.B. Sheremetov: Agent-based Infrastructure of Supply Chain Network Management, in: L.M., Camarinha-Matos, H. Afsarmanesh, and R. Rabelo, eds., E-Business and Virtual Enterprises, Managing Business-to-Business Cooperation (Kluwer Academic Publishers, 2000) 221-232.
7. Contreras, M., Germán, E., Chi, M., and Sheremetov L.: Design and implementation of a FIPA compliant agent platform in .NET, in J. of Object Technology, 3(9), Oct. 2004, Sp. Issue: .NET Tech. Workshop, pp. 5-28. http://www.jot.fm/issues/issue_2004_10/article1
8. D., Dubois, and H. Prade: Mathematics in science and engineering, Vol. 144, (Academic Press Inc., 1980).

9. E. H. Durfee: Distributed problem solving and planning, in: M. Luck, V. Marik, O. Stepankova, and R. Trappl, eds., Multiagent Systems and Applications: Selected tutorial papers from the Ninth ECCAI Advanced Course, ACAI 2001, and AgentLink's 3rd Eur. Agent Systems Summer School, EASSS 2001, LNAI, 2086 (Springer-Verlag, Berlin, 2001) 118-149.
10. M., Ettl, G. E. Feigin, G. Y. Lin and D. D. Yao, A Supply Network Model with Base-Stock Control and Service Requirements, IBM Technical Report, RC 20473, 1996.
11. Evolver Release 4.01. (Palisade, 2001).
12. S. C., Graves, and S. P. Willems, Optimizing Strategic Safety Stock Placement in Supply Chains, Working Paper, 1998.
13. ILOG, Corporate website, URL: http://www.ilog.com, 2005.
14. N.R., Jennings, On agent-based software engineering, Artificial Intelligence, 117 (2000) 277-296.
15. N.R., Jennings, P., Faratin, A.R., Lomuscio, S., Parsons, C. Sierra, and M. Wooldridge, Automated negotiation: prospects, methods and challenges. Int. J. of Group Decision and Negotiation 10:2 (2001) 199-215.
16. Haag, Sales Configuration in Business Processes, IEEE Intelligent Systems and their Applications, 13:4 (1998) 78-85.
17. J.H. Holland, Adaptation, Progress in theoretical biology, 4 (1976) 263-293.
18. J. Koza, Genetic Programming: On the Programming of Computers by Means of Natural Selection, (MIT Press, 1992).
19. Mackworth, Constraint Satisfaction, in: S.C. Shapiro, ed., Encyclopedia of Artificial Intelligence (Wiley-Interscience Publication, New York, 1992), 285-293.
20. M., Mareš, Fuzzy coalition forming, in: Proc. of the Seventh IFSA World Congress, (1997) 1-18.
21. M., Mareš, Fuzzy coalition structures, Fuzzy sets and systems, Elsevier Sciences, 114 (2000) 23-33.
22. H. Merchant, Configurations of International Joint Ventures, MIR (Management International Review), Gabler Verlag, 40:2 (2000) 107-140.
23. H., Raiffa, The art of the Negotiation (Harvard University Press, 1987).
24. Rechenberg, Evolutionsstrategie'94 (Frommann-Holzboog, Stuttgart, 1994).
25. J., Rosenmuller, The theory of games and markets (North Holland, Amsterdam, 1982).
26. A., Roth, Game-theoretic models of bargaining (Cambridge, 1995).
27. O. Shehory and S. Kraus, Feasible Formation of Coalitions Among Autonomous Agents in Non-Super-Additive Environments, Computational Intelligence, 15:3 (1999) 218-251.
28. Sheremetov L. and Romero-Cortés J. Fuzzy Coalition Formation Among Rational Cooperative Agents. Multi-Agent Systems and Applications III, V. Marik, J. Muller, M. Pechoucek (Eds.), LNAI 2691, Springer Verlag, 2003. pp. 268-280
29. Sheremetov L., Contreras M., Smirnov A. Implementation of an Ontology Sharing Mechanism for Multiagent Systems based on Web Services. In Favela, J.; Menasalvas, E.; Chávez, E. (Eds.), LNAI 3034: 54-63, 2004.
30. Smirnov A.V., Pashkin M.P., Chilov N.G., Levashova T.V., Agent-Based Support of Mass Customization for Corporate Knowledge Management, Engineering Applications of Artificial Intelligence. V. 16, Issue 4, June 2003, pp. 349-364.
31. A.V., Smirnov, L. B., Sheremetov, N., Chilov and J. Romero Cortes, Configuring of Supply Chain Networks Based on Constraint Satisfaction, Genetic and Fuzzy Game Theoretic Approaches, in Proc. of the BASYS'02 Conf. (Kluwer Academic Publ., 2002).
32. J. G. A. J., Van der Vorst, Beulens, A. J. M., and P. Van Beek, Modelling and Simulating Multi-Echelon Food Systems, European J. of Operational Research, 122 (2000) 354-366.
33. M., Yokoo, and K. Hirayama, Algorithms for Distributed Constraint Satisfaction: A Review, Autonomous Agents and Multi-Agent Systems, 3:2 (2000) 198-212.

Design, Implementation and Test of Collaborative Strategies in the Supply Chain

Thierry Moyaux [1], Brahim Chaib-draa [1], and Sophie D'Amours [2]

[1] Universit Laval, Dpt. d'Informatique et de Gnie Logiciel, DAMAS & FOR@C
Ville de Qubec G1K 7P4 (Qubec, Canada), {moyaux, chaib}@iad.ift.ulaval.ca
[2] Universit Laval, Dpt. de Gnie Mcanique, FOR@C & CENTOR
Ville de Qubec G1K 7P4 (Qubec, Canada), sophie.damours@gmc.ulaval.ca

Abstract. In general, game theory is used to analyze interactions formally described by an analytical model. In this paper, we describe a methodology to replace the analytical model by a simulation one in order to study more realistic situations. We use this methodology to study how the more-or-less selfishness of agents affects their behaviour. We illustrate our methodology with the case study of a wood supply chain, in which every company is seen as an agent which may use an ordering strategy designed to reduce a phenomenon called the bullwhip effect. To this end, we assume that every agent utility can be split in two parts, a first part representing the direct utility of agents (in practice, their inventory holding cost) and a second part representing agent social consciousness, i.e., their impact on the rest of the multi-agent system (in practice, their backorder cost). We find that company-agents often apply their collaborative strategy at whatever their same level of social consciousness. Our interpretation of this specific case study is that every company is so strongly related with one other, that all should collaborate in our supply chain model. Note that a previous paper outlined this methodology and detailed its application to supply chains; our focus is now on the presentation and the extension of the methodology, rather than on its application to supply chains.

1 Introduction

There is an increasing interest in game theory in the multi-agent community [1]. In particular, some methodologies have applied concepts from game theory to analyze multi-agent simulations [2, 3], even though, in general, game theory is used to analyze the behaviour of agents described with an analytical model. Our ultimate goal is to have a methodology suitable for the analysis of interactions in agent-based simulations that would profit from most concepts from game theory. To this end, we now enhance the methodology we proposed in a previous work [2] to study (i) how the obtained results vary with some change in the utility function of agents, and (ii) how robust the results are with regard to the stochasticity of some parameters.

Let us focus on these two enhancements. Concerning the first point, agents in our previous work had a common fixed way to calculate their utility that was

T. Moyaux et al.: Design, Implementation and Test of Collaborative Strategies in the Supply Chain, Studies in Computational Intelligence (SCI) 28, 247–272 (2006)
www.springerlink.com

based on their direct utility and on their impact on other agents. In contrast, the methodology proposed in the present paper, the way in which utility is calculated is *neither common* to all agents, *nor fixed.* In practice, we continue to assume that agent utility is the addition of two components: the direct utility that implements the individual utility of agents, and a utility that represents the social consciousness of agents in regard to other agents. In [2], the weight of the direct utility was half of the weight of the social utility for every agent; we now change this ratio to verify that agents still have incentives to use the same strategy, or conversely, if agents prefer to change their behaviour when they are more or less socially conscious. The second improvement to our methodology deals with taking *probability distribution* for some parameters into account in replacement of a fixed sequence of numbers, in order to study stochastic models. This means that we generate an instance of the considered distribution, and then, we apply our methodology with this fixed sequence of numbers. This is the same method as in our previous work, except that we repeat the process with other instances of the considered distribution in order to next use statistical tools. Our methodology is described in Section 2.

We illustrate this methodology in the context of the management of a wood supply chain. In other words, we present our methodology as generally as possible, while its application to a specific supply chain demonstrates it. We profit from the first enhancement provided by our methodology to study whether selfish as well as benevolent company-agents should collaborate by sharing information concerning market consumption. This shared information is of importance to companies when they place orders, because this information may be distorted by a phenomenon called the "bullwhip effect" [4] that causes an increase of their inventory level and a decrease in their customer service level, and information sharing is the most often proposed solution to the bullwhip effect [5].

We take the direct utility as agent profit, and social utility as the level at which company-agents take client satisfaction into account in their utility because the goal of the entire supply chain should be to satisfy end-customers. We profit from the second enhancement provided by our methodology to consider a stochastic market consumption pattern. More precisely, we assume that market demand is informally distributed, and we carry out our experiments with ten particular instances of this uniform distribution. We consider only ten instances of the distribution, because, for each of these ten instances, a great number of simulations is required by our methodology to simulate all the combinations of the possible strategies among the considered agents. However, all of these simulations are carried out with the same particular instance of the market consumption in order to build a game in the normal form. We next repeat all these simulations with the nine other considered instances of the market consumption in order to build nine other games. In the ten obtained games, we then change the ratio between the direct and the social utility of agents when we look for Nash equilibria. The experimental results show that the level of social consciousness does not have an impact on the choice of which level of collaboration is chosen by agents. In other words, selfish agents agree to collaborate as often as benevolent agents, which shows that every company-agent in our model is so strongly *dependent* on each other that they should all collaborate. This illustration of our methodology on an issue of supply chain management, namely the bullwhip effect, is detailed in Section 3.

Finally, we discuss how we have applied our methodology in practice to study supply chain management, and we identify some future enhancements on this methodology. Concerning our application to supply chains, we can note that we only consider scenarios in which all companies have the same level of social consciousness, while our methodology is more general and allows us to mix benevolent and selfish agents in a same supply chain, which would allow situations as the Prisoner's Dilemma to occur. We also note that we have only considered ten instances of the uniform distribution for the market consumption pattern, i.e., for the stochastic parameter. In both limitations of the application of our methodology to supply chains, we made these simplifications to save either analysis (first limitation) or simulation (second limitation) time, but we indicate how we expect to manage these two limitations in future work. Concerning the possible enhancements on our methodology, the main one deals with the addition of probability on the use of strategies, which corresponds to the concept of mixed strategies in game theory. Such an addition of probability in our context of supply chains is not as easy as in game theory, but it could be carried out in some contexts other than information sharing decisions in supply chains. The application of our methodology to supply chain management and the possible extensions of our methodology are discussed in Section 4.

2 Methodology

Before the presentation of our methodology, we need to introduce a few notations. For the sake of simplicity, we will continuously illustrate this section with the case study of a supply chain, even if this case study will only be detailed in Section 3.

2.1 Notations

We write $i \in I$ a particular agent, and $|.|$ the cardinal of a set, so that $|I|$ is the number of agents in the considered system. In the case study in Section 3, every agent i represents a company in a wood supply chain, and there are $|I| = 6$ agents. $s^i \in S$ denotes the strategy used by Agent i during an entire run of a simulation, and S is common to all agents, i.e., all agents have the same set of possible strategies, even if they do not use the same strategy during a particular simulation. Simulations are run over fifty weeks during which agents cannot change their strategy s^i. In the case study in Section 3, $S = \{\mu, \beta, \gamma\}$ for every company i to represent three levels of collaboration by information sharing strategies, where:

- μ implements no information sharing, so that placed orders are only based on inventory level. Notice that incoming demand (i.e., orders placed by the direct client) could also have been taken into account here.

- β represents little information sharing, because it allows companies to inform their direct suppliers of the market consumption, i.e., of what is demanded by end-customers. Specifically, every retailer informs its wholesalers of its sales when it places an order, then every wholesaler transmits this information to its suppliers, etc.
- γ implements full information sharing, when retailers broadcast the market consumption (approximated as their sales) to the rest of the supply chain. Such information sharing is referred as information centralization. γ shares the same information as β, i.e., the market consumption information, but instantaneously instead of as slowly as orders.

Agent i's utility is $u^i_{s^i}$ when it uses the strategy s^i, but for the purpose of simplicity we will always write u^i instead of $u^i_{s^i}$. To study social consciousness, we assume that every utility u^i is the convex combination of the direct utility u^i_{dir} and of the social utility u^i_{soc}, so that $u^i = ((1-\epsilon^i)u^i_{\text{dir}} + \epsilon^i u^i_{\text{soc}})/(1-\epsilon^i)$, with $\epsilon^i \in [0;1)$; we divide by $(1-\epsilon^i)$ because u^i_{dir} represents realistic costs in our case study and u^i thus remains a cost (notice that $\epsilon^i = 2/3$ for all i in [2]). This splitting of utility function as direct vs. social utility is similar to others, such as the representation of individual outcome vs. social consciousness [6], or of self vs. group interest [7].

In particular, Glass and Grosz [6] note that computing a measure of social consciousness and combining this measure with individual outcome are two challenges. In our case study, we address these two challenges by defining u^i_{dir} as company-agent's inventory holding cost (thus, $u^i_{\text{dir}} < 0$ because the agents in our simulation cannot earn money), and u^i_{soc} as company-agent's backorder cost (again, $u^i_{\text{soc}} < 0$) to represent the more or less will to satisfy clients; backorders correspond to orders that were not filled in the past and that have to be shipped as soon as possible, or, in other words, there are no lost sales in the simulation model used in our case study because clients wait for product availability, and the backorder cost measures (industrial or end-) customer *dissatisfaction* due to this wait. Since dissatisfaction generates a bad reputation for agents, u^i_{soc} also measures the *reputation* that agents expect to have in the system. Since backorders are measured in our specific case as negative inventory levels, the calculation of u^i_{dir} and u^i_{soc} are similar, and thus, their aggregation into u^i represents a cost. Notice that, $\epsilon^i = 0$ means that Agent i only has a selfish behaviour because u^i_{soc} has no influence on u^i in the previous convex combination, while $\epsilon^i \to 1$ implements a fully benevolent agent because u^i_{dir} becomes negligible in comparison with u^i_{soc}. In the specific of our case study, we can reformulate the question addressed in this paper as:

> At what set of values $\{\epsilon^i\}_{i=1}^{i=|I|}$ do the Nash equilibria switch from an entire non-collaborating supply chain to an entire collaborating supply chain?

In this paper, the proposed methodology allows studying such a question, but the application of this methodology in the case study only addresses this

Algorithm 1 methodology

$\{\{Inv^{i,s}_{init}\}^{i=|I|}_{i=1}\}_{s \in S}$ = optParam $(\{\epsilon^i\}^{i=|I|}_{i=1}, M)$
for each $m \in M$ do
  $\{\{u^i_{dir}, u^i_{soc}, s^i\}^{i=|I|}_{i=1}\}_{s^1 \in S \times \ldots \times s^{|I|} \in S}$ = **simAllConfig**$(\{\{Inv^{i,s}_{init}\}^{|I|}_{i=1}\}_{s \in S}, m)$
  sets of Nash equilibria = anaSimOut $(\{\{u^i_{dir}, u^i_{soc}, s^i\}^{|I|}_{i=1}\}_{s^1 \times \ldots \times s^{|I|}}, \{\epsilon^i\}^{i=|I|}_{i=1})$
returns all sets of the Nash equilibria corresponding to the considered $\{\epsilon^i\}^{i=|I|}_{i=1}$ and the $|M|$ instances of the stochastic parameter.

question for sets $\{\epsilon^i\}^{i=|I|}_{i=1}$ in which $\epsilon^1 = \ldots = \epsilon^{|I|}$. This means that, in the case study, all agents have the same level of selfishness, but that this common level may be different between two of the considered situations. That is, we do not consider the case in which selfish agents cohabit with benevolent ones (i.e., Prisoner's Dilemma), although the presented methodology allows doing so. We make this assumption to simplify the analysis of the outputs of our methodology, but we will try to relax this assumption in future work, as discussed in Section 4. We call ϵ the common value of $\{\epsilon^i\}^{i=|I|}_{i=1}$, so that $\epsilon = \epsilon^i$ for every agent i.

In the previous reformulation of the question addressed in this paper, we specified that we focus on Nash equilibrium as the solution concept. Indeed, this equilibrium is the main solution concept in game theory [8], in particular, because it is the most general one and thus includes the other concepts of equilibrium. A Nash equilibrium is only a stable state of the multi-agent system, that is, no agent has incentives to unilaterally deviate from it. As a result, when we refer to such an equilibrium, we do not understand it as a description of what will always occur, but as what holds when it occurs. The issue of knowing what equilibria should be preferred by agents was addressed in [2]; now, we reconsider the same example of supply chains to study how equilibria change with ϵ.

Next, we write Inv^i_{init} the value of some parameter Inv^i at the beginning of a simulation. In the particular case of our study of a supply chain, Inv^i_{init} will be company-agent i's initial inventory level. Finally, m_A represents one particular instance of a probability distribution for a second parameter m. We write M the set of the considered instances of the probability distribution under study. In our case study, the distribution under study is the uniform distribution of 2×50 integer numbers over the interval $\{11, 12, \ldots, 17\}$, so that each of the 50 integers represents the consumption of one of the 2 markets during one of the 50 weeks of the simulation. We generate once 10 instances of this uniform distribution, so that $M = \{m_A, m_B \ldots m_J\}$.

2.2 Explanation of the Algorithms

With these notations, we can now describe our methodology, as outlined in Algorithms 1, 2, 3, and 4. Algorithm 1 outlines the methodology at the highest level:

(i) Parameters Inv are first optimized by optParam for a specific value of ϵ (Algorithm 1 is general, but as previously stated, we will only consider the sets

Algorithm 2 optParam

inputs : $\{\epsilon^i\}_{i=1}^{i=|I|}, M$
for each strategy $s \in S$ do
  for each considered instance $m \in M$ do
    for each agent $i \in I$ do
      set agent i to use $s^1 = s$
    optimize the initial value of Inv^i for every agent i, so that $\sum_i (u^i_{dir} + u^i_{soc})$ is maximized under m.
    save the set of optimal values $\{Inv^{i,s}_{init}\}_{i=1}^{|I|}$.
  save in a new set $\{Inv^{i,s}_{init}\}_{i=1}^{|I|}$ the average of the values in the $|M|$ sets $\{Inv^{i,s}_{init}\}_{i=1}^{|I|}$ saved in the previous loop.
returns $\{\{Inv^{I,s}_{init}\}_{i=1}^{|I|}\}_{s \in S}$.

$\{\epsilon^i\}_{i=1}^{i=|I|}$ in which $\epsilon^i = \epsilon$ for every company i, with $\epsilon = 1/6, 2/6, 3/6, 4/6$ or $5/6$). (ii) Next, the supply chain is simulated by sim**AllConfig** for all combinations of s^i among the $|I|$ agents with the optimum values of Inv_{init}. (iii) Then, the simulation outcomes are used to build games in the normal form, in which Nash equilibria are next looked for. Depending on $\{\epsilon^i\}_{i=1}^{i=|I|}$, different games can be built here based on the output of sim**AllConfig**. We now describe each step in this process.

Algorithm 2 presents how the optimum values of parameter Inv are found. This method returns the best intial value of Inv depending on the considered agent i and on what common strategy s is used by all agents (we only optimize for s common to all agents in order to save time by greatly reducing the number of optimizations) so that the social welfare of agents is maximized.

We assume here that the social welfare is the sum of all agent utilities $\sum_i (u^i_{dir} + u^i_{soc})$ [9] because this assumption is compatible with our case study in which this sum represents money, but, in general, social welfare can be much harder to define [10]. In brief, Algorithm 2 carries out $|M|$ simulations to find the best value of the parameter Inv and return the average of these $|M|$ values (we thus find values good on average in all instances in M), and repeat this process for every $s \in S$.

As a consequence, the word "optimize" in Algorithm 2 both refers to a simulation and to a parameter optimization.

In our case study, each invocation of optParam returns (i) a set of the six best values of Inv (one value per company-agent) when all agents use β, (ii) a set of the six best values of Inv when all agents use γ, (iii) and the set $\{0, 0, 0, 0, 0, 0\}$ when all agents use □. Concerning □, we use the vector of zeros as optimal parameters, because we proved that $Inv^i_{init} = 0$ is locally optimal for every company i [2]. This local optimality is compatible with the fact that □ is a non-collaborative strategy.

Next, Algorithm 3 uses the values of Inv that are the best on average, and simulates with these average values all the possible combinations of the $|S|$ strate-

Algorithm 3 simAllConfig

inputs : $\{\{Inv^{i,s}_{init}\}_{i=1}^{|I|}\}_{s \in S}, m$
for each $s^1 \in S$ do
  set agent i to use s^1 with Inv^{1,s^1}_{init}.
  for each $s^2 \in S$ do
    set agent i to use s^2 with Inv^{2,s^2}_{init}.
    ⋮
    for each $s^{|I|} \in S$ do
      set agent $|I|$ to use $s^{|I|}$ with $Inv^{1,s^{|I|}}_{init}$.
      simulate the supply chain under instance m.
      save the simulated value of $\{u^i_{dir}, u^i_{soc}\}_{i=1}^{|I|}$ with the corresponding s^i.
returns $\{\{u^i_{dir}, u^i_{soc}, s^i\}_{i=1}^{|I|}\}_{s^1 \in S \times \ldots \times s^{|I|} \in S}$

Algorithm 4 anaSimOut (m)

inputs : $\{\{u^i_{dir}, u^i_{soc}, s^i\}_{i=1}^{|I|}\}_{s^1 \in S \times \ldots \times s^{|I|} \in S}, \{\epsilon^i\}_{i=1}^{i=|I|}$
for each considered set $\{\epsilon^i\}_{i=1}^{i=|I|}$ do
  for each $m \in M$ do
    copy the $|I|^{|S|}$ sets $\{s^i\}_{i=1}^{|I|}$ and $\{(1-\epsilon^i)u^i_{dir} + \epsilon^i u^i_{self}/(1-\epsilon^i)\}_{i=1}^{|I|}$ in a Gambit file.
    apply Gambit method 'EnumPureSolve' to find all Nash equilibria.
    save Nash equilibria in a file.
returns all sets of the Nash equilibria corresponding to the considered $\{\epsilon^i\}_{i=1}^{i=|I|}$ and the $|M|$ instances of the stochastic parameter.

gies among the $|I|$ agents, which is implemented by the imbricated 'for' loops. $|S|^{|I|}$ simulations are performed here. In our case study, we have $|I| = 6$ agents with $|S| = 3$ strategies per agent, and thus, $3^6 = 729$ simulations are performed for each invocation of Algorithm 3. In addition, Algorithm 3 is called $|M| = 10$ times by Algorithm 1.

Finally, Algorithm 4 builds games in the normal form based on the simulation outcomes provided by Algorithm 3.

More precisely, each of the $|S|^{|I|}$ simulations produces $|I|$ utilities u^i_{dir} (one utility per agent) and $|I|$ utilities u^i_{soc}. We apply here $u^i = ((1-\epsilon)u^i_{dir} + \epsilon^i u^i_{soc})/(1-\epsilon)$, and put these $|I|$ utilities u^i in one entry of a $|I|$-dimension matrix. As a result, this matrix contains $|I| \times |S|^{|I|}$ utilities. Let us now illustrate Algorithm 4 with our case study, in which each of the $|S|^{|I|} = 3^6 = 729$ simulations produces $|I| = 6$ utilities u^i_{dir} and $|I| = 6$ utilities u^i_{soc}, that we put under the form u^i in one entry of a 6-dimension matrix. Specifically, this is a $3 \times 3 \times 3 \times 3 \times 3 \times 3$ matrix, which contains $6 \times 3^6 = 4,374$ utilities u^i (each of the 3^6 simulations produce one utility per agent). We obtain a matrix for each of the $|\{\epsilon^i\}_{i=1}^{i=|I|}| \times |M|$ simulations run by Algorithm 1. In our case study, $|\{\epsilon^i\}_{i=1}^{i=|I|}| \times |M| = 5 \times 10 = 50$ matrices.

Every matrix is a game in the normal form that we analyze with Gambit [11]. Gambit is a set of software tools used to analyze games. These tools may be used either with the graphical user interface, or with the Gambit Command Language, which is a powerful way to carry out complex or repetitive operations on games. In our work, we use the latter language for two reasons: it is able to save Nash equilibria in a file (while the graphical interface only displays them), and all operations are automatized by placing the set of instructions to perform in the initialization file of Gambit. In practice, we only have to implement our simulator so that it writes the simulation outcomes in the Gambit format. Roughly, this deals with making the simulator write $|S|^{|I|}$ lines, where each line starts with the $|I|$ individual strategies s^i followed by the $|I|$ individual utilities u^i, and to add the head and the tail to this file to respect the Gambit format, so that this file can be read as a game with $|I|$ and $|S|$ strategies per player by Gambit. Next, the initialization file of Gambit has to open all files/games, find their equilibria, and write these equilibria in a single output file.

3 Case Study: Do Companies in a Supply Chain Agree to Share Demand Information?

We now illustrate our methodology in the context of supply chain management to study when the companies in a supply chain have incentives to share information. This information sharing aims to reduce a phenomenon called the bullwhip effect, which makes the overall supply chain less efficient. We first present the bullwhip effect in Subsection 3.1; next we propose a solution to this effect which is based on information sharing in Subsection 3.2. Since this solution requires information sharing, companies may disagree to use it, and therefore, we use the methodology described in Section 2 to study when companies prefer to use an ordering strategy involving information sharing. The simulation model of the supply chain used by our methodology (cf. Algorithm 3) is presented in Subsection 3.3. In this simulation, companies can use one of the three ordering strategies presented in Subsection 3.4, in which two strategies involve information sharing. Finally, the obtained results are presented in Subsection 3.5, and discussed in Subsection 3.6.

3.1 The Bullwhip Effect: A Problem of Supply Chain Management

The goal of this case study is to illustrate the methodology by studying when companies agree to share demand information in order to reduce the bullwhip effect. To this end, we now present the bullwhip effect (this phenomenon was shortly introduced in Subsection 2.2 of Chapter 1 as one of the problems faced by companies). Figure 1 shows how this effect propagates in a simple supply chain with seven companies: two retailers, two wholesalers, a pulp mill, a sawmill and a forest.

In this figure, each retailer exclusively sells to its customer and buys from its wholesaler, each wholesaler sells to its retailer and buys from its supplier (either the sawmill, or the pulp mill), etc. The ordering patterns of the companies are

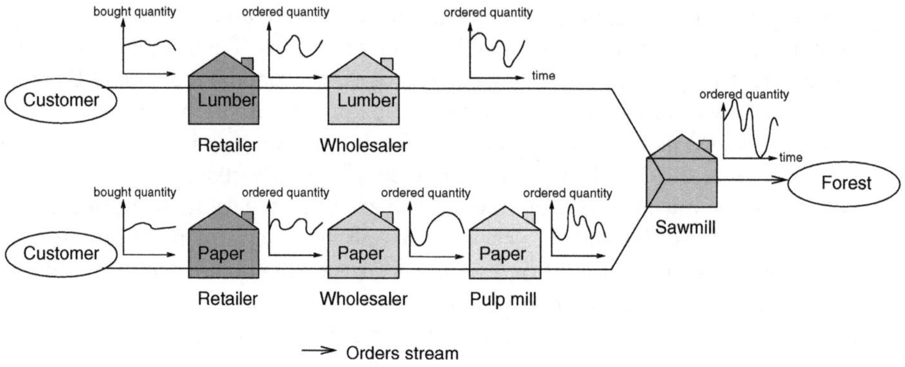

Fig. 1. The bullwhip effect [4, 12]

similar in the way that the variabilities of an upstream site are always greater than those of the downstream site [12]. As a variability, the bullwhip effect is measured by the standard deviation σ of orders.

This phenomenon is not harmful by itself, but because of its consequences. Here are the consequences reported by Carlsson and Fullér [13]:

- *Excessive inventory investments*: Since the bullwhip effect makes the demand more unpredictable, all companies need to safeguard themselves against the variations to avoid stockouts;
- *Poor customer service levels*: Despite the excessive inventory levels mentioned in the first consequence, demand unpredictability may cause stockouts anyway;
- *Lost revenues*: In addition to the poor customer service levels of the second consequence, stockouts may also cause lost revenues;
- *Reduced productivity*: Since revenues are lost (cf. third consequence), operations are less cost efficient;
- *More difficult decison-making*: Decisions-makers react to demand fluctuations and adapt (production and inventory) capacities to meet peak demands;
- *Sub-optimal transportation*: Transportation planning is made more difficult by demand uncertainties induced by the bullwhip effect;
- *Sub-optimal production*: As transportation, a greater demand unpredictability causes missed production schedules.

Carlsson and Fullér [13] noted that these consequences are not due to changes in the demand of end-customers, but only to inefficiencies in the supply chain. As an insight into the importance of these inefficiencies, these same authors claim that the bullwhip effect would cost the Finnish forest products industry 100-200 MFIM (17-34 million euros) per year, the industry having a total turnover of more than 100 BFIM (17 billion euros).

Causes	Proposed solutions	Authors
Demand forecast updating	Information sharing (e.g., VMI, CRP...), echelon-based inventory and leadtime reduction	[4, 12]
Order batching	EDI (Electronic Data Interchange) and Internet technologies	[4, 12] [15]
Price fluctuation	EDLP (Every Day Low Pricing)	[4, 12]
Rationing and shortage gaming	Allocation based on past sales	[4, 12]
Misperception of feedbacks	Giving a better understanding of the supply chain dynamics to managers	[16–20]
Local optimization without global vision	None	[5] [21–24]
Company processes	None	[15]

Table 1. Proposed causes and solutions of the bullwhip effect [14].

Next, some causes and solutions of the bullwhip effect have been identified. Table 1 summarizes them [14]. Lee and his colleagues [4, 12] proposed the first four causes and solutions presented in this table.

- *Demand Forecast Updating:* Companies base their orders on forecasts, which are themselves based on their incoming orders while such forecasts are not perfectly accurate. Therefore, companies order more or less than what they really require to fulfill their demand. In other words, forecasting errors amplifies the variability of orders.
  A solution proposed for this cause is information sharing: each client provides more complete information to its supplier in order to allow the supplier to improve its forecasting. Information sharing is already part of industry practices, such as VMI (Vendor-Managed Inventory), CRP (Continuous Replenishment Program), etc. Two of the three ordering strategies available to the company-agents in our simulation employ such information sharing.
- *Order Batching* (lot sizing in a more general way): Companies discretize orders in order to profit from economies of scales (e.g. transportation with full truck-loads), and therefore, place orders for more or less products than what they actually need.
  The proposed solution for this cause is electronic transactions (e-commerce, EDI...) to reduce transaction costs and thus make companies' orders more frequent and for smaller quantities. Similarly, the size of production batches may be reduced with SMED (Single Minute Exchange of Die), which may next reduce the quantities ordered.

- *Price Fluctuation:* Every client (company or end-customer) profits from promotions by buying more products than what it really requires, and next, buying nothing when the promotion stops because it has enough products in inventory.
  The proposed solution is the EDLP (Every Day Low Pricing) policy, where price is set at the promotion level. However, EDLP also has some drawbacks, e.g., always looking for the lowest price may put a stress on the supply chain that may eventually reduce profits [25].
- *Rationing and Shortage Gaming:* Since every client behaves opportunistically, it overorders when its supplier cannot fulfill its entire demand, e.g., in the case where the supplier has a machine breakdown. Through such behaviour, this client does not hope to receive the quantity that it has ordered, but a lower quantity that matches its actual need.
  Since this behaviour occurs when the supplier allocates shippings in proportion to the ordered amount, it is preferable to allocate the few available products in proportion with the history of past orders.

Other authors have extended Lee and his colleagues' causes to the bullwhip effect:

- *Misperception of Feedback:* Sterman [20] has noted that players in the Beer Game place orders in a non-optimal way because they do not understand the whole dynamics in their supply chain. For example, they do not correctly interpret their incoming orders, and in consequence, smooth their orders when they should order more, because they do not understand that market consumption has increased.
- *Local Optimization without Global Vision:* Several authors [22–24] have noted that companies maximize their own profit without taking into account the effect of their decisions on the rest of the supply chain. In particular, some companies use an ordering scheme, such as the (s, S) policy (in which the company orders for $S - I$ products when inventory level I falls below s) that is the operationalization of this local optimization. It has been formally proven that some of these policies induce the bullwhip effect [5, 16, 21].
- *Company Processes:* Taylor and his colleagues [15] propose two causes to the bullwhip effect: variability in machine reliability and output, and variability in process capability and subsequent product quality. In these two causes, which are summarized as "Company processes" in Table 1, production problems at each workstation are amplified from one workstation to another. This cause recalls that intracompany problems and uncertainties may affect each company's behavior, which in turn may make them change the way they place orders.

In this context, our problem is first to find the way to place orders that reduces the bullwhip effect. Such ordering strategies should be the most efficient for the whole supply chain, which has been checked in a previous paper [14], and we next have to check whether companies agree to use it, which is the topic of this case study. The next subsection presents such an ordering strategy.

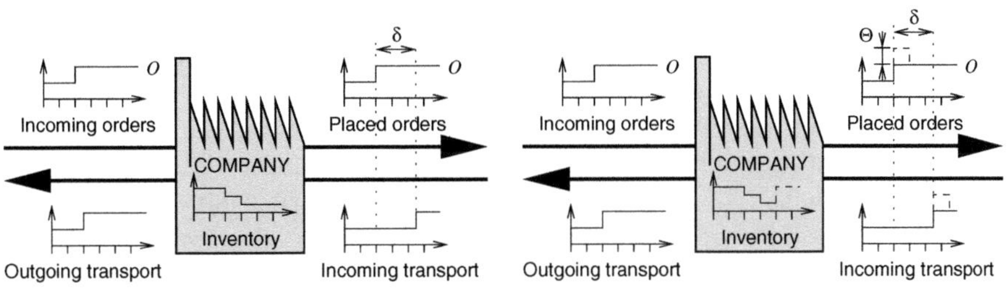

(a) Lot-for-lot ordering policy (b) (O, Θ) orders

Fig. 2. Lot-for-lot ordering policy and (O, Θ) orders

3.2 One Cause and its Solution to the Bullwhip Effect

In this subsection, we present one of the possible refinements of the cause "misperception of feedbacks" in Table 1, then we propose a solution based on information sharing which addresses this specific cause of the bullwhip effect.

To avoid the bullwhip effect, the basic idea is that if companies' orders *follow* their clients' demand with a *lot-for-lot* ordering policy, there is *no bullwhip effect*, but inventories fluctuate greatly. In other words, either there is a bullwhip effect or inventories fluctuate greatly. This assertion is illustrated in Figures 2(a) and 2(b) that represent a company travelled by an ordering and a product stream. In these figures, we assume the company places orders strictly equal to its demand, following a strict lot-for-lot ordering policy. We now detail Figure 2(a) in four points, to show why companies prefer the bullwhip effect, rather than use the lot-for-lot policy.

1. The lot-for-lot ordering policy eliminates the bullwhip effect, because each company has the same ordering pattern as its client and thus, as the market consumption. Therefore, the two curves Incoming orders and Placed orders are identical. Since the bullwhip effect is measured as the standard-deviation of placed orders, we can see that the standard-deviation of each company's orders is exactly the same as the standard-deviation of its client's orders, and therefore, as the standard-deviation of the market consumption.

This first point explains why a lot-for-lot ordering policy eliminates the bullwhip effect, but not why companies do not use this ordering policy. As we shall see, the goal of companies is not to avoid the bullwhip effect, but to have products to carry out their activity:

2. The considered company tries to fulfill its entire demand, and thus, the two curves Incoming orders and Outgoing transport are the same, that is, as many products are shipped as ordered. The curve Outgoing transport in Figure 2(a) is valid as long as no stockouts occur by the considered company.

In short, the first two points say that the three curves Incoming orders, Placed orders and Outgoing transport are similar. The third point below says that the fourth curve also has the same pattern, but with a temporal shift δ:

3. The fourth curve Incoming transport has the same pattern as the three other curves, except that it is delayed by δ in comparison with the three other curves. This curve represents the reception of products by the company. The temporal shift δ corresponds to the ordering and shipping lead times, because items ordered by the company are not immediately received. The problem with lot-for-lot orders is that the inventory is not managed, because the temporal shift δ makes inventory decrease (respectively increase, when we inverse the pattern of "Incoming Orders"). In fact, the company ships more (respectively less) products than it receives during the ordering and shipping lead times δ.
   Note that Incoming transport has the same pattern as the three other curves only when the supplier has no stockouts, because the supplier is assumed to want to fulfill its entire demand, like the considered company.
4. Since every company wants to avoid stockouts (respectively huge inventory), rather than eliminate the bullwhip effect, it does not use the lot-for-lot ordering policy. Instead, it overorders (respectively underorders) to stabilize its inventory, which amplifies the demand variabilities, because the company overorders (respectively underorders) when the demand increases (respectively decreases). This shows that the *bullwhip effect always appears* each time the market consumption has an infinitesimal change, if companies want to *keep a steady inventory.*

Notice that "lead times" refer to the cause of the bullwhip effect that has just been presented. One should also note, that some of the other causes of the bullwhip effect presented in Table 1 induce the bullwhip effect even with a steady demand, while lead times amplify order fluctuations, but do not induce fluctuations when the demand is steady. Specifically, the bullwhip effect amplifies, because if a retailer overorders to stabilize its inventory, a worse phenomenon takes place with its suppliers: since the demand variation is now bigger, their inventories decrease much more, and thus, they must overorder more.

As we can see, the problem is not only to reduce the bullwhip effect, because companies just have to apply the lot-for-lot ordering policy to eliminate this effect, but inventories also have to be managed so that stockouts and high inventory levels are avoided. The solution we proposed in [2] is illustrated in Figures 2(b) and 3. The idea is to have an ordering strategy with a unique order amplification for each change in market consumption. Since companies have to know the market consumption to make this unique order amplification, this solution requires the sharing of demand information. As a result, our solution is the same as the one proposed by Lee and his colleagues to improve demand forecasting updating (cf. "information sharing" in Table 1), because companies have to share their incoming orders information with their suppliers. In fact, companies signal to their suppliers when they over- or underorder. This information sharing is presented in Figures 2(b) and 3, in which each company uses a vector (O, Θ) of two orders, where the actual order is the sum $(O + \Theta)$:

1. orders O follow the lot-for-lot policy to avoid the bullwhip effect;

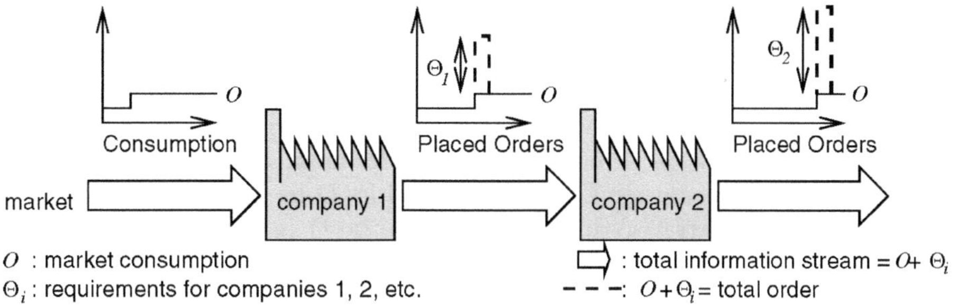

Fig. 3. Information streams cut into two parts

2. orders Θ are used to order more or less products than O to stabilize inventory level.

We now present the two principles ruling the use of O and Θ. As presented in Figures 2(b) and 3, each company using (O, Θ) orders places a vector of orders (O, Θ), instead of a unique number O that englobes and hides these two numbers[3]. In (O, Θ), O is the market consumption transmitted from each company to its supplier(s) with the lot-for-lot policy, and Θ is chosen such that $O + \Theta$ represents what the company needs. As O follows a lot-for-lot policy, the bullwhip effect cannot occur in it. Unfortunately, it may occur in Θ, that is, non-zero Θs may be emitted anytime and anyhow. This is the reason why our (O, Θ) ordering strategies rest on the following two principles, and not only one of them.

First Principle: The lot-for-lot ordering policy eliminates the bullwhip effect, but does not manage inventories.
This first principle rules the way of choosing O in (O, Θ). Lot-for-lot orders means that each company orders what is demanded from it; if its client wants 10 products, the company places an order for 10 products. As previously stated in Figure 2(a), the problem is that the bullwhip effect is eliminated with the lot-for-lot policy, but inventory level is not managed. Therefore, we keep lot-for-lot orders for ruling O, but we add another piece of information Θ to manage inventory level. Note that lot-for-lot orders allow O to share the market consumption information, as illustrated in Figure 3.

Second Principle: Companies should react only once to each market consumption change.
This second principle rules the way of choosing Θ in (O, Θ). Θ is equal to zero all the time, except when the market consumption changes, in which case companies react to this change by sending non-zero Θ, in order to stabilize their inventory to the initial level. The purpose of Θ is to trigger a product

[3] O like Order and Θ like Token, as these two pieces of infor mation were called in our previous papers. Moreover, O and Θ have the advantage of looking si milar; while they have a very si milar meaning: they are both Orders.

wave from the most upstream company when this company receives this Θ. This product wave will increase, or decrease when $\Theta < 0$, each company's inventory as it travels the supply chain down to the retailers.

We now illustrate these two principles with two ordering strategies, namely β and γ. In strategy β, we use the following trick to satisfy the second principle: since the market consumption is transmitted in O by the lot-for-lot policy (cf. first principle), a strategy in which $\Theta = 0$ when O is steady, and $\Theta \neq 0$ when O fluctuates, would satisfy the second principle. If we write i is the considered company, $i-1$ its unique client, and if we assume time is continuous (which is not the case in the simulation in this case study), then Equation 1 describes how the company i would place orders (O_t^i, Θ_t^i) with strategy β at instant t:

$$(O_t^i, \Theta_t^i) = (O_t^{i-1}, \Theta_t^{i-1} + \lambda \frac{dO_t^{i-1}}{dt}) \tag{1}$$

Note that $\frac{dO_t^{i-1}}{dt} \geq 0$ (respectively $\frac{dO_t^{i-1}}{dt} \leq 0$) represents the forecasted inventory decrease (respectively increase) during the ordering and shipping lead times. This quantity has to be overordered (respectively underordered) in order to keep a steady inventory. The constant λ depends on the duration of ordering and shipping lead times. We shall see how we adapt strategy β to our simulation in Subsection 3.4, i.e., after the presentation of the model of this simulation.

Before the presentation of the simulation, let us present another way of implementing our two principles, i.e., strategy γ. We assume in γ that information centralization is used, that is, retailers multicast the market consumption along the whole supply chain. Information sharing with information centralization is much quicker than information sharing with (O, Θ) orders, because the market consumption transmitted in O is as slow as orders, while information centralization is assumed to be instantaneous, reflecting thus the actual market consumption in real-time.

As a result, γ should be more efficient for the entire supply chain than β: as soon as the market consumption changes, non-zero Θs are sent by all companies to satisfy the second principle as quickly as possible. To do so, Θ is proportional to the variation of the market consumption transmitted by retailers. Moreover, companies also base O on the market consumption transmitted by retailers, instead of on incoming O, again in order to react quicker to the market consumption change. If we note again i the considered company, $O_t^{retailer}$ the market consumption (approximated as the demand received by a retailer), and if we assume time is continuous, Equation 2 presents how company i places its orders (O_t^i, Θ_t^i) when there is information centralization. Again, λ is a constant depending on lead times:

$$(O_t^i, \Theta_t^i) = (O_t^{retailer}, \Theta_t^{i-1} + \lambda \frac{dO_t^{retailer}}{dt}) \tag{2}$$

We shall see an adaptation of β and γ to discrete time in Subsection 3.4. But before that, we present the simulation model of a supply chain that we use in this case study.

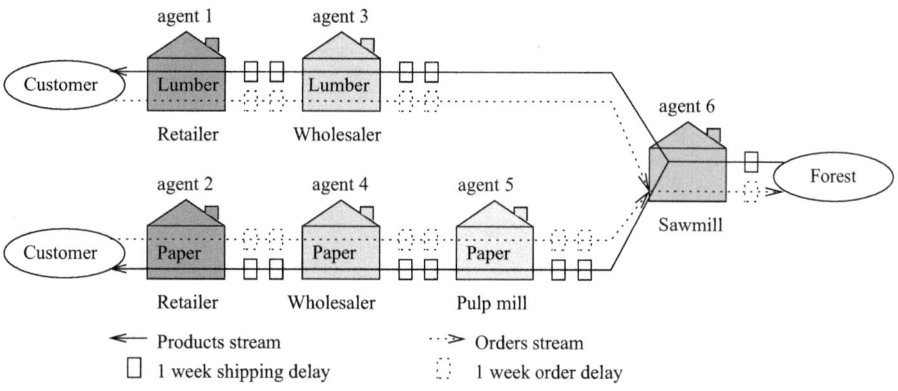

Fig. 4. Structure of the supply chain in our simulation [2]

3.3 Simulation Model

In Section 2, we assumed that the simulator was a source of data without giving additional details about it. Actually, the simulator used by our methodology is the only link with the application field. Of course, the results that our methodology will produce and their interpretation are only specific to this field.

In this context, each agent represents a company that places orders with its supplier in order to be able to fulfill its client's orders. Specifically, we use the same simulator as in our previous work [2, 14]. This simulation model relies on the supply chain structure and on the company model in the Quebec Wood Supply Game. In this model, agent direct utility is the opposite of their inventory holding cost, since agents cannot earn money and only incur costs, and thus, only have a negative direct utility. Similarly, agents' social consciousness is measured as backorders, that are equal to the duration and to the amplitude of stockouts. It is important to understand that we use backorders as the measure for social consciousness, because agents are not obliged to consider them when they make decisions (while they always have to take their inventory holding costs into account, which is the reason why inventory levels measure the direct utility), but the goal of the entire supply chain should be to satisfy end-customers. More technically, u^i_{dir} is the sum of the inventory level over the 50 weeks of a simulation, and u^i_{soc} is the sum of the backorder level over these 50 weeks.

Next, we detail the supply chain structure illustrated in Figure 4.

There are $|I|=6$ company-agents, where the first one represents the Lumber-Retailer ($i = 1$), which receives orders from the lumber market and places orders with the LumberWholesaler ($i = 3$). To fulfill these orders, this latter company places orders to the Sawmill ($i = 6$). Then, the Sawmill places orders to the Forest (which is not an agent because it does not place orders), but the Sawmill has to base its orders on two incoming demands in order to fulfill, as best as possible, the needs of the LumberWholesaler and the PulpMill. In fact, the Sawmill is the only company in our model that has to manage two types of products (lumber

Ordering strategy	Level of collaboration	Type of communication
□	No collaboration	No communication
β	Information sharing with direct neighbours	Point to point
γ	Information centralization	Bulletin board

Table 2. The three ordering strategies.

and paper). Finally, there is a two-week delay to transmit an order and another two-week delay to ship the corresponding products.

3.4 Agent Strategies

To decide how much to order, every company i applies its ordering strategy s^i that has to be chosen in set S. In the case of the Sawmill, we operate as if there were a lumber Sawmill and a paper Sawmill that apply the same ordering strategy s^6, and the orders that are eventually placed to the Forest are the average of the orders that would have been placed by each sub-Sawmill.

For all companies, we have already introduced the fact that we assume $|S| = 3$ possible ordering strategies s^i for each of the six companies (including the Sawmill) and $S = \{\mu, \beta, \gamma\}$ in our case study, but we have not yet detailed μ, β and γ. Every strategy μ, β and γ implements a different level of collaboration in the supply chain, as outlined in Table 2. In comparison with our previous work [2, 14], β and γ are identical, while □ is an enhancement on α to take backorders more efficiently into account by considering inventory position instead of on-hand inventory [4]. Here are more details about μ, β and γ:

- With ordering strategy □, company-agents make their decision on their own and can only rely on their incoming orders and shippings to decide what order to place. The only information transmitted with □ are orders and no additional information is shared to help other agents place their own orders. In practice, □ is an (s, S) policy, which is classic in Inventory Management, and in which a company orders $(S - InvP)$ items when the level of its inventory position $InvP$ falls below s. The inventory position $InvP$ is the sum of on-hand inventory Inv and inventory on order minus backorders. We show in [2] that taking $s = 0$ and S equals to the current incoming order is optimal.
- There is little collaboration with strategy β, because agents now transmit the market consumption information to their supplier with (O, Θ) orders, which

[4] The on-hand inventory Inv represents the products that are actually in inventory, while the inventory position $InvP$ represents what would be in inventory if there were no lead times due to transportation lead times, suppliers' stockouts, etc.

allows every agent to place orders in a more accurate way. As explained in Subsection 3.2, every placed order is a two-dimension vector ($Op, \theta p$), and the quantity ordered is equal to the sum ($Op + \theta p$) of its two elements: the first element corresponds to the market consumption transmitted by each company to its direct supplier, and the second element represents the products to be ordered in more or in less than the market consumption. In our simulation, agents place orders ($Op, \theta p$) = ($Oi, \theta i + 4 * \Delta Oi$), where ($Op, \theta p$) is the two-dimension placed order, ($Oi, \theta i$) is the incoming order, 4 is the value of a parameter depending on the ordering and shipping lead times between two companies (i.e., 2 week ordering lead time plus 2 week shipping lead time), and ΔOi is the variation of Oi since the previous week.

- The highest level of collaboration is implemented in strategy γ, in which retailer-agents write the market consumption information on a bulletin board, and the rest of the agents can read this information (this is information centralization). In practice, γ works as β, except that companies take the first element O of their two-dimension orders equal to the current market consumption, which approximated as retailer's incoming demand $Oi^{retailer}$. Consequently, agents place orders ($Op, \theta p$) = ($Oi^{retailer}$, $\theta i + 2 * \Delta Oi^{retailer}$). Note that the value of the parameter is now 2 rather than 4, because information centralization eliminates the ordering lead time by transmitting information instantaneously.

See [14] for details about these three strategies, about the parameters 2 and 4 depending on lead times in β and γ, and about the efficiency of these strategies for the overall supply chain when all companies use the same strategy.

3.5 Results and Analysis

In this section, we first introduce Table 3 to show the raw data obtained with our methodology for the particular instance m_A of the market demand. Next, we show how we aggregate this data in order to present results for ten instances of the market demand in Table 4.

As just stated, we first introduce the raw data obtained with our methodology. To this end, Table 3 presents the results obtained with the particular instance m_A of a uniform distribution of market demand. The distribution is the same for the ten considered instances m_A to m_J, and it is made of integer numbers uniformally spread over $\{11, 12, \ldots, 17\}$. In particular, the instance m_A in Table 3 corresponds to the two series of 50 numbers $D^{lumber} = \{16, 13, 16, 11, 15, \ldots, 15, 15, 12, 16, 15\}$ and $D^{paper} = \{15, 15, 14, 15, 14, \ldots, 13, 14, 14, 11, 13\}$, where the wth number represents the market demand in Week w, e.g., $D_2^{lumber} = 13$ means that 13 units are demanded by the lumber market in Week 2. We approximate $D^{lumber} = O_i^l$ and $D^{paper} = O_i^l$.

Next, Table 3 enumerates all the Nash equilibria obtained with different values of ϵ.

For example, for $\epsilon = 5/6$, there are 6 Nash equilibria, and the first one is the *strategy profile* $(s^1, s^2, s^3, s^4, s^5, s^6) = (\beta, \mu, \beta, \mu, \mu, \beta)$; in this equilibrium, the

Value of ϵ	Nash equilibria
$\epsilon = 1/6$	$(\gamma,\mu,\gamma,\mu,\gamma,\gamma)$, $(\gamma,\gamma,\gamma,\mu,\mu,\gamma)$
$\epsilon = 2/6$	$(\gamma,\mu,\gamma,\mu,\mu,\beta)$, $(\gamma,\gamma,\gamma,\mu,\mu,\gamma)$
$\epsilon = 3/6$	$(\gamma,\beta,\gamma,\mu,\mu,\beta)$, $(\gamma,\gamma,\gamma,\mu,\gamma,\gamma)$
$\epsilon = 4/6$	$(\gamma,\gamma,\gamma,\gamma,\gamma,\gamma)$
	$(\beta,\mu,\beta,\mu,\mu,\beta)$, $(\beta,\beta,\beta,\beta,\beta,\beta)$
$\epsilon = 5/6$	$(\beta,\gamma,\beta,\gamma,\beta,\gamma)$, $(\beta,\gamma,\beta,\gamma,\gamma,\beta)$
	$(\gamma,\beta,\gamma,\beta,\beta,\beta)$, $(\gamma,\gamma,\gamma,\gamma,\beta,\beta)$

Table 3. Results for Instance m_A.

LumberRetailer (i = 1), the LumberWholesaler (i = 3) and the Sawmill (i = 6) use Strategy β, while all other agents use γ. The interpretation of all the equilibria in Table 3 is that no company has an incentive to unilaterally deviate from these equilibria. This does not mean that these strategy profiles are the best either for the whole supply chain or for every agent. This only means that these profiles are stable for the multi-agent system under this particular value of ϵ and under the particular instance m_A of the market demand. Note that □ occurs 15 times, β 29 times and γ 40 times in overall Table 3, which corresponds respectively to frequencies of 20%, 29% and 51%. Therefore, the highest level of collaboration occurs in half of the found Nash equilibria.

Next, the question addressed in this paper is to determine at which value of ϵ the Nash equilibria switch from a full-collaborating system (that is, most companies collaborate at the highest level, e.g., ($\gamma,\gamma,\gamma,\beta,\gamma,\gamma$)) to a full non-collaborating one (that is, most companies disagree to collaborate, e.g., ($\mu,\beta,\mu,$ □□)). To answer this question, we check if the Nash equilibria with low ϵ (selfish agents) often apply □ (i.e., the non-collaborative strategy), and/or if the Nash equilibria with high ϵ (benevolent agents) frequently use γ (i.e., the highly collaborative strategy). To see this, we calculate the frequency at which every strategy is used depending on the value of ϵ. That is, we calculate the central column 'Instance m_A' in Table 4 based on the numbers in Table 3. This calculation is achieved in the following way. In line '$\epsilon = 5/6$' in Table 3, there are 6 equilibria that count a total of 3 □, 20 β and 13 γ. These three numbers are reported in line $\epsilon = 5/6$ in Table 4 as '3 / 36 = 8% $\rightarrow$ □', which means that 3 of the 36 (=3+20+13) strategies used in a Nash equilibrium are □.

When we look at the central column in Table 4, we cannot make a clear distinction between the equilibria obtained for the five considered value of ϵ, because there is no regularity in the distribution of percentages: we were waiting for □ with a maximum frequency when $\epsilon = 1/6$ (while γ is more frequent than □ in this entry), and for γ with a maximum value when $\epsilon = 5/6$ (which is again not true since β has a greater frequency). On the other hand, the collaborative strategy γ in general occurs the most frequently over most values of ϵ, which indicates that all agents are strongly related in the system whatever the level of selfishness is: if one behaves in a harmful way for the rest of the agents, this agent will also suffer from its incorrect behaviour.

Value of ϵ	Frequency of μ, β and γ under Instance m_A	Average frequency of μ, β and γ over Instances m_A to m_J
$\epsilon = \frac{1}{6}$	$4/12 = 33\% \rightarrow$ □	$27\% \rightarrow$ □
	$0/12 = 0\% \rightarrow \beta$	$9\% \rightarrow \beta$
	$8/12 = 67\% \rightarrow \gamma$	$64\% \rightarrow \gamma$
$\epsilon = \frac{2}{6}$	$5/12 = 42\% \rightarrow$ □	$22\% \rightarrow$ □
	$1/12 = 8\% \rightarrow \beta$	$6\% \rightarrow \beta$
	$6/12 = 50\% \rightarrow \gamma$	$72\% \rightarrow \gamma$
$\epsilon = \frac{3}{6}$	$3/12 = 25\% \rightarrow$ □	$17\% \rightarrow$ □
	$2/12 = 17\% \rightarrow \beta$	$14\% \rightarrow \beta$
	$7/12 = 58\% \rightarrow \gamma$	$69\% \rightarrow \gamma$
$\epsilon = \frac{4}{6}$	$0/6 = 0\% \rightarrow$ □	$9\% \rightarrow$ □
	$0/6 = 0\% \rightarrow \beta$	$16\% \rightarrow \beta$
	$6/6 = 100\% \rightarrow \gamma$	$75\% \rightarrow \gamma$
$\epsilon = \frac{5}{6}$	$3/36 = 8\% \rightarrow$ □	$7\% \rightarrow$ □
	$20/36 = 56\% \rightarrow \beta$	$43\% \rightarrow \beta$
	$13/36 = 36\% \rightarrow \gamma$	$50\% \rightarrow \gamma$

Table 4. Relative frequency of the occurence of μ, β and γ in the incurred Nash equilibria.

This conclusion only holds for the particular instance m_A of the uniformly distributed market consumption. We have repeated for the nine other instances m_B to m_J of market consumption. Their results are in the right column in Table 4. To calculate every frequency in the right column, we take the average of the frequencies obtained with the ten market demands, i.e., these frequencies are not weighted by the number of equilibria: to obtain '7% $\rightarrow$ □' with $\epsilon = 5/6$ in the right column of Table 4, the 8% obtained with m_A has the same importance as the percentages obtained with $m_B, m_B, \ldots, m_J$, even though they do not represent the same quantity of equilibria. We can draw the same conclusions with the right column as with the center column in Table 4, i.e., □ only has low occurrence frequencies. We now detail our findings in the context of the reduction of the bullwhip effect.

3.6 Discussion on the Case Study

We have just seen that the collaborative strategy γ occurs the most frequently, and that its frequency does not depend on the value of ϵ. From the viewpoint of supply chain management, this is an important result, because it indicates (as for m_A) that every agent is strongly related to each other, and therefore, all agents prefer collaboration with β and γ, whatever their value of selfishness is (i.e., for all values of ϵ). In addition, the frequency of □ (i.e., the only considered non-collaborating strategy) depends on the value of ϵ: the frequency of □ regularly decreases from 27% to 7% when the social consciousness ϵ increases.

Consequently, the frequency of □ is intuitive, since companies in the selfish supply chain ($\epsilon = 1/6$) do not collaborate more frequently than companies in the socially conscious supply chain ($\epsilon = 5/6$). These two points mean that, if the supply chain stabilizes on a strategy profile (i.e., on a Nash equilibrium), this strategy profile will probably be made of many γ. In this result, we do not detail the utility incurred by every company, but we noticed in [2] that some equilibria will be avoided because they are Pareto-dominated by another equilibrium, which means that all agents are worse off in the Pareto-dominated equilibria than in another equilibrium.

From the viewpoint of supply chain management, this is an excellent result, because it was possible that each company-agent would have liked the rest of the supply chain to collaborate except itself, because this company-agent would have profited from the effort made by the rest of the supply chain (problem of the free rider). Fortunately, such a scenario does not occur in our simulations, but this is only true when all agents have the same level of self consciousness because we do not consider mixes of selfish and benevolent agents.

The answer to the question addressed in this paper is thus that collaboration holds in the entire supply chain whatever the level of social consciousness is, i.e., for all considered value of ϵ. Our interpretation is that company-agents are strongly dependent on each other. However, the non-collaborating strategy is more frequently chosen when the level of social consciousness is low. What we call dependence is that, if a company-agent disturbs the operation of the supply chain, it may earn a little from this in its direct utility (which represents what is incurred by the agent and which cannot be disregarded), but this agent will also lose a lot because of the supply chain dynamics that will disturb this agent as a feedback of its bad behaviour. As a consequence, benevolent agents ($\epsilon \rightarrow 1$), as selfish agents ($\epsilon = 0$), will collaborate by sharing information, that is, in general, all agents prefer to use Strategy γ (or at least β) rather than □, whatever their selfishness level is. This is a high level result that should lead the design of lower level interaction mechanisms for agents in supply chains [26].

4 Discussion on the Methodology

The previous section has illustrated the methodology proposed in this paper by showing how to study social consciousness in supply chains. We now discuss our methodology and possible improvements. First of all, it would be interesting to be able to run exactly the presented methodology. This means that we should use different values of ϵ^i for our agents, instead of a ϵ common to all agents. As previously stated, we would be able to study if the free rider problem occurs. Unfortunately, this will greatly increase the bulk of outputs produced by our methodology.

To see this, let us consider that ϵ^i may take E values for every agent i. Thus, $E^{|I|}$ combinations of the E possible values of ϵ^i among the $|I|$ agents should be considered. For example, if $E = 3$ (e.g., $\epsilon^i \in \{1/4, 2/4, 3/4\}$) and $|I| = 6$, $3^6 = 729$ cases have to be considered per instance m of M. To analyze this great number of outputs, we could, for example, define some classes of joint strategies , and observe the distribution of the equilibria among these classes depending on

$\{\epsilon^i\}_{i=1}^{i=|I|}$. Such classes of joint strategies may be something like "all the $|I|$ agents use one specific strategy, or $|I| - 1$ agents use one specific strategy except one agent that uses any other strategy", or any other definition in which the classes do not overlap. Finally, a few manipulations are not yet automatized in the implementation of our methodology, which obliges us to reduce the number of instances of $|M|$ to 10, while it should be much greater to get significant results.

Moreover, it would be interesting to include the concept of mixed strategies, which comes from game theory [10]. In fact, we only consider pure strategies in this paper, that is, agents use the same strategy during the entire run (i.e., over 50 weeks) of each of the $|S|^{|I|}$ simulations. If mixed strategies were used, we would use probabilities for all these strategies, e.g., Agent i would use s_1^i for 75% of the time (i.e., over 38 weeks), and s_2^i during the rest of the run of the simulation (i.e., over 12 weeks). Unfortunately, we cannot use mixed strategies for the following three reasons:

1. Reason from supply chain management: Companies cannot switch between collaboration and non-collaboration all the time, because contracts have to be signed to insure the secrecy of the information shared with β and γ, and collaborative devices have to be bought by companies.

The two other reasons can be found in any application field:

2. Dynamics in the simulated system may create some "inertia": Since this phenomenon occurs in our case study, we use it to illustrate what we mean by inertia; when a company switches from Strategy □ to γ, the products previously ordered with □ will still arrive, which has an impact on the supplier's inventory level, and therefore on its future orders. Such a transition does not exist in traditional game theory, and is hard to handle.
   As a result, we are not allowed to calculate the expected outcome of a simulation with mixed strategies based on two simulations with pure strategies because the result would have no meaning in reality. For instance, if the PaperRetailer uses □ 75% of the time and γ during the remaining time, the expected PaperRetailer's outcome is different from 75% of its outcome in simulation in which only □ is used, plus 25% of its outcome in a second simulation in which only γ is applied, e.g., if the PaperRetailer uses □ first, γ next, the outcome will be different than if the PaperRetailer starts with γ, even if the ratio of 75%/25% is the same.
   Therefore, the method used in traditional game theory to calculate expected utilities is not appropriate in the context of our case study, because of the transition time between the two ordering rules that we have just described. Fortunately, this issue of inertia does not occur in all contexts, e.g., in all traditional games studied with game theory [10], such as the "Prisoners' Dilemma", etc. When this problem does not happen, the methodology presented in Section 2 is the same, except that Gambit has to be asked for the Nash equilibria in mixed strategies instead of in pure strategies in Algorithm 4.

3. Algorithmic complexity: We do not know of any algorithms used to determine Nash equilibria in mixed strategies requiring a reasonable time for any game. The determination of Nash equilibria in pure strategies requires the comparison of individual outcomes, while in mixed strategies, it requires the resolution of linear equations, a more complex task. From a practical viewpoint, computing (i.e., finding in a given game) Nash equilibria may also be a problem in pure strategies. For instance, it took around six minutes per game for Gambit to obtain our results, but it is much lower than what was required for mixed strategies because we stopped the computation of some of these same games after a week on a 2.7 GHz PC while the computation was not yet finished! From a theoretical viewpoint, Papadimitriou and Roughgarden [27] presented the finding of a polynomial-time algorithm for computing a Nash equilibrium as the "holy grail". In fact, Papadimitriou [28] thinks that this problem is harder than P and easier than NP-hard.

Let us now compare our methodology with another one that also studied interactions in a supply chain with simulations. This other work was carried out by Wellman and his colleagues [3] who refer to their methodology as "empirical game theory". They used the TAC/SCM (Trading Agent Competition/Supply Chain Management) [29] as data input to analyze the strategic behaviour of agents applying strategies inspired by the ones performed by the opponents in the TAC/SCM 2003 competition (note that Chapter 5 presents RedAgent, which is the winner of TAC/SCM 2003).

The two main differences between the approach of Wellman's team and ours are (i) that their data relies on the more detailed and complex simulation model of the TAC/SCM, in which our methodology would be too long to run (they claim that each simulation instance of the TAC/SCM takes 55 minutes, while our simulator runs within a second; as a result, the TAC/SCM simulator applied to our methodology would require $3^6 * 55$ minutes, i.e., almost a month, to build the game corresponding to one particular market consumption pattern), and (ii) they can consider symmetric games because all agents have the same role in TAC/SCM, while our agents have different roles in the supply chain. Considering symmetric games allows reducing the number of strategy profiles to consider from 729 to 28 when each of 6 agents can choose between 3 strategies.

Finally, readers interested in game theory applied to supply chains may be interested in several other chapters of this book. First, Chapter 4 defines the optimal production quantity and transactions of manufacturers as a Nash equilibrium. Then, Chapter 8 uses marketing science and cooperative game theory to propose a coordination and negotiation mechanism amongst the members of a virtual enterprise. The chapters related to auctions also deal with game theory, and in particular, Chapter 12 relies on mechanism design, which is the contrary of game theory (that is, game theory studies the properties of a system in which agents are given some strategies, while mechanism design searches for the strategies agents should be given so that the system has some properties).

5 Conclusion

In this paper, we presented a methodology to analyze stochastic simulations with game theory, when every utility function can be split in an individual utility and a utility representing the social consciousness of the agent. More precisely, we studied the impact on agents' choice of strategy when these agents consider more or less the part of their utility that represents social consciousness. We illustrated this methodology with a case study of the reduction of the bullwhip effect (i.e., the reduction of the amplification of order variability) in a supply chain. Specifically, we studied whether selfish company-agents would use non-collaborative strategies more frequently than benevolent company-agents. In this context, we assumed that every agent's individual utility is the inventory holding cost, and that agents take their impact on the rest of the system into account as the service delivered to their clients (which is measured as their backorder level, i.e., as the amplitude and duration of their stockouts).

We found that companies collaborate frequently, whatever the level of social consciousness is. We think that this result is due to the fact that companies are strongly dependent on each other in our model. However, non-collaboration is more frequent when the level of social consciousness is low, even if non-collaboration is never the most frequently used strategy.

Finally, we discussed our methodology and pointed out some possible improvements on it. In particular, we noted that the concept of mixed strategies may be difficult to introduce in some contexts. In particular, in our case study, there is some inertia in the supply chain that results in a transition time that forbids the calculation of expected utilities as in traditional game theory. Fortunately, this problem does not occur in all contexts.

6 Acknowledgment

We would like to thank FOR@C, the Research Consortium in E-business in the forest products industry (Université Laval, Québec, Canada), for supporting this research. This work was also partially supported by the National Sciences and Engineering Research Council of Canada (NSERC).

References

1. Parsons, S., Wooldridge, M.: Game theory and decision theory in multi-agent systems. Autonomous Agents and Multi-Agent Systems 5 (2002) 243–254
2. Moyaux, T., Chaib-draa, B., D'Amours, S.: Experimental study of incentives for collaboration in the Québec wood supply game. IEEE Transactions on Engineering Management (2005) (submitted).
3. Wellman, M.P., Estelle, J., Singh, S., Vorobeychik, Y., Kiekintveld, C., Soni, V.: Strategic interactions in a supply chain game. Computational Intelligence 21 (2005) 1–26
4. Lee, H.L., Padmanabhan, V., Whang, S.: Information distortion in a supply chain: The bullwhip effect. Management Science 43 (1997) 546–558

5. Simchi-Levi, D., Kaminsky, P., Simchi-Levi, E.: Designing and managing the supply chain. McGraw-Hill Higher Education (2000)
6. Glass, A., Grosz, B.J.: Socially conscious decision-making. Autonomous Agents and Multi-Agent Systems 6 (2003) 317–339
7. Stirling, W.C., Goodrich, M.A., Packard, D.J.: Satisficing equilibria: A non-classical theory of games and decisions. Autonomous Agents and Multi-Agent Systems 5 (2002) 305–328
8. Lipton, R.J., Markakis, E.: Nash equilibria via polynomial equations. In: Proc. LATIN 2004: Theoretical Informatics, 6th Latin American Symposium. (2004) 413–422
9. Sandholm, T.W.: Distributed rational decision making. In Weiss, G., ed.: Multiagent Systems - A Modern Approach to Distributed Artificial Intelligence. MIT Press, Cambridge (Massachusetts, USA) (1999) 201–258
10. Jehle, G.A., Reny, P.J.: Advanced Microeconomic Theory. Addison Wesley (2000) 2d Edition.
11. McKelvey, R.D., McLennan, A.M., Turocy, T.L.: Gambit: Software tools for game theory, version 0.97.0.4 (2004) http://econweb.tamu.edu/gambit/ (accessed 16 October 2004).
12. Lee, H.L., Padmanabhan, V., Whang, S.: The bullwhip effect in supply chain. Sloan Management Review 38 (1997) 93–102
13. Carlsson, C., Fullér, R.: Soft computing and the bullwhip effect. Economics and Complexity 2 (2000)
14. Moyaux, T., Chaib-draa, B., D'Amours, S.: Information sharing as a coordination mechanism for reducing the bullwhip effect in a supply chain. IEEE Trans. on Systems, Man, and Cybernetics (2005) (to appear).
15. Taylor, D.: Measurement and analysis of demand amplification across the supply chain. The International Journal of Logistics Management 10 (1999) 55–70
16. Daganzo, C.: A theory of supply chains. Springer (2003)
17. Dejonckheere, J., Disney, S.M., Lambrecht, M.R., Towill, D.R.: Transfer function analysis of forecasting induced bullwhip in supply chains. International Journal of Production Economics 78 (2002) 133–144
18. Forrester, J.W.: Industrial dynamics - A major breakthrough for decision-makers. Harvard Business Review 36 (1958) 37–66
19. Forrester, J.W.: Industrial Dynamics. The M.I.T. Press - Massachusetts Institute of Technology - Cambridge (Massachusetts, USA) (1961)
20. Sterman, J.D.: Modeling managerial behavior: Misperceptions of feedback in a dynamic decision making experiment. Management Science 35 (1989) 321–339
21. Chen, F., Drezner, Z., Ryan, J.K., Simchi-Levi, D.: Quantifying the bullwhip effect in a simple supply chain: The impact of forecasting, lead times, and information. Management Science 46 (2000) 436–443
22. Kahn, J.A.: Inventories and the volatility of production. The American Economic Review 77 (1987) 667–679
23. Naish, H.F.: Production smoothing in the linear quadratic inventory model. The Economic Journal 104 (1994) 864–875
24. Shen, Y.: Impact of asymmetric information on inventory policy (2001) Term paper to Dr. Miller for the partial fulfillment of "Production Control".
25. Butman, J.: A pain in the (supply) chain. Harvard Business Review 80 (2002) 31
26. Rosenschein, J.S., Zlotkin, G.: Rules of Encounter: Designing Conventions for Automated Negotiation Among Computers. MIT Press, Cambridge (Massachusetts, USA) (1994)

27. Papadimitriou, C.H., Roughgarden, T.: Computing equilibria in multiplayer games. In: Proc. SODA. (2005) Algorithmic results for the special case of symmetric and other succinctly representable games: correlated equilibria can be found in polynomial time for a broad class that includes all previously known cases.
28. Papadimitriou, C.H.: Algorithms, games, and the internet. In: Proc. STOC 2001 (extended abstract in ICALP 2001). (2001)
29. Arunachalam, R., Sadeh, N.M.: The supply chain trading agent competition. Electronic Commerce Research and Applications 4 (2005) 63–81

MAGNET: A Multi-Agent System using Auctions with Temporal and Precedence Constraints *

John Collins and Maria Gini

Department of Computer Science and Engineering
University of Minnesota
{jcollins,gini}@cs.umn.edu

Abstract. We consider the problem of rational, self-interested, economic agents who must negotiate with each other in order to carry out their plans. Customer agents express their plans in the form of task networks with temporal and precedence constraints. The market runs a combinatorial reverse auction, in which supplier agents submit bids specifying prices for combinations of tasks, along with time windows and duration data that the customer may use to compose a work schedule. The presence of temporal and precedence constraints among the items at auction requires extensions to the standard winner-determination procedures for combinatorial auctions, and the use of the enhanced winner-determination procedure within the context of a real-time negotiation requires that we predict its runtime when planning the negotiation process.

We address two specific issues related to this problem. The first is the need for a market infrastructure to support decision processes. We propose a set of requirements for a market that can support this type of negotiation, and describe an architecture that can meet these requirements. We also describe the high-level design of an agent that can act as a customer in this environment, and discuss the decision behaviors such an agent must implement to maximize its utility. The second issue we consider is the determination of auction winners. We explore and characterize a winner determination method, which is an extension of the bidtree-based Iterative-Deepening A* (IDA*) formulation proposed by Sandholm.

1 Introduction

We believe that much of the commercial potential of the Internet will remain unrealized until a new generation of autonomous systems is developed and deployed. A major problem is that the global connectivity and rapid communication capabilities of the Internet can present an organization with vast numbers

* Work supported in part by the National Science Foundation under grants IIS-0084202 and IIS-0414466

J. Collins and M. Gini: MAGNET: A Multi-Agent System using Auctions with Temporal and Precedence Constraints, Studies in Computational Intelligence (SCI) 28, 273–314 (2006)
www.springerlink.com

of new opportunities, to the point that users are overwhelmed, and conventional automation is insufficient.

Much has been done to enable simple buying and selling over the Internet, and systems exist to help customers and suppliers find each other, such as search engines, vertical industry portals, personalization systems, and recommender engines. However, many business operations are much more complex than the simple buying and selling of individual items. We are interested in situations that require coordinated combinations of goods and services, where there is often some sort of constraint-satisfaction or combinatorial optimization problem that needs to be solved in order to assemble a "deal." Commonly, these extra complications are related to constraints among task and services, and to time limitations. The combinatorics of such situations are not a major problem when an organization is working with small numbers of partners, but can easily become nearly insurmountable when "opened up" to the public Internet.

We envision a new generation of systems that will help organizations and individuals find and exploit opportunities that are otherwise inaccessible or too complex to seriously evaluate. These systems will help potential partners find each other (matchmaking), negotiate mutually beneficial deals (negotiation, evaluation, commitment), and help them monitor the progress of distributed activities (monitoring, dispute resolution). They will operate with variable levels of autonomy, allowing users to delegate or reserve authority as needed, and they will provide users with a market presence and power that is far beyond what is currently achievable with today's telephone, fax, web, and email-based methods. We believe that an important negotiation paradigm among these systems will be market-based combinatorial auctions, with added precedence and temporal constraints.

The Multi-AGent NEgotiation Testbed (MAGNET) project represents a first step in bringing this vision to reality. MAGNET provides a unique capability that allows self-interested agents to negotiate over complex coordinated tasks, with precedence and time constraints, in an auction-based market environment. This paper introduces many of the problems a customer agent must solve in the MAGNET environment, explores in detail the problem of solving the extended combinatorial-auction winner determination problem.

This paper is organized as follows. Section 2 works through a complete interaction scenario with an example problem, describing each of the decision processes a customer agent must implement in order to maximize the expected utility of its principal. Section 3 focuses on one specific decision problem, that of deciding the winners in a MAGNET auction. We describe an optimal tree search formulation algorithm for this problem. Section 4 briefly describes the results of experiments that characterize the performance of our search algorithm. This is important because it is a difficult combinatorial problem, and the negotiation process requires that time be allocated to it before the details of the problem can be known. Section 5 places this work in context with other work in the field. Finally, Section 6 wraps up the discussion and points out a set of additional research topics that must be addressed to further realize the MAGNET vision.

2 Decision Processes in a MAGNET Customer Agent

We focus on negotiation scenarios in which the object of the interaction is to gain agreement on the performance of a set of coordinated tasks that one of the agents desires to complete in order to maximize its own utility. We assume that self-interested agents will cooperate in such a scheme to the extent that they believe it will be profitable for them to do so. After a brief high-level overview of the MAGNET system, we focus on the decision processes that must be implemented by an agent that acts as a customer in the MAGNET environment. We intend that our agents exhibit rational economic behavior. In other words, the agent should always act to maximize the expected utility of its principal.

We will use an example to work through the agent's decisions. Imagine that you own a small vineyard, and that you need to get last autumn's batch of wine bottled and shipped [1]. During the peak bottling season, there is often a shortage of supplies and equipment, and your small operation must lease the equipment and bring on seasonal labor to complete the process. If the wine is to be sold immediately, then labels and cases must also be procured, and shipping resources must be booked. Experience shows that during the Christmas season, wine cases are often in short supply and shipping resources are overbooked.

2.1 Agents and their Environment

Agents may fulfill one or both of two roles with respect to the overall MAGNET architecture, as shown in Figure 1. A *Customer agent* pursues its goals by formulating and presenting *Requests for Quotations* (RFQs) to *Supplier agents* through a *market infrastructure* [1]. An RFQ specifies a task network that includes task descriptions, a precedence network, and temporal constraints that limit task start and completion times. Customer agents attempt to satisfy their goals for the greatest expected profit, and so they will accept bids at the least net cost, where cost factors can include not only bid prices, but also goal completion time, risk factors, and possibly other factors, such as preferences for specific suppliers. More precisely, these agents are attempting to maximize the utility function of some user, as discussed in detail in [2].

A supplier agent attempts to maximize the value of the resources under its control by submitting bids in response to RFQs. A bid specifies what tasks the supplier is able to undertake, when it is available to perform those tasks, how long they will take to complete, and a price. Each bid may specify one or more tasks. Suppliers may submit multiple bids to specify different combinations of tasks, or possibly different time constraints with different prices. For example, a supplier might specify a short duration for some task that requires use of high cost overtime labor, as well as a longer duration at a lower cost using straight-time labor. MAGNET currently supports simple disjunction semantics for bids from the same supplier. This means that if a supplier submits multiple bids, any non-conflicting subset can be accepted. Other bid semantics are possible [3, 4].

[1] This example is taken from the operations of the Weingut W. Ketter winery, Kr□ov, Germany.

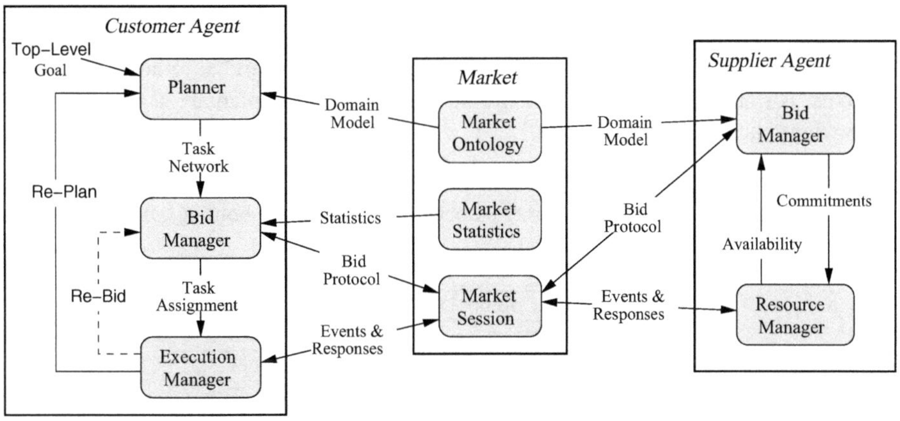

Fig. 1. The MAGNET architecture

2.2 Planning

A transaction (or possibly a series of transactions) starts when the agent or its principal acquires a goal that must be satisfied, or an opportunity arises that, if satisfied, would likely yield a positive payoff. Attributes of the goal might include a payoff and a deadline, or a payoff function that varies over time, either according to a discount rate or some other function.

While it would certainly be possible to integrate a general-purpose planning capability into a MAGNET agent, we expect that in many realistic situations the principal will already have a plan, perhaps based on standard industry practices. Figure 2 shows such a plan, for our winery bottling operation. We shall use this plan to illustrate the decision processes the agent must perform (or provide assistance to its principal in performing).

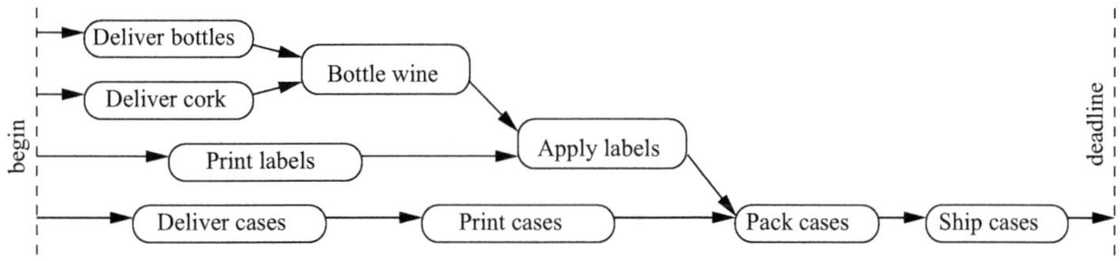

Fig. 2. Plan for the wine-bottling example

Formally, we define a plan $P = (S, V)$ as a task network containing a set of tasks S, and a set of precedence relations V. A precedence relation relates two tasks $s, s' \in \mathcal{S}$ as $s \prec s'$, interpreted as "task s must be completed before task s' can start."

We assume that markets will be sponsored by trade associations and commercial entities, and will therefore be more or less specialized. A consequence of this is that agents must in general deal in multiple markets to accomplish their goals. For our example, we assume that the tasks in our plan are associated with markets as specified in Table 1.

Table 1. Tasks and market associations for the wine-bottling example

Task	Description	Market
s_1	Deliver bottles	Vineyard Services
s_2	Deliver cork	Vineyard Services
s_3	Bottle wine	Vineyard Services
s_4	Print labels	Printing & Graphic Arts
s_5	Apply labels	Vineyard Services
s_6	Print cases	Vineyard Services
s_7	Deliver cases	Vineyard Services
s_8	Pack cases	(none)
s_9	Ship cases	Transport Services

It appears that we will need to deal with 3 different markets, and we will pack the cases ourselves. Or perhaps we'll open a few bottles and invite the village to help out.

So far, our plan is not situated in time, and we have not discussed our expected payoff for completing this plan. In the wine business, the quality of the product depends strongly on time. The wine must be removed from the casks within a 2-week window, and the bottling must be done immediately. For some varieties, the price we can get for our wine is higher if we can ship earlier, given a certain quality level. All the small vineyards in the region are on roughly the same schedule, so competition for resources during the prime bottling period can be intense. Without specifying the exact functions, we assume that the payoff drops off dramatically if we miss the 2-week bottling window, and less dramatically as the shipment date recedes into the future.

This example is admittedly a bid contrived, and it is important not to stretch it too far. We are treating the bottling and labeling operations as atomic – the entire bottling operation must be finished before we can start labeling – even though common-sense would inform us that you would probably want to apply this constraint at the per-bottle level, not the per-batch level. On the other hand, some varieties of wine are aged in the bottles for 6 months or more before the labels are applied.

2.3 Planning the Bidding Process

At this point, the agent has a plan, and it knows which markets it must deal in to complete the plan, the value of completing the plan, and how that value depends on time. The next step is to decide how best to use the markets to maximize its utility. It will do this in two phases. First, the agent generates an overall plan for the bidding process, which may involve multiple RFQs in each of multiple markets. We call this a "bid-process plan". Then a detailed timeline is generated for each RFQ.

The simplest bid-process plan would be to issue a single RFQ in each market, each consisting of the portion of the plan that is relevant to its respective market. If all RFQs are issued simultaneously, and if they are all on the same timeline, then we can combine their bids and solve the combined winner-determination problem in a single step. However, this might not be the optimum strategy. For example:

- We may not have space available to store the cases if we are not ready to pack them when they arrive.
- Our labor costs might be much lower if we can label as we bottle; otherwise, we will need to move the bottles into storage as we bottle, then take them back out to label them.
- Once cases are packed, it is easy for us to store them for a short period. This means that we can allow some slack between the packing and shipping tasks.
- There is a limit to what we are willing to pay to bottle our wine, and there is a limit to the premium we are willing to pay to have the bottling completed earlier.

The agent can represent these issues as additional constraints on the plan, or in some cases as alternative plan components. For example, we could constrain the interval between s_5 (labeling) and s_8 (packing) to a maximum of one day, or we could add an additional storage task between s_3 (bottling) and s_5 that must be performed just in case there is a non-zero delay between the end of s_3 and the start of s_5.

There are many possible alternative actions that the agent can take to deal with these issues. It need not issue RFQs in all markets simultaneously. It need not include all tasks for a given market in a single RFQ. Indeed, dividing the plan into multiple RFQs can be an important way to reduce scheduling uncertainty. For example, we might want to have a firm completion date for the bottling and labeling steps before we order the cases.

Market statistics can be used to support these decisions. For example, if we knew that resources were readily available for the steps up through the labeling process (tasks $s_1 \ldots s_5$), we could include the case delivery and printing steps (tasks s_6 and s_7) in the same RFQ. This could be advantageous if suppliers were more likely to bid or likely to bid lower prices if they could bid on more of the business in a single interaction. In other words, some suppliers might be willing to offer a discount if we agree to purchase both bottles and cases from them, but if we negotiate these two steps in separate RFQs, we eliminate the ability to find out about such discounts.

We should note that suppliers can either help or hinder the customer in this process, depending on the supplier's motivations. For example, the supplier can help the customer mitigate issues like the constraint between bottling and packing. For example, if a supplier knew about this constraint, it could offer both tasks at appropriate times, or it could give the customer the needed scheduling flexibility by offering the case delivery over a broad time window or with multiple bids with a range of time windows. In some domains this could result in higher costs, due to the large speculative resource reservations the supplier would have to commit to in order to support its bids. On the other hand, if a supplier saw an RFQ consisting of s_6 and s_7, it would know that the customer had likely already made commitments for the earlier tasks, since nobody wants cases printed if they aren't bottling. If the supplier also knew that there would be little competition within the customer's specified time window, it could inflate its prices, knowing that the customer would have little choice.

The bid-process plan that results from this decision process is a network of negotiation tasks and decision points. Figure 3 shows a possible bid-process plan for our wine-bottling example.

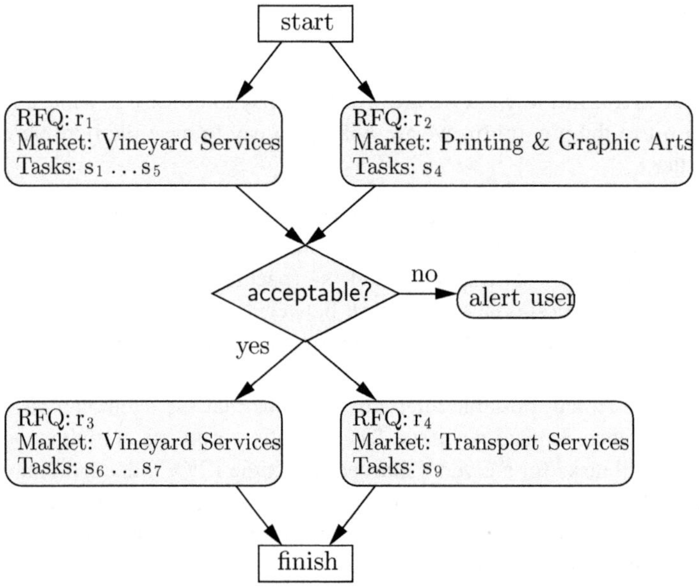

Fig. 3. Bid-process plan for the wine-bottling exa mple

Once we have a bid-process plan, we know what markets we will enter, and how we want to divide up the bidding process. We must then schedule the bid-process plan, and allocate time within each RFQ/bidding interaction. These two scheduling problems may need to be solved together if the bid-process plan contains multiple steps and it is important to finish it in minimum time. Each RFQ step needs to start at a particular time, or when a particular event occurs or some condition becomes true. For example, if the rules of the market require deposits to be paid when bids are awarded, the customer may be motivated to move RFQ steps as late as possible, other factors being equal. On the other hand, if resources such as our bottling and labeling steps are expected to be in short supply, the agent may wish to gain commitments for them as early as possible in order to optimize its own schedule and payoff. We assume these decisions can be supported by market statistics, the agent's own experience, and/or the agent's principal.

Each RFQ must also be allocated enough time to cover the necessary deliberation processes on both the customer and supplier sides. Some of these processes may be automated, and some may involve user interaction. The timeline in Figure 4 shows an abstract view of the progress of a single negotiation. At the beginning of the process, the customer agent must allocate deliberation time to itself to compose its RFQ[2], to the supplier for bid preparation, and to itself again for the bid evaluation process. Two of these time points, the bid deadline and the bid award deadline, must be communicated to suppliers as part of the RFQ. The bid deadline is the latest time a supplier may submit a bid, and the bid award deadline is the earliest time a supplier may expire a bid. The interval between these two time points is available to the customer to determine the winners of the auction.

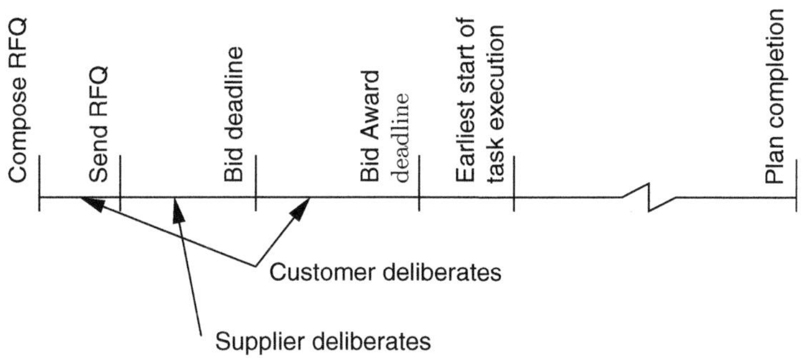

Fig. 4. Typical timeline for a single RFQ

In general, it is expected that bid prices will be lower if suppliers have more time to prepare bids, and more time and schedule flexibility in the execution phase. Minimizing the delay between the bid deadline and the award deadline will also minimize the supplier's opportunity cost, and would therefore be expected to reduce bid prices. On the other hand, the customer's ability to find a good set of bids is dependent on the time allocated to bid evaluation, and if a user is making the final decision on bid awards, she may want to run multiple bid-evaluation cycles with some additional think time. We are interested in the performance of the winner determination process precisely because it takes place

[2] This may be a significant combinatorial problem – see for example [5].

within a window of time that must be determined ahead of time, before bids are received, and because we expect better overall results, in terms of maximizing the agent's utility, if we can maximize the amount of time available to suppliers while minimizing the time required for customer deliberation. These time intervals can be overlapped to some extent, but doing so can create opportunities for strategic manipulation of the customer by the suppliers, as discussed in [6].

The process for setting these time intervals could be handled as a non-linear optimization problem, although it may be necessary to settle for an approximation. This could consist of estimating the minimum time required for the customer's processes, and allocating the remainder of the available time to the suppliers, up to some reasonable limit.

2.4 Composing a Request for Quotes

At this point in the agent's decision process, we have the information needed to compose one or more RFQs, we know when to submit them, and we presumably know what to do if they fail (if we fail to receive a bid set that covers all the task in the RFQ, for example). The next step is to set the time windows for tasks in the individual RFQs, and submit them to their respective markets.

Formally, an RFQ $r = (\mathcal{S}_r, \mathcal{V}_r, \mathcal{W}_r, \tau)$ contains a subset $\mathcal{S}_r$ of the tasks in the task network $\mathcal{P}$, with their precedence relations $\mathcal{V}_r$, the task time windows $\mathcal{W}_r$ specifying constraints on when each task may be started and completed, and the RFQ timeline τ containing at least the bid deadline and bid award deadline. As pointed out earlier, there might be elements of the task network $\mathcal{P}$ that are not included in the RFQ. For each task $s \in \mathcal{S}_r$ in the RFQ the bid manager must specify a time window $w \in \mathcal{W}_r$, consisting of an earliest start time $t_{es}(s,r)$ and a latest finish time $t_{lf}(s,r)$, and a set of precedence relationships $\mathcal{V}_r = \{\text{for each } s, s' \in \mathcal{S}_r, s' \prec s\}$, associating s with each of the other tasks $s' \in \mathcal{S}_r$ whose completion must precede the start of s.

The principal outcome of the RFQ-generation process is a set of values for the early-start and late-finish times for the time windows $\mathcal{W}_r$ in the RFQ. We obtain a crude first approximation using the Critical Path (CPM) algorithm [7], after making some assumptions about the durations of tasks, and about the earliest start time for tasks that have no predecessors in the RFQ (the *root tasks* $\mathcal{S}_R$) and the latest finish times for tasks that have no successors in the RFQ (the *leaf tasks* $\mathcal{S}_L$). Market mean-duration statistics can be used for the task durations. Overall start and finish times for the tasks in the RFQ may come from the bid-process plan, or we may already have commitments that constrain them as a result of other activities. For this discussion, we assume a continuous-time domain, although we realize that many real domains effectively work on a discrete-time basis. Indeed, it is very likely that some of our wine bottling activities would typically be quoted in whole-day increments. We also ignore calendar issues such as overtime/straight time, weekends, holidays, time zones, etc.

The critical path algorithm walks the directed graph of tasks and precedence constraints, forward from the early-start times of the root tasks to compute the earliest start $t_{es}(s)$ and finish $t_{ef}(s)$ times for each task $s \in \mathcal{S}_r$, and then backward from the late-finish times of the leaf tasks to compute the latest finish $t_{lf}(s)$ and start $t_{ls}(s)$ times for each task. The minimum duration of the entire task network specified by the RFQ, defined as $\max_{s' \in \mathcal{S}_L}(t_{ef}(s')) - \min_{s \in \mathcal{S}_R}(t_{es}(s))$, is called the makespan of the task network. The smallest slack in any leaf task $\min_{s \in \mathcal{S}_L}(t_{lf}(s) - t_{ef}(s))$ is called the total slack of the task network within the RFQ. All tasks s for which $t_{lf}(s) - t_{ef}(s) =$ total-slack are called critical tasks. Paths in the graph through critical tasks are called critical paths.

Some situations will be more complex than this. This can happen when there are constraints that are not captured in the precedence network of the RFQ. For example, some non-leaf task may have successors that are already committed but are outside the RFQ. The CPM algorithm is still applicable, but the definition of critical tasks and critical paths becomes more complex.

Figure 5 shows the result of running the CPM algorithm on the tasks of RFQ r_1 from our bid-process plan. We are assuming task durations as given in the individual "task boxes." We observe several problems immediately. The most obvious is that it is likely that many bids returned in response to this RFQ would conflict with one another because they would fail to combine feasibly. For example, if I had a bid for the label printing task s_5 for days 5-7, then the only bids I could accept for the labeling task s_4 would be those that had a late start time at least as late as day 7. If the bids for s_4 were evenly distributed across the indicated time windows, and if all of them specified the same 4-day duration, then only 1/3 of those bids could be considered. In general, we want to allow time windows to overlap, but excessive overlap is almost certainly counterproductive. We will revisit this issue shortly.

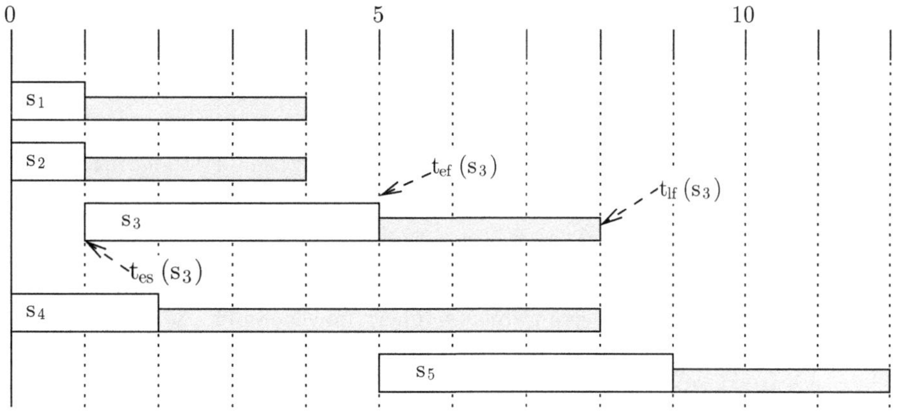

Fig. 5. Initial time allocations for tasks in RFQ r_1. Only the $t_{es}(s)$ and $t_{lf}(s)$ times are actually specified in the RFQ

Once we have initial estimates from the CPM algorithm, there are several issues to be resolved, as described in the following sections.

Setting the total slack The plan may have a hard deadline, which may be set by a user or determined by existing commitments for tasks that cannot be started until tasks in the current RFQ are complete. Otherwise, in the normal case, the bid-process plan is expected to set the time limits for the RFQ.

It would be interesting to find a way to use the market to dynamically derive a schedule that maximizes the customer's payoff. This would require cooperation of bidders, and could be quite costly. Parkes and Ungar [8] have done something like this in a restricted domain, but it's hard to see how to apply it to the more generalized MAGNET domain.

Task ordering For any pair of tasks in the plan that could potentially be executed in parallel, we may have a choice of handling them in parallel, or in either sequential order. For example, in our wine-bottling example, we could choose to acquire the bottles before buying the corks. This example is a bit contrived, perhaps, but if there is uncertainty over the ability to complete tasks which could cause the plan to be abandoned, then (given some straightforward assumptions such as payments being due when work is completed) the agent's financial exposure can be affected by task ordering. If a risky task is scheduled ahead of a "safe" task, then if the risky task fails we can abandon the plan without having to pay for the safe task. Babanov [5] has worked out in detail how to use task completion probabilities and discount rates in an expected-utility framework to maximize the probabilistic "certain payoff" for an agent with a given risk-aversion coefficient.

For some tasks, linearizing the schedule will extend the plan's makespan, and this must be taken into account in terms of changes to the ultimate payoff. Note that in many cases the agent may have flexibility in both the start time and the completion time of the schedule. This would presumably be true of our wine-bottling example.

Allocating time to individual tasks Once we have made decisions about the overall time available and about task ordering, the CPM algorithm gives us a set of preliminary time windows. In most cases, this will not produce the best results, for several reasons:

Resource availability – In most markets, services will vary in terms of availability and resource requirements. There may be only a few dozen portable bottling and labeling machines in the region, while corks may be stored in a warehouse ready for shipping. There is a high probability that one could receive several bids for delivery of corks on one specific day, but a much lower probability that one could find even one bid for a 6-day bottling job for a specific 6-day period. More likely one would have to allow some flexibility in the timing of the bottling operation in order to receive usable bids.

Lead-time effects – In many industries, suppliers have resources on the payroll that must be paid for whether their services are sold or not. In these cases, suppliers will typically attempt to "book" commitments for their resources into the future. In our example, the chances of finding a print shop to produce our labels tomorrow is probably much lower than the chances of finding shops to print them next month. This means that, at least for some types of services, one must allow more scheduling flexibility to attract short lead time bids than for longer lead times. We should also expect to pay more for shorter lead times.

Task-duration variability – Some services are very standardized (delivering corks, printing 5000 labels), while others may be highly variable, either because they rely on human creativity (software development) or the weather (bridge construction), or because different suppliers use different processes, different equipment, or different staffing levels (wine bottling). These two types of variability can usually be differentiated by the level of predictability; suppliers that uses a predictable process with variable staffing levels are likely to be able to deliver on time on a regular basis, while services that are inherently unpredictable will tend to exhibit frequent deviations from the predictions specified in bids[3].

For services that exhibit a high variability in duration, as specified in bids, the customer's strategy may depend on whether a large number of bidders is expected, and whether there is a correlation between bid price and quoted task duration. If a large number of bidders is expected, then the customer may be able to allocate a below-average time window to the task, in the expectation that there will be some suppliers at the lower end of the distribution who will be able to perform within the specified window. On the other hand, if few bidders are expected, a larger than average time window may be required in order to achieve a reasonable probability of receiving at least one usable bid.

Excessive allocations to non-critical tasks – One obvious problem with the time allocations from the CPM algorithm as shown in Figure 5 is that non-critical tasks (tasks not on the critical path) are allocated too much time, causing unnecessary overlap in their time windows. All other things being equal, we are likely to be better off if RFQ time windows do not overlap, because we will have fewer infeasible bid combinations.

Trading off feasibility for flexibility In general we expect more bidders, and lower bid prices, if we offer suppliers more flexibility in scheduling their resources by specifying wider time windows. On the other hand, if we define RFQ time windows with excessive overlap, a significant proportion of bid combinations will be unusable due to schedule infeasibility. The intuition is that there will be some realistic market situations where the customer is better off allowing RFQ time windows to overlap to some degree, if we take into account price, plan completion time, and probability of successful plan completion (which requires at minimum a set of bids that covers the task set and can be composed into

[3] Whether the market or customers would be able to observe these deviations may depend on market rules and incentives, such as whether a supplier can be paid early by delivering early.

a feasible schedule). This means that the winner-determination procedure must handle schedule infeasibilities among bids.

Figure 6 shows a possible updated set of RFQ time windows for our wine-bottling example, taking into account the factors we have discussed. We have shortened the time windows for tasks s_1 and s_2, because we believe that bottles and corks are readily available, and can be delivered when needed. There is no advantage to allowing more time for these tasks. Market data tells us that bottling services are somewhat more difficult to schedule than labeling services, and so we have specified a wider time window for task s_3 than for s_4. Our deadline is such that the value of completing the work a day or two earlier is higher than the potential loss of having to reject some conflicting bids. We also know from market data that a large fraction of suppliers of the bottling crews can also provide the labeling service, and so the risk of schedule infeasibility will be reduced if we receive bids for both bottling and labeling. Finally, there is plenty of time available for the non-critical label-printing task s_5 without needing to overlap its time window with its successor task s_4.

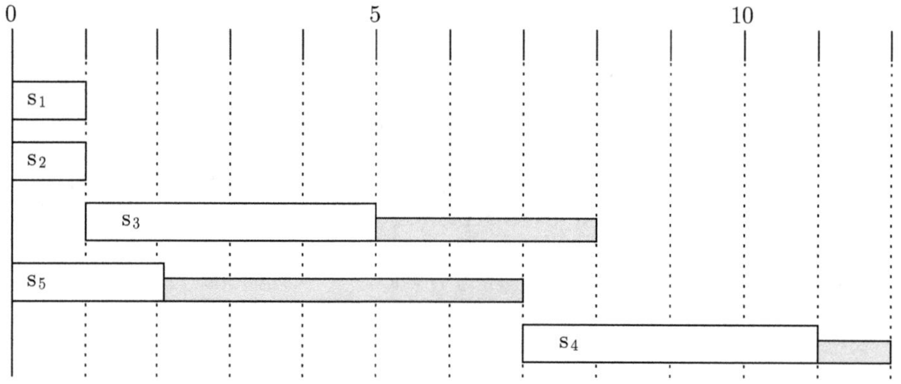

Fig. 6. Revised time allocations for tasks in RFQ r_1

2.5 Evaluating Bids

Once an RFQ is issued and the bids are returned, the agent must decide which bids to accept. The bidding process is an extended combinatorial auction, because bids can specify multiple tasks, and there are additional constraints the bids must meet (the precedence constraints) other than just covering the tasks. The winner-determination process must choose a set of bids that maximize the agent's utility, covers all tasks in the associated RFQ, and forms a feasible schedule.

Formal description of the winner-determination problem Each bid represents an offer to execute some subset of the tasks specified in the RFQ, for a specified price, within specified time windows. Formally, a bid $b = (r, \mathcal{S}_b, \mathcal{W}_b, c_b)$ consists of a subset $\mathcal{S}_b \in \mathcal{S}_r$ of the tasks specified in the corresponding RFQ r, a set of time windows $\mathcal{W}_b$, and an overall cost c_b. Each time window $w_s \in \mathcal{W}_b$ specifies for a task s an earliest start time $t_{es}(s, b)$, a latest start time $t_{ls}(s, b)$, and a task duration $d(s, b)$.

It is a requirement of the protocol that the time window parameters in a bid b are within the time windows specified in the RFQ, or $t_{es}(s, b) \geq t_{es}(s, r)$ and $(t_{ls}(s, b) + d(s, b)) \leq t_{lf}(s, r)$ for a given task s and RFQ r. This requirement may be relaxed, although it is not clear why a supplier agent would want to expose resource availability information beyond that required to respond to a particular bid. For bids that specify multiple tasks, it is also a requirement that the time windows in the bids be internally feasible. In other words, for any bid b, if for any two of its tasks $(s_i, s_j) \in \mathcal{S}_b$ there is a precedence relation $s_i \prec s_j$ specified in the RFQ, then it is required that $t_{es}(s_i, b) + d(s_i, b) \leq t_{ls}(s_j, b)$.

A solution to the bid-evaluation problem is defined as a complete mapping $\mathcal{S} \rightarrow \mathcal{B}$ of tasks to bids in which each task in the corresponding RFQ is mapped to exactly one bid, and that is consistent with the temporal and precedence constraints on the tasks as expressed in the RFQ and the mapped bids.

Figure 7 shows a very small example of the problem the bid evaluator must solve. As noted before, there is scant availability of bottling equipment and crews, so we have provided an ample time window for that activity. At the same time, we have allowed some overlap between the bottling and labeling tasks, perhaps because we believed this would attract a large number of bidders with a wide variation in lead times and lower prices. Bid 1 indicates this bottling service is available from day 3 through day 7 only, and will take the full 5 days, but the price is very good. Similarly, bid 2 offers labeling from day 7 through day 10 only, again for a good price. Unfortunately, we can't use these two bids together because of the schedule infeasibility between them. Bid 3 offers bottling for any 3-day period from day 2 through day 7, at a higher price. We can use this bid with bid 2 if we start on day 4, but if we start earlier we will have to handle the unlabeled bottles somehow. Finally, bid 4 offers both the bottling and labeling services, but the price is higher and we would finish a day later than if we accepted bids 2 and 3.

Evaluation criteria We have discussed the winner-determination problem in terms of price, task coverage, and schedule feasibility. In many situations, there are other factors that can be at least as important as price. For example, we might know (although the agent might not know) that the bottling machine being offered in bid 3 is prone to breakdown, or that it tends to spill a lot of wine. We might have a long-term contract with one of the suppliers, Hermann, that gives us a good price on fertilizer only if we buy a certain quantity of corks from him every year. We might also know that one of the local printers tends to miss his time estimates on a regular basis, but his prices are often worth the

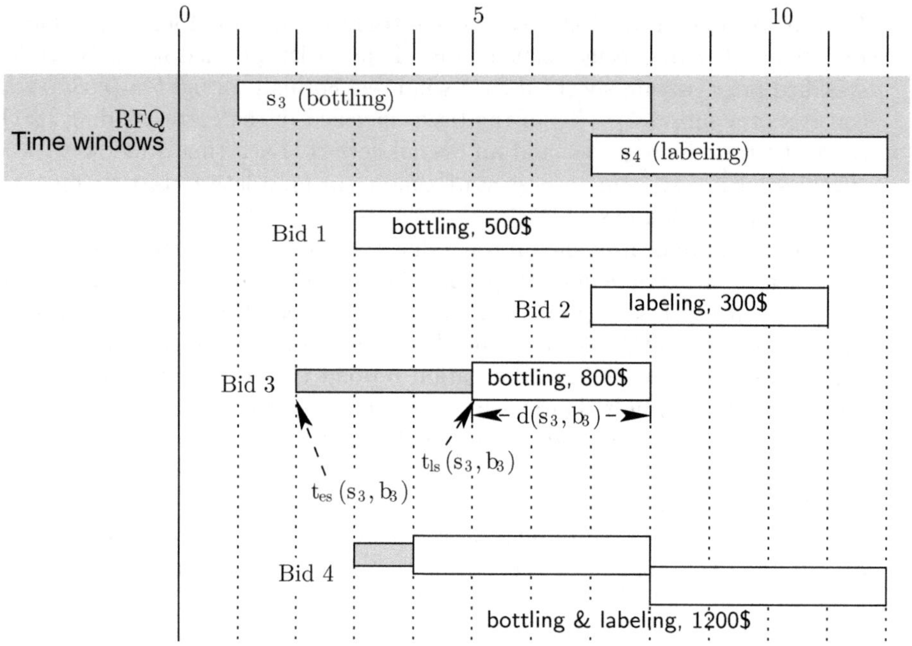

Fig. 7. Bid Example

hassle, as long as we build some slack into the schedule when we award a bid to him.

Many of these factors can be expressed as additional constraints on the winner-determination problem, and some can be expressed as cost factors. These constraints can be as simple as "don't use bid b_3" or more complex, as in "if Hermann bids on corks, and if a solution using his bid is no more than 10% more costly than a solution without his bid, then award the bid to Hermann." Some of them can be handled by preprocessing, some must be handled within the winner-determination process, and some will require running it twice and comparing results.

Mixed-initiative approaches There are many environments in which an automated agent is unlikely to be given the authority to make unsupervised commitments on behalf a person or organization. In these situations, we expect that many of the decision processes we discuss here will be used as decision-support tools for a human decision-maker, rather than as elements of a completely autonomous agent. The decision to award bids is one that directly creates commitment, and so it is a prime candidate for user interaction. We have constructed an early prototype of such an interface. It allows a user to view bids, add simple bid inclusion and exclusion constraints, and run one of the winner-determination search methods. Bids may be graphically overlaid on the RFQ, and both the

RFQ and bid time windows are displayed in contrasting colors on a Gantt-chart display.

Effective interactive use of the bid-evaluation functions of an agent require the ability to visualize the plan and bids, to visualize bids in groups with constraint violations highlighted, and to add and update constraints. The winner-determination solver must be accessible and its results presented in an understandable way, and there must be a capability to generate multiple alternative solutions and compare them.

2.6 Awarding Bids

The result of the winner-determination process is a (possibly empty) mapping $\mathcal{S} \rightarrow \mathcal{B}$ of tasks to bids. We assume that the bids in this mapping meet the criteria of the winner-determination process: they cover the tasks in the RFQ and can be composed into a feasible schedule, and they maximize the agent's or user's expected utility. However, we cannot just award the winning bids. In general, a bid b contains one or more offers of services for tasks s, each with a duration $d(s, b)$ within a time window $w(s, b) > d(s, b)$. The price assumes that the customer will specify, as part of the bid award, a specific start time for each activity. Otherwise, the supplier would have to maintain its resource reservation until some indefinite future time when the customer would specify a start time. This would create a disincentive for suppliers to specify large time windows, raise prices, and complicate the customer's scheduling problem.

This means that the customer must build a final work schedule before awarding bids. We will defer to the next section the issue of dealing with schedule changes as work progresses. This scheduling activity represents another opportunity to maximize the customer's expected utility. In general, the customer's utility at this point is maximized by appropriate distribution of slack in the schedule, and possibly also by deferring task execution in order to defer payment for completion.

3 Solving the MAGNET Winner-Determination Problem

We now focus on the MAGNET winner-determination problem, originally introduced in Section 2.5. Earlier we have described both an Integer Programming formulation [9] and a simulated annealing framework for solving this problem [10] for this problem. Here we will focus on an application of the A* method. The algorithm presented here solves the winner-determination problem under assumption of a fixed payoff, and does not deal with a payoff that depends on completion time.

The winner determination problem for combinatorial auctions has been shown to be $\mathcal{NP}$-complete and inapproximable [11]. This result clearly applies to the MAGNET winner determination problem, since we simply apply an additional set of (temporal) constraints to the basic combinatorial auction problem, and we don't allow free disposal. In fact, because the additional constraints create additional bid-to-bid dependencies, and because bids can vary in both price and in time specifications, the bid-domination and partitioning methods used by others to simplify the problem (for example, see [12]) cannot be applied in the MAGNET case.

Sandholm has shown that there can be no polynomial-time solution, nor even a polynomial-time bounded approximation [12], so we must accept exponential complexity. We will see in Section 4 that we can determine probability distributions for search time, based on problem size metrics, and we can use those empirically-determined distributions in our deliberation scheduling process.

Sandholm described an approach to solving the standard combinatorial auction winner-determination problem [12] using an iterative-deepening A* formulation. Although many of his optimizations, such as the elimination of dominated bids and partitioning of the problem, cannot be easily applied to the MAGNET problem, we have adapted the basic structure of Sandholm's formulation, and we have improved upon it by specifying a means to minimize the mean branching factor in the generated search tree.

In general, tree search methods are useful when the problem at hand can be characterized by a solution path in a tree that starts at an initial node (root) and progresses through a series of expansions to a final node that meets the solution criteria. Each expansion generates successors (children) of some existing node, expansions continuing until a final node is found. The questions of which node is chosen for expansion, and how the search tree is represented, lead to many different search methods. In the A* method, the node chosen for expansion is the one with the "best" evaluation[4], and the search tree is typically kept in memory in the form of a sorted queue. A* uses an evaluation function

$$f(N) = g(N) + h(N)$$

for a node N, where $g(N)$ is the cost of the path from initial node N_0 to node N, and $h(N)$ is an estimate of the remaining cost to a solution node. If $h(N)$ is a strict lower bound on the remaining cost (upper bound for a maximization problem), we call it an admissible heuristic and A* is complete and optimal; that is, it is guaranteed to find a solution with the lowest evaluation, if any solutions exist, and it is guaranteed to terminate eventually if no solutions exist.

We describe a basic A* formulation of the MAGNET winner-determination problem, and then we show how this formulation can be adapted to a depth-first iterative-deepening model [13] to reduce or eliminate memory limitations.

3.1 Bidtree Framework

For a basic introduction to the A* algorithm, see [14], or another textbook on Artificial Intelligence. Our formulation depends on two structures which must be prepared before the search can run. The first is the bidtree introduced by

[4] lowest for a minimization problem, highest for a maximization problem.

Sandholm, and the second is the bid-bucket, a container for the set of bids that cover the same task set.

A bidtree is a binary tree that allows lookup of bids based on item content. The bidtree is used to determine the order in which bids are considered during the search, and to ensure that each bid combination is tested at most once. In Sandholm's formulation, the collection of bids into groups that cover the same item sets supports the discard of dominated bids, with the result that each leaf in the bidtree contains one bid. However, because our precedence constraints create dependencies among bids in different buckets, bid domination is a much more complex issue in the MAGNET problem domain. Therefore, we use bid-buckets at the leaves rather than individual bids.

The principal purpose of the bidtree is to support content-based lookup of bids. Suppose we have a plan S with tasks s_m, $m = 1..4$. Further suppose that we have received a set of bids b_n, $n = 1..10$, with the following contents: $b_1 : \{s_1, s_2\}$, $b_2 : \{s_2, s_3\}$, $b_3 : \{s_1, s_4\}$, $b_4 : \{s_3, s_4\}$, $b_5 : \{s_2\}$, $b_6 : \{s_1, s_2, s_4\}$, $b_7 : \{s_4\}$, $b_8 : \{s_2, s_4\}$, $b_9 : \{s_1, s_2\}$, $b_{10} : \{s_2, s_4\}$. Figure 8 shows a bidtree we might construct for this problem. Each node corresponds to a task. One branch, labeled in, leads to bids that include the task, and the other branch, labeled out, leads to bids that do not.

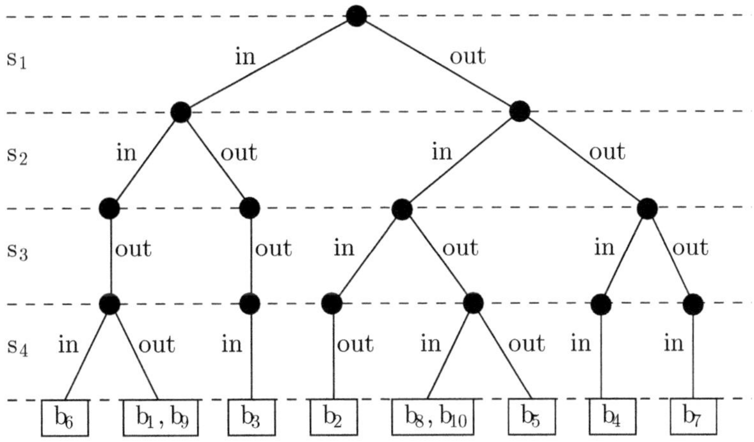

Fig. 8. Example bidtree, lexical task order

We use the bidtree by querying it for bid-buckets. A query consists of a mask, a vector of values whose successive entries correspond to the "levels" in the bidtree. Each entry in the vector may take on one of three values, {in, out, any}. A query is processed by walking the bidtree from its root as we traverse the vector. If an entry in the mask vector is in, then the in branch is taken at the corresponding level of the tree, similarly with out. If an entry is any, then both branches are taken at the corresponding level of the bidtree. So, for example, if we used a mask of [in, any, any, in], the bidtree in Figure 8 would return the bid-buckets containing $\{b_6\}$ and $\{b_3\}$.

A bid-bucket is a container for a set of bids that cover the same task set. In addition to the bid set, the bid-bucket structure stores the list of other bid-buckets whose bids conflict with its own (where we use "conflicts" to mean that they cover overlapping task sets). This recognizes the fact that all bids with the same task set will have the same conflict set.

In order to support computation of the heuristic function, we use a somewhat different problem formulation for A* and IDA* than we used for the IP formulation described in [9]. In that formulation, we were minimizing the sum of the costs of the selected bids. In this formulation, we minimize the cost of each of the tasks, given a set of bid assignments. This allows for straightforward computation of the A* heuristic function $f(N)$ for a given node N in the search tree. We first define

$$f(N) = g(S_m(N)) + h(S_u(N))$$

where $S_m(N)$ is the set of tasks that are mapped to bids in node N, while $S_u(N) = S_r \setminus \mathcal{S}_m(N)$ is the set of tasks that are not mapped to any bids in the same node. We then define

$$g(S_m(N)) = \sum_{j \mid s_j \in \mathcal{S}_m} \frac{c(b_j)}{n(b_j)}$$

where b_j is the bid mapped to task s_j, $c(b_j)$ is the total cost of b_j, $n(b_j)$ is the number of tasks in b_j, and

$$h(S_u(N)) = \sum_{j \mid s_j \in \mathcal{S}_u} \frac{c(b_j^*)}{n(b_j^*)}$$

where b_j^* is the "usable" bid for task s_j that has the lowest cost/task. By "usable," we mean that the bid b_j^* includes s_j, and does not conflict (in the sense of having overlapping task sets) with any of the bids b_j already mapped in node N.

Note that, unlike the case with the IP solver, the definition of $g(S_m(N))$ can be expanded to include other factors, such as risk estimates or penalties for inadequate slack in the schedule, and these factors can be non-linear. The only requirement is that any such additional factor must increase the value of $g(S_m(N))$, and not decrease it, because otherwise the admissibility of the heuristic will be compromised, and we no longer would have a complete search method.

3.2 A* Formulation

Now that we have described the bidtree and bid-bucket, we can explain our optimal tree search formulation. The algorithm is given in Figure 9.

The principal difference between this formulation and the "standard" A* search formulation (see, for example, [14]), is that nodes are left on the queue

```
1  Procedure A*_search
2  Inputs:
3      {S, V}: the task network to be assigned
4      B: the set of bids, represented as a bidtree
5  Output:
6      N_opt : the node having a mapping M (N_opt) = S → B
           of tasks to bids with an opti mal evaluation, if one exists
7  Process:
8      Q ← priority _queue, sorted by node evaluation f (N)
9      N_0 ← empty node
10     mask ← {in, any, any, cdots}
11     B^c_N0 ← bidtree _query (B, mask)
         "B^c_N0 is a set of bids (in the form of a set of bid buckets), containing
         the bids that can be used to expand N_0"
12     insert (Q, N_0)
13     loop
14         if e mpty(Q) then return failure
15         N ← first (Q)
16         if solution (N) then return N
17         N′ ← astar _expand(N) "see Figure 10"
18         if N′ = null then re move_front (Q) "Remove nodes that fail to expand"
19         else if feasible (N′) then insert (Q, N′)
```

Fig. 9. Bidtree-based A* search algorithm

(line 15) until they cannot be expanded further, and only a single expansion is tried (line 17) at each iteration. This is to avoid expending unnecessary effort evaluating nodes.

The expansion of a parent node N to produce a child node N' (line 17 in Figure 9) using the bidtree is shown in Figure 10. Here we see the reason to keep track of the buckets for the candidate-bid set of a node. In line 16, we use the mask for a new node to retrieve a set of bid-buckets. In line 18, we see that if the result is empty, or if there is some unallocated task for which no usable bid remains, we can go back to the parent node and just dump the whole bucket that contains the candidate we are testing.

In line 17 of Figure 10, we must find the minimum-cost "usable" bids for all unallocated tasks S_u (tasks not in the union of the task sets of $B_{N'}$), as discussed earlier. One way (not necessarily the most efficient way) to find the set of usable bids is to query the bidtree using the mask that was generated in line 14, changing the single in entry to any. If there is any unallocated task that is not covered by some bid in the resulting set, then we can discard node N' because it cannot lead to a solution (line 22). Because all other bids in the same bidtree leaf node with the candidate bid b_x will produce the same bidtree mask and the same usable-bid set, we can also discard all other bids in that leaf node from the candidate set of the parent node N.

```
1  Procedure astar_expand
2  Inputs:
3      N: the node to be expanded
4  Output:
5      N′: a new node with exactly one additional bid, or null
6  Process:
7      buckets ← ∅
8      while buckets = ∅ do
9          if B_N^c = ∅ then return null "B_N^c is set of candidate bids for node N"
10         b_x ← choose (B_N^c) "pick a bid from the set of candidates"
11         B_N^c ← B_N^c − b_x "remove the chosen bid from the set"
12         N′ ← new node
13         B_N′ ← B_N + b_x "B_N′ is the set of bids in node N′"
14         S_u ← unallocated_tasks (N′) "tasks not covered by any bid b ∈ B_N′"
15         mask ← create_mask (B′_N)
               "for each task that is covered by a bid in B_N′, set the corresponding
               entry to out. Then find the first task in s ∈ S_u (the task in S_u with
               the minimum index in the bidtree) and set its entry to in. Set the
               remaining entries to any"
16         buckets ← bidtree_query (B, mask)
17         B_u ← ∀s ∈ S_u, minimum_usable_bid (s) "see the narrative"
18         if (solution (N′))
               ∨((buckets ≠ ∅) ∧ (¬∃s ∈ S_u | minimum_usable_bid (s) = null))
19         then
20             B_N′^c ← buckets "candidates for N′"
21         else
22             remove (B_N^c, bucket (b_x))
                   "all bids in the bucket containing b_x in node N will produce the
                   same mask and therefore an empty candidate set or a task that
                   cannot be covered by any usable bid"
23     end while
24     g(N′) ← Σ_{b∈B_N′} c_b
25     h(N′) ← Σ_{b∈B_u} avg_cost (b)
26     return N′
```

Fig. 10. Bidtree-based node-expansion algorithm

This implementation is very time-efficient but A* fails to scale to large problems because of the need to keep in the queue all nodes that have not been fully expanded. Limiting the queue length destroys the optimality and completeness guarantees. Some improvement in memory usage can be achieved by setting an upper bound once the first solution is found in line 18 of Figure 10. Once an upper bound flimit exists, then any node N for which $f(N) > flimit$ can be safely discarded, including nodes already on the queue. Unfortunately, this helps only on the margin; there will be a very small number of problems for which the resulting reduction in maximum queue size will be sufficient to convert a failed or incomplete search into a complete one. We address this in the next section.

One of the design decisions that must be made when implementing a bidtree-based search is how to order the tasks (or items, in the case of a standard combinatorial auction) when building the bidtree. It turns out that this decision can have a major impact on the size of the tree that must be searched, and therefore on performance and predictability. As we have shown in [15], the tasks should be ordered such that the tasks with higher numbers of bids come ahead of tasks with lower numbers of bids. This ordering is exploited in line 18 of Figure 10, where bid conflicts are detected.

3.3 Iterative Deepening A*

Iterative Deepening A* (IDA*) [13] is a variant of A* that uses the same two functions g and h in a depth-first search, and which keeps in memory only the current path from the root to a particular node. In each iteration of IDA*, search depth is limited by a threshold value flimit on the evaluation function f (N).

```
Procedure  IDA*_search
Inputs:
    {S, V}: the task network to be assigned
    B: the set of bids, represented as a bidtree
Output:
    N_opt : the node having a mapping M (N_opt) = S → B of tasks to bids
            with an optimal evaluation, if one exists
Process:
    N_0 ← empty node
    g(N_0) ← 0
    h(N_0) ← Σ_{s∈S} avg_cost (minimumbid (s))
    flimit ← f (N_0)
    mask ← {in, any, any, ···}
    best_node ← null
    while (best_node = null) ∧ (flimit ≠ ∞) do
        B^c_{N_0} ← bidtree _query (B, mask)
            "B^c_{N_0} is a set of bids (in the form of a set of bid buckets),
            containing the bids that can be used to expand N_0. We have to
            repeat this for every iteration."
        new_limit ← dfs _contour (N_0)
        if best_node = null then
            flimit ← max(new_limit, z · flimit) "see narrative in this section"
    end while
    return best_node
```

Fig. 11. Bidtree-based Iterative Deepening A* search algorithm: top level

We show in Figure 11 a version of IDA* that uses the same bidtree and node structure as the A* algorithm. The recursive core of the algorithm is shown in Figure 12. This search algorithm uses the same node expansion algorithm as we used for the A* search, shown in Figure 10 above.

```
1  Procedure dfs_contour
2  Inputs:
3      N: a node
4  Output:
5      new_limit: a candidate for the flimit value for the next contour. This is
                  either the first f(N) value seen that is larger than flimit, or
                  the value of the best solution node found. If it is determined
                  that no solution is possible, then we return ∞.
6  Process:
7      new_limit ← f(N)
8      if new_limit > flimit then "enforce contour limit"
9          return new_limit
10     if solution(N) then "switch to branch-and-bound"
11         flimit ← new_limit "set new upper bound"
12         best_node ← N
13         return new_limit
14     nextL ← ∞
15     while B_N^c ≠ ∅ do
16         N′ ← astar_expand(N)
17         if (N′ ≠ null) ∧ (feasible(N′)) then
18             new_limit ← dfs_contour(N′)
19             nextL ← min(nextL, new_limit)
20     end while
21     return nextL
```

Fig. 12. Bidtree-based Iterative Deepening A* search algorithm: depth-first contour

There are three issues to note in this algorithm:

- The tuning parameter z shown in line 19 of Figure 11 is a positive number > 1. This controls the amount of additional depth explored in each iteration of the main loop that starts on line 15. Experimentation shows that a good value is 1.15, and that it is not very sensitive (performance falls off noticeably with $z < 1.1$ or $z > 1.2$).
- Whenever a solution is found, the value of flimit is updated in line 11 of Figure 12. This follows the usage in Sandholm [12], and limits exploration to nodes (and solutions) that are better than the best found so far.
- We test nodes for feasibility in line 17 of Figure 12 to prevent consideration and further expansion of nodes that cannot possibly lead to a solution.

4 Search Performance

The winner determination problem must be solved within the confines of a time-limited negotiation scenario. More significantly, the agent must allocate time to the winner-determination process when it sets the negotiation timeline prior to issuing an RFQ. Failure to solve the winner-determination problem within the allocated time will result in failure of the negotiation process.

Our goal is to measure the probability distributions of search times across a range of easily-measured (or easily-estimated) problem metrics. We prefer metrics that can be estimated prior to issuing an RFQ, since that is when deliberation scheduling must be done. This will allow us to schedule the winner-determination deliberation with a known level of confidence in finding a solution.

We have chosen four problem size and complexity metrics for evaluation.

Task count – This is simply the number of tasks m in a task network, and can be directly measured.

Bid count – The number of bids submitted n. Alternatively, the number of bids/task. We will assume that this value can be estimated from market statistics, given the task network composition and possibly other data such as lead time or allowable schedule slack.

Bid size – The mean number of tasks specified in each bid $\overline{|\mathcal{S}_i|}, i = 1 \ldots n$. We assume this can also be estimated from market statistics. We can think of a "specialist market" as being one in which most bidders bid on only one or a few task types, while a "generalist market" is one in which many bidders will bid on large chunks of a plan.

Plan complexity – The mean size of the precedence set of a task in the task network. This can be directly measured in the task network.

Some of these problem-size parameters can be controlled independently, and some are not as independent as we might like. For example, when we increase the number of tasks in the task network, we also need to increase the number of bids and/or the number of tasks/bid if we wish to retain the same number of bids for each task. For that reason, we will use bids/task and bid size/task network size when it is important to consider independent variables.

In the remainder of this chapter, we report on a set of experiments that give us the necessary probability distribution data for the IDA* winner-determination method. There are clearly limits on its scalability, because the problem has unavoidable exponential complexity. We shall observe runtime characteristics with exponential tails, so there will be unpredictable situations where good solutions will not be found within any fixed time limit. This is just another way of saying that 100% probability of finding a solution within a predetermined time limit is not achievable. Instead, our goal is to develop the data that will allow us to convert a desired probability of success to a time allocation, given the necessary problem-size metrics.

4.1 Experimental Setup

The experimental setup consists of a plan generator to produce randomly-generated task networks, a bid generator to produce randomly-generated bids for each task network, and a bid evaluator to perform the winner-determination search. The winner-determination process is instrumented to measure both elapsed time and the number of steps performed.

The problem-generation process requires a stream of random numbers. In order to be able to repeat test conditions, we maintain 2 separate streams, one for generating plans and RFQs, and one for generating bids Each can be initialized with a seed, so we have the ability to repeat the same plan sets with different bid sets.

All experiments were run using a dedicated 1800 MHz Intel/Linux machine. Timings are given in wall-clock time. The MAGNET system (including the IP preprocessor and the IDA* solver) is written in Java, and the IP solver is `lp` solve , written in C and available from ftp://ftp.ics.ele.tue.nl/pub/lp_solve/ .

Customer: Generate plan For these experiments, plans are randomly-generated task networks, with a number of controllable parameters. Task Network variables include:

Number of tasks

Mix of task types – Task types are characterized by expected duration d_e, duration variability $\sigma(d)$, average price c_e, and price variability $\sigma(c)$. Both duration and price are normally distributed, positive values (the distributions are truncated at 0). For all experiments, the task types and their properties are as given in Table 2.

Branch factor – This controls the average number of precedence relationships generated per task, which we call "fan-in." For example, if a particular task has two predecessors, then the fan-in value for that task is 2. As the plan is built, each new task s_j is linked to each of the previous tasks $s_{j'}, j' = 1 \ldots (j-1)$ with a probability p_l of (branchFactor/(taskCount − 1)). The completed network is filtered to remove most redundant precedence relations, so the final number is lower than the initial number generated.

Table 2. Task types and relative proportions used in performance experiments

Name	Proportion	Duration d_e	Duration $\sigma(d)$	Price c_e	Price $\sigma(c)$	Resource Availability
short	40%	2.0	0.4	500	0.2	0.8
medium	40%	3.0	0.2	1000	0.3	0.7
long	20%	8.0	0.3	1500	0.3	0.4

As an example, Figure 13 shows the task network for the first of the problems used in the bid-count test series described in Section 4.2 below. Keep in mind that duration values at this point in the process are expected values. Actual durations cannot be known until bids are awarded.

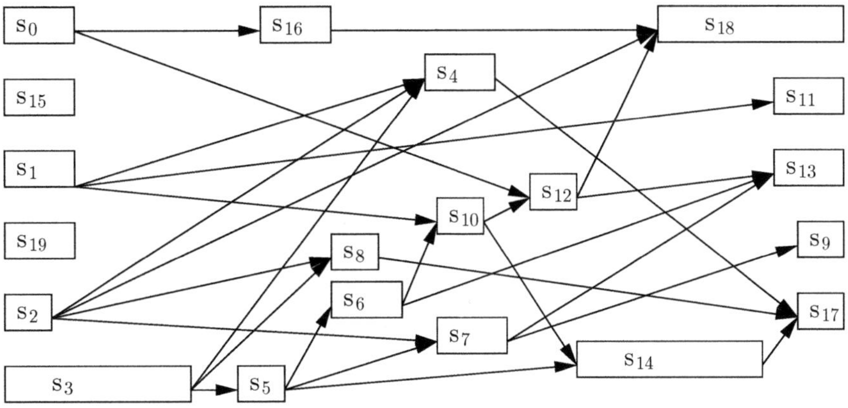

Fig. 13. Task network for Proble m #1. Box widths indicate relative task durations

Custo mer: Construct and issue a request for quotes The output of the planning step is simply a task network, made up of tasks for which we have some statistical data. Before we can ask for bids, we must add some constraints to communicate our desires more clearly to potential bidders, and we must consider the availability of bidders and their resources. The goal is to maximize the "usefulness" and minimize the cost of the bids we receive. Ultimately, we would expect to use Babanov's method based on Expected Utility to do this [5], but for these experiments we use a simpler approach that generates bids with well-defined statistics that are adequate to exercise our winner-determination solvers. We illustrate the RFQ generation process with the example task network shown in Figure 14.

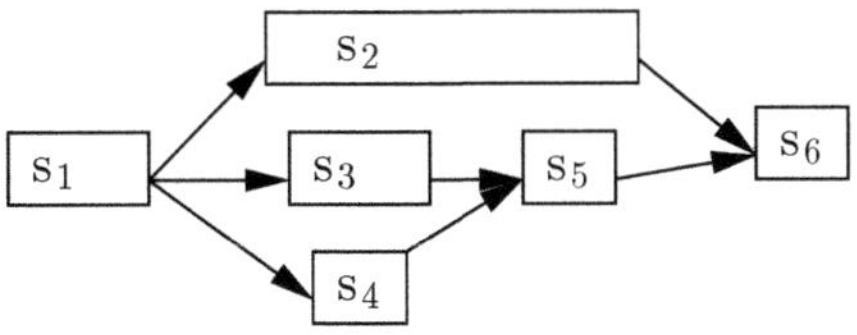

Fig. 14. Example task network for RFQ illustration

For these experiments, the RFQ time windows W_r are computed as follows:

1. Compute the "expected makespan" d_m for the entire task network. The makespan is simply the sum of the expected durations of the "critical" tasks,

or the tasks that are on the critical path assuming all tasks are assigned their expected durations.

2. Add some amount of "plan slack" ζ to the task network. This simply relaxes the plan deadline by some fraction. If the start time of the plan is t_0, then the earliest possible completion time, assuming all tasks are completed within their expected durations, is $t_m = t_0 + d_m$. After applying plan slack, we have $t_{goal} = t_0 + \zeta d_m$.
3. Determine final time windows for individual tasks. This is done in two steps. The first step is to set the minimum time allocation $d(s_j)$ for each task s_j to some value $d_{min}(s_j) = \eta d_e(s_j)$ where $\eta <= \zeta$. Then we run the CPM algorithm with a start time of t_0, and a deadline of t_{goal}. Figure 15 shows the resulting start and finish times for the task network of Figure 14, using $\zeta = 1.2$ and $\eta = 1.15$.

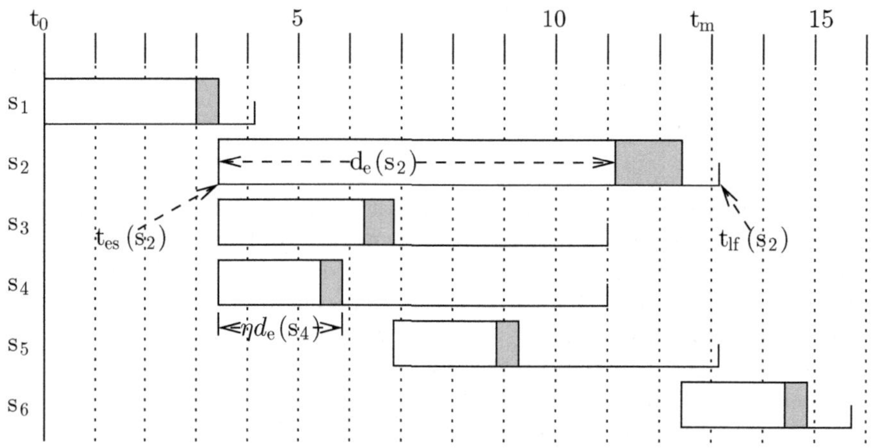

Fig. 15. Gantt chart for task network of Figure 14 with $\zeta = 1.2$, $\eta = 1.15$

There is no doubt that this procedure generates time windows that are sub-optimal, because of the large overlaps produced for noncritical tasks. However, it is very adequate for our experimental purposes, because it exercises the feasibility testing by ensuring that virtually all problems will have some infeasibilities among bids.

Supplier: Generate bids For these experiments, we used a test component that masquerades as an entire community of supplier agents. Each time a new RFQ is passed to it, it attempts to generate a specified number of bids.

For a given RFQ r, individual bids b_i are generated as follows:

1. Select a task s_j at random from S_r, and attempt to generate bid parameters for s_j.

2. Use the task-type parameters to generate a supplier time window
$$w_j(b_l) = \{t_{es}(s_j, b_l), t_{ls}(s_j, b_l), d(s_j, b_l)\}$$
for each task. We first generate the duration $d(s_j, b_l)$ using a normal distribution with mean and standard deviation equal to the expected duration and variability from the task type. If the resulting duration
$$d(s_j, b_l) > (t_{lf}(s_j, r) - t_{es}(s_j, r))$$
then it cannot be used and the attempt is abandoned. Otherwise, an early start $t_{es}(s_j, b_l)$ is generated from a uniform distribution over the interval
$$[t_{es}(s_j, r), (t_{lf}(s_j, r) - d(s_j, b_l))],$$
and a late start over the remaining slack in the RFQ time window
$$[t_{es}(s_j, b_l), (t_{lf}(s_j, r) - d(s_j, b_l))].$$
Actually, this step is slightly more complex after the first task-bid is generated, because of the need to generate bids with internal feasibility. We simply tighten up each RFQ time window with the maximum early finish of preceding tasks and the minimum late start of succeeding tasks.
3. If we successfully generate a time window, we call it a "valid task specification" and add the task s_j and its time window $w_j(b_l)$ to the bid. If a valid task specification was generated, then with probability p_{link}, each predecessor and successor link from task s_j is followed to choose additional tasks $s_{j'}$ to add to the bid, and so on recursively.
4. To complete the process for a single bid, we determine a cost for the overall bid and return it.
5. Because valid task specifications are not always achieved, some attempts to generate bids will fail, and so the number of bids actually generated is nearly always somewhat smaller than the target number. We can optionally check the resulting bid-set before it is returned, and test for coverage. If coverage is not achieved, then for each missing task, we make an additional attempt to generate a bid, with the missing task selected as a starting point in Step 1 rather than a random task. This is useful because returned bid-sets that do not cover all tasks will not exercise our winner-determination solvers. All experiments reported in this chapter used this feature.

The resulting bids specify "contiguous" sets of tasks, and each bid is guaranteed to be internally feasible.

4.2 Characterizing the Iterative Deepening A* Solver

In this section we examine the performance of the bidtree-based IDA* solver that was described in Section 3. We want to examine the scalability and predictability of this algorithm, and to compare it with the IP solver.

Each problem set consists of 100 problems, with randomly-generated task networks and randomly-generated bids as described above. Our problem generator has a large number of parameters; we kept all of them constant except for Task Count, Bid Count, Bid Size, and Network Complexity (branch factor).

Bid count experiment Here we examine the scalability of the IDA* method as the number of bids is varied over a 4:1 range, with task count and bid size held constant. Each row in Table 3 represents 100 problems, each with 30 tasks and varying numbers of bids. The same set of 100 task networks is used in each row; only the bid sets are varied. In the table, the "Bid Size" column gives the average size of bids (number of tasks per bid). The "# Solved" column gives the number of problems solved out of 100 (not all 100 problems were necessarily solvable), and "Mean Time" gives the mean search time in milliseconds. The meaning of the remaining 3 columns will become clear shortly.

Table 3. Bid Count experiment for the IDA* solver

Task Count	Bid Count	Bid Size	# Solved (of 100)	Mean Time (ms)	m	v	χ^2 (9 dof)
30	80	7.45	64	45.4	15.9	1.35	2.36
30	106	7.45	82	106	48.5	1.33	1.48
30	133	7.53	99	894	163	1.96	7.33
30	159	7.55	100	2180	500	2.36	3.50
30	186	7.48	100	34,700	1410	3.03	1.18
30	213	7.38	100	22,000	3070	3.33	2.92
30	239	7.46	99	137,000	9230	4.80	0.83
30	265	7.35	100	152,000	16,000	4.55	1.60

While the data in Table 3 show average performance and give some indication of variability, we are really more interested in knowing the probability that a solution can be determined in a given amount of time. For that purpose, we show in Figure 16 the complete runtime distributions for these problem sets. For each curve, we show actual observations along with a lognormal distribution that minimizes the χ^2 metric[5]. In Table 3, the last 3 columns give the parameters of the inferred lognormal density function. The m column is the median value (the value of $\log m$ is the mean of the log of the distribution), and the v column is the standard deviation of the log of the distribution. We also give χ^2 values, computed by the equiprobable method [16]. For 9 degrees of freedom, $\chi^2 = 3.5$ corresponds to roughly a 95% confidence that the data fit the hypothesized distribution. In general, a large value of χ^2 shows up in a plot such as Figure 16 as a very obvious failure of the observations to lie atop the inferred distribution curve.

It seems clear that the 186-bid row in Table 3 contains a large outlier. Figure 16 shows the observed and inferred distributions for all of the problem sets in this experiment. We can see visually that the lognormal distributions provide a good approximation for this data. For 9 dof, the 95% confidence point

[5] The χ^2 metric is determined by dividing the inferred density function into a number of equal areas, and counting the number of observations that fall into each of those zones. The χ^2 value is a measure of the deviation from a "perfect fit".

is approximately 3.5, so all but one of these sets is at or above a 95% match to the given lognormal distribution. This gives us confidence that we can base predictions on these inferred distributions. The single outlier in the 186-bid set can be seen in the plot at about $3 \cdot 10^6$ msec.

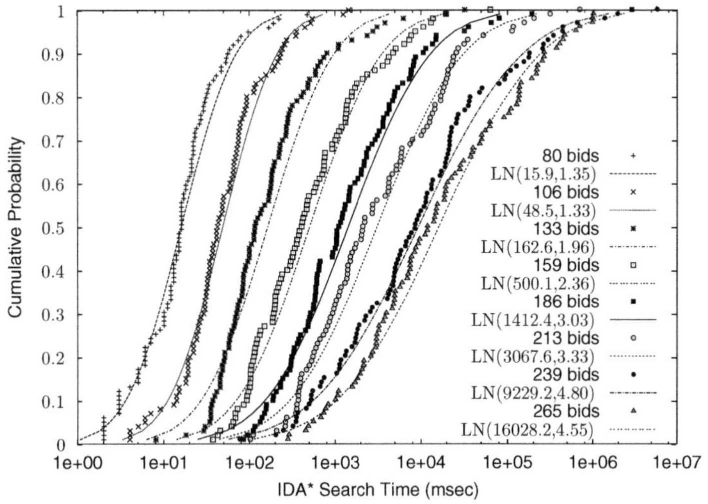

Fig. 16. IDA* run-time distributions for proble ms with 30 tasks, bid count ranging from 80 bids to 265 bids, bid size approx. 7.5 tasks/bid

Bid size experi ment We now turn to the bid-size dimension of problem variability. We expect that larger average bid sizes will result in shorter search times, because the number of bids in a solution will be smaller. This leads to fewer combinations in the search space, and fewer precedence constraints that must be generated and solved (since bids are required to be internally feasible). One way to see this is to look at an upper bound on the number of possible solutions as bid size varies. A simple upper bound is the number of combinations of bids $C = n!/(n-(m/b))!$ where n is the number of bids, m is the number of tasks, and b is the mean number of tasks/bid. Sandholm [12, 17] avoids this with small bid sizes by partitioning the bids into non-overlapping subsets. This works because small bids have fewer overlaps, and because the items in a simple combinatorial auction are arbitrarily separable, there being no constraints among them. We cannot use this approach because of the presence of precedence constraints that connect the items at auction (the tasks).

Our problem generator uses a somewhat inexact method to "influence" the sizes of bids. As described earlier, the bid-generation process generates "contiguous" bids by choosing a starting point in the task network, and then recursively following predecessor and successor links with some probability. In this series, we have varied that probability from a low of 0.2 to a high of 0.9. Bid size is also influenced by the sizes of time windows, since the "success" in generating a bid for a particular task is a function of both the size of the time window specified in the RFQ, and the simulated "resource availability" recorded for the type of each task. The task networks in these problem sets are the same as the ones used in the bid-count experiment in the previous section.

For this experiment, we generated 100 problems with 30 tasks and 150 "bid attempts". We then varied the probability that the bidder will follow precedence links in generating multiple-task bids from 0.3 to 0.9. Table 4 gives the raw data and the parameters for the inferred lognormal distributions for this experiment. All but the 7.53 tasks/bid and the 11.5 tasks/bid sets fit the lognormal distribution quite well, as evidenced by the χ^2 values.

Table 4. Bid Size experiment for the IDA* solver

Task Count	Bid Count	Bid Size	# Solved (of 100)	Mean Time (ms)	m	v	χ^2 (9 dof)
30	134	2.43	96	3,120,000	21,900	7.76	1.13
30	133	3.81	92	800,000	1300	5.38	1.66
30	133	5.58	97	2030	314	2.61	1.13
30	133	7.53	99	936	163	1.98	7.10
30	133	9.00	90	249	110	1.34	3.82
30	133	10.30	91	121	60.2	0.874	3.17
30	133	11.50	92	69.0	48.4	0.483	7.63

Figure 17 shows the observed and inferred runtime distributions for selected sets from this experiment. Clearly, both difficulty and variability rise significantly as bid size is reduced.

Task count experiment The next set of experiments examines scalability of the search process as the number of tasks varies, with a (nearly) constant ratio of bids to tasks (the ratio varies somewhat due to the random nature of the bid-generation process). We again generated 100 problems for each set, varying the task count over a 10:1 range from 5 tasks to 50 tasks, with the number of "bid attempts" at about 3.5 bids/task. Table 5 gives the raw data and the parameters for the inferred lognormal distributions for this experiment. These data do not fit the "inferred" lognormal distributions as well as the data in the previous section, for reasons that are unclear. In the first two sets that appears to be because of quantizing error (the clock resolution is only 1 ms, and the mean search time for the 5-task set was just 2.12 ms). For the other poor fits in the 35-task and 50-task sets, the errors seem unlikely to strongly impact our ability to use the 95% point as an estimator, since there is a crossover between the data and the lognormal curve in that region.

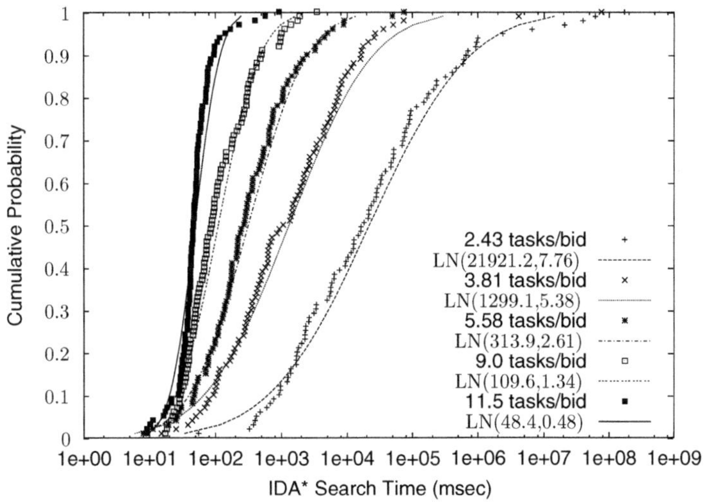

Fig. 17. IDA* run-time distributions for proble ms with 30 tasks, 133 bids, bid size ranging fro m 2.4 tasks to 11.5 tasks/bid

Table 5. Task Count experi ment for the IDA* solver

Task Count	Bid Count	Bid Size	# Solved (of 100)	Mean Time (ms)	m	v	χ^2 (9 dof)
5	13.5	2.10	94	2.12	1.81	0.499	31.7
10	29.0	3.27	94	4.78	3.01	0.716	9.27
15	45.4	4.46	90	9.22	5.93	0.664	4.02
20	61.0	5.45	85	22.4	12.2	1.18	3.97
25	77.7	6.30	85	31.1	19.0	0.874	0.76
30	93.4	7.40	80	81.4	27.3	0.955	4.03
35	111	8.42	74	914	47.0	1.96	8.55
40	127	9.65	68	149	61.1	1.35	2.99
45	142	11.2	66	146	74.7	1.19	1.44
50	160	12.2	61	597	98.9	1.61	6.38

Figure 18 shows the observed and inferred runtime distributions for selected sets from this experiment.

Task network co mplexity experi ment We have examined variability due to the number of tasks, the number of bids, and the number of tasks in an auction. There are many other potential sources of variability, some of which can be measured or estimated prior to submitting an RFQ, and some of which clearly cannot (for example, the number of bids in an optimal solution). One parameter that is easy to measure is the density of precedence constraints in the

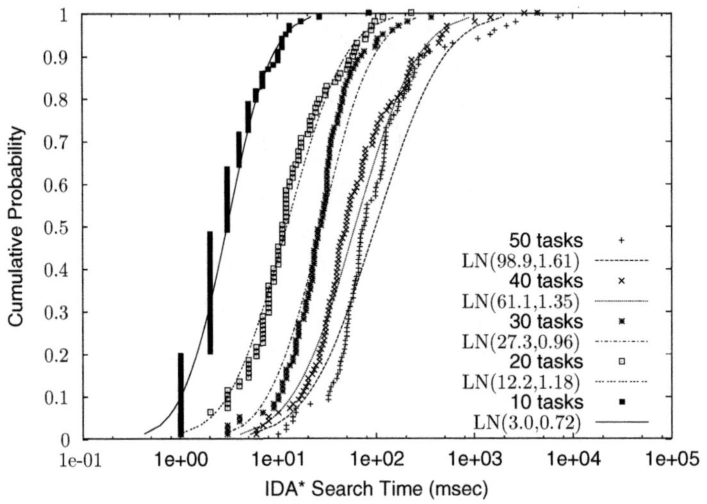

Fig. 18. IDA* run-time distributions for proble ms with 10 to 50 tasks, about 3.1 bids/task

task network. However, given the way we generate plans and bids, it is not easy to build experiments that control this parameter independently.

In all experiments reported so far, the "Branch Factor" (see the definition in Section 4.1) has been set to a value of 2.4. The example task network shown in Figure 13 was built with this value. In this experiment, the branch factor is varied from 1.0 to 4.0 in increments of 1.0. Because root tasks have no predecessors, and because of the elimination of redundant precedence links, the actual number of precedence relationships per task is generally much lower than the branch factor. The results are shown in Table 6, where the first column, labeled "Fan In," gives the actual mean number of precedence links per task. There were 35 tasks in each set of 100 problems.

Table 6. Task network co mplexity experi ment for the IDA* solver.

Fan In	Bid Count	Bid Size	# Solved (of 100)	Mean Time (ms)	σ	m	v	χ^2 (9 dof)
0.489	112	2.01	86	76100	516,000	505	6.74	0.966
0.876	111	5.06	100	427	2000	75.8	2.15	0.974
1.371	109	11.9	100	30.5	22.8	22.9	0.632	0.972
1.735	109	16.0	100	20.7	14.8	15.9	0.552	0.930
1.735	109	2.65	56	365,000	2,044,000	1975	9.85	1.000

As is evident from the table, the way bids are generated in our experimental setup causes bid count to go down as the number of precedence links is reduced. This is likely to cause a conflation of the network complexity factor with bid count effects. The probability distributions for this set are shown in Figure 19.

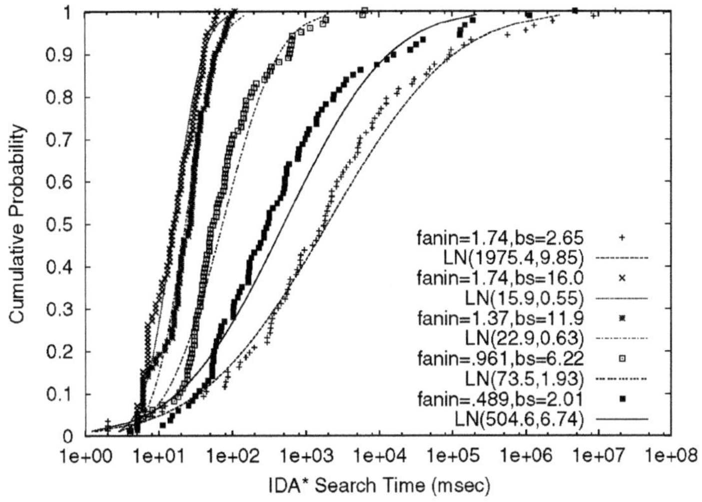

Fig. 19. Observed and inferred runti me distributions for the IDA* solver across a range of task network co mplexity values

5 Related Work

This work draws from several fields. In Computer Science, it is related to work in artificial intelligence and autonomous agents. In Economics, it draws from auction theory and expected utility theory. From Operations Research, we draw from work in combinatorial optimization.

5.1 Multi-Agent Negotiation

MAGNET proposes using an auction paradigm to support problem-solving interactions among autonomous, self-interested, heterogeneous agents. Several other approaches to multi-agent problem-solving have been proposed. Some of them use a "market" abstraction, and some do not.

Rosenschein and Zlotkin [18] show how the behavior of agents can be influenced by the set of rules system designers choose for their agents' environment. In their study the agents are homogeneous and there are no side payments. In other words, the goal is to share the work, in a more or less "equitable" fashion, but not to have agents pay other agents for work. They also assume that each agent has sufficient resources to handle all the tasks, while we assume the contrary.

In Sandholm's TRACONET system [19, 20], agents redistribute work among themselves using a contracting mechanism. Sandholm considers agreements involving explicit payments, but he also assumes that the agents are homogeneous – they have equivalent capabilities, and any agent can handle any task. MAGNET agents are heterogeneous, and in general do not have the resources or capabilities to carry out the tasks necessary to meet their own goals without assistance from others.

Planning systems (see, for instance, [21–23]) assume multiple agents that operate independently. However, in those systems the agents are explicitly cooperative, and all work toward the achievement of a shared goal. MAGNET agents are trying to achieve their own goals and to maximize their own profits; there is no global or shared goal.

Solving problems using markets and auctions MAGNET uses an auction-based negotiation style because auctions have the right economic and motivational properties to support "reasonable" resource allocations among heterogeneous, self-interested agents. However, MAGNET uses the auction approach not only to allocate resources, but also to solve constrained scheduling problems.

A set of auction-based protocols for decentralized resource-allocation and scheduling problems is proposed in [24]. The analysis assumes that the items in the market are individual discrete time slots for a single resource, although there is a brief analysis of the use of the Generalized Vickrey Auctions [25] to allow for combinatorial bidding. A combinatorial auction mechanism for dynamic creation of supply chains was proposed and analyzed in [26]. This system deals with the constraints that are represented by a multi-level supply-chain graph, but does not deal with temporal and precedence constraints among tasks. MAGNET agents must deal with multiple resources and continuous time, but we do not currently deal explicitly with multi-level supply chains[6]

Several proposed bidding languages for combinatorial auctions allow bidders to express constraints, for example [27, 4]. However, these approaches only allow bidders to communicate constraints to the bid-taker (suppliers to the customer, in the MAGNET scenario), while MAGNET needs to communicate constraints in both directions.

Infrastructure support for negotiation Markets play an essential role in the economy [28], and market-based architectures are a popular choice for multiple agents (see, for instance, [29–32] and our own MAGMA architecture [33]). Most market architectures limit the interactions of agents to manual negotiations, direct agent-to-agent negotiation [20, 34], or some form of auction [35].

[6] Individual MAGNET agents can deal with multi-level supply chains by subcontracting, but this requires that the initial time allocation provide sufficient slack for the extra negotiation cycles.

The Michigan Internet AuctionBot [35] is a very interesting system, in that it is highly configurable, able to handle a wide variety of auction rules. It is the basis for the ongoing Trading Agent Competition [36], which has stimulated interesting research on bidding behavior in autonomous agents, such as [37].

Matchmaking, the process of making connections among agents that request services and agents that provide services, will be an important issue in a large community of MAGNET agents. The process is usually done using one or more intermediaries, called middle-agents [38]. Sycara et al. [39] present a language that can be used by agents to describe their capabilities and algorithms to use it for matching agents over the Web. Our system casts the Market in the role of matchmaker.

The MAGNET market infrastructure depends on an Ontology to describe services that can be traded and the terms of discourse among agents. There has been considerable attention to development of detailed ontologies for describing business and industrial domains (see, for instance, [40, 41]).

5.2 Combinatorial Auctions

Determining the winners of a combinatorial auction [42] is an NP-complete problem, equivalent to the weighted bin-packing problem. A good overview of the problem and approaches to solving it is [43]. Dynamic programming [44] works well for small sets of bids, but it does not scale well, and it imposes significant restrictions on the bids. Sandholm [12, 45] relaxes some of the restrictions and presents an algorithm for optimal selection of combinatorial bids, but his bids specify only a price and a set of items. Hoos and Boutilier[46] describe a stochastic local search approach to solving combinatorial auctions, and characterize its performance with a focus on time-limited situations. A key element of their approach involves ranking bids according to expected revenue; it's very hard to see how this could be adapted to the MAGNET domain with temporal and precedence constraints, and without free disposal[7]. Andersson et al. [47] describe an Integer Programming approach to the winner determination problem in combinatorial auctions. Nisan [27] extends this model to handle richer bidding languages for combinatorial auctions, and we have extended it to handle the MAGNET situation in [9]. More recently, Sandholm [45] has described an improved winner-determination algorithm called BOB that uses a combination of linear programming and branch-and-bound techniques. It is not clear how this technique could be extended to deal with the temporal constraints in the MAGNET problem, although the bid-graph structure may be of value.

One of the problems with combinatorial auctions is that they are nearly always run in a single round sealed-bid format, and this is the format MAGNET uses. Parkes and Ungar [48] have shown how to organize multiple-round combinatorial auctions. Another problem is that the individual items in a combinatorial auction are individual items; there is no notion of quantity. MAGNET

[7] Under the "free disposal" assumption, the goal is to maximize revenue even if this means failing to allocate all the items at auction.

will eventually need to address this. This limitation is overcome in [49] for simple items without side constraints. The addition of precedence constraints would seriously complicate their procedure, and it has not yet been attempted.

5.3 Deliberation scheduling

The principal reason we are interested in search performance is because the search is embedded in a real-time negotiation scenario, and time must be allocated to it before bids are received, and therefore before the exact dimensions of the problem are known. In [50], deliberation scheduling is done with the aid of anytime and contract algorithms, and performance profiles. An anytime algorithm is one that produces a continuously-improving result given additional time, and a contract algorithm is one that produces a result of a given quality level in a given amount of time, but may not improve given additional time. The best winner-determination algorithms we know of for the MAGNET problem have marginal anytime characteristics, and we know of no applicable contract-type algorithms. In fact, [12] presents an inapproximability result for the winner-determination problem, leading us to believe that there may not be an acceptable contract algorithm.

One way to think about deliberation scheduling is to assign the time required for deliberation a cost, and then to balance the cost of deliberation against the expected benefit to be gained by the results of the deliberation. This is the approach taken in [51]. However, much of this analysis assumes that there is a "default" action or state that can be used or attained without spending the deliberation effort, and that there is a clear relationship between the time spent in deliberation and the quantifiable quality of the result. In the MAGNET case, the alternative to deliberation is to do nothing.

6 Conclusions

We have examined the problem of rational economic agents who must negotiate among themselves in a market environment in order to acquire the resources needed to accomplish their goals. We are interested in agents that are self-interested and heterogeneous, and we assume that a plan to achieve an agent's goal may be described in the form of a task network, containing task descriptions, precedence relationships among tasks, and time limits for individual tasks. Negotiation among agents is carried out by holding combinatorial reverse auctions in a marketplace, in which a customer agent offers a task network in the form of a request for quotes (RFQ). Supplier agents may then place bids on portions of the task network, each bid specifying the tasks they are interested in undertaking, durations and time limits for those tasks, and a price for the bid as a whole. The presence of temporal and precedence constraints among the items at auction requires extensions to the standard winner-determination procedures for combinatorial auctions, and the use of the enhanced winner-determination procedure within the context of a real-time negotiation requires us to be able to predict its runtime when planning the negotiation process.

There are a number of real-world business scenarios where such a capability would be of value. These include flexible manufacturing, mass customization, travel arrangement, logistics and international shipping, health care resource management, and large-scale systems management. Each of these areas is characterized by limited capabilities and suboptimal performance, due at least in part to the limits imposed by human problem-solving capabilities. In each of these areas, a general ability to coordinate plans among multiple independent suppliers would be of benefit, but does not exist or is not used effectively because of an inability to solve the resulting combinatorial problems. The use of extended combinatorial auctions such as we propose is one approach to solving these problems. There are many difficulties yet to be overcome before this vision can be realized, however, not the least of which is that such auction-based markets would not be effective without wide adoption of new technology across an industry, and a willingness to delegate at least some level of autonomy and authority to that new technology.

We have designed and implemented a testbed, which we call MAGNET for Multi-AGent NEgotiation Testbed, to begin exploring and testing this vision. It includes a customer agent, a rudimentary market infrastructure, and a simple simulation of a population of supplier agents. The customer agent implementation is designed so that virtually all behaviors can be specified and implemented in terms of responses to events. Events can be external occurrences, internal state changes, or the arrival of a particular point in time. The MAGNET software package is available to the research community under an open-source license.

When a goal arises, the agent and its principal must develop a plan, in the form of a task network Once a plan is available, a bid-process plan must be developed to guide the negotiation process. The bid-process plan specifies which tasks are to be offered in which markets, allocates time to the bidding process and to the plan execution, and may split the bidding into phases in order to mitigate risk. For each bidding step in the bid-process plan, time must be allocated to the customer to compose its RFQ, to the supplier to compose bids, and to the customer to evaluate bids. For each auction episode specified in the bid-process plan, a RFQ must be composed. The RFQ specifies a subset of tasks in the task network, and for each task, it specifies a time window within which that task must be accomplished. The setting of time windows is critical, because it influences the likelihood that bidders will bid, the prices bidders are likely to charge, and the difficulty of the resulting winner-determination process. If the time windows specified in the RFQ allow task precedence relationships to be violated, then the winner-determination process will need to choose a set of bids that can be composed into a feasible schedule. Once the RFQ has been issued and bids received, the agent must determine winners. We have described an optimal algorithm for determining winners based on an IDA* framework.

Because the winner-determination problem must be solved within a predetermined period of time, it is important to have a clear idea of how much time to allocate to it, and to know what parameters to use in predicting its run time. We therefore conducted a series of experiments to characterize the performance of our winner-determination method over a variety of problem sizes and shapes. The goal was to determine probability distributions, so that we could allocate the time necessary to achieve a known probability of solving the problem. We found that the lognormal distribution was a good model of search performance.

Much work remains to be done before the vision of the MAGNET project is fully realized. Some of that work, particularly with respect to the supplier agent and its decision processes, is already under way by other members of the team.

With respect to the customer agent, many of the decision processes outlined in Section 2 still need to be worked out and tested. The present work has resulted in models for the auction winner-determination problem and the time that must be allocated to it. For the remainder of the decisions, we need models that will maximize the expected utility of the agent or its principal. These include composing the plan, developing the bid-process plan, allocating time to the deliberation processes of the customer and suppliers, balancing negotiation time against plan execution time, setting the time windows in the RFQ, scheduling the work in preparation for awarding bids, and dealing with unexpected events during plan execution. Babanov et al. [5, 52] have addressed the problem of setting time windows in the customer's RFQ.

The language we currently use for plans and bids treats tasks as simple atomic objects, without attributes. There are many real-world problems in which attributes are important, both for specifying tasks and for expressing offers in bids. Examples include colors, quantities, dimensions, and quality attributes. In addition, many real-world operations operate on a "flow" basis. This includes the wine-making example we used in Chapter 2, in which the precedence between filling bottles and applying labels would normally be applied bottle-by-bottle, and not at the batch level. In addition, the expressivity of our bidding language is limited. A number of proposals have been made for more expressive bidding languages in combinatorial auctions [27, 4]. Bidding can also be done with oracles, which are functions passed from bidder to customer that can be evaluated to produce bid conditions. Some features of a more expressive bidding language would likely have minimal impact on the winner-determination process (parameterized quality values, for example), while others, including the use of oracles, could require wholesale re-invention.

References

1. Collins, J., Tsvetovat, M., Mobasher, B., Gini, M.: MAGNET: A multi-agent contracting system for plan execution. In: Proc. of SIGMAN, AAAI Press (1998) 63–68
2. Collins, J., Bilot, C., Gini, M., Mobasher, B.: Mixed-initiative decision support in agent-based automated contracting. In: Proc. of the Fourth Int'l Conf. on Autonomous Agents. (2000) 247–254
3. Nisan, N.: Bidding and allocation in combinatorial auctions. Technical report, Institute of Computer Science, Hebrew University (2000)

4. Boutilier, C., Hoos, H.H.: Bidding languages for combinatorial auctions. In: Proc. of the 17th Joint Conf. on Artificial Intelligence. (2001) 1211–1217
5. Babanov, A., Collins, J., Gini, M.: Asking the right question: Risk and expectation in multi-agent contracting. Artificial Intelligence for Engineering Design, Analysis and Manufacturing 17 (2003) 173–186
6. Collins, J., Jamison, S., Gini, M., Mobasher, B.: Temporal strategies in a multi-agent contracting protocol. In: AAAI-97 Workshop on AI in Electronic Commerce. (1997)
7. Hillier, F.S., Lieberman, G.J.: Introduction to Operations Research. McGraw-Hill (1990)
8. Parkes, D.C., Ungar, L.H.: An auction-based method for decentralized train scheduling. In: Proc. of the Fifth Int'l Conf. on Autonomous Agents, Montreal, Quebec, ACM Press (2001) 43–50
9. Collins, J., Gini, M.: An integer programming formulation of the bid evaluation problem for coordinated tasks. In Dietrich, B., Vohra, R.V., eds.: Mathematics of the Internet: E-Auction and Markets. Volume 127 of IMA Volumes in Mathematics and its Applications. Springer-Verlag, New York (2001) 59–74
10. Collins, J., Bilot, C., Gini, M., Mobasher, B.: Decision processes in agent-based automated contracting. IEEE Internet Computing 5 (2001) 61–72
11. Sandholm, T.: An algorithm for winner determination in combinatorial auctions. In: Proc. of the 16th Joint Conf. on Artificial Intelligence. (1999) 524–547
12. Sandholm, T.: Algorithm for optimal winner determination in combinatorial auctions. Artificial Intelligence 135 (2002) 1–54
13. Korf, R.E.: Depth-first iterative deepening: An optimal admissible tree search. Artificial Intelligence 27 (1985) 97–109
14. Russell, S.J., Norvig, P.: Artificial Intelligence: A Modern Approach. Prentice-Hall, Upper Saddle River, NJ (1995)
15. Collins, J., Demir, G., Gini, M.: Bidtree ordering in IDA* combinatorial auction winner-determination with side constraints. In Padget, J., Shehory, O., Parkes, D., Sadeh, N., Walsh, W., eds.: Agent Mediated Electronic Commerce IV. Volume LNAI2531., Springer-Verlag (2002) 17–33
16. Law, A.M., Kelton, W.D.: Simulation Modeling and Analysis. Second edn. McGraw-Hill (1991)
17. Sandholm, T., Suri, S., Gilpin, A., Levine, D.: CABOB: A fast optimal algorithm for combinatorial auctions. In: Proc. of the 17th Joint Conf. on Artificial Intelligence, Seattle, WA, USA (2001) 1102–1108
18. Rosenschein, J.S., Zlotkin, G.: Rules of Encounter. MIT Press, Cambridge, MA (1994)
19. Sandholm, T.W., Lesser, V.: On automated contracting in multi-enterprise manufacturing. In: Distributed Enterprise: Advanced Systems and Tools, Edinburgh, Scotland (1995) 33–42
20. Sandholm, T.W.: Negotiation Among Self-Interested Computationally Limited Agents. PhD thesis, Department of Computer Science, University of Massachusetts at Amherst (1996)
21. Pollack, M.E.: Planning in dynamic environments: The DIPART system. In Tate, A., ed.: Advanced Planning Technology. AAAI Press (1996)
22. Wilkins, D.E., Myers, K.L.: A multiagent planning architecture. In: Proc. Int'l Conf. on AI Planning Systems. (1998) 154–162
23. Cox, J.S., Durfee, E.H., Bartold, T.: A distributed framework for solving the multiagent plan coordination problem. In: Autonomous Agents and Multi-Agent Systems. (2005) 821–827

24. Wellman, M.P., Walsh, W.E., Wurman, P.R., MacKie-Mason, J.K.: Auction protocols for decentralized scheduling. Games and Economic Behavior 35 (2001) 271–303
25. Varian, H.R., MacKie-Mason, J.K.: Generalized vickrey auctions. Technical report, University of Michigan (1995)
26. Walsh, W.E., Wellman, M., Ygge, F.: Combinatorial auctions for supply chain formation. In: Proc. of ACM Conf on Electronic Commerce (EC'00). (2000) 260–269
27. Nisan, N.: Bidding and allocation in combinatorial auctions. In: Proc. of ACM Conf on Electronic Commerce (EC'00), Minneapolis, Minnesota, ACM SIGecom, ACM Press (2000) 1–12
28. Bakos, Y.: The emerging role of electronic marketplaces on the Internet. Comm. of the ACM 41 (1998) 33–42
29. Chavez, A., Maes, P.: Kasbah: An agent marketplace for buying and selling goods. In: Proc. of the First Int'l Conf. on the Practical Application of Intelligent Agents and Multi-Agent Technology, London, UK, Practical Application Company (1996) 75–90
30. Rodriguez, J.A., Noriega, P., Sierra, C., Padget, J.: FM96.5 - A Java-based electronic auction house. In: Second Int'l Conf on The Practical Application of Intelligent Agents and Multi-Agent Technology (PAAM'97), London (1997) 207–224
31. Sycara, K., Pannu, A.S.: The RETSINA multiagent system: towards integrating planning, execution, and information gathering. In: Proc. of the Second Int'l Conf. on Autonomous Agents. (1998) 350–351
32. Wellman, M.P., Wurman, P.R.: Market-aware agents for a multiagent world. Robotics and Autonomous Systems 24 (1998) 115–125
33. Tsvetovatyy, M., Gini, M., Mobasher, B., Wieckowski, Z.: MAGMA: An agent-based virtual market for electronic commerce. Journal of Applied Artificial Intelligence 11 (1997) 501–524
34. Faratin, P., Sierra, C., Jennings, N.R.: Negotiation decision functions for autonomous agents. Int. Journal of Robotics and Autonomous Systems 24 (1997) 159–182
35. Wurman, P.R., Wellman, M.P., Walsh, W.E.: The Michigan Internet AuctionBot: A configurable auction server for human and software agents. In: Second Int'l Conf. on Autonomous Agents. (1998) 301–308
36. Collins, J., Arunachalam, R., Sadeh, N., Ericsson, J., Finne, N., Janson, S.: The supply chain management game for the 2005 trading agent competition. Technical Report CMU-ISRI-04-139, Carnegie Mellon University, Pittsburgh, PA 15213 (2004)
37. Stone, P., Schnapire, R.E., Csirik, M.L.L.J.A., McAllester, D.: ATTac-2001: A learning, autonomous bidding agent. Submitted to the Eigtheenth National Conference on Artificial Intelligence (AAAI-2002) (2002)
38. Sycara, K., Decker, K., Williamson, M.: Middle-agents for the Internet. In: Proc. of the 15th Joint Conf. on Artificial Intelligence. (1997) 578–583
39. Sycara, K., Klusch, M., Widoff, S., Lu, J.: Dynamic service matchmaking among agents in open information environments. SIGMOD Record (ACM Special Interests Group on Management of Data) 28 (1999) 47–53
40. Fox, M., Barbuceanu, M., Teigen, R.: Agent-oriented supply-chain management. The International Journal of Flexible Manufacturing Systems 12 (2000) 165–188
41. Chandra, C., Tumanyan, A.: Supply chain system taxonomy: development and application. In: Proc. 12th Annual Industrial Engineering Research Conference IERC-2003, Portland, Oregon, USA (2003)

42. McAfee, R., McMillan, P.J.: Auctions and bidding. Journal of Economic Literature 25 (1987) 699–738
43. de Vries, S., Vohra, R.: Combinatorial auctions: a survey. Technical report, Technische Universität München (2001)
44. Rothkopf, M.H., Pekeč, A., Harstad, R.M.: Computationally manageable combinatorial auctions. Management Science 44 (1998) 1131–1147
45. Sandholm, T., Suri, S.: Bob: Improved winner determination in combinatorial auctions and generalizations. Artificial Intelligence 145 (2003) 33–58
46. Hoos, H.H., Boutilier, C.: Solving combinatorial auctions using stochastic local search. In: Proc. of the Seventeen Nat'l Conf. on Artificial Intelligence. (2000) 22–29
47. Andersson, A., Tenhunen, M., Ygge, F.: Integer programming for combinatorial auction winner determination. In: Proc. of 4th Int'l Conf on Multi-Agent Systems. (2000) 39–46
48. Parkes, D.C., Ungar, L.H.: Iterative combinatorial auctions: Theory and practice. In: Proc. of the Seventeen Nat'l Conf. on Artificial Intelligence. (2000) 74–81
49. Leyton-Brown, K., Shoham, Y., Tennenholtz, M.: An algorithm for multi-unit combinatorial auctions. In: Proc. of the Seventeen Nat'l Conf. on Artificial Intelligence, Austin, Texas (2000)
50. Greenwald, L., Dean, T.: Anticipating computational demands when solving time-critical decision-making problems. In Goldberg, K., Halperin, D., Latombe, J., Wilson, R., eds.: The Algorithmic Foundations of Robotics. A. K. Peters, Boston, MA (1995)
51. Boddy, M., Dean, T.: Decision-theoretic deliberation scheduling for problem solving in time-constrained environments. Artificial Intelligence 67 (1994) 245–286
52. Babanov, A., Collins, J., Gini, M.: Harnessing the search for rational bid schedules with stochastic search and domain-specific heuristics. In: Proc. of the Third Int'l Conf. on Autonomous Agents and Multi-Agent Systems, New York, AAAI Press (2004) 269–276

Incentive Compatible Supply Chain Auctions *

Moshe Babaioff[1] and William E. Walsh[2]

[1] The School of Information Management and Systems (SIMS),
University of California at Berkeley, Berkeley, CA 94720-4600
moshe@sims.berkeley.edu **
[2] CombineNet, Inc.
Fifteen 27th St., Pittsburgh, PA, USA
wwalsh@combinenet.com ***

Abstract. We address supply chain formation—the problem of coordinating the buying and selling of goods in multiple markets across a supply chain—as a mechanism design problem. Our basic assumption is that each agent has private information about its cost of providing a service or about its value for purchasing a set of goods it desires. Because effective negotiation strategies in the presence of such private information can be difficult to design, we focus on incentive compatible mechanisms. In such mechanisms, the agents' dominant strategy is to simply report their private information truthfully. Unfortunately, with two-sided negotiation, characteristic of supply chains, it is impossible to simultaneously achieve three desirable properties with incentive compatibility: perfect efficiency, budget balance, and individual rationality. In this chapter we address this impossibility and describe auctions that are incentive compatible, individually rational and budget balanced, and guarantee high, but not perfect efficiency. We describe auctions for various assumptions regarding the supply chain topologies and the private information the agents have about their technologies. We discuss economic properties of the auctions, as well as computational efficiency and distributed implementation of the auctions.

1 Introduction

Supply chain formation is the problem of determining the production and exchange relationships across a supply chain. Whereas typical research in supply chain management focuses on optimizing production and delivery in a fixed supply chain structure, we are concerned with establishment of supply chain relationships in response to current needs, costs, and resource availability. These individual relationships cannot be established in isolation because a functioning supply chain requires a complete sequence of production through the supply chain. As business relationships become ever more flexible and dynamic, there is an increasing need to automate this supply chain formation process. Automated supply chain formation is being recognized as an important

* This chapter contains material previously published in Decision Support Systems [4].

** The research for this chapter was performed while the first author was a student at the Hebrew University of Jerusalem.

*** This research for this chapter was performed while the second author was at IBM T. J. Watson Research Center.

M. Babaioff and W.E. Walsh: Incentive Compatible Supply Chain Auctions, Studies in Computational Intelligence (SCI) 28, 315–350 (2006)
www.springerlink.com

research challenge [18, 22], and was the subject of the Trading Agent Competition for Supply Chain Management (TAC SCM) [19, 17], for the last several years.

Because procurement and supply contracts in a supply chain can involve significant production commitments and large monetary exchanges, it is important for an agent to negotiate effectively on behalf of a business. However, strategic analysis can be very complex when agents must negotiate contracts for outputs and multiple inputs simultaneously across a supply chain. Fortunately, careful design of the negotiation mechanism can simplify the agents' strategic problem enormously. We can effectively engineer away the agents' strategic problem by designing an auction to be ***incentive compatible*** (IC), in which case an agent's dominant strategy is to simply report its private information truthfully. Other properties are also important in a business setting. An auction should be ***individually rational*** (IR), that is, no agent would pay more than its valuation for the goods it receives when it bids truthfully. The auction should be ***budget balanced*** (BB) (the auction house does not lose money), else there would typically be little incentive to run the auction. Additionally, it is desirable that the auction be ***efficient*** (maximize total agent value) to ensure that all gains from trade are realized.

To date, there has been relatively little mechanism design work that meets the needs of automated supply chain formation. To address the problem of individual rationality, much recent effort has focused on combinatorial auctions [7] which, by allowing agents to place indivisible bids for bundles of goods, ensure that agents do not buy partial bundles of no value. Much of this work has been on one-sided auctions. To address the two-sided negotiation necessary in a supply chain, Walsh et al. [23] analyzed an IR and budget-balanced auction that avoids negotiation miscoordination by allowing combinatorial bids across the supply chain. They found the strategic analysis challenging, and were able to derive Bayes-Nash equilibria for only restricted network topologies [21]. In contrast, the well-known Vickrey-Clarke-Groves (VCG) auction [5, 9, 20] (also called the Generalized Vickrey Auction [12]) is incentive compatible and efficient, but not BB with the two-sided bidding needed in a supply chain. Myerson and Satterthwaite showed that, in two-sided negotiation, it is, unfortunately, impossible to simultaneously achieve perfect efficiency, BB, and IR from an IC mechanism [14].

Because the Myerson-Satterthwaite theorem precludes the possibility of an auction that guarantees individual rationality, budget balance, and efficiency with incentive compatibility in ours or any two-sided supply chain model, the best one can do is to obtain an approximation of these desirable economic properties. One approach, taken by Parkes et al. [15], relaxes both incentive compatibility and efficiency. Parkes et al. describe a general methodology for two-sided combinatorial auctions whereby prices are computed that minimize the distance to VCG prices, according to some metric, subject to the budget balance constraint. Their informal argument is that, if prices are not far from VCG prices, then agents will have minimal incentive to strategize and report their values untruthfully. They show empirically, for some distance metrics, that it is possible to obtain high efficiency with prices that deviate little from VCG prices. Note that, although they describe their two-sided combinatorial auction model differently than the supply chain model, it is easy to see that their approach also applies to the present model.

Despite the impossibility theorem, it *is* possible to attain incentive compatibility with any two of the three desirable properties (efficiency, individual rationality, budget balance) in an auction for supply chain formation. In particular, one can ensure IC, IR, and BB, but achieve only approximately optimal efficiency. One approach that has been used is to intentionally reduce efficiency by throwing away profitable trades, while maintaining the other properties. If this approach seems misguided, we note that the Myerson-Satterthwaite theorem actually states more strongly that the three properties cannot be obtained even in Bayes-Nash equilibrium. Thus, since efficiency loss is inevitable in supply chain formation (assuming BB and IR), one may choose to focus on simplifying the agents' strategic problem by ensuring IC. Still, it is important that we do not ignore efficiency altogether, for a highly inefficient auction would likely be unacceptable for business negotiations. Indeed, a trivial way to get IC, IR, and BB is to perform no allocation, which is clearly unacceptable.

In the next section we outline the supply chain formation problem, and in subsequent sections we discuss incentive compatible, individually rational, budget balanced, and highly efficient auctions for a series of progressively more general subclasses of the supply chain model. In Section 3 we review McAfee's double auction and a variant Trade Reduction auction that forms the basis for our subsequent auctions. In Section 4 we describe a generalization of the Trade Reduction auction to linear supply chains. In Section 5 we describe a further generalization of the Trade Reduction auction to supply chains with the unique manufacturing technologies property. In addition to showing the auction is IC, BB, IR, and highly efficient, we describe computationally efficient centralized and distributed algorithms for the auction. We analyze the economic properties for the case when the auction knows the supply chain structure and also discuss how to generalize the auction to the case when the structure is unknown. In Section 6 we generalize the Trade Reduction auction further to give IC and IR for auctions without the unique manufacturing technologies property. Unfortunately, the auction is not BB in this case, and we do not know of any auction with all the desirable economic properties for supply chains at this level of generality. We review the results from the chapter and discuss open problems in in Section 7.

2 Supply Chain Formation Problem

In this section we describe the supply chain formation problem. First we describe our model of agent preferences, then we define allocations in a supply chain. Our goal in subsequent sections is then to define rules that choose an allocation and set prices in a way that support desirable economic properties. Note that we are concerned with the one-time problem of computing the allocation. Readers interested in managing the dynamics of existing supply chains may refer to Kjenstad [10] for an extensive review.

2.1 Supply Chain Model

Before describing the formal details, we illustrate a supply chain with a stylized example in a small lemonade industry, as shown in Figure 1. The figure shows in a ***supply chain graph*** how the lemon juice and lemonade can be manufactured from lemons and sugar by agents in the supply chain. In the figure, an oval indicates a good in the supply chain. A box indicates a market, which is a set of agents who desire exactly the same set of input and output goods. The arrows indicate the input/output relationships between the agents and the goods. The goods are traded in discrete quantities, and with each good we indicate its discretization. For each market, the quantity of inputs needed by one agent are indicated next to the respective arrows. We assume that an agent can provide one unit (at the appropriate discretization) of its output good but may require multiple units of an input good. Borrowing a term from Lehmann et al. [11], we say the agents are single minded to identify the property that each agent has a single bundle of input and output goods that is of interest to that agent. This can often be a reasonable assumption, for companies typically have an established way to produce a product.

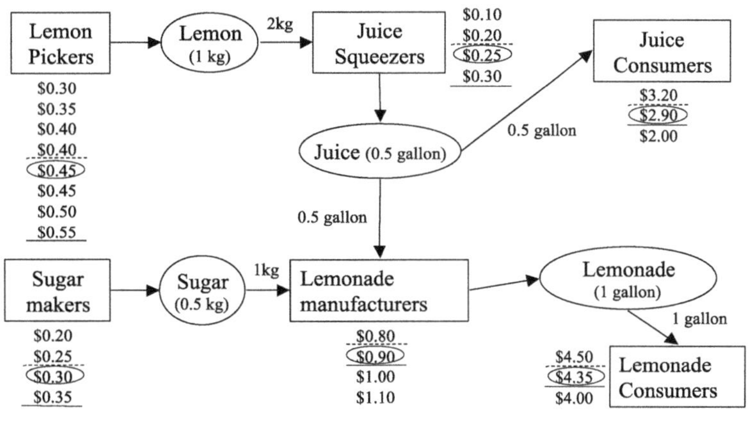

Fig. 1. An example supply chain graph in the lemonade industry

An agent with an output is a producer, and an agent with only inputs is a consumer. For instance, a lemonade manufacturer (a producer) requires 1kg of sugar and 0.5 gallons of juice to produce one gallon of lemonade, and a lemonade consumer wishes to buy 1 gallon of lemonade. A consumer obtains a monetary value from acquiring its bundle of interest, and a producer incurs a monetary cost from producing a good. The values and costs of the individual agents are indicated in a list adjacent to each market (observe that values are sorted from highest to lowest and costs are sorted from lowest to highest).

The formal model we describe subsumes the linear supply chain model described by Babaioff and Nisan [3] but is subsumed by the model described by Walsh and Wellman [22].

Formally, we have a set A of agents and a set G of goods, with agents indicated by integers in $[1, \ldots, |A|]$ and goods indicated by integers in $[1, \ldots, |G|]$. A bundle $q = (q^1, \ldots, q^{|G|})$ indicates the quantity q^g of each good g exchanged by an agent. Positive quantity indicates acquisition of a good (input), and negative quantity indicates provision of a good (output). We restrict our attention to quantities $q^g \in \{-1, 0\} \cup \mathbf{Z}^+$. In other words, we consider agents that can require multiple units of an input, but produce at most one unit of an output. We further restrict our attention to ***single output agents*** that supply at most a single unit of a single good. That is $q^g = -1$ for at most one good g. When comparing quantities in bundles of goods, we assert $\tilde{q} \geq q$ when $\tilde{q}^g \geq q^g$ for all g, and assert $\tilde{q} > q$ when $\tilde{q} \geq q$ and $\tilde{q}^k > q^k$ for some good k.

Agent i has a ***valuation function*** V_i that assigns a value to any bundle q, and $V_i(q) \in \{-\infty, \mathbf{Z}\}$. Agent i obtains utility $U_i(q, M) = V_i(q) - M$, for exchanging bundle q and paying M monetary units. We assume that the agents are rational and try to maximize their utility over all possible outcomes. We refer to $V_i(q)$ as agent i's ***value*** for the bundle of goods q, and we denote the vector of all agents' valuation functions by $\mathbf{V} = (V_1, \ldots, V_{|A|})$. We interpret negative values as ***costs*** (e.g., cost of production, or opportunity cost of providing a good). We assume the valuation functions are normalized at $V_i(\mathbf{0}) = 0$ and the value is ***weakly monotonic***[3] in the quantity of goods, that is $V_i(\tilde{q}_i) \geq V_i(q)$ for all $\tilde{q}_i$ such that $\tilde{q}_i > q$. When $V_i(q) = -\infty$, we say that the bundle q is ***infeasible*** for agent i, and when $V_i(q) \in \mathbf{Z}$ we say the bundle is ***feasible*** for the agent.

We assume that each producer has only one ***technology*** it can use to manufacture an output good, and that each consumer has one specific set of goods it likes to buy. This means that agents are ***single minded*** in that each agent i has a unique ***bundle of interest*** $\check{q}_i$ that it tries to trade. The composition of a bundle of interest and an agent's valuation thereof, depend on the class of the agent, as we detail below. We assume for all agents i that $V_i(\check{q}_i) \in \mathbf{Z}$. For convenience, we subsequently denote $V_i(\check{q}_i)$ as $\check{v}_i$. We denote as $\check{\mathbf{v}}$ the vector of agents' values and denote as $\check{\mathbf{q}}$ the vector of agents' bundles of interest. The ***market*** $K(i)$ of agent i is the set of agents with exactly the same bundle of interest, formally defined as $K(i) = \{j \mid \check{q}_j = \check{q}_i\}$.

There are two classes of agents, defined by further constraints on agents' bundles of interest and values. A ***consumer*** i obtains non-negative value ($\check{v}_i \geq 0$) for acquiring all goods in its bundle of interest ($\check{q}_i > \mathbf{0}$), but cannot produce any good. The consumer's value $V_i(q)$ for bundle q is such that:

- If $q \geq \check{q}_i$, then $V_i(q) = \check{v}_i$ (single minded and weakly monotonic).
- Else, if $q^k < 0$ for some good k, then $V_i(q) = -\infty$ (a consumer cannot feasibly produce any good).
- Otherwise $V_i(q) = 0$ (a consumer has zero value for any feasible bundle not containing its bundle of interest).

A ***producer*** i can produce a single unit of a single output from a specific (possibly empty) set of inputs, while incurring a cost: $\check{v}_i \leq 0$ and $\check{q}_i^g = -1$ for exactly one good g. A producer cannot feasibly produce its output without all inputs, nor can it feasibly produce any other output. The producer's value $V_i(q)$ for bundle q is such that:

- If $q^g = \check{q}_i^g = -1$ and $q \geq \check{q}_i$, then $V_i(q) = \check{v}_i$ (the producer bears a cost of $\check{v}_i$ if it manufactures its output, it is single minded, and has a weakly monotonic valuation).
- Else, if $q^g = \check{q}_i^g = -1$ and $q^k < \check{q}_i^k$ where $\check{q}_i^k > 0$ for some good k, then $V_i(q) = -\infty$ (a producer needs all inputs to feasibly produce its output).
- Else, if $q^k < 0$ where $\check{q}_i^k \geq 0$ for some good k, then $V_i(q) = -\infty$ (a producer can feasibly produce only one good).

[3] Weak monotonicity is equivalent to free disposal for agents.

- Otherwise $V_i(q) = 0$ ($q \geq 0$ and a producer has zero cost if it does not produce any good).

We denote by *CM* the set of consumer markets and *PM* the set of producer markets.

The relationship between markets and goods can be represented as a supply chain graph, as illustrated in Figure 1 and described above. We assume that any graph is directed acyclic, but can have undirected cycles. The market structure defines the supply chain topology.

Definition 1 (supply chain topology). *A **supply chain topology** is a set of markets.*

2.2 Allocations

Given a set of agents, we want to determine the production and exchange of goods that constructs a supply chain. An ***allocation*** $\mathbf{q}$ specifies how much of each good is bought and sold by each agent. Let the allocation of good g to agent i be q_i^g, with $q_i^g > 0$ meaning that i buys q_i^g units of g, and $q_i^g < 0$ meaning that i sells $|q_i^g|$ units of g in the allocation. Allocation $\mathbf{q}$ is ***feasible*** if and only if each agent is feasible and each good is in ***material balance***, that is $\sum_{i \in A} q_i^g = 0$ for each good g.

Throughout this paper, we will consider only allocations that give an agent either all or none of its bundle of interest. Since each agent has one bundle of interest, it will be convenient to identify an allocation $\mathbf{q}$ by the set of agents A' that receive their bundle in the allocation: $A' = \bigcup_{i \in A | q_i \neq \emptyset} i$. We refer to the agents that receive their bundle of interest as the ***winning*** agents. The ***trade size*** $T_m(A')$ in market m of allocation A' is the number of agents in $A' \cap m$. We say that there is ***trade*** in market m in the allocation A', if the trade size in market m is positive.

The ***value*** $\mathbf{V}(A')$ ***of an allocation*** A' is the sum the agent values in A': $\mathbf{V}(A') \equiv \sum_{j \in A'} \check{v}_j$. The value of an allocation A' excluding the value of agent i is $\mathbf{V}_{-i}(A') \equiv \sum_{j \in A', j \neq i} \check{v}_j$. In the auctions below, the true values are not known, so the allocation values are computed with respect to the values reported in the agents' bids. When it is necessary to specify the values explicitly, we denote the value of allocation A' with respect to specific values v as $\mathbf{V}^v(A')$. An allocation $\breve{A}$ is ***efficient*** (optimal) if it is feasible and maximizes the value over all feasible allocations. The ***efficiency*** of allocation A' is $\frac{\mathbf{V}(A')}{\mathbf{V}(\breve{A})}$.

The efficient allocation $\breve{A}$ for the supply chain graph shown in Figure 1 has value \$7.90, and contains the agents whose costs and values are specified above the solid line in each market. The reader can verify that all goods are in material balance and that each agent in $\breve{A}$ receives its bundle of interest. For instance, each of the two lemonade manufacturers in $\breve{A}$ require 1kg of sugar to produce its output, and there are four sugar makers in $\breve{A}$ to provide the 2kg required in total. Similarly, there are two lemonade consumers to buy each of the 1 gallons of lemonade produced by the lemonade manufacturers.

3 Two-Sided, Single-Good

A double auction (DA) is designed for the most simple form of supply chain, one with a single good in which each producer requires no input goods to create its output. In a such a supply chain, a single good g is traded between *sellers* (producers with no input good and a single output) and *buyers* (consumers with a single input). Seller i has a single unit of g ($q^g = -1$) and has some non-negative cost s_i if it sells its unit ($-s_i = \check{v}_i < 0$). Buyer i has a non-negative value b_i ($b_i = \check{v}_i > 0$) if it receives one unit of good g ($q^g = 1$). The set of feasible allocations is the set of materially balanced allocations, namely the allocations with the same number of sellers and buyer. The ***trade quantity*** is the number of trading sellers, which equals the number of trading buyers.

Assuming that the agents are truthful, the efficient allocation is easily found by the following procedure. First the ***supply and demand curves*** are constructed, by sorting the supply bids from low cost to high cost ($s_1 \leq s_2 \leq \ldots$), and the demand bids from high value to low value ($b_1 \geq b_2 \geq \ldots$). Using the two sorted lists, ***optimal trade size*** l is the maximal trade size such that $b_l \geq s_l$. Trading l units between the l highest value buyers and the l lowest cost sellers maximizes the efficiency. Any additional trade has a negative contribution to the efficiency since $b_m < s_m$ for any $m > l$, by definition.

The VCG double auction (***VCG DA***) selects the efficient allocation (with respect to the reported values, which we assume to be the true values, as the auction is incentive compatible), hence the trade quantity is the optimal trade quantity l. The l winning buyers pay $p_B = max(s_l, b_{l+1})$ and the l winning sellers receive $p_S = min(s_{l+1}, b_l)$. VCG mechanisms [5, 9, 20] in general, and the VCG DA mechanism as a special case, are efficient, IR and IC. Incentive compatibility leads to maximal efficiency since agents are truthful, and the trade is efficient with respect to their bids. The VCG DA has budget deficit since $p_B \leq p_S$.

Myerson and Satterthwaite [14] have shown that if individual rationality (also called "participation constraints") is guaranteed, an incentive compatible mechanism that always achieves the efficient outcome leads to a budget deficit (even for a more general model and weaker solution concept than we consider). McAfee [13] suggested a DA rule that is IR and IC, and achieves budget balance at the price of slightly sub-optimal efficiency, for sometimes an efficient trade is discarded. By ***McAfee's DA*** rule, if a suggested clearing price $p = \frac{s_{l+1}+b_{l+1}}{2}$ is accepted by the first l buyers and first l sellers (that is, $p \in [s_l, b_l]$) then l units of the good g are traded at the price of p. Otherwise, $l-1$ units of the good are traded, each trading buyer pays $p_B = b_l$, and each trading seller receives $p_S = s_l$. Since $b_l \geq s_l$ this auction is BB.

A simplified version of McAfee's DA rule is the ***Trade Reduction (TR) DA*** [3]. In this auction the least efficient trade is never executed. Thus, $l-1$ units of the good are always traded (unless $l = 0$), and each trading buyer pays $p_B = b_l$, and each trading seller receives $p_S = s_l$. We call the l-th agent of each class the ***price-bounding agent***, and the value of this agent the ***price-bounding value*** (b_l for the buyers, and s_l for the sellers).[4]

The TR double auction is IR, IC, and BB, like McAfee's auction, but it's expected efficiency is clearly lower. However, the worst-case efficiency of both auctions is the same. In a worst-case analysis, we must assume that in both auctions, the least efficient

[4] While in TR double auction (and the linear TR auction we present in the next section) the price "bounding" value is the actual price and not only a bound, this is not the case for the more general supply chain models. For consistency we use the same term also for the simpler auctions.

trade is not performed. The ***efficiency competitive ratio*** $\text{Ratio}^{AUC}(\check{\mathbf{v}})$ of an auction AUC is the fraction of the social welfare it guarantees when the agents play rationally. Both auctions achieve a competitive ratio of $\frac{l-1}{l}$, since only the unit with the lowest gain from trade might not be traded. This bound is tight in the sense that it is actually achieved when all units have the same gain from trade. Note that the efficiency competitive ratio is a function of the true values, and the fraction $\frac{l-1}{l}$ depends on the trade size in the efficient allocation with respect to the true values. Observe that as the optimal trade size grows, the competitive ratio converges to 1 (perfect efficiency). Furthermore, as we show in subsequent sections, the TR DA generalizes to give IR, IC, BB, and high efficiency for more general classes of supply chains. In contrast, similar generalizations of McAfee's DA do not maintain IC [3].

Now we briefly explain how to prove that the above auctions are IR and IC. The two-sided, single good model considered above is a ***single parameter*** model, that is, the only private information of each agent can be represented by a single number. An allocation rule for single parameter domain is ***bid monotonic*** if it ensures that no winning agent becomes a loser by improving its bid (decreasing the cost for a seller, or increasing the value for a buyer), when other bids are fixed. If the allocation rule is bid monotonic, then for each agent there must exist a ***critical value***, such that the agent wins if it bids above this value, and loses if it bids below this value. It is well known (see e.g. [3]) that for normalized mechanisms (losers pay zero), a mechanism is incentive compatible and individually rational if and only if the allocation rule is bid monotonic, and the payments of the winners are the critical values for winning.

We now use this characterization to prove that the Trade Reduction DA is IR and IC. First we observe that the allocation rule is indeed bid monotonic. If an agent wins the auction, then it is one of the highest-valued $l-1$ agents of its class (sellers or buyers), and clearly by improving its bid, the allocation will not change and the agent will remain a winner. What is the critical value for winning? If a winning buyer bids above b_l, it wins as there will be $l-1$ winning buyers and it will be one of them. If it bids below b_l, then the optimal trade size must be at most l, and it is no longer one of the first $l-1$ buyers, thus it loses. We conclude that b_l is the critical value for a buyer to win, and since the payments are set to be the critical values, the mechanism is indeed IR and IC for buyers. A similar argument holds for sellers.

We subsequently use the above characterization of truthful mechanisms to prove the properties of more general supply chain mechanisms for models in which the only private information is the value for winning.

4 Linear Supply Chains

Having considered the simple case of a two-sided, single-good supply chain, we present a result of Babaioff and Nisan [3] for linear supply chains, which we later extend to more general supply chain models. Babaioff and Nisan consider a linear supply chain model, in which any unit of an initial good can be converted to a unit of the final good through a sequence of unit to unit conversions of intermediate goods. Because a producer has an exposure problem due to acting in two different markets (the market for its input good and the market for its output good), Babaioff and Nisan suggest organizing markets for manufacturing capacity, rather than for the goods themselves. For a given market of this type, all producers with the same technology (the same input and output) compete to sell their manufacturing capacity in the market. Babaioff and Nisan design protocols for exchange of information between the sequence of markets for manufacturing services along a single linear supply chain. These protocols allow each of these markets to function separately, while the information exchanged ensures highly efficient global behavior across the supply chain. Each market that forms a link in the supply chain operates as a double auction market, where the bids on one side of the double auction come from bidders in the corresponding segment of the industry, and the bids on the other side are synthetically generated by the protocol to express the combined information from all other links in the chain. The double auctions in each of the markets can be of several types, and the paper studies several variants of incentive compatible double auctions, comparing them in terms of their efficiency and market revenue.

More formally, in a linear supply chain there is an linear order on the set of goods G. Each agent i is of one of the following $|G|+1$ classes. An ***initial supplier*** can supply one unit of the first good ($q^1 = -1$, and for any other good $g \neq 1$ it holds that $q^g = 0$) and has a non-negative cost s_i ($-\check{v}_i = s_i \geq 0$) if it trades. A ***converter*** of one unit of good $r \in \{1,\ldots,|G|-1\}$ to one unit of good $r+1$ ($q^r = -1$, $q^{r+1} = 1$, and for any other good $g \notin \{r,\ r+1\}$ it holds that $q^g = 0$) has a non-negative cost s_i ($-\check{v}_i = s_i \geq 0$) if it converts a unit. A ***consumer*** has a non-negative value b_i ($\check{v}_i = b_i \geq 0$) if it receives one unit of the final good ($q^{|G|} = 1$, and for any other good $g \neq |G|$ it holds that $q^g = 0$). A ***market*** is the set of agents of the same class (same input and output).

Babaioff and Nisan suggest two protocols that can be used to conduct an auction in a chain of markets in a distributed manner: ***The Symmetric Protocol*** and ***The Pivot Protocol***. Both protocols run on servers connected by a network with a linear chain topology. Each server represents one market and receives bids from only one class of agents (suppliers, consumers or converters from some specified good to the next good in the chain). Each of the conversion markets is connected with bi-directional communication channels to the market that supplies its input good and to the market which demands its output good. Each of the markets, after receiving bids from the appropriate segment of the industry, sorts the bids from high value to low value. This process constructs the supply curves in the initial suppliers' and converters' markets, and the demand curve at the consumers' market. A ***supply curve*** S is a vector of values, sorted from high to low value. The S^1 vector is the supply curve of the initial good, and the values are the bids from the initial supplier. For any $r \in [1,\ldots|G|-1]$, $S^{r\to r+1}$ is the supply curve for the r to $r+1$ conversion, and the values are the bids from the converters of r to $r+1$. $D^{|G|}$ is the demand curve for the final good, and the values are the bids from the consumers.

We focus on the execution of the Pivot Protocol using the Trade Reduction DA rule (see Babaioff and Nisan [3] for details about the general protocol and its properties). We call the resulting mechanism the ***linear TR auction***. In the first stage, the Pivot Protocol constructs the supply curve for the consumers' market. The supply curve for the consumers' market can be calculated by $S^{|G|} = S^1 + \sum_{g=1}^{|G|-1} S^{g\to g+1}$, performed as a vector sum. The protocol performs this calculation as follows. First the initial suppliers' market sends S^1 forward to the first conversion market. Each conversion market from good r to good $r+1$ receives the supply curve $S^r = S^1 + \sum_{g=1}^{r-1} S^{g \to g+1}$ for good r, adds its supply curve $S^{r \to r+1}$, and send the new supply curve of good $r+1$ to the next market. Finally, the supply curve of the last good arrives to the consumers' market.

Next, the consumers' market applies the Trade Reduction DA rule, and finds the optimal trade size l, and the actual trade size $l-1$. The trade size $l-1$ is propagated backwards to all other markets. In each market, the first $l-1$ agents trade. As with the TR DA, payments are computed as the price-bounding value, the bid of the price-bounding agent. Now though we have a different price-bounding value for each market. In any given market, the price-bounding agent is the ***reduced*** agent, which is the agent that is in the efficient allocation that does not trade in the TR allocation (the l-th agent). For agent i, we denote the price-bounding value as PBV_i which is the bid of the price-bounding agent in its market.

The linear Trade Reduction auction is IR and IC [3]. The proof shows that the mechanism created by the above protocol is bid monotonic, and the price bounding value in each market is indeed the critical value for a trading agent in the market. The efficient allocation has l agents trading in each market, while the linear TR auction has only $l-1$. Thus, in each market there is one agent that is reduced from the efficient allocation. The trading agents can be partitioned to $l-1$ ***procurements sets***, each of which includes one agent from each market, such that each set has a higher value then the reduced procurement set (the reduced procurement set contains the l-th agent in each market). This implies that the linear TR auction has an efficiency competitive ratio of $\frac{l-1}{l}$.

Finally, we note that the linear TR auction is budget balanced, as the total payment of each procurement set equals to the non-negative gain from trade of the reduced procurement set. Specifically, given a procurement set of trading agents, the consumer pays $D_l^{|G|}$ (the bid of the l-th highest value consumer), the initial supplier receives S_l^1, and each converter that converters good r to good $r+1$ receives $S_l^{r \to r+1}$. The sum of payments of each procurement set is thus $D_l^{|G|} - S_l^1 - \sum_{g=1}^{|G|-1} S_l^{g \to g+1}$. This is exactly the gain from trade of the procurement set that includes the l-th agent in each market, and it equals to $D_l^{|G|} - S_l^{|G|}$, which by the Trade Reduction DA rule, must be non-negative.

5 Unique Manufacturing Technology Supply Chains

Now we relax the the linear restriction and consider supply chains with the ***unique manufacturing technologies (UMT)*** property, in which there is only one market that produces any good. Formally, this means that if two producers have the same output, then they also use the same technology, which means that they have the same bundle of interest. Note that, although a good can be made in only one way, there can be multiple producers in any market, and multiple markets that require the good as an input.

Next we describe a generalization of the linear Trade reduction auction for UMT supply chains in a way that maintains IC, IR, BB, and high efficiency. The rules consist of an allocation rule, defined in Section 5.1 and a payment rule, defined in Section 5.2.

5.1 UMT Trade Reduction Auction Allocation

The auction is one-shot and sealed-bid. Each agent reports a bundle of interest q_i and a value v_i, either of which may not be truthful. The auction then computes an allocation, which assigns, for each agent, either its reported bundle of interest or the empty bundle. It also computes payments to be made by each agent. Conceptually, the auction first computes an optimal allocation, based on the reported values, and uses this to compute a TR allocation and the agents' payments, as with the previous TR auctions. To ensure incentive compatibility and budget balance, the auction removes some beneficial trades from the optimal allocation.

Define the ***bid-optimal allocation*** $A^*(\mathbf{v})$, as the feasible allocation that maximizes the sum of values with respect to the reported bids $\mathbf{v}$ (in the case that the reported bids are the true values, the bid-optimal allocation is efficient). We now present a definition of a Trade Reduction allocation that generalizes the TR double auction and linear TR auction to our present combinatorial framework.

Definition 2 (Generalized Trade Reduction allocation). *The **Generalized Trade Reduction allocation** $A^{TR}(\mathbf{v})$ is a feasible allocation with maximal value with respect to the bids, that satisfies:*

- *$A^{TR}(\mathbf{v})$ is a sub allocation of the bid-optimal allocation $A^*(\mathbf{v})$. That is, $A^{TR}(\mathbf{v}) \subseteq A^*(\mathbf{v})$.*
- *$A^{TR}(\mathbf{v})$ has strictly fewer winners than the bid-optimal allocation $A^*(\mathbf{v})$ in each market with non-zero trade in $A^*(\mathbf{v})$. That is, for any market m if $T_m(A^*) > 0$ then $T_m(A^{TR}) < T_m(A^*)$, where $T_m(A^*)$ and $T_m(A^{TR})$ are the trade sizes in market m in the bid-optimal and the reduction allocations, respectively.*

For simplicity, we henceforth refer to the Generalized Trade Reduction allocation as simply the Trade Reduction (TR) allocation.

We say that the agents in $A^*(\mathbf{v})$ but not in $A^{TR}(\mathbf{v})$ are ***reduced***. (We explicitly include the bid valuations here because it will be necessary to refer to different sets of bid valuations for computing the payments and proving incentive compatibility. To simply the notation we denote $A^* = A^*(\mathbf{v})$ and $A^{TR} = A^{TR}(\mathbf{v})$).

In the definition, we implicitly assume that there is exactly one bid-optimal allocation and exactly one Trade Reduction allocation that satisfy the equations. In general, we need a rule to break ties between multiple bid-optimal and TR allocations. It can be shown that the auction is not incentive compatible if we break ties between alternate bid-optimal allocations in favor of the one that gives the maximum bid-value TR allocation. However, we maintain IC if we break ties randomly, independent of reported valuations. To simply the exposition, in this chapter we assume that there are no two allocation with the same value (for details about an efficient implementation of the randomized tie breaking rule, see [4]).

If the true values shown in Figure 1 are reported to the auction, then $\mathbf{V}(A^*) = \mathbf{V}(\breve{A}) = \$7.90$ and A^* contains all agents with values and costs above the solid lines. All agents above the dashed lines are in the Trade Reduction allocation A^{TR} and $\mathbf{V}(A^{TR}) = \$4.70$, giving an efficiency of 0.59. The TR rules require that we remove at least one agent from A^* for each market (all markets have winners), hence we reduce one agent from each of the following markets: juice consumers and lemonade consumers. This implies that the inputs consumed by the reduced consumers should not be manufactured (as this adds extra cost and creates goods that are not consumed). Thus, we propagate the fact that some of the goods should not be manufactured backwards in the supply chain, and as we soon see, this also ensures that there will be at least one reduced agent in each market. We should remove a lemonade manufacturer that produces the lemonade consumed by the reduced lemonade consumer. Now, since one juice consumer and one lemonade manufacturer are reduced, we should remove two agents from the juice squeezers' market to maintain material balance of juice. Because each juice squeezer require 2kg of lemons, but each lemon picker provides only 1kg of lemons, we should remove four agents from the lemon pickers' market to maintain material balance of lemons. Similarly, we should remove two agents from the sugar markers' market to maintain material balance of sugar.

An equivalent, and perhaps more illuminating way of defining the Trade Reduction allocation is described below. We begin with an important definition.

Definition 3 (procurement set). *A **procurement set** $S\{A'\}$ in allocation A' is a set of agents constituting a non-empty feasible allocation that contains no other non-empty feasible allocation.*

This definition of procurement set generalizes the notion of procurement set we used for the linear TR auction.

For example, in Figure 1, the following constitutes one procurement set: the juice consumer bidding \$2.90, the juice squeezer bidding \$0.25 and the two lemon pickers bidding \$0.50 and \$0.55. Note that since a producer can produce exactly one unit of one good, any procurement set has exactly one consumer. Clearly, any non-empty feasible allocation can be partitioned into procurement sets. In Figure 1, the reduced agents (indicated by bids between the solid and dashed lines) can be partitioned into two procurement sets, one including the reduced juice consumer and one including the reduced lemonade consumer.

The following lemma provides an alternate view of the Trade Reduction allocation. As we show in Section 5.5, it also allows us to construct an efficient algorithm for computing the TR allocation from the bid-optimal allocation.

Lemma 1. *The Trade Reduction allocation A^{TR} can be obtained from the bid-optimal allocation A^* by reducing a disjoint set of procurement sets as follows. For each consumer market m with trade in A^*, reduce exactly one procurement set containing one consumer in m. In each market, the agents reduced must have no higher value than the agents in A^{TR}.*

Proof. See [4].

Corollary 1 (to Lemma 1). *Given A^*, the number of agents in each market in A^{TR} is uniquely defined and the number of agents reduced from each market is uniquely defined, and both can be calculated in polynomial time. Thus, the A^{TR} allocation is uniquely defined and can be calculated in polynomial time, by taking the appropriate number of highest value agents in each market.*

5.2 UMT Trade Reduction Auction Payments

Having defined the TR allocation rule, we next describe the payment scheme for the TR auction, for UMT supply chains. Since the TR auction could reduce *multiple* agents in a market (in contrast to the TR DA and linear TR auctions which always reduce only one agent), we generalize the notion of ***price-bounding agent*** in a market to refer to the *highest* value agent removed from the efficient allocation at the market. The ***price-bounding value*** for agent i, denoted PBV_i, is then the value reported by the price bounding agent in the market of agent i. (Note that these definitions generalize the more specific definitions presented for the special cases of two-sided auction for a single good, and a linear supply chain). In Figure 1, the values reported by price-bounding agents are just below the dashed lines and circumscribed by ovals.

Although PBV_i payments give incentive compatibility for linear supply chains, they do not give IC in our more general model, as demonstrated in Figure 2. If agent A1 bids \$13.00, as indicated, it does not win because it is not in the bid-optimal allocation. A1 has an incentive to bid any value above \$20.00 because then it would win but pay only $PBV_{\mathrm{A1}} = \$7.00$.

To obtain an IC payment rule, we first denote as $\mathbf{v} \equiv (v_i, \mathbf{v}_{-i})$ the vector of values reported by all agents, where $\mathbf{v}_{-i}$ is the vector of values reported by all agents except i. Let $A^*(\mathbf{v})$ be the bid-optimal allocation with respect to $\mathbf{v}$, and $A^*(\mathbf{v}_{-i})$ be the bid-optimal allocation with respect to $\mathbf{v}_{-i}$.

The ***Vickrey-Clark-Groves (VCG) payment*** from agent i with respect to the bids $\mathbf{v}$ is defined as

$$VCG_i(\mathbf{v}) \equiv \mathbf{V}(A^*(\mathbf{v}_{-i})) - \mathbf{V}_{-i}(A^*(\mathbf{v})) \tag{1}$$

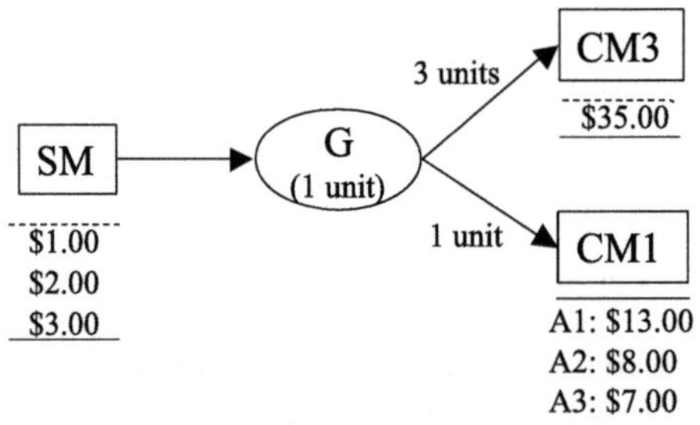

Fig. 2. Agent A1 does not win the Trade Reduction auction if it reports its true value of \$13.00. It would win and pay \$7.00 if it reports any value above \$20.00

Intuitively, $VCG_i(\mathbf{v})$ is the "harm" done by agent i to the other agents by bidding v_i. Observe that $VCG_i(\mathbf{v}) \leq v_i$ and that $VCG_i(\mathbf{v}) = 0$ if i is not in a bid-optimal allocation, thus VCG is individually rational. Consider agent A1 in Figure 2. If it bids as shown in the figure, it is not in the bid-optimal allocation and $VCG_i(\mathbf{v}) = 0$. If instead it bids \$100, then A1 is in the bid-optimal allocation and $VCG_{\mathrm{A1}} = 29 - 9 = 20$. Observe that i would be in the bid-optimal allocation if it bids any value above \$20.

As mentioned above, the VCG auction is not budget balanced for supply chains, but we can modify it slightly to achieve BB for UMT supply chains by using the Trade Reduction allocation and the payment we now define. The payment rule is that losing agents pay zero and each winning agent $i \in A^{TR}$ pays

$$\max\{PBV_i, VCG_i(\mathbf{v})\} \tag{2}$$

This payment rule is a strict generalization of the payment rule described for linear supply chains. For linear structures $VCG_i(\mathbf{v}) < PBV_i$, hence (2) is always PBV_i.

5.3 Auction Properties with the Known Single-Minded Model

We first consider the properties of the auction for UMT supply chains for a ***known single-minded*** model of agent utility. We say "known" because we assume that it is common knowledge that the auction correctly knows the bundle of interest of all agents, but an agent's monetary valuation for its bundle of interest is private and independent of other agents' values. The "known" assumption can be plausible in established industries where production technologies are well known.

Under this model, we call the auction KSM-UMT-TR (Known Single-Minded, Unique Manufacturing Technologies, Trade Reduction). Because the auction knows the bundles of interest, each agent only reports its valuation v_i, which may or may not be its true value $\check{v}_i$, to the auction. Under the KSM model, an auction is incentive compatible (IC) if and only if each agent's dominant strategy is to report its true valuation for its desired bundle.

In this section we show certain desirable economic properties of the KSM-UMT-TR auction. First, we show that the auction is incentive compatible, individually rational, and budget balanced. Then, we show that it has a good competitive ratio for efficiency. In the next section, we enumerate the properties that do and do not hold if we relax the unique manufacturing technologies constraint.

Incentive Compatibility, Individual Rationality, and Budget Balance The main theorem we prove for the KSM-UMT-TR mechanism is:

Theorem 1. *The KSM-UMT-TR auction produces a feasible allocation, and is incentive compatible in dominant strategies, individually rational, and budget balanced.*

Proof concept. The allocation is feasible by construction. We give the complete proof for incentive compatibility (Lemma 6) and individual rationality (Lemma 7) in Section A of the appendix. Below is an outline of the proof concept. As we noted earlier, it has been shown that, for known single-minded agents, an auction is incentive compatible and individually rational, if and only it is bid-monotonic (i.e., an agent that would win by reporting value a x, would also win by reporting any value $x' > x$) and trading agents pay their ***critical value*** for winning. The critical value is the value such that if an agent bids above the value it will win the auction, and if it bids below the value it will lose. We show in the appendix that VTR_i is indeed the critical value for the KSM-UMT-TR allocation rule, hence the auction is IC. For IR, it is easy to see that both VCG and PBV payments are lower than a winning agent's reported value. Hence if an agent reports truthfully, it will pay no more than its value.

We give the complete proof of budget balance, which is rather technical, in Section B of the appendix. Here we outline the proof concept. The budget balance proof shows that each individual procurement set is BB. It is possible to construct a one-to-one correspondence between agents in a trading procurement set, to agents in are reduced procurement set. The reduced procurement set has a non-negative value, and each agent in the trading procurement set pay at least the value of the corresponding reduced agent in the reduced procurement set (as it pays at least the PBV, which is at least the value of the corresponding reduced agent). □

Efficiency Analysis We have established that the Known Single-Minded, Unique Manufacturing Technologies, Trade Reduction auction is incentive compatible, individually rational, and budget balanced, but we also want acceptable efficiency. In this section, we establish a good worst-case bound on the efficiency of the auction. This bound is such that, in a fixed topology with the UMT property, the allocation converges to perfect efficiency as the number of trades in the consumer markets grows. More precisely, we show that the efficiency competitive ratio of the KSM-UMT-TR auction, denoted by $\text{Ratio}^{\text{KSM-UMT-TR}}(\check{\mathbf{v}})$, is the minimal ratio of trade sizes of the TR allocation and the efficient allocation, over all consumers' markets with trade in the efficient allocation.

We denote by CM^* the set of consumer markets with non-zero trade size in $\breve{A}$, the efficient allocation with respect to values $\check{\mathbf{v}}$.

Theorem 2. *For any values $\check{\mathbf{v}}$, when agents play according to their dominant strategies, the KSM-UMT-TR auction achieves at least the following fraction of the maximal social welfare:*

$$Ratio^{KSM\text{-}UMT\text{-}TR}(\check{\mathbf{v}}) = \min_{m \in CM^*} \frac{T_m(\breve{A}) - 1}{T_m(\breve{A})},$$

where we define the minimum to be 1 *if* $\breve{A} = \emptyset$.

Proof concept. Here we present the basic intuition for the theorem. Refer to Section B of the appendix for the technical details of the proof. Consider a simple supply chain with one consumer market wanting a single unit of a single good from one producer market. The Trade Reduction allocation reduces the lowest value buyer and seller, that is, the lowest value procurement set. If the trade size in the optimal allocation is $n > 0$, then the trade size in the TR allocation is $n - 1$. The efficiency is the lowest when all procurement sets (seller-buyer pairs) have the same value, in which case the efficiency is $(n-1)/n$.

More generally, in a supply chain with unique manufacturing technologies property, any two consumers of the same market belong to the same topology of procurement sets. That is, the same number of the same goods need to be produced for each consumer. Again, the lowest value procurement set is reduced in each consumer market. The consumer market m with the lowest trade size has the largest fraction of agents reduced. If the procurement sets of m also constitutes most of the value in A^*, then the greatest value will be lost here and this reduction would dictate the efficiency loss. So, the guaranteed efficiency is the minimal over all consumer markets, of the ratio between the trade size in the TR allocation and the trade size in the optimal allocation. □

Note that Theorem 2 gives a worst case bound which holds for *any* valuations of the agents, and in particular it holds if we assume any underlying distribution of valuations. The bound is dependent only on the number of trades in the optimal allocation.

Recall that the efficiency of A^{TR} for Figure 1 is 0.59. In this supply chain, there are two trades in each consumer market in $\breve{A}$, giving us $\text{Ratio}^{\text{KSM-UMT-TR}}(\breve{\mathbf{v}}) = 1/2$, which is indeed less than the actual efficiency. Typically, our auction achieves higher efficiency than the competitive ratio. The efficiency can be significantly higher when there is a large difference between the value of the agents in the auction allocation and the value of the agents reduced (recall that the low-valued agents are reduced). For instance, consider a supply chain with two markets: a producer market m_1 with no inputs and an output desired by consumers in market m_2. If $T_{m_2}(\breve{A}) = 2$, then $\text{Ratio}^{\text{KSM-UMT-TR}}(\breve{\mathbf{v}}) = 1/2$. But if both producers in m_1 have a value of 0, c_1 is the highest-value consumer, and c_2 is the second-highest-value consumer in m_2, then $\text{Eff}^{\text{KSM-UMT-TR}}(\breve{\mathbf{v}}) = \breve{v}_{c_1}/(\breve{v}_{c_1} + \breve{v}_{c_2})$. Clearly then, $\text{Eff}^{\text{KSM-UMT-TR}}(\breve{\mathbf{v}}) \to 1$ as $\breve{v}_{c_1}/\breve{v}_{c_2} \to \infty$.

Nevertheless, the competitive ratio is a tight worst-case bound, in the following sense. Given an optimal allocation, there exists a set of bids supporting the allocation that give efficiency arbitrarily close to the competitive ratio.

Theorem 3. *Let $\breve{A}$ be the efficient allocation for agents A and some set of values. Then for any $\varepsilon > 0$, there exists a vector of values $\breve{\mathbf{v}}$ for agents A with the same optimal allocation and the same TR allocation, such that the TR allocation achieves only the fraction*

$$\min_{m \in CM^*} \frac{T_m(\breve{A}) - 1}{T_m(\breve{A})} + \varepsilon$$

of the maximal social welfare.

Proof. See [4].

The Myerson-Satterthwaite impossibility theorem [14] (discussed in Section 1) holds, in particular, for the case of a single producer with no inputs wishing to sell one good to a single consumer (bilateral trade). In this case, the impossibility theorem implies that there is no mechanism which is budget balanced, individually rational, incentive compatible, and in which the agents always trade if it is efficient to do so. Thus, for any mechanism, there are some values for which there is efficient trade, but it is not conducted by the mechanism. With this in mind, and using reasoning similar to that in the proof of Theorem 3, we can conclude that, when the consumer's market has only one consumer, no auction can have better than a zero efficiency competitive ratio (as in the worst-case there is no trade conducted by the mechanism, even though an efficient trade exists). Thus, KSM-UMT-TR gives the best possible competitive ratio in this case.

5.4 Auctions for the Unknown Single-Minded Model

In many situations it may not be reasonable to assume that an auction knows the technology that each agent uses, thus it does not know the bundle of interest of each of the agents. Now we consider the case where both an agent's bundle of interest and monetary valuation are private and independent of other agents. With this model, which we call the ***Unknown Single-Minded*** (USM) model, the auctions must elicit the bundle of interest information from the agents. An auction is ***incentive compatible in dominant strategies*** if and only if each agent maximizes its utility by reporting its bundle of interest, and its valuation thereof, truthfully, regardless of what the other agents report (truth telling is a dominant strategy). For the USM model, we sometimes need to use a weaker solution concept. An auction is ***ex post Nash incentive compatible*** if and only if each agent maximizes its utility by reporting its bundle of interest, and its valuation thereof, truthfully, given that all other agents also do so (independent of the distribution of the values of the other agents).

Auction Mechanisms using Bundles Conversion Consider the Trade Reduction auction under the USM model, in which each agent i reports a value v_i and bundle of interest q_i, either of which may not be true. Unfortunately, the auction is not generally incentive compatible because, due to weak monotonicity of preferences, an agent may be able to gain by (untruthfully) reporting a bundle that contains its bundle of interest. For instance, consider the case in which we have some consumer a with $\breve{q}_a^g = 1$ for good g only, and we have another consumer b with the same bundle of interest except that $\breve{q}_b^k = 1$ for some good k such that $\breve{q}_a^k = 0$. Assume that a is the only agent in its true market. If a bids truthfully, it gets reduced if it is in the optimal allocation, hence gets zero utility. Assume further that $\breve{v}_a > \breve{v}_b$. Then if b is winning in its own market, a would win by reporting the bundle $q_a = \breve{q}_b$ with value $V_a(\breve{q}_b)$ to the auction. Since $V_a(\breve{q}_a) = V_a(\breve{q}_b)$, agent a would obtain a higher utility by misreporting its bundle of interest than by reporting truthfully.

Since agents cannot obtain positive value by bidding untruthfully for bundles that are not supersets of their bundle of interest, the agents' motivation to untruthfully bid for supersets is the root cause of lack of incentive compatibility for the TR auction under the USM model. We formalize a variant of this notion by introducing another class of auctions based on the Trade Reduction auction. In the next sections we present specific auctions from this class that remove the incentive for agents to untruthfully bid for larger bundles, giving incentive compatibility.

A ***USM-UMT-TR auction*** (Unknown Single-Minded, Unique Manufacturing Technologies, Trade Reduction auction) accepts bids from all agents, then applies a publicly known ***conversion function*** F to the bundles reported in the bids, to get a ***converted*** bundle for each agent. The auction then applies the Trade Reduction auction rules on the converted bundles with the submitted values. The output of F is the converted bundle for agent i and is denoted by $F_i(q_i, \mathbf{q}_{-i})$. The first argument q_i is the bundle reported by i and the second argument $\mathbf{q}_{-i}$ is the vector of bundles reported by all agents other than i. Because we do not want F to discriminate between agents, we require that for any bundle q, vectors of bundles $\mathbf{q}_{-i}, \mathbf{q}_{-j}$, and agents i and j, $F_i(q, \mathbf{q}_{-i}) = F_j(q, \mathbf{q}_{-j})$ when $\mathbf{q}_{-i}$ equals $\mathbf{q}_{-j}$ as *sets* of bundles (without considering the order of the bundles). In order to make sure that the valuation of the bundle of interest after the conversion remains the same, we require that for any agent i and for any $\mathbf{q}_{-i}$, a conversion function F must also satisfy the following conditions for any bundle q_i:

1. $q_i \leq F_i(q_i, \mathbf{q}_{-i})$ (the converted bundle is a superset of the reported bundle).
2. The original and converted bundles have the same output: for any good g, $F_i(q_i, \mathbf{q}_{-i})^g = -1$ if and only if $q_i^g = -1$ (a producer remains a producer with the same output).

We can effectively remove an agent from the auction by adding to the converted bundle of the agent a demand for some null good $\bar{g}$ which is not produced by any agent, we effectively remove the agent from the auction.

The properties we require from a conversion function ensure that any agent i has the same valuations for its desired bundle and for the conversion, that is $V_i(\check{q}_i) = V_i(F_i(\check{q}_i, \mathbf{q}_{-i}))$. This implies that the auction is individually rational. Additionally, the TR auction is budget balanced (even if agents are not truthful).

Observation 4 *For any conversion function, the USM-UMT-TR auction is individually rational and budget balanced.*

Next, we characterize conversion functions that create incentive compatible auctions for the Unknown Single-Minded case (for a general characterization of truthful mechanisms in the more general model of Unknown "Single Value" domain[5] see [2]).

Theorem 5. *A USM-UMT-TR auction is incentive compatible in dominant strategies if and only if, for any agent i, any* $\mathbf{q}_{-i}$ *and any* $\mathbf{v}_{-i}$*, i cannot increase its utility by reporting any bundle* q_i *with the same output as* $\check{q}_i$*, such that* $q_i \neq \check{q}_i$ *and* $\check{q}_i \leq F_i(q_i, \mathbf{q}_{-i})$.

Proof. See [4].

Note that in particular, the auction must ensure that the agent cannot increase its utility by reporting a bundle that contains its desired bundle, since after conversion it still contains its desired bundle.

Theorem 6. *A USM-UMT-TR auction is ex post Nash incentive compatible if and only if, for any agent i, any* $\check{\mathbf{q}}$*, and any* $\check{\mathbf{v}}$*, i cannot increase its utility by reporting any bundle* q_i *with the same output as* $\check{q}_i$*, such that* $q_i \neq \check{q}_i$ *and* $\check{q}_i \leq F_i(q_i, \mathbf{q}_{-i})$*, when all other agents report truthfully.*

Proof. See [4].

Ex Post* Nash Incentive Compatibility by Removing Bids** Consider a particular USM-UMT-TR auction we call the ***USM-UMT-TR-RB (for USM-UMT-TR Remove Bigger) auction, with the conversion function F defined as follows:

1. For any producer i we have $F_i(q_i, \mathbf{q}_{-i}) = q_i$.
2. For any consumer i, if there exists an agent j such that $q_i > q_j$, then $F_i(q_i, \mathbf{q}_{-i}) = q_i \cup \bar{g}$ (the bid is removed), otherwise $F_i(q_i, \mathbf{q}_{-i}) = q_i$.

[5] In a "Single Value" domain the outcome space is partitioned to desired and undesired outcomes, and each of the desired outcomes has the same private value for the agent. In an Unknown Single Value domain, the partition is private information.

At first glance it might appear that, because consumer bundle q_i is removed when $q_i > q_j$ for another consumer j, the auction satisfies Theorem 6 for consumers. However, if j reports q_i instead of q_j, and there is no other reported bundle q_k such that $q_i > q_k$, then j is not removed and it might be able to increase its utility by this manipulation. We can guarantee that all such q_i bids get removed if there exists a reported bundle q_k other than q_j such that $q_i > q_k$. If this holds, then the auction satisfies Theorem 6.

Theorem 7. *The USM-UMT-TR-RB auction is* ex post *Nash incentive compatible, if, for any consumer j, if there is a consumer i such that $\check{q}_i > \check{q}_j$ then there exists a consumer $k \neq j$ such that $\check{q}_i > \check{q}_k$.*

Proof. See [4].

Note that the condition of the theorem above always holds if there are at least two agents using the same technology, for each technology used (in each market the are at least two agents). Also note that a consumer would be forced out of the auction if another consumer has a strictly smaller bundle of interest (which corresponds to a better technology in the sense that it requires fewer inputs). This could happen even if the consumer that is forced out has a very high value, potentially causing the efficiency to be very low. In fact, the efficiency can be arbitrarily close to zero, hence, in general, we can give no positive competitive ratio for efficiency. However, if there are no consumers i and j such that $\check{q}_i > \check{q}_j$, then Theorem 7 holds and there is a good competitive ratio for efficiency.

Observation 8 *If there are no consumers i and j such that $\check{q}_i > \check{q}_j$, then the USM-UMT-TR-RB auction is* ex post *Nash incentive compatible, no agent is removed by the conversion function, and the efficiency competitive ratio from Theorem 2 holds for the USM-UMT-TR-RB auction.*

Dominant Strategies Incentive Compatibility by Merging Markets We can ensure incentive compatibility in dominant strategies by *merging* markets, rather than removing bids. The ***USM-UMT-TR-Merge*** auction is a USM-UMT-TR auction with a conversion function F that sets the converted bundle for agent i to have the maximum number of units of each input of any reported bundle with the same output as reported by i. Formally, for agent i, let O_i be the set of agents with the same output as i, then for each good g, the number of units of g in the converted bundle of i is $F_i^g(q_i, \mathbf{q}_{-i}) = \max_{j \in O_i} q_j^g$. Observe that all producers with the same reported output are merged into a single market and all consumers are merged into a single market.

Theorem 9. *The USM-UMT-TR-Merge auction is incentive compatible in dominant strategies.*

Proof. See [4].

Recall that, by the unique manufacturing technologies property assumption, there is only one technology to manufacture any good (at most one market with any specific output good). Thus, since producers bid truthfully in USM-UMT-TR-Merge, no markets of producers will actually be merged (although markets for consumers might be merged). Still, although no producer markets are actually merged, the merging rule is necessary to ensure incentive compatibility in *dominant* strategies for the producers.

In general, it is ambiguous whether the USM-UMT-TR-Merge auction would give higher or lower efficiency than the Trade Reduction auction for agents with privately known bundles and values, with agents reporting truthfully. If all true consumer markets contain only one consumer, then there would be no trade without merging, hence merging could not make the allocation worse and might improve it. But if consumer markets contain multiple consumers, then merging markets could increase the costs of an allocation, giving it a lower value than without merging. In fact, efficiency could be arbitrarily close to zero, hence we can establish no positive competitive ratio for efficiency.

With an additional assumption on the consumers' bundle of interest, we can avoid merging consumers markets and ensure high efficiency. We say that the supply chain has the ***k-Input Consumers*** property, if for some fixed integer k (publicly known), each consumer desires exactly k units of all goods in total. That is, for any consumer i, $\sum_{g\in G}\breve{q}_i^g = k$. If for some fixed k, the k-input consumers property holds, we can gain incentive compatibility without merging any consumer markets. In the ***USM-UMT-TR-Merge-kIC*** (USM-UMT-TR-Merge k-Input Consumers) auction, F_i is defined for producers as in USM-UMT-TR-Merge. No consumer markets are merges, but the auction rejects all consumer bids for other than k units in total. Formally, for each consumer i, if $\sum_g \breve{q}_i^g = k$ then $F_i(q_i, \mathbf{q}_{-i}) = \breve{q}_i$, otherwise $F_i(q_i, \mathbf{q}_{-i}) = \breve{q}_i \cup \bar{g}$ (the bid is removed). With the k-input consumers property, no consumer can feasibly misrepresent itself as any other consumer. Since no merging is actually performed, and no agent is removed when the agents bid truthfully, our competitive ratio bound holds for the USM-UMT-TR-Merge-kIC auction.

Observation 10 *When the k-Input Consumers property holds, the USM-UMT-TR-Merge-kIC auction is incentive compatible in dominant strategies, and the efficiency competitive ratio from Theorem 2 holds.*

Proof. See [4].

5.5 Computational Complexity of the UMT Auction

As we show in [4], computing both the allocation and auction payment are NP-hard (i.e., computationally intractable) for UMT supply chains. However, it is possible to compute the allocation A^{TR} from the bid-optimal allocation A^* in polynomial time as follows: for each consumer market m with trade in A^* we remove one trade, and propagate the reduction in demand for the input goods backwards in the supply chain. In fact, we calculate the number of traders in each market in Trade Reduction auction and from it we calculate the TR allocation, according to Corollary 1. Procedure `Trade-Reduction`(A^*) (Figure 3) implements the algorithm by computing the trade size $T_m(A^{TR})$ for each market m in A^{TR}. Procedure `Producer-Market-Trade-Sizes` (Figure 4) actually computes and returns the trade sizes for the producers in reverse topological order (ordered from the consumers). Its argument $\{t_m\}_{m\in CM}$ is a ***configuration***, which specifies, for each consumer market m, the trade size in m. When the trade size $T_m(A^{TR})$ is computed for market m, the $T_m(A^{TR})$ highest value agents are included in A^{TR}. In the procedures, we denote as I_g the set of markets that desire good g as an input. Recall that we denote as $\breve{q}_n^g$ the number of items of good g desired by each agent in market n, so $\breve{q}_n^g \cdot t_n$ is the total number of items of good g desired by all agents in market n.

Theorem 11. *Procedure* `Trade-Reduction`(A^*) *computes the Trade Reduction allocation* A^{TR} *from* A^* *in time polynomial in the number of agents* $|A|$.

Proof. See [4].

From the above theorem, it is clear that if we can calculate the VCG mechanism (allocation and payments) then from the VCG allocation and payments we can calculate the Trade Reduction allocation and payments in time polynomial in the number of agents.

Observation 12 *The Trade Reduction mechanism can be calculated from the VCG mechanism in time polynomial in the number of agents.*

```
Trade-Reduction(A*)
   FOR EACH market m, Set T_m(A^TR) ← 0.
   FOR EACH consumer market m ∈ CM
      IF T_m(A*) > 0 THEN set T_m(A^TR) ← T_m(A*) − 1.
   {T_m(A^TR)}_{m∈PM} ←
      Producer-Market-Trade-Sizes({T_m(A^TR)}_{m∈CM}).
   A^TR ← ∅.
   FOR EACH market m, add the T_m(A^TR) highest value agents to A^TR.
   RETURN A^TR.
```

Fig. 3. Procedure `Trade-Reduction`

```
Producer-Market-Trade-Sizes({t_m}_{m∈CM})
   For each producer market m ∈ PM producing good g, in reverse
         topological order
      For each market n in I_g, set t_m ← t_m + q̆_n^g · t_n.
   RETURN {t_m}_{m∈PM}.
```

Fig. 4. Procedure `Producer-Market-Trade-Sizes`

Still, because computing the bid-optimal allocation is NP-hard, the Trade Reduction auction remains NP-hard in general. However, for our auction we can exploit the structure of a unique manufacturing technologies supply chain to compute the bid-optimal and the Trade Reduction allocations in polynomial time, if certain constraints hold on the structure of consumer preferences. If we can enumerate all configurations of trade sizes in the consumers' markets in polynomial time (in the number of agents) then we can calculate the auction in polynomial time. Denote as N_{max} the maximal number of consumers in any consumer market.

Theorem 13. *If there are constants* c,k *such that* ${N_{max}}^{|CM|} \leq c|A|^k$ *then the Trade Reduction auction is polynomial-time computable in* $|A|$.

Proof concept. For brevity, we present only the proof concept. For any given configuration, we can apply procedure `Producer-Market-Trade-Sizes` on the configuration and calculate the size of trade in every market in a feasible allocation with this configuration (this can be done time polynomial in $|A|$). Given the trade sizes, we simply pick the highest value agents in each market, which can clearly be done in time polynomial in $|A|$. Thus, for a fixed configuration, we can find the highest value feasible allocation in time polynomial in $|A|$.

The number of configurations is at most ${N_{max}}^{|CM|}$, which is polynomial in $|A|$ by the assumption that ${N_{max}}^{|CM|} \leq c|A|^k$. Therefore, finding the optimal allocation can be done by enumerating all configurations and picking the highest value feasible allocation in time polynomial in $|A|$. Given the bid-optimal allocation, we can then compute the Trade Reduction allocation in polynomial time using the `Trade-Reduction` procedure, as described above. Finally, because calculating payments requires computing a polynomial number of optimal and TR allocations, these calculations are also polynomial-time computable in $|A|$. □

Note that $N_{max} \leq |A|$, therefore if the number of consumer markets is bounded by some (fixed) constant k, then ${N_{max}}^{|CM|} \leq |A|^k$ and by the theorem the algorithm is polynomial-time computable in $|A|$. So we have the following corollary.

Corollary 2. *For the family of supply chains with the number of consumer markets bounded by a fixed constant, the Trade Reduction auction is polynomial-time computable in* $|A|$.

Also note that if the maximal number of consumers in a market N_{max} is bounded by a constant k_1, and there is a constant k_2 such that $|CM| \leq k_2 \log |A|$ (the number consumer market is logarithmic in the number of agents) then ${N_{max}}^{|CM|} \leq k_1^{k_2 \log |A|} \leq |A|^{k_2 \log k_1}$ and the algorithm is polynomial-time computable in $|A|$.

For some supply chains it is quite reasonable to assume that each good is manufactured from at least two units of its input goods. In this case the number of agents in a procurement set grows exponentially with the depth of the supply chain. In such supply chains, the maximal number of consumers in a market that can feasibly be in an allocation is indeed significantly smaller then the total number of agents, and we can reasonably expect the assumptions of Theorem 13 to hold.

5.6 Distributed Implementation

The Trade Reduction auction algorithm described in Theorem 13 can also be implemented as a distributed protocol between markets (for a general overview of Distributed Algorithmic Mechanism Design, we refer the reader to Feigenbaum and Shenker [8]), generalizing the protocol presented in Babaioff and Nisan [3]. Again, for a fixed number of consumer markets for example, this protocol will run in time polynomial in the number of agents.

Each agent communicates with a mediator representing its market. Each market communicates with its input and output markets, and consumer markets also communicate with a single coordinator. The mediators and coordinator are assumed to be obedient and run the protocol as specified (only the agents are strategic).

The protocol works as follows. Each agent first sends its bid to its respective mediator. When a consumer market mediator receives all bids from its consumers, it sends the *number* of bids in that market to the coordinator. When the coordinator receives the numbers of bids from each consumer mediator, it enumerates all configurations, each specifying the trade size for each consumer market. Then, for each configuration, it sends the respective trade size to each consumer mediator. Recall that we denote the set of markets that desire good g as an input by I_g. When a mediator receives the number of units it needs to produce for each market in I_g (a consumer mediator treats the coordinator as demanding a virtual output from it) it performs `Market-Trade-Size-Propagate` (Figure 5). This procedure sums the number of output items demanded to determine the number of winners in the market m. It then propagates to each input market of m the number of input units needed by m from that input market. In the following, we denote by Υ_m the set of markets that produce input goods used by market m.

Market-Trade-Size-Propagate()
  WHEN $\breve{q}_n^g \cdot t_n$ is received from each market n in I_g, Set $t_m \leftarrow \sum_{n \in I_g} \breve{q}_n^g \cdot t_n$.
  IF m has no input
    THEN perform Market-Value-Propagate().
  ELSE BEGIN
    FOR EACH market $n \in \Upsilon_m$ producing good k, send $\breve{q}_m^k \cdot t_m$ to n.
    WHEN a value V_n is received from each market $n \in \Upsilon_n$
      Perform Market-Value-Propagate().
  END

Fig. 5. Procedure `Market-Trade-Size-Propagate` executed by the mediator for market m with output good g

When the trade sizes are computed, producers with no inputs initiate `Market-Value-Propagate` (Figure 6) propagates back to the coordinator the total value of the agents that will trade. The coordinator receives all the values from the consumer markets and sum them to find the total value of the allocation.

The entire protocol described above is performed for each configuration, and the coordinator chooses the configuration with the highest value. Then, to determine the Trade Reduction allocation, the coordinator subtracts one from the optimal trade size for each consumer market and sends these trade sizes to the consumer mediators. The same protocol then computes the TR allocation in one round.

```
Market-Value-Propagate()
    Set V_m ← the sum of the bid values of the t_m highest agents in m.
    FOR EACH market n ∈ Υ_m, Set V_m ← V_m + V_n.
    Send V_m to one arbitrary market in I_g and send 0 to all other
        markets in I_g.
```

Fig. 6. Procedure `Market-Value-Propagate` executed by the mediator for market m with output good g

By the same argument as in Theorem 13 and from the above protocol we conclude the following observation.

Observation 14 *If there are constants c,k such that $N_{max}{}^{|CM|} \leq c|A|^k$ then the auction can be implemented as a distributed protocol with running time polynomial in $|A|$.*

6 Supply Chains without the Unique Manufacturing Technologies Constraint

The payment rule described for UMT supply chains is not incentive compatible when the UMT property does not hold. However, the following generalization of the payment rule is incentive compatible for non-UMT supply chains. By this payment rule, losing agents pay zero and a winning agent pays:

$$VTR_i(\mathbf{v}) \equiv \mathbf{V}(A^{TR}(VCG_i(\mathbf{v}), \mathbf{v}_{-i})) - \mathbf{V}_{-i}(A^{TR}(\mathbf{v})) \tag{3}$$

where $A^{TR}(VCG_i(\mathbf{v}), \mathbf{v}_{-i})$ is the auction allocation obtained when the bid of i is replaced by $VCG_i(\mathbf{v})$. For UMT supply chains (and in particular, for linear supply chains and double auctions), this payment rule is exactly the payment rule that we described before, that is $VTR_i(\mathbf{v}) = \max\{PBV_i, VCG_i(\mathbf{v})\}$.

The auction is incentive compatible and individually rational for non-UMT supply chains, as we prove these properties for the general $VTR_i(\mathbf{v})$ payment rule. However, the Trade Reduction auction is not budget balanced if the UMT property does not hold, as shown in Figure 7. There, $VTR = PBV$ for all agents, and it is easy to see that the total payment is $34 - 3 \cdot 13 < 0$.

Observation 15 *The Trade Reduction auction is incentive compatible and individually rational, but not necessarily budget balanced, if the unique manufacturing technologies property does not hold.*

It is possible to regain budget balance by adding an explicit budget balance constraint to the TR auction, but this at the expense of incentive compatibility.

The competitive ratio for Theorem 2 does not hold when the UMT property does not hold. Consider a topology with two markets that supply the same good g. Market 1 has one agent that can produce g with a low cost L from some zero cost good k, and Market 2 has two agents that can each produce g with a high cost H. There are three agents, each with value $H + 1$ in a consumer market that desire g. In the efficient allocation, all three

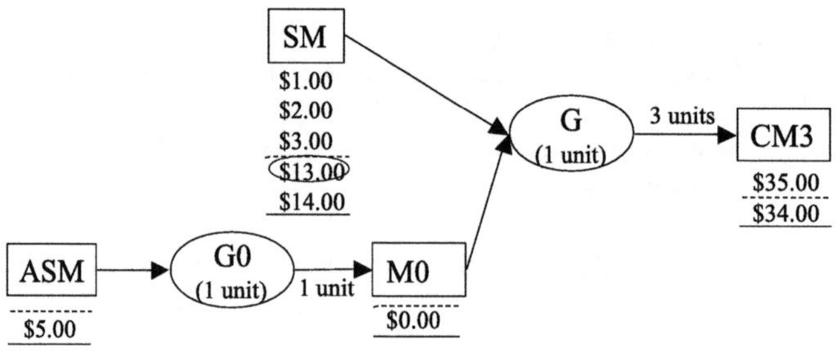

Fig. 7. A supply chain without the unique manufacturing technologies property and for which Trade Reduction auction is not budget balanced

items will be traded and the value of the efficient allocation is $3*(H+1)-(L+2*H)=(H-L+3)$. In the TR allocation one agent in each market will be reduced and the allocation value is $H+1-H=1$. The efficiency is then $1/(H-L+3)$, which can be arbitrarily close to zero as H grows. This violates the efficiency ratio of $2/3$ from the theorem for this non-UMT supply chain.

The auction is still NP-hard when the UMT property might not hold, as we have proven the hardness for the restricted case that the UMT property has to hold. Unfortunately, because Lemma 1 (and hence Corollary 1) depends on the UMT property, the centralized and decentralized polynomial (given constraints on the consumers) algorithms cannot be used for non-UMT supply chains.

7 Discussion and Open Problems

We have presented auctions for supply chain formation that are incentive compatible, individually rational, and budget balanced. We are not aware of any auctions, other than the Trade Reduction auctions, with these properties and with comparably high efficiency for as broad a class of supply chain topologies we consider. Nevertheless, we believe there may be further opportunities for improving efficiency of the Trade Reduction auction while maintaining the properties. Our present approach relies on the existence of multiple agents with the same bundles of interest to obtain high efficiency. We hope that methods can be developed to lessen the dependence. It is also our hope that mechanisms can be developed to obtain incentive compatibility and budget balance with higher efficiency in the unknown single minded model.

We are not aware of effective mechanisms for broader classes of supply chains. Of particular interest is the loosening of the single minded restriction. For instance, OR preferences allow an agent to specify that it would accept one or more bundles specified in a set of bundles, each with a different value, and that it values the set of bundles it receives as the sum of the values of the bundles. XOR preferences allow an agent to specify that it wants exactly one bundle from a specified set, each with a different value, and that it values the set of bundles as the maximal value of any bundle it receives. These extensions would allow agents (companies) to express, for instance, different values/costs for different quantities and alternate production technologies. Consider the following obvious variant of our auction to allow OR or XOR bids. We change the auction to allow agents to place OR or XOR bids, and include the OR and XOR constraints in the auction. We also change the VTR_i payments so that i's payment does not depend on its own bids. With these changes, an agent can manipulate the allocation in its favor by changing one of its bids, thus violating incentive compatibility. Consider the case in Figure 2 where the true preferences of A1 contain XOR components $13 in market CM1 and $35 in market CM3. If A1 bids truthfully, it will win none of its bids. If instead, A1 bids less than $28 in the CM3 market, it will win one unit of the good in market CM1 and pay less than $13, giving it a positive utility. We get the same phenomenon if the bid is OR instead or XOR. In either case, the auction is not incentive compatible.

Further interesting additions to the supply chain model would include more realistic features of supply chains, including multiple unit production, time-based preferences (e.g., deadlines and production lead time) and non-commodity multi-attribute goods. We expect these to impose interesting challenges in designing mechanisms with desirable economic properties.

In a supply chain, agents may not wish to reveal their full valuations outright in the single-shot auction we developed. Indeed, if agents are not single minded, it could be infeasible for agents to communicate their entire preferences. To address this problem, iterative combinatorial auctions have been studied by others [1] [16]. Extending these results to a supply chain model might increase the viability of auctions for real-world supply chains [6].

Acknowledgments

The first author was supported in this research by grants from the Israeli Academy of Sciences and the USA-Israel Bi-national Science Foundation while he was a graduate student at the Hebrew University of Jerusalem.

References

1. Lawrence M. Ausubel and Paul Milgrom. Ascending auctions with package bidding. Technical report, University of Maryland and Stanford University, 2001.
2. Moshe Babaioff, Ron Lavi, and Elan Pavlov. Mechanism design for single-value domains. In *Twentieth National Conference on Artificial Intelligence*, pages 241–247, 2005.
3. Moshe Babaioff and Noam Nisan. Concurrent auctions across the supply chain. *Journal of Artificial Intelligence Research*, 21:595–629, 2004. Preliminary version in *The 3rd ACM Conference on Electronic Commerce*, pages 1-10, 2001.
4. Moshe Babaioff and William E. Walsh. Incentive-compatible, budget-balanced, yet highly efficient auctions for supply chain formation. *Decision Support Systems*, 39:123–149, March 2005. Preliminary version in *The 4th ACM Conference on Electronic Commerce*, pages 64-75, 2003.
5. E. H. Clarke. Multipart pricing of public goods. *Public Choice*, 11:17–33, Fall 1971.
6. Peter Cramton. Applications of auction theory: Ascending auctions. *European Economic Review*, 42:745–756, 1998.
7. Sven de Vries and Rakesh Vohra. Combinatorial auctions: A survey. *INFORMS Journal on Computing*, 15(3):284–309, 2003.

8. Joan Feigenbaum and Scott Shenker. Distributed algorithmic mechanism design: Recent results and future directions. In *6th International Workshop on Discrete Algorithms and Methods for Mobile Computing and Communications*, pages 1–13, 2002.
9. Theodore Groves. Incentives in teams. *Econometrica*, pages 617–631, 1973.
10. Dag Kjenstad. *Coordinated Supply Chain Scheduling*. PhD thesis, Norwegian University of Science and Technology, 1998.
11. Daniel Lehmann, Liadan Ita O'Callaghan, and Yoav Shoham. Truth revelation in approximately efficient combinatorial auctions. *Journal of the ACM*, 49(5):1–26, 2002.
12. Jeffrey K. MacKie-Mason and Hal R. Varian. Generalized Vickrey auctions. Technical report, Dept. of Economics, Univ. of Michigan, July 1994.
13. R. Preston McAfee. A dominant strategy double auction. *Journal of Economic Theory*, 56:434–450, 1992.
14. Roger B. Myerson and Mark A. Satterthwaite. Efficient mechanisms for bilateral trading. *Journal of Economic Theory*, 29:265–281, 1983.
15. David C. Parkes, Jayant Kalagnanam, and Marta Eso. Achieving budget-balance with Vickrey-based payment schemes in exchanges. In *Seventeenth International Joint Conference on Artificial Intelligence*, pages 1161–1168, 2001.
16. David C. Parkes and Lyle H. Ungar. Iterative combinatorial auctions: Theory and practice. In *Seventeenth National Conference on Artificial Intelligence*, pages 74–81, 2000.
17. Norman Sadeh, Raghu Arunachalam, Joakim Eriksson, Niclas Finne, and Sverker Janon. TAC-03: A supply chain trading competition. *AI Magazine*, 24(1):92–93, 2003.
18. Norman M. Sadeh, David W. Hildum, Dag Kjenstad, and Allen Tseng. MASCOT: An agent-based architecture for dynamic supply chain creation and coordination in the internet economy. *Production Planning and Control*, 12(3), 2001.
19. The Trading Agent Competition website. `http://tac.eecs.umich.edu`.
20. William Vickrey. Counterspeculation, auctions, and competitive sealed tenders. *Journal of Finance*, 16:8–37, 1961.
21. William E. Walsh. *Market Protocols for Decentralized Supply Chain Formation*. PhD thesis, University of Michigan, 2001.
22. William E. Walsh and Michael P. Wellman. Decentralized supply chain formation: A market protocol and competitive equilibrium analysis. *Journal of Artificial Intelligence Research*, 19:513–567, 2003.
23. William E. Walsh, Michael P. Wellman, and Fredrik Ygge. Combinatorial auctions for supply chain formation. In *Second ACM Conference on Electronic Commerce*, pages 260–269, 2000.

A Proof of Incentive Compatibility and Individual Rationality of the KSM-UMT-TR Auction

In this section we prove that the KSM-UMT-TR auction is individually rational and incentive compatible. We first present several supporting lemmas and give the final results in Lemma 6 and Lemma 7. Recall that for simplicity of the exposure we assume that there is a single bid-optimal allocation and single TR allocation with respect to any set of reported values.

We prove the lemmas for the more generalized payment rule (3) which we describe for non-UMT supply chains, thus also proving IC and IR for that case. Recall that this payment is $VTR_i(\mathbf{v}) \equiv \mathbf{V}(A^{TR}(VCG_i(\mathbf{v}), \mathbf{v}_{-i})) - \mathbf{V}_{-i}(A^{TR}(\mathbf{v}))$. As we proved in [4], this rule is equivalent to $\max\{PBV_i, VCG_i(\mathbf{v})\}$ for UMT supply chains.

In the following, we denote by $\mathbf{v}^1$ the set of reported values when i bids v_i^1, that is $\mathbf{v}^1 = (v_i^1, \mathbf{v}_{-i})$. Similarly $\mathbf{v}^2 = (v_i^2, \mathbf{v}_{-i})$.

Lemma 2. *If $i \in A^*(\mathbf{v}^1)$ and $i \in A^*(\mathbf{v}^2)$ for some agent i, then $VCG_i(\mathbf{v}^1) = VCG_i(\mathbf{v}^2)$ and $A^*(\mathbf{v}^2) = A^*(\mathbf{v}^1)$.*

Proof. That $VCG_i(\mathbf{v}^1) = VCG_i(\mathbf{v}^2)$ follows directly from the well-known fact that VCG payments are incentive compatible when agents receive the bid-optimal allocation (if $VCG_i(\mathbf{v}^1) \neq VCG_i(\mathbf{v}^2)$ then i could manipulate its payment by changing its bid, in which case VCG payments would not be incentive compatible). It follows then that $A^*(\mathbf{v}^2) = A^*(\mathbf{v}^1)$ because there is only one optimal allocation.

Denote as A_i^* the optimal allocation containing i. When such an allocation exists, we define $VCG_i = \mathbf{V}(A^*(\mathbf{v}_{-i})) - \mathbf{V}_{-i}(A_i^*)$. ($A_i^*$ is uniquely defined by Lemma 2.)

Lemma 3. *If there exists a feasible allocation containing i, agent i is in $A^*(\mathbf{v})$ if $v_i > VCG_i$ and i is not in $A^*(\mathbf{v})$ if $v_i < VCG_i$.*

Proof. If $v_i > VCG_i$ then $v_i > VCG_i = \mathbf{V}(A^*(\mathbf{v}_{-i})) - \mathbf{V}_{-i}(A_i^*)$, giving us $\mathbf{V}(A^*(\mathbf{v})) = v_i + \mathbf{V}_{-i}(A_i^*) > \mathbf{V}(A^*(\mathbf{v}_{-i}))$. Thus A_i^* must be optimal, hence $i \in A^*(\mathbf{v})$ when $v_i > VCG_i$. If $v_i < VCG_i$ then $v_i < \mathbf{V}(A^*(\mathbf{v}_{-i})) - \mathbf{V}_{-i}(A_i^*)$. Thus A_i^* is not optimal, hence $i \notin A^*(\mathbf{v})$ when $v_i < VCG_i$.

Lemma 4. *If $i \in A^{TR}(\mathbf{v}^1)$ and $i \in A^{TR}(\mathbf{v}^2)$ for some agent i, then $VTR_i(\mathbf{v}^1) = VTR_i(\mathbf{v}^2)$ and $A^{TR}(\mathbf{v}^1) = A^{TR}(\mathbf{v}^2)$.*

Proof. Assume, wlog, that $v_i^2 > v_i^1$. By Lemma 2, $VCG_i = VCG_i(\mathbf{v}^1) = VCG_i(\mathbf{v}^2)$, hence $\mathbf{V}(A^{TR}(VCG_i(\mathbf{v}^1), \mathbf{v}_{-i})) = \mathbf{V}(A^{TR}(VCG_i(\mathbf{v}^2), \mathbf{v}_{-i}))$. It remains to show that $\mathbf{V}_{-i}(A^{TR}(\mathbf{v}^2)) = \mathbf{V}_{-i}(A^{TR}(\mathbf{v}^1))$, to prove $VTR_i(\mathbf{v}^1) = VTR_i(\mathbf{v}^2)$.

Since i is in the TR allocation, it is also in the bid optimal allocation. So by Lemma 3, $A^*(\mathbf{v}^2) = A^*(\mathbf{v}^1)$, hence both TR allocations are chosen from the same bid optimal allocation. Then, since $A^{TR}(\mathbf{v}^1)$ clearly satisfies all the auction constraints when i bids v_i^2, and since $A^{TR}(\mathbf{v}^2)$ is the optimal TR allocation when i bids v_i^2

$$\mathbf{V}(A^{TR}(\mathbf{v}^2)) \geq \mathbf{V}^{\mathbf{v}^2}(A^{TR}(\mathbf{v}^1)) = \mathbf{V}(A^{TR}(\mathbf{v}^1)) - v_i^1 + v_i^2.$$

Subtracting v_i^2 from both sides, we have $\mathbf{V}_{-i}(A^{TR}(\mathbf{v}^2)) \geq \mathbf{V}_{-i}(A^{TR}(\mathbf{v}^1))$. Now we need to show that $\mathbf{V}_{-i}(A^{TR}(\mathbf{v}^2)) \leq \mathbf{V}_{-i}(A^{TR}(\mathbf{v}^1))$.

Assume, contrary to which we wish to prove, that $\mathbf{V}_{-i}(A^{TR}(\mathbf{v}^2)) > \mathbf{V}_{-i}(A^{TR}(\mathbf{v}^1))$. If $A^{TR}(\mathbf{v}^2)$ satisfies all the auction constraints when i bids v_i^1 then since $A^{TR}(\mathbf{v}^1)$ is optimal with respect to $\mathbf{v}^1$

$$\mathbf{V}(A^{TR}(\mathbf{v}^1)) \geq \mathbf{V}^{\mathbf{v}^1}(A^{TR}(\mathbf{v}^2)) = \mathbf{V}(A^{TR}(\mathbf{v}^2)) - v_i^2 + v_i^1.$$

Subtracting v_i^1 from both sides gives us $\mathbf{V}_{-i}(A^{TR}(\mathbf{v}^1)) \geq \mathbf{V}_{-i}(A^{TR}(\mathbf{v}^2))$ which is a contradiction. If, on the other hand, $A^{TR}(\mathbf{v}^2)$ does not satisfy all the auction constraints, then the only constraint that could be violated is that $v_j > v_i^1$, where j is the price bounding agent in $K(i) \cup A^{TR}(\mathbf{v}^2)$. Consider allocation $A' = (A^{TR}(\mathbf{v}^2) \setminus \{i\}) \cup \{j\}$. A' satisfies all the auction constraints when i bids v_i^1. So

$$\begin{aligned}\mathbf{V}(A^{TR}(\mathbf{v}^1)) \geq \mathbf{V}^{\mathbf{v}^1}(A') &= \mathbf{V}(A^{TR}(\mathbf{v}^2)) - v_i^2 + v_j \\ &> \mathbf{V}(A^{TR}(\mathbf{v}^2)) - v_i^2 + v_i^1.\end{aligned}$$

Subtracting v_i^1 from both sides gives us $\mathbf{V}_{-i}(A^{TR}(\mathbf{v}^1)) \geq \mathbf{V}_{-i}(A^{TR}(\mathbf{v}^2))$, contradicting our assumption. Thus $\mathbf{V}_{-i}(A^{TR}(\mathbf{v}^2)) = \mathbf{V}_{-i}(A^{TR}(\mathbf{v}^1))$, giving us $VTR_i(\mathbf{v}^1) = VTR_i(\mathbf{v}^2)$. Also, since there is only one TR allocation, we have $A^{TR}(\mathbf{v}^1) = A^{TR}(\mathbf{v}^2)$.

Denote as A_i^{TR} the optimal TR allocation containing agent i. When such an allocation exists, we define $VTR_i = \mathbf{V}(A^{TR}(\mathbf{v}^{VCG_i})) - \mathbf{V}_{-i}(A_i^{TR})$. ($VTR_i$ is uniquely defined by Lemma 4.)

Lemma 5. *If there exists a feasible TR allocation containing i, agent i wins the TR auction if* $v_i > VTR_i$*, and loses the TR auction if* $v_i < VTR_i$.

Proof. First, we establish that $VTR_i \geq VCG_i$. Assume, to the contrary, that $VTR_i < VCG_i$, then $\mathbf{V}(A^{TR}(\mathbf{v}^{VCG_i})) - \mathbf{V}_{-i}(A_i^{TR}) < VCG_i$. Thus $\mathbf{V}(A^{TR}(\mathbf{v}^{VCG_i})) < \mathbf{V}_{-i}(A_i^{TR}) + VCG_i$. Because $A^{TR}(\mathbf{v}^{VCG_i})$ is optimal when i bids VCG_i, we have $\mathbf{V}(A^{TR}(\mathbf{v}^{VCG_i})) \geq \mathbf{V}_{-i}(A_i^{TR}) + VCG_i$, which is a contradiction.

Now, we prove that if $v_i > VTR_i$ then $i \in A^{TR}(\mathbf{v})$. By the above, $v_i > VTR_i \geq VCG_i$, so $i \in A^*(\mathbf{v})$ by Lemma 3. Then, since $v_i > VTR_i = \mathbf{V}(A^{TR}(\mathbf{v}^{VCG_i})) - \mathbf{V}_{-i}(A_i^{TR})$ it follows that $\mathbf{V}(A_i^{TR}) > \mathbf{V}(A^{TR}(\mathbf{v}^{VCG_i}))$. Assume, contrary to which we wish to prove, that i loses the auction, giving us $\mathbf{V}(A_i^{TR}) \leq \mathbf{V}(A^{TR}(\mathbf{v}))$. Since $v_i > VCG_i$, i must also lose with bid VCG_i, hence $\mathbf{V}(A^{TR}(\mathbf{v})) = \mathbf{V}(A^{TR}(\mathbf{v}^{VCG_i}))$. It follows that $\mathbf{V}(A_i^{TR}) \leq \mathbf{V}(A^{TR}(\mathbf{v}^{VCG_i}))$, which is a contradiction.

Finally, we prove that if $v_i < VTR_i$ then $i \notin A^{TR}(\mathbf{v})$. By Lemma 3, if $v_i < VCG_i$ then $i \notin A^*(\mathbf{v})$ and therefore $i \notin A^{TR}(\mathbf{v})$. Now, consider the case where $VCG_i \leq v_i < VTR_i$. Assume, contrary to which we wish to prove, that $i \in A^{TR}(\mathbf{v})$. Then $A^{TR}(\mathbf{v}) = A_i^{TR}$, hence $v_i < VTR_i = \mathbf{V}(A^{TR}(\mathbf{v}^{VCG_i})) - \mathbf{V}_{-i}(A_i^{TR})$, giving us $v_i + \mathbf{V}_{-i}(A_i^{TR}) < \mathbf{V}(A^{TR}(\mathbf{v}^{VCG_i}))$. Also, since $VCG_i \leq v_i$, we have $\mathbf{V}(A^{TR}(\mathbf{v}^{VCG_i})) \leq \mathbf{V}(A^{TR}(\mathbf{v}))$. Therefore

$$\mathbf{V}(A^{TR}(\mathbf{v})) = v_i + \mathbf{V}_{-i}(A_i^{TR}) < \mathbf{V}(A^{TR}(\mathbf{v}^{VCG_i})) \leq \mathbf{V}(A^{TR}(\mathbf{v}))$$

which is a contradiction.

Lemma 6. *The KSM-UMT-TR auction is individually rational.*

Proof. We must prove that an agent receives non-negative utility from bidding truthfully. If i loses, it pays zero and has zero utility. If i wins by bidding truthfully, then by Lemma 5, $\breve{v}_i \geq VTR_i$, hence its utility is $\breve{v}_i - P_i = \breve{v}_i - VTR_i \geq 0$

Lemma 7. *The KSM-UMT-TR auction is incentive compatible in dominant strategies.*

Proof. Consider the case in which agent i wins the auction by bidding its true value. If i bids untruthfully and loses, then it gets zero utility, which by Lemma 6 cannot be better than its utility with a truthful bid. If i bids untruthfully and wins the auction, then by Lemma 4 its payment, and hence it's utility remains the same.

Now consider the case in which i loses the auction by bidding truthfully. Its utility is zero and $\breve{v}_i \leq VTR_i$ by Lemma 5. If i bids untruthfully and loses, its utility remains zero. If i bids untruthfully and wins, its utility is $\breve{v}_i - P_i = \breve{v}_i - VTR_i \leq 0$.

In both cases, we have shown that an agent cannot improve its utility by bidding truthfully, thus proving the lemma.

B Proof of the Efficiency Competitive Ratio and Budget Balance of the KSM-UMT-TR Auction

In this section we prove the efficiency competitive ratio (Theorem 2) and budget balance (Lemma 10) properties of the KSM-UMT-TR auction. Before doing so, we present a number of definitions and lemmas necessary for the proofs.

We denote by $S\{A'\}(m)$ the number of agents in market m in the procurement set $S\{A'\}$ and denote by $Markets(S\{A'\})$ the set of markets $m \in M$ such that $S\{A'\}(m) \neq 0$. We denote by $V(S\{A'\})$ the sum of valuations of all agents in procurement set $S\{A'\}$, so $V(S\{A'\}) = \sum_{i \in S\{A'\}} v_i$.

Definition 4 (procurement set topology). *Two procurement sets $S\{A'\}_1$, $S\{A'\}_2$ of the allocation A' are of the same* ***procurement set topology*** $\mathbf{S}$*, if for every market m, $S\{A'\}_1(m) = S\{A'\}_2(m)$. In this case we write $S\{A'\}_1, S\{A'\}_2 \in \mathbf{S}$. We say that procurement set topology* $\mathbf{S}$ *is* ***in the allocation*** *A' if there exist a procurement set $S\{A'\}$ of the topology* $\mathbf{S}$ *in A'. We denote as $\hat{\mathbf{S}}(\mathbf{A}')$ the set of all procurement set topologies* $\mathbf{S}$ *that are in A'.*

We denote as $\hat{\mathbf{S}}_T$ the set of all procurement set topologies that are possible with supply chain topology T.

We note that procurement sets topologies are independent of the actual allocation, given the topology of the supply chain. For a fixed supply chain topology, the set of procurement set topologies that are possible in that supply chain topology is fixed.

For any procurement set topology $\mathbf{S} \in \hat{\mathbf{S}}_\mathbf{T}$, we denote as $\mathbf{S}(m)$ the number of winners in market m in any procurement set of topology $\mathbf{S}$. $\mathbf{S}(m)$ is well defined. Since a procurement set topology is minimal, there is exactly one consumer market m for which $\mathbf{S}(m) > 0$. For that market $\mathbf{S}(m) = 1$, and we call m the ***consumer market of the procurement set topology*** $\mathbf{S}$.

For any procurement set topology $\mathbf{S} \in \hat{\mathbf{S}}_\mathbf{T}$, let $N^*(\mathbf{S})$ be the maximal number of disjoint procurement sets with the same topology $\mathbf{S}$ in the optimal allocation $\breve{A}$. For any procurement set topology $\mathbf{S} \in \hat{\mathbf{S}}(\breve{\mathbf{A}})$, we have $N^*(\mathbf{S}) > 0$, and for any topology $\mathbf{S} \notin \hat{\mathbf{S}}(\breve{\mathbf{A}})$ we have $N^*(\mathbf{S}) = 0$. Similarly, let $N^{TR}(\mathbf{S})$ be the maximal number of disjoint procurement sets with the same topology $\mathbf{S}$ in the Trade Reduction allocation.

For any market m we define $T_m(\breve{A})$ to be the number of winning agents (trade size) in market m in the efficient allocation $\breve{A}$, and $R_m(A^{TR})$ to be the number of reduced agents in market m (winning agents in $\breve{A}$ that are losers in the Trade Reduction allocation in market m). The size of trade in market m in the Trade Reduction allocation is therefore $T_m(A^{TR}) = T_m(\breve{A}) - R_m(A^{TR})$.

We denote as CM^* the set of consumer markets with non zero trade size in the efficient allocation.

Lemma 8. *Let $\breve{\mathbf{v}}$ be any vector of agent values. Let $\hat{\mathbf{S}}(\breve{A})$ be the set of procurement set topologies in the efficient allocation $\breve{A}$ for $\breve{\mathbf{v}}$. Then:*

1. *For any consumer market $m \in CM^*$, there is only one procurement set topology in $\hat{\mathbf{S}}(\breve{A})$ for which m is the consumer market. So we can mark the consumer markets by m_i and the procurement set topologies by $\mathbf{S}^i$ for $i = 1, \ldots, |CM^*|$.*
2. *For each procurement set topology $\mathbf{S}^i$, $N^*(\mathbf{S}^i) = N^{TR}(\mathbf{S}^i) + 1$.*
3. *For each procurement set topology $\mathbf{S}^i$ and its consumer market m_i, $N^*(\mathbf{S}^i) = T_{m_i}(\breve{A})$ and $N^{TR}(\mathbf{S}^i) = T_{m_i}(A^{TR}) = T_{m_i}(\breve{A}) - 1$.*
4. *For each market m, the trade sizes in market m in the optimal and TR allocations are:*

$$T_m(\breve{A}) = \sum_{i=1}^{|CM^*|} \mathbf{S}^i(m) T_{m_i}(\breve{A}) \quad and \quad T_m(A^{TR}) = \sum_{i=1}^{|CM^*|} \mathbf{S}^i(m)(T_{m_i}(\breve{A}) - 1)$$

5. *The allocations $\breve{A}$ and A^{TR} can be partitioned to procurement sets in the following way.*

$$\breve{A} = \bigcup_{i=1}^{|CM^*|} \bigcup_{j=1}^{N^*(\mathbf{S}^i)} S^i_j \quad and \quad A^{TR} = \bigcup_{i=1}^{|CM^*|} \bigcup_{j=1}^{N^*(\mathbf{S}^i)-1} S^i_j$$

where S^i_j is the j procurement set of topology $\mathbf{S}^i$, $\bigcup_{i=1}^{|CM^*|} S^i_{N^*(\mathbf{S}^i)}$ are the reduced agents and $V(S^i_{N^*(\mathbf{S}^i)}) \leq V(S^i_j)$ for all $j = 1, \ldots, N^*(\mathbf{S}^i)$.

Proof. Proof of (1). The unique manufacturing technologies property of the supply chain causes that there is only one procurement set topology that has an agent in market m as we show below. The proof is by induction on the markets in reverse topological order. If a procurement set topology has one agent in consumer market m, it uniquely defines the number of goods that are needed to satisfy this agent, and since a single market produces each good, it uniquely sets the number of winners in each of those markets. This can be continued till the number of agents in each market is uniquely specified, hence there is only one procurement set topology for which m is the consumer market. Hence we have proven (1).

Proof of (2). Let $\mathbf{S} \in \hat{\mathbf{S}}(\breve{\mathbf{A}})$ be the unique procurement set topology of any consumer market $m \in CM^*$. We now show that $N^*(\mathbf{S}) = N^{TR}(\mathbf{S}) + 1$. From the above we conclude that in order to reduce one agent in market m and maintain feasibility, a procurement set of topology $\mathbf{S}$ must be reduced in the Trade Reduction allocation, therefore $N^*(\mathbf{S}) \geq N^{TR}(\mathbf{S}) + 1$.

It remains to prove that $N^*(\mathbf{S}) \leq N^{TR}(\mathbf{S}) + 1$. Assume that this is not true, then for some $\mathbf{S} \in \hat{\mathbf{S}}(\breve{\mathbf{A}})$, $N^*(\mathbf{S}) > N^{TR}(\mathbf{S}) + 1$, or equivalently $N^*(\mathbf{S}) \geq N^{TR}(\mathbf{S}) + 2$. This means that at least two procurement sets of the same topology $\mathbf{S}$ are reduced. Below we show that for any procurement set S of topology $\mathbf{S} \in \hat{\mathbf{S}}(\breve{\mathbf{A}})$, we have $V(S) > 0$. Since $V(S) > 0$ for any such procurement set S, we can add one of the reduced procurement sets of topology $\mathbf{S}$ and increase the value of the allocation while ensuring that every market with trade has at least one reduced agent. This is true since we still have one procurement set of topology **S** reduced and both procurement sets share the same set of markets. Thus, we can add one of the reduced procurement sets while maintaining the constraints on a TR allocation, contradicting the requirement that KSM-UMT-TR maximizes the allocation value, subject to the constraints. Therefore the assumption is not true, and we have proven that $N^*(\mathbf{S}) \leq N^{TR}(\mathbf{S})+1$ and therefore $N^*(\mathbf{S}) = N^{TR}(\mathbf{S})+1$ and we have proven (2).

We now show that for any procurement set S of topology $\mathbf{S} \in \hat{\mathbf{S}}(\breve{\mathbf{A}})$, we have $V(S) > 0$ from the point of view of the KSM-UMT-TR auction. To see this, assume to the contrary that $V(S) < 0$. Then, by removing this procurement set we increase the efficient allocation value, which is a contradiction. We can ensure that $V(S) = 0$ by modifying the bids slightly, as described in [4]. There, it is assumed that the bids are integral, and for each agent with unique integer index i, the value 2^{-i} is added to the agent's bid.

Proof of (3). Any procurement set topology **S** contains a single agent in its consumer market m, and by feasibility of the allocation, all inputs to each of the consumers must be manufactured, therefore $N^*(\mathbf{S}) = T_m(\breve{A})$ and $N^{TR}(\mathbf{S}) = T_m(A^{TR})$. Since we have shown that $N^*(\mathbf{S}) = N^{TR}(\mathbf{S})+1$, then $N^{TR}(\mathbf{S}) = N^*(\mathbf{S})-1 = T_m(\breve{A})-1 = T_m(A^{TR})$ and we have proven (3).

Proof of (4). The trade size $T_m(\breve{A})$ for each market m is as stated in the lemma from the following observation. Feasibility of the allocation $\breve{A}$ dictates that if the trade size in each consumer market m_i is $T_{m_i}(\breve{A})$, then for each market m a set of agents of size $\mathbf{S}^i(m)T_{m_i}(\breve{A})$ must be in the allocation (and these sets must be disjoint since each producer manufacturers a single item of one good). The trade size in market m is the sum of these quantities over all consumer markets. A similar argument works for the TR trade sizes. We have proven (4).

Proof of (5). From all the above we conclude that

$$\breve{A} = \bigcup_{i=1}^{|CM^*|} \bigcup_{j=1}^{N^*(\mathbf{S}^i)} S^i_j \quad and \quad A^{TR} = \bigcup_{i=1}^{|CM^*|} \bigcup_{j=1}^{N^*(\mathbf{S}^i)-1} S^i_j$$

where S^i_j is the j procurement set of topology $\mathbf{S}^i$. $\bigcup_{i=1}^{|CM^*|} S^i_{N^*(\mathbf{S}^i)}$ are the reduced agents and $V(S^i_{N^*(\mathbf{S}^i)}) \leq V(S^i_j)$ for all $j = 1, \ldots, N^*(\mathbf{S}^i)$. This is since the reduced agents are the lowest value agents in each market, so any reduced procurement set has a lower value (as sum of the agents value) then any procurement set of the same topology which is in the optimal allocation. We have seen that one procurement set of each consumer market must be reduced, and therefore all the reduced agents can be partitioned to one procurement set of each consumer market. We have proven (5).

Lemma 9. *For any winning agent i, $VTR_i \geq PBV_i$.*

Proof. Assume, to the contrary, that $VTR_i < PBV_i$. If i bids any value v_i such that $VTR_i < v_i$, then i wins the auction by Lemma 5. In particular, i wins if it bids $VTR_i < v_i < PBV_i$. But by the definition of PBV_i, $v_i \geq PBV_i$, which is a contradiction.

Lemma 10. *Let $\breve{\mathbf{v}}$ be any vector of agent values. The KSM-UMT-TR allocation for $\breve{\mathbf{v}}$ with KSM-UMT-TR payments is budget balanced.*

Proof. We denote by $V(S^i_j)$ the sum of valuations of all agents in procurement set S^i_j, and by $P(S^i_j)$ the sum of payments from all those agents. We denote by $Pay(A^{TR})$ the sum of payments of all agents in the Trade Reduction allocation.

We must show that $Pay(A^{TR}) \geq 0$ to prove that A^{TR} with KSM-UMT-TR payments is budget-balanced. By Lemma 8 the Trade Reduction allocation is

$$A^{TR} = \bigcup_{i=1}^{|CM^*|} \bigcup_{j=1}^{N^*(\mathbf{S}^i)-1} S^i_j$$

therefore $Pay(A^{TR}) = \sum_{i=1}^{|CM^*|} \sum_{j=1}^{N^*(\mathbf{S}^i)-1} P(S^i_j)$. So it is sufficient to show that $P(S^i_j) \geq 0$ for $i = 1, \ldots, |CM^*|$ and $j = 1, \ldots, N^*(\mathbf{S}^i) - 1$, to prove the lemma.

We can build a one-to-one mapping of agents from procurement set S^i_j to agents from procurement set $S^i_{N^*(\mathbf{S}^i)}$, since both procurement sets are of the same topology and have the same number of agents in each market.

Since $S^i_{N^*(\mathbf{S}^i)}$ is in the efficient allocation, it must be that $V(S^i_{N^*(\mathbf{S}^i)}) \geq 0$, else it could be removed from the efficient allocation to get a better allocation, which is a contradiction.

By Lemma 9 the payment P_k from each agent k in S^i_j is at least as high as the PBV_k. This agent has the highest value of all reduced agents in k's market. In particular PBV_k is higher then the valuation of the agent that agent k is mapped to in $S^i_{N^*(\mathbf{S}^i)}$. Hence, by summing over all agents in the procurement set, we conclude that $P(S^i_j) \geq \sum_{k \in S^i_j} PBV_k \geq V(S^i_{N^*(\mathbf{S}^i)}) \geq 0$, which is what we wanted to prove.

We need some additional definitions to carry on with our proofs.

Definition 5 (allocation partition). *An* ***allocation partition*** $P^{A'}$ *of a feasible allocation* A' *is a partition* $P^{A'}_1, P^{A'}_2, \ldots, P^{A'}_k$ *of the agents in* A'*. The* ***size of the partition*** *is* k*. For any set* $P^{A'}_i$*, the* ***value****,* $V(P^{A'}_i)$*, of the set is* $\sum_{i \in P^{A'}_i} \breve{v}_i$

We call an allocation A^{FTR} a ***feasible reduction allocation*** if it satisfies all constraints for a Trade Reduction allocation, except that it possibly does not maximize value.

Definition 6 (good partition pair). *Given vector of agent values* $\breve{\mathbf{v}}$*, with efficient allocation* $\breve{A}$ *and a feasible reduction allocation* A^{FTR}*, we say that the allocations have a* ***good partition pair*** P^*, P^{tr} *if there exists a partition* P^* *for the efficient allocation* $\breve{A}$ *of size* k*, and a partition* P^{tr} *for a feasible reduction allocation* A^{FTR} *of size* k*, such that for any* $i = 1, \ldots, k$*:*

- $V(P^*_i) \geq V(P^{tr}_i) \geq 0.$
- $\frac{V(P^{tr}_i)}{V(P^*_i)} \geq \min_{\mathbf{S} \in \hat{\mathbf{S}}(\breve{\mathbf{A}})} \frac{N^{TR}(\mathbf{S})}{N^*(\mathbf{S})}.$

For valuations with a good partition pair we can bound the efficiency of KSM-UMT-TR in the following way:

Lemma 11. *Given vector of agent values* $\check{\mathbf{v}}$*, with non-empty efficient allocation* $\check{A}$ *which has a good partition pair* P^*, P^{tr}*, we have:*

$$Eff^{KSM-UMT-TR}(\check{\mathbf{v}}) = \frac{V(A^{TR})}{V(\check{A})} \geq \min_{\mathbf{S}\in\hat{\mathbf{S}}(\check{\mathbf{A}})} \frac{N^{TR}(\mathbf{S})}{N^*(\mathbf{S})}$$

Proof. Let the good partition pair P^*, P^{tr} have size k. Since P^* is a partition of the efficient allocation, $V(\check{A}) = \sum_{i=1}^{k} V(P_i^*)$, and since P^{tr} is a partition of a feasible reduction allocation, $V(A^{TR}) \geq V(A^{FTR}) = \sum_{i=1}^{k} V(P_i^{tr})$. Therefore

$$Eff^{KSM-UMT-TR}(\check{\mathbf{v}}) = \frac{V(A^{TR})}{V(\check{A})} \geq \frac{\sum_{i=1}^{k} V(P_i^{tr})}{\sum_{i=1}^{k} V(P_i^*)}.$$

Since P^*, P^{tr} is a good partition pair, for every $i = 1,\ldots,k$ it is true that $V(P_i^*) \geq V(P_i^{tr}) \geq 0$. Therefore, we can apply Lemma 12 to get

$$Eff^{KSM-UMT-TR}(\check{\mathbf{v}}) \geq \frac{\sum_{i=1}^{k} V(P_i^{tr})}{\sum_{i=1}^{k} V(P_i^*)} \geq \min_{i=1}^{k} \frac{V(P_i^{tr})}{V(P_i^*)}$$

Since P^*, P^{tr} is a good partition pair, for every $i = 1,\ldots,k$ it is true that $\frac{V(P_i^{tr})}{V(P_i^*)} \geq \min_{\mathbf{S}\in\hat{\mathbf{S}}(\check{\mathbf{A}})} \frac{N^{TR}(\mathbf{S})}{N^*(\mathbf{S})}$, therefore

$$Eff^{KSM-UMT-TR}(\check{\mathbf{v}}) \geq \min_{i=1}^{k} \frac{V(P_i^{tr})}{V(P_i^*)} \geq \min_{i=1}^{k} \left(\min_{\mathbf{S}\in\hat{\mathbf{S}}(\check{\mathbf{A}})} \frac{N^{TR}(\mathbf{S})}{N^*(\mathbf{S})} \right)$$
$$= \min_{\mathbf{S}\in\hat{\mathbf{S}}(\check{\mathbf{A}})} \frac{N^{TR}(\mathbf{S})}{N^*(\mathbf{S})}$$

Lemma 12. *For any set of indexes m and pairs* R_m *and* O_m *such that* $0 \leq R_m \leq O_m$ *it is true that*

$$\frac{\sum_m R_m}{\sum_m O_m} \geq \min_m \left(\frac{R_m}{O_m} \right)$$

Proof. Let k be the index of elements that minimize the ratio $\frac{R_m}{O_m}$. For every m $\frac{R_m}{O_m} \geq \frac{R_k}{O_k}$, therefore for every m, $O_k * R_m \geq R_k * O_m$. Summing over m we get $O_k * (\sum_m R_m) \geq R_k * (\sum_m O_m)$. Hence, $\frac{\sum_m R_m}{\sum_m O_m} \geq \frac{R_k}{O_k} = \min_m \frac{R_m}{O_m}$, which is what we wanted to prove.

From Lemma 11 we conclude that, if the efficient allocation has a good partition pair and there is no procurement set topology with a single procurement set of this topology in the efficient allocation, then we get a competitive ratio of at least $1/2$.

Lemma 13. *Let* $\check{\mathbf{v}}$ *be any vector of agents values with efficient allocation* $\check{A}$*.* $\check{A}$ *has a good partition pair.*

Proof. By Lemma 8 the efficient allocation is constructed from procurement set topologies $\mathbf{S}^i$ for $i = 1,\ldots,|CM^*|$ such that

$$\check{A} = \bigcup_{i=1}^{|CM^*|} \bigcup_{j=1}^{N^*(\mathbf{S}^i)} S_j^i \text{ and } A^{TR} = \bigcup_{i=1}^{|CM^*|} \bigcup_{j=1}^{N^*(\mathbf{S}^i)-1} S_j^i$$

where S_j^i is the j procurement set of topology $\mathbf{S}^i$, and $S_{N^*(\mathbf{S}^i)}^i$ is the lowest valuation agents of all agents in procurement sets of topology $\mathbf{S}^i$. Observe that there are $N^{TR}(\mathbf{S}^i) = N^*(\mathbf{S}^i) - 1$ procurement sets of the i topology in the Trade Reduction allocation A^{TR}.

Let $P_i^* = \bigcup_{j=1}^{N^*(\mathbf{S}^i)} S_j^i$, and let $P_i^{tr} = \bigcup_{j=1}^{N^*(\mathbf{S}^i)-1} S_j^i$. We need to show that the two requirements for a good partition pair hold.

- $V(P_i^*) \geq V(P_i^{tr}) \geq 0$: Since $S_{N^*(\mathbf{S}^i)}^i$ is a procurement set, it has a non-negative value. Hence, $V(P_i^*) - V(P_i^{tr}) = V(S_{N^*(\mathbf{S}^i)}^i) \geq 0$. Since every procurement set has non-negative value, we also have $V(P_i^{tr}) \geq 0$.
- $\frac{V(P_i^{tr})}{V(P_i^*)} \geq \min_{\mathbf{S}\in\hat{\mathbf{S}}(\breve{\mathbf{A}})} \frac{N^{TR}(\mathbf{S})}{N^*(\mathbf{S})}$: First observe that $\frac{V(P_i^{tr})}{V(P_i^*)} = \frac{\sum_{j=1}^{N^*(\mathbf{S}^i)-1} V(S_j^i)}{\sum_{j=1}^{N^*(\mathbf{S}^i)} V(S_j^i)}$. Hence, by applying Lemma 14 we get

$$\frac{V(P_i^{tr})}{V(P_i^*)} \geq \frac{N^*(\mathbf{S}^i)-1}{N^*(\mathbf{S}^i)} = \frac{N^{TR}(\mathbf{S}^i)}{N^*(\mathbf{S}^i)} \geq \min_{\mathbf{S}\in\hat{\mathbf{S}}(\breve{\mathbf{A}})} \frac{N^{TR}(\mathbf{S})}{N^*(\mathbf{S})}.$$

Lemma 14. *Let $n \in \mathbf{Z}^+$, $m \in \{1,\dots,n\}$, and $X_i \in \mathbf{R}^+$ for all $i \in \{1,\dots,n\}$. If $X_i \geq X_m$ for all $i < m$ and $X_i \leq X_m$ for all $i > m$, then*

$$\frac{\sum_{i=1}^m X_i}{\sum_{i=1}^n X_i} \geq \frac{m}{n}$$

Proof. The proof is by induction on n for any fixed m. For any n such that $n \geq m$ we prove the claim by induction on n.

If $n = m$ the claim is true since we have 1 on both sides of the inequality. Now assume that we have proven the claim for some n_0 such that $n_0 \geq m$, to prove the claim for $n_0 + 1$. By the induction hypothesis,

$$\frac{\sum_{i=1}^m X_i}{\sum_{i=1}^{n_0} X_i} \geq \frac{m}{n_0},$$

hence $n_0 \sum_{i=1}^m X_i \geq m \sum_{i=1}^{n_0} X_i$.

Since $X_i \geq X_m \geq X_{n_0+1}$ for all $i \leq m$, we have $\sum_{i=1}^m X_i \geq m X_{n_0+1}$. Using the induction hypothesis we get by summation

$$n_0 \sum_{i=1}^m X_i + \sum_{i=1}^m X_i \geq m \sum_{i=1}^{n_0} X_i + m X_{n_0+1}$$

therefore

$$\frac{\sum_{i=1}^m X_i}{\sum_{i=1}^{n_0+1} X_i} = \frac{\sum_{i=1}^m X_i}{\sum_{i=1}^{n_0} X_i + X_{n_0+1}} \geq \frac{m}{n_0+1}$$

which is what we wanted to prove.

Finally, we are ready to prove the theorem.

Theorem 2 *Let $\check{\mathbf{v}}$ be any vector of agents values. The following is an efficiency competitive ratio function for the KSM-UMT-TR auction:*

$$Ratio^{KSM-UMT-TR}(\check{\mathbf{v}}) = \min_{m \in CM^*} \frac{T_m(\breve{A}) - 1}{T_m(\breve{A})}$$

if $\breve{A} \neq \emptyset$ and

$$Ratio^{KSM-UMT-TR}(\check{\mathbf{v}}) = 1$$

if $\breve{A} = \emptyset$.

Proof. The second component of the competitive ratio is true by definition, hence we prove the first component. By Lemma 13 $\check{\mathbf{v}}$ has a good partition pair. By applying Lemma 11, the efficiency of KSM-UMT-TR satisfies the following:

$$Eff^{KSM-UMT-TR}(\check{\mathbf{v}}) = \frac{V(A^{TR})}{V(\breve{A})} \geq \min_{\mathbf{S} \in \hat{\mathbf{S}}(\breve{\mathbf{A}})} \frac{N^{TR}(\mathbf{S})}{N^*(\mathbf{S})}$$

From Lemma 8 we know that there is a one-to-one mapping of procurement set topologies in the efficient allocation to consumer markets with non-zero trade. If procurement set topology $\mathbf{S}$ is mapped to market m, then $N^{TR}(\mathbf{S}) = T_m(A^{TR}) = T_m(\breve{A}) - 1$ and $N^*(\mathbf{S}) = T_m(\breve{A})$.

So we conclude that

$$Eff^{KSM-UMT-TR}(\check{\mathbf{v}}) = \frac{V(A^{TR})}{V(\breve{A})} \geq \min_{\mathbf{S} \in \hat{\mathbf{S}}(\breve{\mathbf{A}})} \frac{N^{TR}(\mathbf{S})}{N^*(\mathbf{S})} = \min_{m \in CM^*} \frac{T_m(\breve{A}) - 1}{T_m(\breve{A})}.$$

Therefore

$$Ratio^{KSM-UMT-TR}(\check{\mathbf{v}}) = \min_{m \in CM^*} \frac{T_m(\breve{A}) - 1}{T_m(\breve{A})} \leq Eff^{\text{KSM-UMT-TR}}(\check{\mathbf{v}}),$$

which is what we wanted to prove.

Supply Chain Coordination by Means of Automated Negotiations Between Autonomous Agents

Andreas Fink

Helmut-Schmidt-University Hamburg,
Department of Economics and Organisational Sciences, Chair of Information Systems,
Holstenhofweg 85, 22043 Hamburg, Germany,
andreas.fink@hsu-hamburg.de, http://ifi.hsu-hh.de/fink

Abstract. This chapter is concerned with coordination problems between firms which collaborate within supply chains. We consider coordination problems at the level of operations management and assume that potential decisions of the involved decision making units can be completely specified as formal contracts. The elements of the contract space are common knowledge, whereas information asymmetry prevails regarding the preferences of the involved decision making units. We investigate an automated negotiation mechanism, with firms or decision making units represented as software agents, for determining mutually beneficial contracts. Taking into account asymmetric information and opportunistic behavior, a mediator agent supports the negotiation process by randomly generating new candidate contracts. In order to approach the Pareto frontier as far as possible we propose a verifiable mechanism which forces the involved agents to behave cooperatively to some extent. The negotiation approach is analyzed by computational experiments for bilateral as well as multilateral supply chain sequencing scenarios. The results show that automated negotiations may provide an effective and efficient mechanism for coordinating decisions within supply chains.

1 Introduction

Research on solving coordination problems within supply chain management is often set up under the paradigm of centralized and hierarchical planning; see, e. g., de Kok and Graves (2003), Erengüc et al. (1999), Sarmiento and Nagi (1999), Shapiro (2001), Stadtler and Kilger (2004), and Thomas and Griffin (1996). However, the coordination of logistics processes (and the related informational and financial flows) between different organizational units within a firm as well as encompassing more than one firm typically involves several autonomous and self-interested decision making units. On the one hand, related research has mainly been dealing with strategic and tactical problems in connection with stochastic models. In particular, research has been studying incentive policies (such as price discounts or penalty costs) which may be set up in specific

A. Fink: *Supply Chain Coordination by Means of Automated Negotiations Between Autonomous Agents*, Studies in Computational Intelligence (SCI) **28**, 351–372 (2006)
www.springerlink.com

supply contracts in order to direct the behavior of the involved decision making units in such a way that the performance of some hypothetical centralized planning is approached; see, e. g., Cachon (2003), Chen et al. (2001), Klastorin et al. (2002), Moses and Seshadri (2000), and Tsay et al. (1999). On the other hand, there seems to be a lack of research that deals with short-term coordination problems with deterministic data (i. e., operational issues in deterministic environments) in connection with autonomous decision making units and information asymmetry; see, e. g., Thomas and Griffin (1996) and Hall and Potts (2003).

This chapter investigates coordination mechanisms for decision problems with autonomous decision making units (agents). In particular, we consider coordination problems at the level of operations management and assume that potential decisions and corresponding forms of cooperation among decision making units can be completely specified as formal contracts. That is, contracts represent common plans concerning inter-agent decision variables. This leads to contract search spaces and necessitates mechanisms for determining mutually beneficial contracts that all involved decision making units agree on. The contribution of this chapter is based on a discussion of feasible decision making procedures and related criteria for the design of coordination mechanisms and the assessment of the resulting outcome. We describe, extend, and analyze a generic automated negotiation scheme that provides such a mechanism. The remainder of this chapter is organized as follows: In the subsequent section, we discuss the characteristics of the considered coordination problems and state respective assumptions and goals. In Sect. 3, we describe a generic automated negotiation procedure for the coordination of autonomous decision making units. This approach is evaluated by computational experiments for exemplary applications within the domain of supply chain management, which includes bilateral as well as multilateral coordination scenarios (Sect. 4). We end up with a summary and discuss requirements for further research.

2 The Coordination Problem

We consider the coordination of autonomous agents whose decisions and resulting actions are interdependent. We assume that the agents have gained agreement about the concurrent decision space in the form of a formal contract space. (Achieving such an understanding may itself be regarded as a higher level coordination problem, which is not in the scope of this chapter.) The restriction to coordination problems on the level of operations management makes it reasonable to presuppose deterministic data about the actual elements of the decision space, notwithstanding private preference information. While the agents generally pursue different objectives, we do not suppose any exogenous imbalance of power that might enable some agent to directly impose decisions on other agents. We aim for a generic coordination mechanism with regard to any contract space that fulfills some minimum requirements as discussed below. The instantiation of such a generic coordination mechanism for a specific contract space leads to a specific coordination mechanism for a specific coordination problem.

In order to illustrate the considered kind of coordination problem and corresponding contract spaces we give three examples: First, if production facilities of different firms are tightly coupled in connection with just-in-time manufacturing, the contract space may consist of the set of feasible production (delivery) schedules at the coupling point. Second, a haulage contractor which serves different firms may need to coordinate pickup and delivery transactions with various clients; the corresponding contract space is made up of the options for vehicle routing and resulting pickup and delivery points in time. Third, firms which collaborate within a construction project may pursue different objectives such as the maximization of the net present value of respective subsets of project activities; the resulting contract space includes all feasible project schedules. We assume that it is individually rational for an agent to cooperate with other agents in the considered coordination scenarios. That is, resulting contracts must be beneficial for individual agents since only collaborations where participation is generally useful for all involved agents are viable in the long run.

Interdependencies between the decisions of the involved agents generally result in potential gains from coordinated decision making. For example, a firm may be able to bring forward some piece of production activity at no or minor costs, which might enable a haulage contractor and/or a manufacturer at the next stage to lower their costs. There has been recent research on the theoretical benefits of coordinated decision making for specific types of coordination problems; see, e. g., Agnetis et al. (2001, 2004, 2005), Dawande et al. (2005), Hall and Potts (2003), and Lee and Chen (2001). Such win-win opportunities are put at risk if no sensibly-designed coordination mechanism is available.

While we assume common knowledge about the contract space, information asymmetry prevails regarding the individual quality (utility) of a contract for the involved decision making units (which thus can be conceived as black boxes). In general, there is a conflict of interest. From the viewpoint of methodological individualism the purposely rational decision maker selfishly seeks for maximizing its utility. In pursuing their interests agents may behave insincerely. Thus it does not appear appropriate to simply presuppose that autonomous agents provide truthful information. In addition, agents may aim to reveal as little information as possible about their preferences to give others no arguments to skim potential gains. Consequently, benevolent collaborative planning cannot be taken for granted.

This leads to the following formal model of the coordination problem: There is a set of agents $a_1, \ldots, a_m$. These agents face a concurrent decision space (concerning inter-agent decision variables), which may be represented as a formal contract space $C = \{c \,|\, c \text{ is a feasible contract}\}$. While common knowledge about the elements of the contract space is assumed, each agent $i = 1, \ldots, m$ has private information about its preferences, which constitute complete transitive preference relations on C. On the assumption of measurable preference value differences, preferences can be represented by private utility functions $f_i : C \rightarrow \mathbb{R}$ which are unique up to positive affine transformations; see, e. g., French (1988).

We seek for a generic coordination mechanism that eventually leads to a mutually beneficial contract that all involved decision making units agree on. In general, no contract may be imposed on an agent without its approval. Two criteria appear fundamental for the acceptance of coordination mechanisms and resulting contracts by the involved agents, namely, some notion of *fairness* of the decision making procedure and some measure of the *quality* of the outcome. We note that fairness has not necessarily to be regarded as a normative concept in its own right, but may constitute a prerequisite for the approval of the coordination mechanism by the involved agents. Drawing on ideas of decision theory (French (1988)) and cooperative game theory (Nash (1950, 1953)), some basic axiomatic properties are regarded as desirable for the design of a coordination mechanism, in particular:

- *Invariance to equivalent utility representations:* The outcome of the coordination mechanism should be invariant to positive affine transformations of the utility functions $f_i'(c) = af_i(c) + b$ with $a > 0$, since the numerical representation of preferences is arbitrary concerning scale and origin.
- *Symmetry:* The agents (or their labeling) should be interchangeable so that that all properties that are relevant for the outcome are captured in the individual preferences.
- *Pareto optimality:* With respect to the resulting contract, there should be no other contract that improves the utility of some agent without deteriorating the utility of any other agent.

The former two properties constitute essential requirements concerning fairness, which may entail, to some extent, an implicit selection among the hypothetical set of Pareto optimal contracts. The latter property is a central criterion for the quality of the outcome, which may be assessed by measuring the deviation from the Pareto frontier (the set of Pareto optimal solutions). We note that in general the Pareto frontier is not known because of, first, private preference information and, second, the complexity of contract spaces. Thus any mechanism that would rely on such knowledge is infeasible.

Further criteria for the assessment of coordination mechanisms are computational tractability, simplicity, and low requirements for information revelation (Sandholm (1999)). That is, all involved agents should be enabled to develop and implement a reasonable strategy without the need to expose individual utility functions. Concerning strategy design and stability, a coordination mechanism preferably comes with a unique Nash equilibrium regarding the strategies of the involved agents. Ideally, the mechanism induces incentive compatibility, i. e., an agent should not be able to profit from untruthful utterances provided that this conforms to the rules of the coordination mechanism.

Within the domain of supply chain management, recent research has been concerned with the design and application of negotiation-based procedures with side-payments for different types of problems; see, e. g., Dudek and Stadtler (2005) and Sucky (2005). Within such approaches the involved agents may make alternating offers concerning changes to common plans in connection with compensation schemes which include side payments. This raises the crucial problem of inducing incentive-compatible behavior. In particular, one might have to assume some level of global information about individual preferences and/or to assume, or build up, some level of trust between autonomous decision making units which might be expected to collaborate benevolently to some extent. However, to implicitly presuppose some level of sincere and cooperative behavior would beg the main question that is addressed in this chapter. Furthermore, we strive for a generic mechanism which does not rely on interpersonal comparisons of utility. This is also motivated by the concern that reasonable monetary evaluations of decision alternatives may not be available, as decision making units often follow subordinate performance indicators and related operational objectives (such as meeting given due dates or minimizing production flow times).

To sum up the discussion: We seek for a generic coordination mechanism that serves for determining a contract out of a complex formal contract space, under consideration of selfish decision making units with strictly private information about their preferences. As one cannot presuppose that the agents provide truthful information, sincere behavior should be entailed by the design of the coordination mechanism. This mechanism should be fair; in particular, it should provide an outcome that is unaffected by positive affine transformations of utility functions and any aspects outside the preference relations. We do not assume any measure of social welfare but aim for approaching the Pareto frontier as far as possible (i. e., minimization of some measure of the deviation from the Pareto frontier). Requirements for information revelation should be kept to a minimum. Furthermore, the coordination mechanism should be computationally tractable and simple. It should be generic, i. e., it should be possible to apply it to any kind of coordination problem that can be represented on the basis of formal contract spaces. The instantiation of the generic coordination mechanism to specific coordination problems should be straightforward and robust in the meaning that its effectiveness should not depend crucially on the particular design of problem-specific adaptations.

3 The Negotiation Protocol

Heuristic search methods have been proven as an effective means for determining high-quality solutions out of complex search spaces that are related to $\mathcal{NP}$-hard single-objective optimization problems. This includes metaheuristics such as iterated local search, simulated annealing, tabu search, and genetic/evolutionary algorithms; see, e. g., Burke and Kendall (2006) and Glover and Kochenberger (2003). Metaheuristics have also been extended for multi-objective problems with information symmetry; see, e. g., Gandibleux et al. (2005), Jaszkiewicz (2004), and Ulungu et al. (1999). Klein et al. (2003a,b) describe novel ideas for adapting heuristic search strategies for coordination problems with two autonomous decision making units. In this concept each iteration of the search process is modeled as an elementary negotiation among the decision making units about which step to take. This results in an iterated negotiation process in the form of a concurrent search in a complex contract search space. Because of the size of the contract space automation of negotiation-based procedures appears indispensable. Complete automation can be achieved on the basis of representing decision making units as autonomous software agents, which adhere to a formal negotiation protocol on the basis of initial instructions from their respective principals. Resulting negotiation processes constitute a system of interacting agents (multiagent system); see, e. g., D'Inverno and Luck (2004), Jennings et al. (2001), Kraus (2001), and Rosenschein and Zlotkin (1994).

Aiming for a mechanism and corresponding incentives for generating mutually beneficial outcomes, we propose the application of an integrative negotiation procedure, in the role of a collaborative problem solving mechanism, which supports the exploitation of win-win opportunities under consideration of complex contract spaces and selfish decision making units. We build on the approach presented in Klein et al. (2003a,b) and Fink (2004).

A negotiation protocol sets the rules of a negotiation process. This includes the definition of feasible interaction sequences in connection with the rights and obligations of the involved agents. The multilateral negotiation protocol examined in this chapter is described in Fig. 1. Starting from a random initial contract that serves as the first active contract (on which all agents agree on), a mediator agent generates in each round a candidate contract which is evaluated, and then accepted or rejected, by each agent. The candidate contract becomes the new active contract that the agents agree on after the particular round if and only if all agents have signaled acceptance. The contract eventually agreed on when the termination criterion is fulfilled, e. g. after R rounds, constitutes the final outcome of the negotiation process.

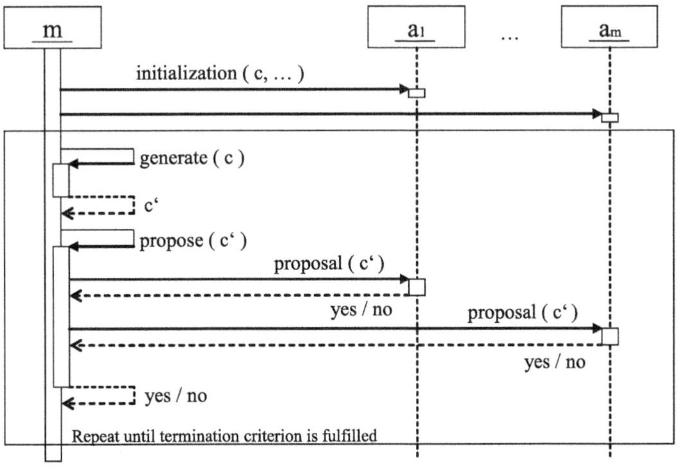

Fig. 1. General negotiation protocol

This negotiation protocol is generic concerning the contract space that represents the actual coordination problem; hence the adaptation for some specific contract space partly involves problem-specific definitions. The overall effectiveness of the negotiation protocol depends on two essential variation points, which are discussed in the subsequent sections. First, there is the issue of generating new candidate contracts. Second, we argue about reasonable acceptance criteria for the agents and conditions that may be imposed on the corresponding acceptance decisions.

3.1 Contract Generation

The mediator agent supports the negotiation process without pursuing particular interests. Besides the contract space, the mediator has neither any knowledge about the individual preferences nor any special trust relationships with the participating agents. The actions of the mediator are transparent; it may be regarded as a plain piece of observable software. In case of doubts about the impartiality of the mediator, an unbiased generation of candidate contracts can be guaranteed by the application of a fully observable contract generation process. This can be achieved by using some public random number generator which feeds an open source algorithm which generates candidate contracts.

As the mediator cannot assess the quality of contracts, the generation of candidate contracts is essentially by chance. A simple option is the random selection of candidate contracts out of the contract space according to a uniform probability distribution. However, if one presumes that the negotiation process successively produces active contracts of higher quality, it appears favorable to incrementally introduce small changes to the active contract. Such incremental changes correspond to the selection of random neighbor solutions in the terms of local search, or random mutations in the terms of evolutionary algorithms. In the remainder of this chapter, we assume that the mediator randomly selects candidate contracts c' from a set $N(c)$, where $N : C \rightarrow 2^C$ defines some neighborhood relation.

More intelligent options for generating candidate contracts are conceivable. In particular, the mediator may learn by observing the negotiation process, trying to infer certain regularities in order to focus the generation procedure accordingly. For example, some attributes or attribute combinations might preferably be set to specific values or candidate contracts might be built by recombining fragments of former active contracts. Moreover, within an adapted negotiation protocol agents might propose candidate contracts alternately.

3.2 Acceptance Criteria

In each round of the negotiation process each agent must decide whether to accept or reject the candidate contract. For such decisions an agent does not necessarily need a complete explicit representation of the contract space and the utility function, but must only be capable of assessing potential contracts relative to each other. The natural behavior of a selfish agent is to myopically accept a candidate contract c' if and only if it is not worse than the currently active contract c. This *greedy acceptance criterion* A_i^g of an agent a_i is defined as follows:

$$A_i^g(c, c') := \begin{cases} \text{accept} & \text{if } f_i(c') \geq f_i(c) \\ \text{reject} & \text{otherwise} \end{cases}$$

The tightened version of this criterion requires a strict utility improvement.

For single-objective local search methods it is well known that a search process may get stuck rather soon in a local optimum if utility deteriorations are not allowed. This risk is even more evident in negotiation scenarios with several agents behaving greedily in combination. In order to prevent the negotiation process from getting stuck in its early stages, Klein et al. (2003a,b) propose that an agent applies the so called Metropolis acceptance criterion of simulated annealing (Metropolis et al. (1953), Kirkpatrick et al. (1983)). The resulting probabilistic *cooperative acceptance criterion* is defined as follows:

$$P(A_i^k(c, c') = \text{accept}) := \begin{cases} 1 & \text{if } f_i(c') \geq f_i(c) \\ e^{(f_i(c') - f_i(c))/T} & \text{otherwise} \end{cases}$$

This probabilistic decision rule allows utility deteriorations depending on the actual utility deterioration in connection with a positive control parameter T (temperature). The temperature may be gradually reduced in the course of the negotiation process. As a result, the probability of accepting candidate contracts decreases with an increasing utility deterioration as well as in the course of the negotiation process.

As the combination of agents that all apply the cooperative acceptance criterion presumably leads to a better outcome than the combination of greedy agents (which is supported by the experimental results presented in Sect. 4), the question arises how to induce the agents to behave accordingly. Because of information asymmetry concerning preferences and utility functions agents cannot be advised to apply a particular acceptance criterion, but one is restricted to rules that are verifiable. That is, related obligations can only be imposed in terms of the public behavior which is constituted by the dichotomic acceptance decisions uttered by the agents in each round. Consequently, the negotiation protocol may prescribe nothing else but conditions that must be fulfilled by the individual acceptance decision sequences $D(i, k) =< d_{i,1}, d_{i,2}, \ldots, d_{i,k}]$ for each agent $i = 1, \ldots, m$ and each round $k = 1, \ldots, R$, with $d_{i,j} \in \{\text{accept}, \text{reject}\}$ for $j = 1, \ldots, k$. While it is not reasonable to prescribe acceptance decisions for particular rounds in detail, conditions may be defined with regard to some aggregate measure function which subsumes a number of rounds. We propose the definition of mandatory acceptance ratios within specific phases of the negotiation process in order to induce acceptance decision sequences which resemble those of the application of the cooperative acceptance criterion. Fairness basically entails imposing the same acceptance ratios on all agents. (In case of some exogenous imbalance of power between the agents it is also conceivable to prescribe different acceptance ratios.) These acceptance ratios should decrease in the course of the negotiation process (in accordance with effective simulated annealing procedures). In what follows we describe a mechanism that conforms to these requirements.

The decision sequence that results from the application of the cooperative acceptance criterion depends on the temperature parameter. In order to enable achieving some acceptance ratio (probability) p, given some arbitrary utility function and search space topology, the agents need a general mechanism for determining temperature values that conform to given mandatory acceptance ratios. (This crucial point is not elaborated by Klein et al. (2003a,b). Even the majority of the research literature on simulated annealing neglects this problem and usually restricts to applying some simple rules of thumb.) If a given mandatory acceptance ratio is already reached by the utility improving candidate contracts, an agent does not need to accept any candidate contract with a deteriorating utility (i. e., T may be set to an infinitesimally small value). In the remainder we consider the case that an agent must additionally accept some deteriorating candidate contracts.

If one presupposes a density function $g(x)$ that represents the probability distribution of the utility variation x that an agent is expected to encounter during the negotiation process (with positive values corresponding to utility deteriorations), the following equation must be fulfilled by a corresponding temperature value T, $T > 0$, depending upon a given acceptance ratio p:

$$\int_{-\infty}^{0} g(x)dx + \int_{0}^{\infty} g(x)e^{-x/T}dx = p$$

However, such analytic probability distribution functions are usually not available. The following equation results when considering the utility variation for the neighborhood of a particular active contract c (with f representing the utility function of some agent):

$$\left(\sum_{\substack{c' \in N(c) \\ f(c') \geq f(c)}} 1 + \sum_{\substack{c' \in N(c) \\ f(c') < f(c)}} e^{(f(c') - f(c))/T} \right) / |N(c)| = p$$

This utility variation generally differs for different active contracts, and is thus not stationary in the course of the negotiation process. Therefore, we propose a simulative approach in order to approximately find out the utility variation to be encountered, which should enable agents to behave in such a way that a given acceptance ratio p is met. The agents privately simulate, separately from each other, the negotiation process in respective trial runs, and record the actual utility variations observed. Such a trial run is to resemble a phase of the negotiation process. A trial run may take, e. g., 10% of the rounds of the complete negotiation process. In each round of a trial run a candidate contract is generated randomly. An agent takes a candidate contract with a non-deteriorating utility variation as the new active contract with probability p^{m-1}. (Note that there is a common mandatory acceptance ratio p which each of the m agents must comply with.) A trial run of r rounds results in a number s of non-deteriorating candidate contracts and a multi-set of (positive) values $\{\Delta_1, \Delta_2, \ldots, \Delta_{r-s}\}$ that correspond to the $r - s$ deteriorating candidate contracts. With this the expected value of the number of individually accepted deteriorating candidate contracts can be calculated, which results in the following equation:

$$\left(s + \sum_{k=1}^{r-s} e^{-\Delta_k / T} \right) / r = p$$

The unique temperature value T that conforms to this equation can be calculated by an iterative approximation procedure.

This simulation-based procedure has two limitations which partly affect the course of the search process within a trial run (i. e., the simulated trajectory in the contract search space), and thus makes it a heuristic procedure. First, the preferences and thus the acceptance decisions of the agents may depend on each other, but because of information asymmetry there seems to be no other option than to assume statistical independence. Second, if an agent only accepts candidate contracts with a non-deteriorating utility variation it usually fails to provide for the mandatory acceptance ratio in simulation mode. To resolve this point an agent can repeat the trial run multiple times with applying the temperature value determined in the preceding run until the temperature converges to a stable value.

Now that each agent is provided with a mechanism for the approximate determination of an appropriate temperature value that brings about an imposed acceptance ratio, one may specify mandatory acceptance ratios as part of the negotiation protocol. We employ the following scheme (Fink (2004)): Two parameters p_1 and p_2, $p_1 > p_2$, define the ratio of the number of candidate contracts that shall be accepted by an agent within the beginning phase and within the final phase of the negotiation process, respectively, where each phase may take, e. g., 10% of the overall number of rounds of the negotiation process. With the mechanism described above each agent $i = 1, \ldots, m$ is enabled to approximately determine corresponding temperature values $T_{i,1}$ and $T_{i,2}$. We propose to imitate a common practice of simulated annealing procedures concerning the reduction of acceptance ratios in connection with geometrically decreasing temperature values. That is, the temperature is successively multiplied by a parameter α, $0 < \alpha < 1$, which can be calculated depending on the initial and final temperature parameters and the rounds of the negotiation process.

The described mechanism is applicable for any coordination problem that conforms to the abstract model stated in Sect. 2. The involved agents may pursue arbitrary private objectives on top of some common contract space. The quality of the resulting outcome will be analyzed in the next section. The mechanism provides fairness as it enforces by design all agents to behave cooperatively (i. e., to accept the same number of candidate contracts). In particular, the mechanism is invariant with regard to positive affine transformations of utility functions. Note that the constant term b is irrelevant, as only utility variations are considered, and the effect of the multiplier a is balanced by means of the temperature parameter.

The application of the cooperative acceptance criterion in connection with the described mechanism for determining temperature values may represent incentive compatible behavior, since otherwise an agent would have to partly accept worse contracts with a higher probability. However, a formal proof of such a conjecture would require demonstrating that the Metropolis criterion is the best conceivable probabilistic acceptance criterion in connection with the used negotiation protocol.

4 Computational Experiments

We analyze the proposed negotiation-based coordination mechanism by means of computational experiments. First, we want to support the basic hypothesis that the combination of agents that all behave cooperatively (in the meaning as defined in the previous section) usually provides a Pareto superior outcome in comparison to the combination of greedy agents. Furthermore, we aim for insights concerning the quality and sensitivity of results depending upon the parameter values. Note that the configuration of an experiment includes (in addition to the fixed general negotiation protocol, the contract space with the contract generation mechanism, and the agents involved) the parameters R (number of rounds), p_1 and p_2 (initial and final acceptance ratio).

As an exemplary application we consider the problem of matching production decisions in two-stage supply chains with a fixed set of participants; see Fig. 2. We suppose that some firm P requires for its production process parts which are produced and supplied by one other firm S in Scenario A (i. e., two decision making units), or parts of two different types which are produced and supplied by two respective firms S_1 and S_2 in Scenario B (i. e., three decision making units). We assume a strong coupling of the production processes in connection with a just-in-time (JIT) approach and focus on the operations management problem of regularly determining delivery schedules. That is, the question to be answered repeatedly is about the job sequence at the coupling point between the production stages. This results in a contract space which includes all permutations of an actual set of n jobs. Conceivable features such as batch deliveries or re-sequencing are neglected. Such additional degrees of freedom would affect, on the basis of some alignment of local plans to inter-agent decision variables, the particular utility functions without fundamentally changing the coordination problem.

Given a set of jobs $J = \{1, \ldots, n\}$, the contract space C includes all job sequences, i. e., permutations $\Pi = [\pi(1), \ldots, \pi(n)>$, with $\pi(j)$ denoting the job that takes the j-th position in the schedule. In order to generate candidate contracts we employ the neighborhood (move) definition of shifting one job to another position in the job sequence.

We assume that the supplier firms aim for cost efficient production schedules, while firm P is interested in reducing the eventual completion times of the jobs according to a customer point of view. In particular, let the utility functions of the supplier firms S, S_1 and S_2 be defined as $f(\Pi) := \sum_{j=1}^{n-1} \tau_{\pi(j),\pi(j+1)}$,

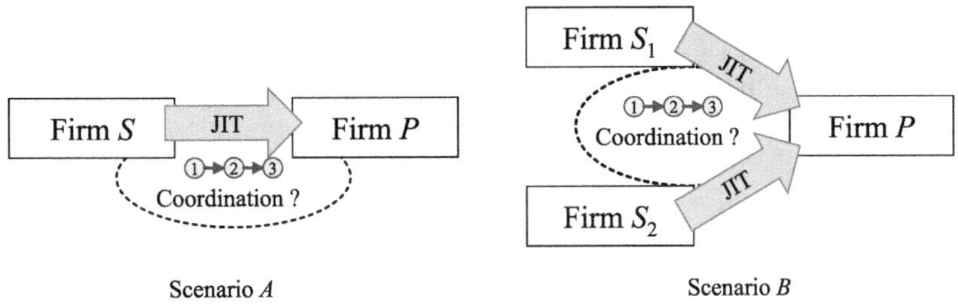

Fig. 2. Coordinating production sequences in two-stage supply chains

which may be interpreted as the sum of sequence-dependent setup costs (given supplier-specific setup cost data τ_{jk}, $1 \leq j, k \leq n$, $j \neq k$). For firm P we consider a production system with a continuous flow-shop characteristic and the objective of minimizing the average completion time of the jobs; see, e. g., Fink and Voß (2003). Note that the related single-objective optimization problems for each firm are $\mathcal{NP}$-hard (Garey and Johnson (1979)).

We use 30 problem instances ta061-ta090, each with 100 jobs, from the benchmark scheduling data set of Taillard (1993). For firm P, these problem instances are treated as continuous flow-shop scheduling problems. For the supplier firms, minimum delay values between the start of successive jobs (which would guarantee continuous production for each job on all production stages) are calculated and taken as setup costs. Each negotiation experiment is made up of 30 so called collaboration instances that result from matching different problem instances: For Scenario A, we use the combinations $(i + j, i + ((j + 2) \bmod 10))$ for $i \in \{61, 71, 81\}$ and $j = 0, \ldots, 9$, with the tuple values denoting the instance numbers from the benchmark data set of Taillard (1993) to be used for the firms S and P, respectively. For Scenario B, we use the combinations $(i{+}j, i{+}((j{+}1) \bmod 10), i{+}((j{+}2) \bmod 10))$ for $i \in \{61, 71, 81\}$ and $j = 0, \ldots, 9$ for the firms S_1, S_2 and P, respectively.

In order to account for random effects each negotiation procedure is repeated ten times with different seed values for the pseudo-random value generator, which affects the initial contracts (random permutations), the generation of candidate contracts as well as the responses from the probabilistic acceptance criteria. We generally report average deviations over all problem / collaboration instances and repetitions.

On the one hand, we are interested in deviations from baseline results for each firm (in the meaning of lower bounds), which are determined on the assumption that a firm can decide alone about the contract (the job sequence). Such reference results are approximately determined by means of the application of state of the art metaheuristics (iterated local search, simulated annealing, reactive tabu search, each for a maximum computation time of 10 seconds, see, e. g., Fink and Voß (2003)). For each problem instance, we use the best result obtained, as documented in Table 1, as the baseline for computing individual average percentage deviations. On the other hand, in order to assess the loss of efficiency due to information asymmetry in connection with selfishly acting decision making units, we apply a multi-objective simulated annealing heuristic (MOSA, Ulungu et al. (1999)) which provides an approximation of the corresponding set of Pareto optimal contracts. In particular, we randomly generated 200 different weight sets and repeatedly applied an adaptation of the simulated annealing procedure of Johnson et al. (1989), which incrementally generates an approximation of the set of Pareto optimal solutions; in each run the actual weight set is used for computing weighted sums of normalized criteria value differences for random neighbor solutions.

Table 1. Reference results: best objective function values obtained for single-objective problems

	setup costs	flow time
ta061	6211	308052
ta062	6061	302386
ta063	5942	295239
ta064	5825	278811
ta065	6022	292757
ta066	5923	290819
ta067	6065	300068
ta068	5966	291859
ta069	6241	307650
ta070	6225	301942
ta071	7583	412700
ta072	7439	394562
ta073	7634	406160
ta074	7857	422301
ta075	7561	400175
ta076	7371	391359
ta077	7443	394179
ta078	7518	402025
ta079	7736	416833
ta080	7683	410372
ta081	9802	562150
ta082	9674	563923
ta083	9676	562404
ta084	9770	562918
ta085	9565	556311
ta086	9691	562253
ta087	9885	574102
ta088	9960	578119
ta089	9779	564803
ta090	9922	572798

This enables, first, the assessment of the obtained utility of a contract for each agent by means of average deviations from approximated single-objective optima, and, second, the illustration of the deviations of contracts from the approximated Pareto frontier. Note that one cannot generally suppose a pre-defined metric for measuring the deviation of the negotiation outcome from the Pareto frontier as a scalar value. As a result of the ambiguous numerical representation of preferences concerning scale and origin the comparison of respective values between the agents may be meaningless. Thus we take into account particular deviations for the decision making units as a tuple.

The negotiation protocol and the reasoning of the agents have been implemented in C++. In particular, we defined an application framework with generic components which represent the general negotiation protocol described above. These components can be applied for a given coordination problem by means of instantiations concerning, e. g., specific contract spaces and acceptance strategies. Computational experiments were being performed using a Pentium IV processor with 1.8 GHz. Computation times are of minor importance as the considered bilateral and trilateral negotiation processes have been taking not more than between one and two seconds for 100,000 rounds.

4.1 Scenario A: Two Decision Making Units

In preliminary experiments we found that parameter values such as $p_1 = 0.2$ und $p_2 = 0.04$ on average provided reasonable outcomes (Fink (2004)), so we first report results for this parameter setting. Second, we analyze the effect of varying these values. Table 2 shows the individual average deviations from approximated single-objective optima, for negotiations of 100,000 rounds, depending on the agents behaving greedily or cooperatively as described above.

Table 2. Negotiation results for scenario A (100,000 rounds)

Average deviations 100,000 rounds $p_1 = 0.2$; $p_2 = 0.04$		Firm / agent 2 (flow time) greedy f_1	 f_2	 cooperative f_1	 f_2
Firm / agent 1	greedy	23.0%	23.9%	11.3%	36.4%
(setup costs)	cooperative	38.9%	10.0%	19.7%	15.1%

Comparing the four principal alternatives the outcome due to two cooperative agents is Pareto efficient; even more, it is Pareto superior in comparison to the combination of two greedy agents. If a cooperative agent negotiates with a greedy agent the latter one clearly profits from exploiting the willingness of the opposite agent to give in. Anyway, the corresponding outcome due to the combination of two agents behaving differently is Pareto efficient.

Table 3 shows the results of extending the negotiation to 1,000,000 rounds. First, as may be expected this does not alter the outcome for the combination of two greedy agents, which presumably get stuck rather soon when greedily negotiating with each other. Second, the extension enables two cooperative agents to achieve even better results than before. Third, giving a greedy agent more time to exploit a cooperative agent widens the gap between the respective results. Extending the negotiations to 10,000,000 rounds generally supports this interpretation (Table 4).

Table 3. Negotiation results for scenario A (1,000,000 rounds)

Average deviations 1,000,000 rounds $p_1 = 0.2$; $p_2 = 0.04$		Firm / agent 2 (flow time) greedy		cooperative	
		f_1	f_2	f_1	f_2
Firm / agent 1 (setup costs)	greedy	23.0%	23.9%	7.4%	42.9%
	cooperative	48.4%	6.2%	18.5%	11.6%

Table 4. Negotiation results for scenario A (10,000,000 rounds)

Average deviations 10,000,000 rounds $p_1 = 0.2$; $p_2 = 0.04$		Firm / agent 2 (flow time) greedy		cooperative	
		f_1	f_2	f_1	f_2
Firm / agent 1 (setup costs)	greedy	23.0%	23.9%	6.0%	47.9%
	cooperative	54.2%	5.2%	17.8%	10.8%

Supposing common knowledge about the ordinal relations of the utility values that result for the involved agents as shown in Tables 2–4, in connection with the usual assumptions of non-cooperative game theory, the shown outcome would resemble the classic prisoner's dilemma (Axelrod (1984)) if an agent can freely decide about its acceptance strategy (Klein et al. (2003a)). This follows from a straightforward application of the dominance criterion. That is, the combination of two agents that behave greedily, and thus the only outcome that is Pareto dominated, represents the unique Nash equilibrium (because of the risk of benevolent behavior being exploited). This unfortunate situation can be overcome by imposing verifiable rules that force the agents to accept given ratios of candidate contracts as described above. This mechanism appears necessary and sufficient for resolving the prisoner's dilemma and achieving a mutually beneficial outcome. (Klein et al. (2003a,b) propose to overcome the prisoner's dilemma by letting the mediator decide about the acceptance of candidate contracts. Among other complications, this adaptation of the negotiation protocol requires that the agents reveal further information to enable the mediator to assess the quality of a candidate contract.)

While the results documented in Tables 2–4 serve for evaluating the relative effect of the combination of agents that follow different strategies, the other main question is about the deviation from the Pareto frontier. That is, we are interested in the degradation that is the consequence of applying the negotiation-based approach under consideration of selfishly acting decision making units with private preference information. We approximately determined the set of Pareto optimal contracts by means of a multi-objective simulated annealing heuristic as described above. Figure 3 shows corresponding results for a particular collaboration instance. This outcome appears representative for other collaboration instances as we generally obtained very similar results. The utility of the negotiation outcomes shown in Fig. 3, for the four typical strategy combinations, resulted from negotiations with 1,000,000 rounds. It appears that the applied negotiation approach with two agents that follow a cooperative acceptance strategy leads to a high-quality outcome very close to the Pareto frontier. This has been confirmed by comparing to the results obtained by the application of metaheuristics as described above for an hypothetical social welfare function as defined by the mean of the average percentage deviation from the reference results of Table 1.

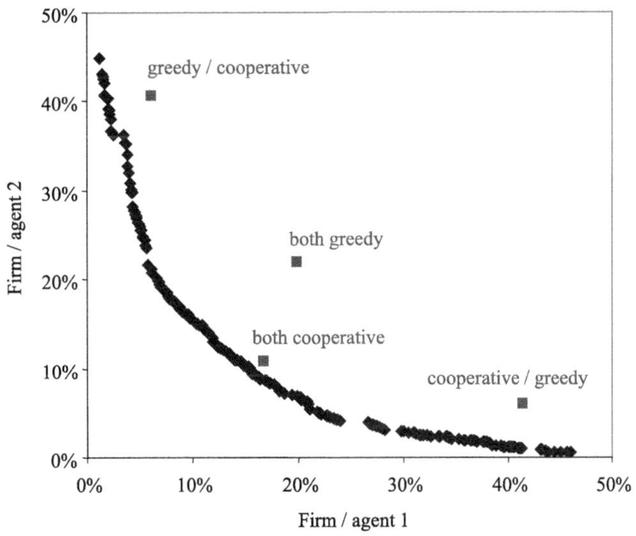

Fig. 3. Approaching the Pareto frontier depending upon the strategies of the agents

We examined the effect of varying the parameter values p_1 and p_2 with both agents behaving cooperatively by performing respective negotiation experiments (again for the whole set of collaboration instances as described above). Figure 4 shows the results obtained by combining different parameter values. In this case we use the mean of the average percentage deviations as a scalar indicator for the quality of the outcome. For the considered scenarios, the parameter value combinations (0.2, 0.02) and (0.1, 0.02) lead to the best average results for negotiations of 1,000,000 (a) and 5,000,000 (b) rounds, respectively. In general, it seems that the results are less sensitive to changing p_1 in comparison to p_2. (Similar results have been obtained by experiments for a different type of coordination problem as described in Fink (2004).) The encountered variations of the solution quality are modest in contrast to the overall gain due to the application of the negotiation mechanism with two agents that follow a cooperative acceptance strategy.

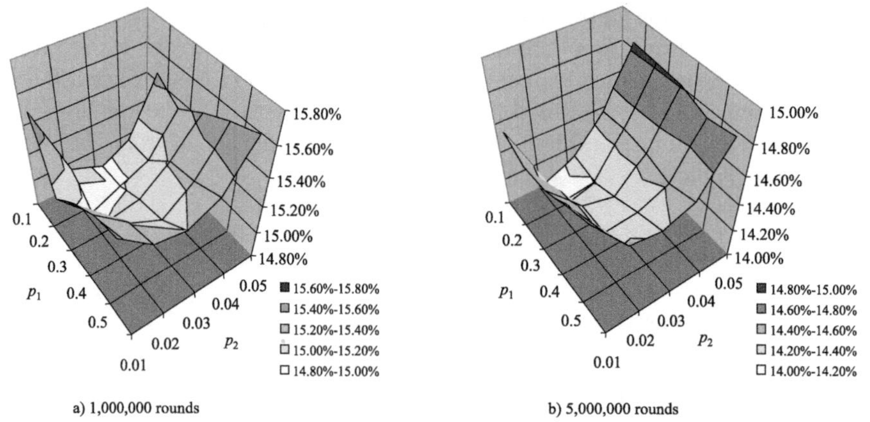

Fig. 4. Influence of the parameter values p_1 and p_2 on the solution quality

4.2 Scenario B: Three Decision Making Units

In this section we want to validate and generalize the results from above by considering the scenario with three agents (Scenario B). In Tables 5 and 6 we report the outcome for negotiations of 100,000 rounds and 1,000,000 rounds, respectively. As is to be expected, the resulting average deviations are larger than before, since a common contract takes into account the conflicting preferences of three instead of two agents.

The outcome that results from the combination of three cooperative agents is Pareto efficient and dominates the outcome that results from the combination of three greedy agents. If one agent behaves greedily against two cooperative agents, the greedy agent considerably profits at the cost of the other ones. If two agents behave greedily in combination with one cooperative agent, the former ones achieve good results while the benevolent behavior is exploited. Overall, all outcomes are Pareto efficient but the one with all agents behaving greedily. When using the mean of the average deviations to measure the quality of a strategy combination by a scalar value, the combination of all agents behaving cooperatively clearly turns out as the best result. Furthermore, the quality of the outcome is improved by extending the negotiation (mean values of 27.2% vs. 23.5% for 100,000 rounds and 1,000,000 rounds, respectively).

Table 5. Negotiation results for scenario B (100,000 rounds)

Average deviations 100,000 rounds $p_1 = 0.2$; $p_2 = 0.04$				Firm / agent 3 (flow time) greedy f_1	greedy f_2	greedy f_3	cooperative f_1	cooperative f_2	cooperative f_3
Firm / agent 1 (setup costs)	greedy	Firm / agent 2 (setup costs)	greedy	34.8%	35.0%	35.8%	28.8%	28.8%	44.0%
			cooper.	27.5%	45.0%	28.2%	18.1%	40.7%	38.1%
	cooper.		greedy	45.3%	27.3%	28.2%	40.9%	18.2%	38.3%
			cooper.	39.6%	40.0%	16.9%	28.8%	28.2%	24.5%

Table 6. Negotiation results for scenario B (1,000,000 rounds)

Average deviations 1,000,000 rounds $p_1 = 0.2$; $p_2 = 0.04$				Firm / agent 3 (flow time) greedy f_1	greedy f_2	greedy f_3	cooperative f_1	cooperative f_2	cooperative f_3
Firm / agent 1 (setup costs)	greedy	Firm / agent 2 (setup costs)	greedy	34.8%	35.0%	35.8%	26.2%	26.1%	47.6%
			cooper.	24.6%	50.0%	25.1%	10.8%	46.3%	41.7%
	cooper.		greedy	49.9%	24.6%	25.3%	45.7%	11.0%	42.0%
			cooper.	45.9%	45.9%	9.3%	25.9%	26.0%	18.5%

Figure 5 shows the effect of varying the parameter values p_1 and p_2 with both agents behaving cooperatively. Best results have been obtained by the parameter value combination (0.2, 0.04) for negotiations of 1,000,000 (a) as well as of 5,000,000 (b) rounds. The modest sensitivity of the solution quality may allow generating a high quality outcome without requiring costly calibration experiments.

To summarize, the results achieved for Scenario B generally support the conclusions drawn for Scenario A in the previous section.

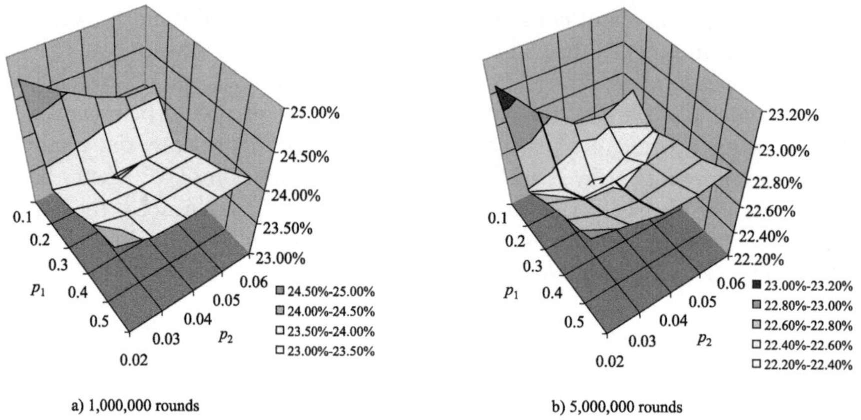

Fig. 5. Influence of the parameter values p_1 and p_2 on the solution quality

5 Conclusions

In this chapter we have described, extended, and analyzed a generic automated negotiation scheme that provides a mechanism for coordinating autonomous and selfishly acting decision making units with private preference information. The mechanism takes into account strict restrictions of distributed decision situations. It is applicable for any coordination problem where the set of feasible outcomes can be represented formally. We have described effective and verifiable rules as part of the negotiation protocol, which force the agents to behave cooperatively in the meaning of accepting certain ratios of candidate contracts. The experimental results, which have been obtained for computationally complex bilateral and trilateral coordination problems within the domain of supply chain management, give evidence of the potential of the fully automated negotiation mechanism to generate mutually beneficial contracts very close to the Pareto frontier.

The described coordination mechanism comes with several advantages: It does not need to presuppose any benevolence of the involved agents. Requirements for information revelation are kept to a minimum, as the mechanism neither requires the exposure of cardinal preference information nor builds on some critique on specific contract features by the agents. This is in contrast to the expectation (e. g., Jennings et al. (2001)) that simple negotiation processes which are only based on a random generation of potential contracts and minor information revelation might be inefficient. The mechanism is computationally tractable and can be adapted for specific kinds of coordination problems straightforwardly. It does not require any exploitation of domain-specific knowledge. Furthermore the mechanism appears quite robust given that the quality of the results is only slightly affected by the actual setting of the parameter values.

There are several requirements and options for further research. The applied scheme for setting mandatory acceptance ratios may be elaborated and examined in more detail. There is the question about the incentive compatibility of following the scheme described above for deriving individual acceptance decisions; that is, one may think about whether making decisions in other ways can lead to better individual results while still conforming to the rules of the negotiation process concerning mandatory acceptance ratios. The mechanism should be applied for other application scenarios to gain more insights concerning the generality of the experimental findings reported in this chapter. From a practical point of view, standardization and formalization of typical contract spaces may be useful, which may involve the definition of related ontologies.

It seems of interest to examine situations where the preferences of the agents partly depend on each other. In case that the mechanism applied to some coordination problem obtains results of unsatisfactory quality concerning the deviation from the Pareto frontier, options for extending the negotiation protocol may be followed (e. g., by generating candidate contracts more intelligently by means of learning from the course of the negotiation as discussed above).

The application of the mechanism within market matching scenarios is also conceivable. Within this context the negotiation mechanism can be used as a building block to compute affinity measures of all potential partners on a marketplace by provisional negotiation experiments, which can be used to generate efficient assignments.

Furthermore, the negotiation mechanism may be applied in connection with coordination problems with an online characteristic and a rolling planning horizon. Considering the coordination problem as an online problem with quasi-instantaneous updates of the elements of the contract space and/or the individual preferences may be handled in a straightforward way.

References

Agnetis, A., P. Detti, C. Meloni, and D. Pacciarelli (2001). Set-up coordination between two stages of a supply chain. Annals of Operations Research 107, 15–32.

Agnetis, A., N.G. Hall, and D. Pacciarelli (2005). Supply chain scheduling: sequence coordination. Discrete Applied Mathematics, to appear.

Agnetis, A., P.B. Mirchandani, D. Pacciarelli, and A. Pacifici (2004). Scheduling problems with two competing users. Operations Research 52, 229–242.

Axelrod, R. (1984). The Evolution of Cooperation. Basic Books, New York.

Burke, E.K. and G. Kendall (Eds.) (2006). Search Methodologies: Introductory Tutorials in Optimization and Decision Support Techniques. Springer, Berlin.

Cachon, G.P. (2003). Supply chain contracts. In A. de Kok and S. Graves (Eds.), Supply Chain Management: Design, Coordination and Operation (Handbooks in Operations Research and Management Science: Volume 11), pp. 229–339. Elsevier, Amsterdam.

Chen, F., A. Federgruen, and Y.S. Zheng (2001). Coordination mechanisms for a distributed system with one supplier and multiple retailers. Management Science 47, 693–708.

Dawande, M., H.N. Geismar, N.G. Hall, and C. Sriskandarajah (2005). Supply chain scheduling: Distribution systems. Working paper.

de Kok, A.G. and S.C. Graves (Eds.) (2003). Supply Chain Management: Design, Coordination and Operation (Handbooks in Operations Research and Management Science: Volume 11). Elsevier, Amsterdam.

D'Inverno, M. and M. Luck (2004). Understanding Agent Systems Springer, Berlin Heidelberg.

Dudek, G. and H. Stadtler (2005). Negotiation-based collaborative planning between supply chains partners. European Journal of Operational Research 163, 668–687.

Erengüc, S.S., N.C. Simpson, and A.J. Vakharia (1999). Integrated production/distribution planning in supply chains: an invited review. European Journal of Operational Research 115, 219–236.

Fink, A. (2004). Supply chain coordination by means of automated negotiations. In Proceedings of the 37th Hawaii International Conference on System Sciences (HICSS-37). IEEE.

Fink, A. and S. Voß (2003). Solving the continuous flow-shop scheduling problem by metaheuristics. European Journal of Operational Research 151, 400–414.

French, S. (1988). Decision Theory: An Introduction to the Mathematics of Rationality. Ellis Horwood, Chichester.

Gandibleux, X., M. Sevaux, K. Sörensen, and V. T'kindt (Eds.) (2005). Metaheuristics for Multiobjective Optimisation Springer, Berlin.

Garey, M.R. and D.S. Johnson (1979). Computers and Intractability– A Guide to the Theory of NP-completeness. Freeman, New York.

Glover, F. and G.A. Kochenberger (Eds.) (2003). Handbook of Metaheuristics Kluwer, Boston.

Hall, N.G. and C.N. Potts (2003). Supply chain scheduling: batching and delivery. Operations Research 51, 566–584.

Jaszkiewicz, A. (2004). A comparative study of multiple objective metaheuristics on the bi-objective set covering problem and the pareto memetic algorithm. Annals of Operations Research 131, 135–158.

Jennings, N.R., P. Faratin, A.R. Lomuscio, S. Parsons, M. Woolridge, and C. Sierra (2001). Automated negotiation: Prospects, methods and challenges. Group Decision and Negotiation 10, 199–215.

Johnson, D.S., C.R. Aragon, L.A. McGeoch, and C. Schevon (1989). Optimization by simulated annealing: An experimental evaluation; part i, graph partitioning. Operations Research 37, 865–892.

Kirkpatrick, S., C.D. Gelatt Jr., and M.P. Vecchi (1983). Optimization by simulated annealing. Science 220, 671680.

Klastorin, T.D., K. Moinzadeh, and J. Son (2002). Coordinating orders in supply chains through price discounts. IIE Transactions 34, 679–689.

Klein, M., P. Faratin, H. Sayama, and Y. Bar-Yam (2003a). Negotiating complex contracts. Group Decision and Negotiation 12, 111–125.

Klein, M., P. Faratin, H. Sayama, and Y. Bar-Yam (2003b). Protocols for negotiating complex contracts. IEEE Intelligent Systems 18 (6), 32–38.

Kraus, S. (2001). Strategic Negotiation in Multiagent Environments MIT Press, Cambridge MA.

Lee, C.-Y. and Z.-L. Chen (2001). Machine scheduling with transportation considerations. Journal of Scheduling 4, 3–24.

Metropolis, N., A.W. Rosenbluth, M.N. Rosenbluth, A.H. Teller, and E. Teller (1953). Equation of state calculations by fast computing machines. Journal of Chemical Physics 21, 1087–1092.

Moses, M. and S. Seshadri (2000). Policy mechanisms for supply chain coordination. IIE Transactions 32, 245–262.

Nash, J.F. (1950). The bargaining problem. Econometrica 18, 155–162.

Nash, J.F. (1953). Two-person cooperative bargaining games. Econometrica 21, 128–140.

Rosenschein, J.S. and G. Zlotkin (1994). Rules of Encounter – Designing Conventions for Automated Negotiation among Computers. MIT Press, Cambridge MA.

Sandholm, T.W. (1999). Distributed and rational decision making. In G. Weiss (Ed.), Multiagent Systems: A Modern Approach to Distributed Artificial Intelligence, pp. 201–258. MIT Press, Cambridge MA.

Sarmiento, A.M. and R. Nagi (1999). A review of integrated analysis of production-distribution systems. IIE Transactions 31, 1061–1074.

Shapiro, J.F. (2001). Modeling the Supply Chain Duxbury, Pacific Grove.

Stadtler, H. and C. Kilger (Eds.) (2004). Supply Chain Management and Advanced Planning - Concepts, Models, Software and Case Studies (third ed.). Springer, Berlin.

Sucky, E. (2005). A bargaining model with asymmetric information for a single supplier-single buyer problem. European Journal of Operational Research, to appear.

Taillard, E.D. (1993). Benchmarks for basic scheduling instances. European Journal of Operational Research 64, 278–285.

Thomas, D.J. and P.M. Griffin (1996). Coordinated supply chain management. European Journal of Operational Research 94, 1–15.

Tsay, A.A., S. Nahmias, and N. Agrawal (1999). Modeling supply chain contracts: A review. In S. Tayur, R. Ganeshan, and M. Magazine (Eds.), Quantitative Models for Supply Chain Management, pp. 299–336. Kluwer, Boston.

Ulungu, E.L., J. Teghem, P. Fortemps, and D. Tuyttens (1999). MOSA method: A tool for solving multiobjective combinatorial optimization problems. Journal of Multi-Criteria Decision Analysis 8, 221–236.

Business Process Support in a Seaport Automobile Terminal — a Multi-Agent Based Appro ach

Torsten Fischer and Hermann Gehring

Froschkönigweg 16a, 53797 Lohmar
Torsten.Fischer@edfischer.de

Abstract. A multi-agent based approach for supporting the planning of transshipments of imported vehicles via a seaport automobile terminal is presented. The logistics supply chain for vehicle transport is analyzed from the point of view of terminal operator by focusing on the corresponding business processes. On this basis the planning problem is subdivided into three tasks, the quay management, the allocation of parking areas for the temporary storage of vehicles, and the allocation of drivers to the vehicles that have to be moved in the terminal area. The problem of storage allocation is often combined with a departure time estimation for the incoming vehicles – here carried out by a learning classifier system. For supporting these planning tasks a multi-agent system (MAS) has been developed. In the MAS the planning tasks are assigned to different agent types, which are coordinated by a central coordinator agent. The MAS was subjected to a test using practically relevant problem instances. The derived test results show a robust behavior of the MAS with regard to varying problem data.

1. Introduction and Problem Description

In Western Europe, a considerable number of transshipment ports for the import and export of vehicles established themselves in the growing transshipment market, e.g. Le Havre, Rotterdam, Antwerp, Bremerhaven, etc. Every year some million vehicles are loaded and unloaded in the automobile terminals of these seaports. In recent years, the terminals were strongly affected by the globalization trend being observed in the automotive industry. In this regard, automobile manufactures focus more on their core competence, while typical logistic services, such as the transportation or (interim) storage of vehicles are increasingly assigned to specialized logistic providers. Consequently, the rise of the automobile production as well as the relocation of production plants all over the world lead to an increasing demand for logistic services, especially transportation and storage services. According to appropriate studies and prognoses, this development will also continue over the next couple of years (cf. [10], [5]).

The following article focuses on the vehicle transshipment over a specified large European automobile terminal. However, most of the considerations of this paper can also be transferred to the conditions in other seaport automobile terminals. In the

T. Fischer and H. Gehring: Business Process Support in a Seapor t Automobile Terminal— a Multi-Agent Based Approach, Studies in Computational Intelligence (SCI) 28, 373–394 (2006)
www.springerlink.com

following, the focus is reduced to the importation of vehicles, because the importation presents some special challenges. In the meantime, in most of the European seaports there is a premium on available storage area – and this cannot be extended. Since the vehicles are frequently not delivered directly to the retailers or customers, the vehicles often have to be stored in the terminal parking area for a longer time until they are finally delivered. As a result of intensified transshipment processes, slot bottlenecks are occurring more and more frequently. In addition, a clear trend towards longer storage periods for imported vehicles can also be observed. This places an additional load on the already restricted storage locations. Beyond, the exact departure time of the incoming vehicles is often unknown at the point of their delivery. Against this background, it is difficult to realize an efficient storage planning in the dynamic environment of vehicle transshipment.

Most of the western European seaport automobile terminals are strongly affected by an increasing demand for logistic services because the majority of finished vehicles imported into Europe are handled via terminals. In spite of this development, the competition between the western European seaports has been aggravated in the last few years caused by a fast increasing number of transshipment capacities and the re-enter of new seaports into the vehicle transshipment market. To cope with the growing pressure of competition, the seaport operators are forced to cooperate with the other logistic providers being involved in the supply chain for vehicle transfer, such as carriers or shipping company owners. Further, the automobile terminal operators have assumed some necessary measures in order to rationalize the underlying workflows and to improve the quality of the logistic services. Special importance is always attached to the planning and processing of the logistic tasks in the seaport automobile terminals which are focused in this paper.

The remainder of the contribution is organized as follows: In section 2 the business processes in the considered seaport automobile terminal is described and critically analyzed. In order to reduce identified deficiencies, a mathematical vehicle transshipment problem is formulated in section 3. The model is embedded in a multi-agent system for supporting integrated storage allocation and deployment scheduling which is described in section 4. In section 5 the proposed multi-agent system is evaluated using randomly generated but practically relevant test instances and in section 6 the paper is summarized.

2. Business Processes in a Seaport Automobile Terminal

The economic success of a seaport automobile hub is affected by various issues concerning the relations to automobile manufacturers. A high calling frequency, e.g., is very important for automobile manufacturers, since they can serve then their markets via a seaport hub on a quick and regular basis. The manufactures also require smooth transport from the point of origin to the point of destination. The seaport with its handling, storage, and service functions is the key link. It has to offer rapid accessibility form the sea as well as good hinterland connections via rail and road. To clarify the terminal's central position, the underlying interorganisational logistic supply chain for vehicle transfer is considered in a first step. Then, the import processes in a typical seaport automobile terminal are described in more detail and analysed via an appropriate method of business process modeling.

2.1 Logistics Supply Chain for Vehicle Transport

The vehicle transfer from European or intercontinental production plants to the retailers or customers is usually handled by different logistic providers. The corresponding vehicle transportation and storage tasks can be represented as an interorganisational logistic supply chain according to their chronological classification as shown in Fig. 1. The tasks focused in this paper are dark highlighted.

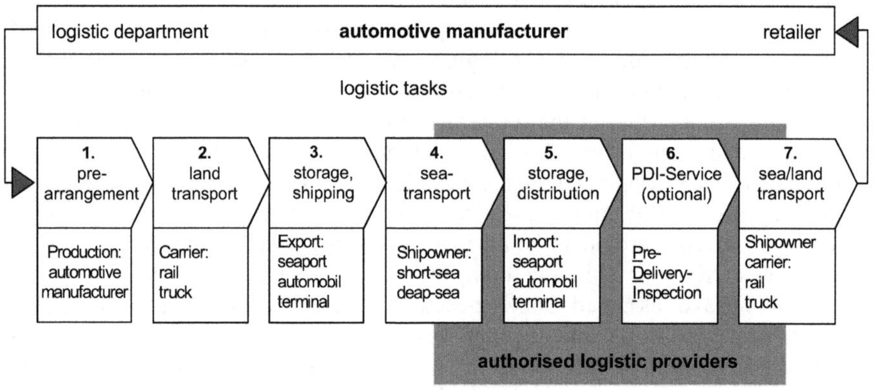

Fig. 1. Logistic supply chain for vehicle transport via seaport automobile terminals

The logistics supply chain is activated by a central logistic department of the automotive manufacturer working as a client representative. First, the logistic department of the automotive manufacturer prearranges the vehicle transport at the production plant and hands a corresponding order over to a carrier (task 1). Then, the authorized carrier accomplishes the vehicle transfer from the production plant to the export automobile terminal by truck or rail (task 2). At the export terminal the vehicles are managed by the responsible terminal operator, who organizes the interim storage and the shipping of the vehicles (task 3). In the next step, the short or deep sea vehicle transport to the import automobile terminal is carried out by a shipping company owner (task 4). The actually focused process of the vehicle transshipment via a seaport automobile terminal follows (task 5). The vehicles arriving by sea are unloaded in the quay area, then stored in the terminal's parking areas and finally removed from storage to be delivered. Prior to delivery to the retailers or customers, the vehicles are often transferred to an import forwarding company (task 6) which is located in the rear area of the terminal. These enterprises provide for an optional pre-delivery inspection, carry out repairs or retrofit the vehicles according to individual buyer wishes. They also take care of customs clearance and delivery to the dealer network. Finally the vehicles are transferred to the automotive retailers, e.g. the dealer network, or even the customers (task 7). This process can take place either by sea via feeder ship or by land via truck or rail. The end of the supply chain is composed by the automotive retailer or even the customer.

2.2 Representation of the Underlying Business Processes

Because of the growing pressure of competition, the seaport terminal operators are forced to assume some necessary measures in order to optimize the planning and the efficiency of the execution of the logistic tasks. In this situation the business process modelling approach seems to be an appropriate method for the description, analysis and design of business processes in seaport automobile terminals. An overview of the respective business processes in the considered automobile terminal is given in Fig.2.

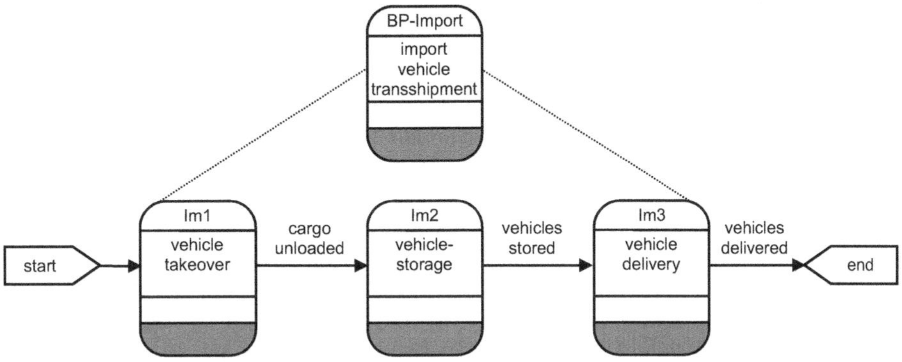

Fig. 2. Overview business process "import vehicle transshipment"

According to Fig. 2, the business process of import vehicle transshipment (BP-Import) is divided up into three sub-processes. In the following, these sub-processes are explained focusing the related planning processes (cf. [6]).

In practice, the administration of order, planning, and storage data is usually carried out via computer aided information systems. Here it is assumed that, in accordance with the situation in the terminal, the following information systems are available. The port information system PORT-INFO manages the quay data, and the terminal information system CAR-INFO manages the order and storage data as well as the personal data. The customer order database is administrated by a classical ERP-system. According to the focus of this contribution, however, a detailed description of these systems can be omitted.

2.2.1 Business Process "Vehicle Takeover" (Im1)

The business process "vehicle takeover" (Im1) extends from the notification of an incoming ship over the quay management to the unloading of the delivered vehicles. In the considered seaport automobile terminal, this process is subdivided into five sub-processes. These are:

1. notification of an incoming ship (Im1.1),
2. quay arrangement (Im1.2),
3. incoming buffer allocation (Im1.3),
4. deployment scheduling (Im1.4), and
5. vehicle unloading (Im1.5).

The ship notification (Im1.1) initializes the transshipment process, which includes, from the point of view of a terminal operator, especially the vehicle takeover (Im1). Within the scope of the notification the terminal operator is given all relevant information about the delivered vehicles, such as the number of vehicles intended for the seaport, a specific stowage plan, and all order data according to the transshipment being available at that time. On the basis of the submitted data, the operator's ship-agent assigns a free quay to the incoming ship by manual planning (Im1.2). The planning results created by the ship agent are fed into the port information system PORT-INFO and also inserted in the ship documents. To realize a fast and cost-efficient unloading process, the vehicles are unloaded in the quay area first, which has to be cleared in a relatively short time. Therefore, an incoming buffer has to be allocated to the vehicles arriving by sea (Im1.3). The incoming buffer store allocation is carried out by a storage operator in comprehension with the terminal information system CAR-INFO. The planning results are also documented in a storage plan generated by the CAR-INFO system. The system functionality is, however, limited to the retrieval and the analysis of relevant order, storage, and personal data. If sufficient capacities are assured at an incoming buffer store, the personal operator carries out the deployment scheduling for the unloading process (Im1.4) and inserts the planning results into the CAR-INFO system. Finally, a shift plan is generated by the information system CAR-INFO. In the end, the vehicles are unloaded by drivers (Im1.5) on the basis of the ship documents and shift plans.

2.2.2. Business Process "Vehicle Storage" (Im2)

After the vehicles are unloaded into an incoming buffer store, they are taken into the terminal's parking areas. These are large paved parking places for storing the incoming vehicles until they are delivered to the carriers or shipping companies. During the whole transshipment process, the incoming vehicles are handled in groups. A group consists of vehicles of the same or similar type that are produced by the same manufacturer and may meet an additional grouping criterion, e.g., all vehicles of a group are to be delivered to the same automobile trader. Moreover, the forming of vehicle groups simplifies the planning of the vehicle transshipment and of the operational tasks. The associated business process "vehicle storage" (Im2) can also be divided up into three sub-processes:

1. storage allocation (Im2.1),
2. deployment scheduling (Im2.2), and
3. insertion of vehicles(Im2.3).

Before the vehicles are stored in the parking areas, some planning steps have to be carried out. In a first step, the storage operator chooses an adequate storage location in the terminal for every vehicle group (Im2.1). For this purpose, the storage and order data are called from the CAR-INFO system and a storage plan is generated manually. In the next step, the personal operator carries out the deployment scheduling for the resulting storage processes (Im2.2). The deployment scheduling contains a temporarily coarse grained and fine grained scheduling. The planning results are secured in the shift plan. According to the storage and shift plan, the vehicles are inserted into the terminal parking areas by drivers (Im2.3).

2.2.3 Business Process "Vehicle Delivery" (Im3)

Starting from their actual position the vehicles are finally removed from storage to be delivered. This activity is modeled via the corresponding business process "vehicle delivery" (Im3). Vehicles that are transported further by a so called feeder ship are moved directly to the waterfront transshipment areas while those to be transferred to the hinterland are placed in the rear areas of the terminal. This process can be described as follows:

1. vehicle call (Im3.1),
2. deployment scheduling according to the removal processes (Im3.2),
3. vehicle removal (Im3.3),
4. deployment scheduling according to the loading processes (Im3.4), and
5. vehicle loading (Im3.5).

The vehicle calls for delivery arrive at least two days before the delivery date (Im3.1). This short time period requires the implementation of a fast and efficient loading process: The customer service desk generates the lists of the vehicles to be delivered by the CAR-INFO system. Before the vehicles are finally loaded onto a carrier for transportation, they are moved to an outgoing buffer store. In contrast to the other storage processes, the destination of a vehicle removal process is well-defined because every type of carrier has its own specific outgoing buffer store. Hence, only a deployment scheduling process (Im3.2) has to be considered to plan the vehicle removal processes (Im3.3). The same can also be assumed for the loading process (Im3.4 and Im3.5). In both cases the planning results are also stored in the CAR-INFO system and handed out to the drivers as a shift plan. The business process is finalized by the operational loading process through the deployed drivers of the terminal.

2.3 Description of the Planning Processes

According to the business process analysis carried out in section 2.2, every vehicle moving process is prepared by one or even more preceding planning processes. These are:

1. quay arrangement,
2. incoming buffer store allocation,
3. storage allocation, and
4. personal deployment.

The quay arrangement is carried out manually by the ship agent of the seaport accessing the information system PORT-INFO. On the basis of the submitted data the terminal operator's ship-agent assigns a free quay to the incoming ship by manual planning. If there is no free quay at the time of the ship's arrival, the ship has to wait outside the seaport as long as a free quay is available at the seaside of the automobile terminal. An imaginable evasion to a quay at another terminal, i.e. the container terminal, is not possible because of the missing landside required infrastructure for the vehicle unloading processes. Form an economic point of view, every idle time to incoming ships is very cost-intensive. Therefore, the ship-agent is instructed to minimize those idle times as far as possible by a foresighted and efficient planning with regard to the customer satisfaction.

The planning process of the incoming buffer store allocation is provided by a storage operator. All vehicles to be unloaded from an incoming car carrier ship have to be assigned to the same incoming buffer because of organizational reasons. To simplify the respective planning tasks, the storage operator tries to handle those vehicles in the same groups as in the vehicle storage process. For unloading, the operator checks the availability of free storage positions at each buffer at the time of the ship arrival. If a sufficient number of free parking spaces can be identified at least at one quay buffer, the storage operator chooses one of the identified quay buffers for unloading on the basis of his individual experiences. Otherwise, the storage operator arranges the clearance of an incoming buffer until the notified ship arrives at the quay to avoid cost intensive delays. The selection of an appropriate buffer store for clearance also bases on the individual experience of the storage operator. All these manual planning processes are carried out in comprehension with the terminal information system CAR-INFO. The main objective of a storage operator with respect to the incoming buffer store allocation consists in the minimization of the required waiting times of the incoming ships.

The storage allocation is also carried out manually by the terminal's storage operator on the basis of several vehicle and storage lists generated by the information system CAR-INFO. The car manufacturers, however, often do not specify the departure time of a vehicle group at the time of their delivery, so that an exact storage time can not be identified. In this situation, the responsible storage operator estimates the departure time on the basis of his personal experience. The estimated departure time is then considered in the following planning processes. The concept of handling vehicles in groups implies that a vehicle group is stored in an interconnected part of a parking area. The storage operator has therefore to search for a free interconnected area of sufficient capacity for the whole storage period. If such an area is not available for an unloaded vehicle group, a trial is made to extend one of the free areas. An extension can be achieved by relocation, i.e., by the transfer of a vehicle group stored in the immediate neighborhood of this free area to one of the other free areas. If the trial fails, the vehicle group has to be stored in the supplementary car park, which has a less well-reinforced surface and is further away from the quays than the other regular parking areas. In the terminal vehicles are moved by drivers organized in gangs. The terminal operations are therefore very cost-intensive. Hence, the main objective of the storage operator with respect to the storage allocation is to minimize the effort or the costs, respectively, for the vehicle moving processes. In case of a deflection between the estimated and the finally realized departure time estimation, the resulting storage allocation has to be corrected and adapted to reality. This often causes extensive re-planning processes.

The deployment scheduling focuses the coarse and fine grained scheduling of the vehicle moving processes as well as the allocation of drivers. According to the business process analysis, this planning process goes ahead to every moving process. The terminal's own drivers are employed shift-wise. This organizational concept also influences the underlying planning process. The coarse grained deployment scheduling is carried out by a shift co-ordinator, while the terminal's shift supervisors are responsible for the fine grained scheduling. The whole procedure consists of the following steps showing the interaction between the shift co-ordinator and the shift supervisors: In a first step, the shift co-ordinator determines the personnel expenses for every upcoming vehicle moving process. The expense is usually measured in so-called driver-time cycles denoting a time unit of a given length, e.g. fifteen or thirty minutes. In a second step, the shift co-ordinator assigns every moving task to an appropriate shift. The coordination between the shift co-ordinator and the concerned shift supervisors is conducted bilateral in the context of a daily coordination meeting. Then, the responsible shift supervisor executes the fine grained scheduling in the corresponding shift. He determines a fix starting and end time for each moving process and assigns an adequate number of drivers for moving the vehicles. In the case of personnel bottlenecks, the permanently occupied drivers can be supplemented by additional drivers hired by the shift or by the hour from a port-wide workforce pool. In these cases, the transshipment costs per vehicle increase in the same way as for the storage of vehicles in the supplementary car park. The main objective pursued by the shift supervisors is a cost-efficient execution of all assigned vehicle moving tasks with a constant workload all over the shift.

2.4 Analysis of Critical Points

Referring to the analysis of the actual state, the operation of vehicle moving processes cause high personnel costs. These costs are an essential part of the operating expenses of a typical seaport automobile terminal. Form this point of view, the profit of the terminal operator largely depends on a cost-efficient execution of the vehicle moving processes. Therefore cost-oriented planning processes have to be established. The considered actual planning processes, however, show some critical points:

- time-consuming planning processes with a high intensity of communication and coordination.
- limitation of foresighted and comprehensible planning decisions due to the associated planning effort.
- inefficient and inflexible storage allocation.
- no decision support according to strategic questions, such as the increase of the terminal's storage or driver capacities.

Furthermore, these critical points are fortified by the dynamic, fast changing environment in which the planning of the vehicle transshipment is embedded. In the actual situation a manual planning approach assisted by the information systems called PORT-INFO and CAR-INFO is applied. Both information systems only support the retrieval and the analysis of relevant order, storage, and personal data, i.e., an active computer based planning support is not available.

In order to reduce or eliminate these deficiencies, modern planning approaches like multi-agent systems seem to be promising. In the following, such a multi-agent based planning support system will be presented.

3. Vehicle Transshipment Optimization Problem

The development of a multi-agent based planning support system includes always the formulation of a more or less simplified mathematical planning or optimization problem reflecting the relevant planning processes. Here, the resulting vehicle transshipment optimization problem is characterized by a high degree of complexity. Taking advantage of an approach by Mattfeld und Kopfer (cf. [10]), this problem is first described as a planning problem comprising the sub-problems quay management, storage allocation and deployment scheduling (cf. [5]) and then completed with another sub-problem, the estimation of the prospective departure time (cf. [4]).

The quay management problem, which aims to pass unloaded vehicles through the quay area as quickly as possible, is an independent sub-problem. According to the business process analysis, this sub-problem covers the quay arrangement and the incoming buffer store allocation planning processes. In contrast, storage allocation, deployment scheduling and estimation of prospective departure time are closely interwoven. For this reason they form the focus of this paper, which will introduce a multi-agent system for the integrated handling of these sub-problems. To make the focus clear, the handling of vehicles is subsequently described in more detail.

According to the business process analysis, the vehicles are operated in groups and the relevant planning data are known beforehand. In exceptional cases, however, the departure time is not announced by car manufacturers up to the time of the delivery at the terminal. In these cases the prospective departure time has to be estimated to enable a timely planning. The handling of a vehicle group follows a sequence of tasks, the so-called task chain. There exist three types of task chains which can be run through by a group of vehicles:

- type 1: unloading – insertion into storage – removal from storage – loading,
- type 2: unloading – insertion into storage – relocation – removal from storage – loading,
- type 3: unloading – insertion into supplementary car park – removal from supplementary car park – loading.

The execution of each task of a task chain is restricted by a time window. The time windows for tasks are limited by e.g. the docking and sailing times of ships, the arrival of other ships, the arrival of delivery carriers etc. With respect to the planning process the following details apply:

- The port buffer to be used for an unloaded group of vehicles is determined in the preceding storage plan. The outgoing buffer is fixed by the carrier or the customer.

- Every moving task for a group of vehicles has to be carried out by the assigned gang of drivers and completed within an eight-hour shift. For purposes of planning and control of terminal operations, the amount of work covered by a moving task is usually measured in driver-time cycles of a given length. A driver-time cycle is the amount of work which has to be invested to move one vehicle by one driver over a given time span.
- The order of processing the vehicle groups is primarily determined by the sequence in which the vehicle groups are unloaded into the port buffer. This sequence is also applied to the insertion of the vehicle groups into storage, since control of the vehicle stream and clearing of the port buffer are then facilitated.
- For organizational reasons it is practical to carry out the task with gangs comprising constant numbers of drivers. Here, a gang consists of a minimum number of two drivers.

The problem of integrated storage allocation and deployment scheduling can now be formulated as follows (cf. [3] and [5]): let be given a set I of vehicle groups $i, i \in I$, which have to be operated during a given planning period subdivided into eight-hour shifts. Let c_{1i}, $i \in I$, be the cost in driver time-cycles for operating the i-th vehicle group if task chain type 1 is executed, c_{2i}, $i \in I$, the corresponding cost in the case of task chain type 2 and c_{3i}, $i \in I$, the cost of task chain type 3. Let x_{1i}, x_{2i} and x_{3i} be binary decision variables which take the value 1 if the task chain type 1, type 2 and type 3 is executed for the i-th vehicle group and the value 0 otherwise. Since only one of the three task chains can be run through, the cost c_i of operating the i-th vehicle group is: $c_i = x_{1i} \cdot c_{1i} + x_{2i} \cdot c_{2i} + x_{3i} \cdot c_{3i}$, where $x_{1i} + x_{2i} + x_{3i} = 1$, $i \in I$.

Since the planning process aims at the minimization and the balancing of the required drivers, the minimization of the total number of driver-time cycles, TN, that have to be scheduled for transshipping the set I of vehicle groups in the planning period is used as the primary objective function:

$$\min \; TN = \sum_{i \in I} c_i \tag{1}$$

The secondary objective function is then the minimization of the squares of the deviations, SD, between the maximum and the mean driver number over all shifts:

$$\min \; SD = \sum_{t=1}^{T} (fm_t - fa_t)^2 . \tag{2}$$

Here T denotes the number of successive eight-hour shifts t, $t = 1, \ldots, T$, in the planning period, while fm_t denotes the maximum and fa_t the mean number of drivers deployed in the t-th shift.

4. Multi-Agent System for Planning Support

In recent years, multi-agent systems (MAS) have been recommended for supporting complex decision-making problems that can be decomposed into interdependent sub-problems (cf. [1] [8]). According to the literature, MAS are applied to areas ranging from production and transportation planning (cf. [13] [2]) through to business process management (cf. [8]). A multi-agent based approach for supporting the vehicle transshipment problem has not yet been presented by other authors. The MAS for supporting personnel deployment scheduling in a software firm described by Struve (cf. [12]) cannot be transferred to the given situation because the problem structure is different. The MAS proposed here is described in two steps, the first step concerns the architecture and the second the operation of the individual agents in the MAS.

4.1 Representation of the Planning Problem via Autonomous Agents

Considering the organizational structure of the considered automobile terminal, the sub-problems of departure time estimation, storage allocation and deployment scheduling are handled by different decision-makers. An analogous distribution of decision competences can be used in a decision support system for the integrated planning problem considered here.

A distribution of decision competences according to the general idea of agent systems is achieved by the use of a specific type of agent for each of the problems of departure time estimation, storage allocation and deployment scheduling. For the departure time estimation the so-called departure time estimator agent (DEA) is used and for the storage allocation to the so-called area agent import (AAI). In contrast to the departure time estimation and the storage allocation, the shift-based deployment scheduling requires a decentralized decision-making over the time. For this reason, deployment scheduling is carried out by a number of S so-called shift agents (SA_s, s = 1, ... , S). Since a shift agent always takes over the planning for exactly one shift, only a number of $n < N$ of shift agents is concurrently active. A further agent, the so-called planning coordinator agent (PCA), is responsible for the coordination of the local planning activities of the individual agents in the sense of the objective functions (2) and (3). The tasks of the PCA depend on the types of agents to be co-ordinated. The overall architecture of the MAS is shown in Fig.3.

Since the planning competences of the DEA and AAI are rather limited, the coordination between PCA and these agents is reduced to a type of client/server relation. While the PCA takes over the role of a client, AAI and DEA are each in the role of a server. For each vehicle group the following steps are carried out: First, the problem data initially transmitted to the PCA are checked. If the departure time for a vehicle group is unknown, PCA submits the respective estimation problem to the DEA, who estimates the departure time and returns the solution back to PCA. Using this solution, the PCA submits then a concrete storage allocation problem to AAI. Otherwise, if all required data are available at the time of planning PCA submits the storage allocation problem directly to the AAI. Finally, the AAI returns the solution to the PCA.

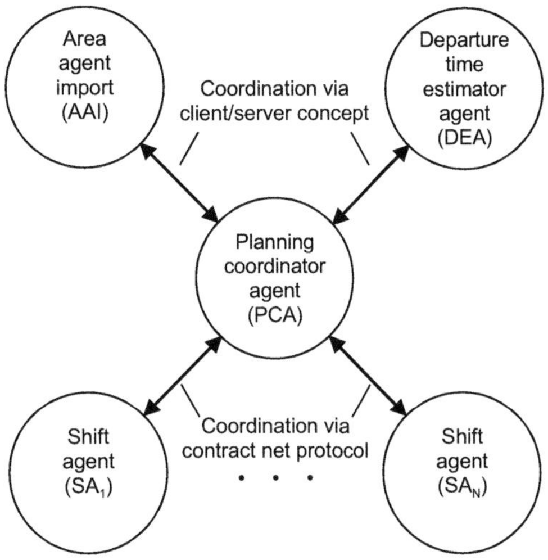

Fig. 3. Architecture of the multi-agent system

A different situation is given for the deployment scheduling for a vehicle group, because each task of the associated task chain can often be carried out in different shifts. Accordingly, an auction mechanism based on the contract network protocol (cf. [9]) seems to be suitable for selecting a shift agent. The contract network protocol used here is oriented to the coordinator variant model introduced by Zelewski for eliminating the coordination deficiencies that occur with decentralized forms of coordination (cf. [11]). Consequently, the PCA assumes the role of the central coordinator, who, for the purpose of managing the flow of vehicle groups, concludes contracts with the AAI as the provider of parking areas and with the shift agents as the providers of driving capacities.

The architecture of the system shown in Fig.3 considers the three typical features of a MAS (cf. [5], [6]). First, the decentralized structure of the underlying planning problem is modeled in conformity with reality. Second, the introduced agents are autonomous and pursue different goals. Third, a mechanism for coordinating the local planning processes is used; this mechanism is embedded in the PCA. The conceptual and procedural realization of these features is specified in the following.

4.2 Operation of the Multi-Agent-System

The operation of the MAS starts with some initializations and ends with some evaluations performed by PCA. In between, five activity phases of the MAS can be distinguished, which are executed until all vehicle groups have been processed. In detail these are:

1. estimation of the departure time for a vehicle group by DEA (if necessary),
2. deployment scheduling (coarse-grained scheduling) for a vehicle group by a SA_S,
3. storage allocation for a vehicle group by AAI,

4. deployment scheduling (fine-grained scheduling) for each task to be executed for a vehicle group by SA_s, and
5. update of the so-called parking area reservation list by PCA (if necessary).

In the following, each phase is described in more detail form the point of view of the associated agent.

4.2.1 Departure Time Estimation

A departure time estimation is carried out by DEA in phase 1 only if the departure time of a vehicle group is not known at the date of the arrival at the terminal. Otherwise this phase is skipped and the system switches over to phase 2.

The environment for estimation is characterized by a high degree of dynamics and uncertainty, because the individual demand strategies of the car manufacturers are difficult to forecast. For this reason, a dynamic estimation method has to be developed which is also able to recognize mutations in the planning periphery and to adapt its own planning strategy to new circumstances. Traditional methods and heuristics often bare significant shortfalls with respect to these requirements. Thus, specialised learning classifier systems (LCS) have been recommended for supporting such complex decision-making problems in dynamic environments in the last few years (cf. [7], [9]).

A typical LCS is an induction self-learning system based on a set of logical rules called classifiers. Each rule or classifier, respectively, has the structure "if <condition> then <action>". A LCS is usually being optimized by using an evolutionary algorithm, mostly a genetic algorithm, as learning rule. During the learning process the rule priorities, the so-called strengths, are changed. In case of a success, current and previous activated rules are encouraged. In case of a failure, these rules are devaluated. In order to adapt the planning process to varying challenges and problem data, an evolutionary method is also used for generating new rules.

Because of the characteristics of the underlying estimation problem, a learning classifier approach seems to be applicable here. The LCS considered here, however, is based on a slightly modified architecture. The evolutionary methods mentioned above are only used for generating new rules but not as so-called learning rules. The resulting one-sided direction of the classifier population towards a few but precise estimating classifiers seems to be an advantage in this special case, because the car manufactures usually apply a more or less continuous demand strategy. In the following, the functionality of the LCS is described in more detail.

The underlying rule types for the departure time estimation are represented by classifiers being composed of a condition part and an action part (cf. [11]). Fig. 4 shows the problem representation for the underlying estimation problem.

In the condition part, the relevant order parameters – here the car manufacturer and the final destination – are encoded. The number and the specification of these parameters are not limited. The control parameters are encoded in the action part of the classifier. As to the departure time estimation, these are the storage time denoting the estimated value for the type of vehicle specified in the condition part and the classifier strength describing the fitness of the classifier.

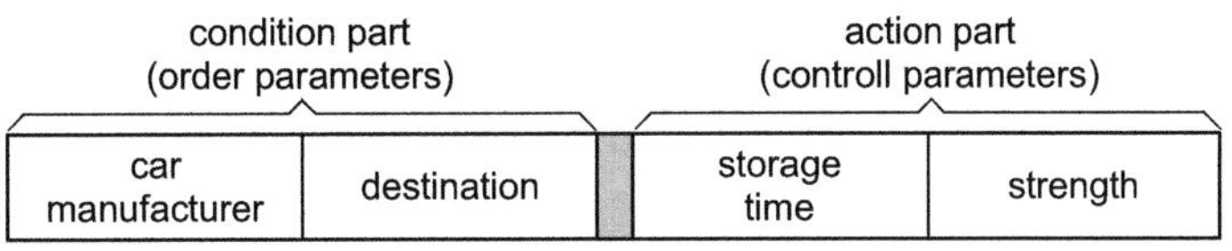

Fig. 4. Estimation rule representation via classifier

In order to estimate a departure time for each vehicle group, the following steps are executed:

1. identification of the relevant classifiers,
2. selection of a classifier for estimating,
3. departure time estimation, and
4. evaluation of the estimation.

In the first step, the order parameters of a vehicle group are compared with the condition part of every classifier. All classifiers with a condition part corresponding exactly to the order parameters are selected. If no corresponding condition part is found a new classifier is created with the parameters of the requested vehicle group and finally selected. The average values of the storage time and of the strength value are then used as (initial) values of the respective parameters over all existing classifiers.

In the second step, the selection of an appropriate classifier is carried out by an auction mechanism. Therefore, all selected classifiers submit an offer for departure time estimation. The offer with highest classifier strength is finally selected.

In the third step the estimation is carried out and the result is finally sent to the PCA.

In order to optimize the estimation, the estimated values are evaluated in a fourth step after the final delivery date is announced. In case of an exact estimation or even a marginal overestimation, the strength of the selected classifier is increased. On the other hand, a strong overestimation of thirty or even more shifts causes a decrease. In contrast, an underestimation always leads to a reduction of the classifier strength because an underestimation causes expensive vehicle relocations in general. An underestimation of less than two shifts leads to a slight reduction, an underestimation of more than two shifts to a significant reduction of the strength. In order to eliminate weak classifiers without chances of being selected for estimating, the strength of all classifiers is reduced by a so called life tax at each estimating process. In all cases of an overestimation and underestimation the DEA also corrects estimated departure times (increase or reduction) towards the new circumstances.

4.2.2 Deployment Scheduling (Coarse Grained Scheduling)

In the second phase, the deployment scheduling (coarse grained scheduling) for a vehicle group is performed on the level of individual tasks. For this purpose, the three interaction steps of the applied contract net protocol are executed for each task. These steps are: invitation to bid, submission of offer, and order.

As to the invitation to bid, the PCA first determines the shifts that overlap wholly or partly with the time window of the task. The respective shift agents are then supplied with the commission data. In the second step, each of the contacted shift agents returns an offer to the PCA. In the third step, the PCA selects the most favorable shift for carrying out the individual task. If the task time window overlaps with one shift only, then this shift is selected. Otherwise, all offers are evaluated by means of a lexicographical function. The shift is selected which is, in decreasing order, characterized by the longest overlapping period, the largest number of free driver-time cycles for regular drivers, the largest number of free driver-time cycles for hired drivers, the smallest number driver-time cycles for hired drivers, and the lowest shift index.

4.2.3 Storage Allocation

In the third phase, the storage allocation for a vehicle group is carried out in two steps. In the first step, PCA submits data on the respective vehicle group to the AAI. On the basis of these data AAI determines a parking area for the vehicle group and returns his decision to PCA. To determine a parking area, the AAI has the option of two algorithms:

- simple heuristic procedure (HP) or
- genetic algorithm "multi-sequence gene" (MSG), cf. [3] and [5].

The central concept used by the heuristic HP is a “multidimensional priority list” consisting of several sub-lists. For each combination of a port buffer at the quay side with an outgoing buffer at the delivery side a corresponding sub-list is included in the multidimensional priority list. A sub-list contains the corresponding de facto transportation times for a vehicle in minutes for all transportation paths connecting a given port buffer via one of the regular parking areas with a given outgoing buffer. Each entry in a sub-list consists of a parking area identifier and a time coefficient. The entries are sorted with increasing order of the coefficient values.

The multidimensional priority list is used to determine a parking area for a vehicle group as follows.

1. First, the heuristic attempts to place the vehicle group in a single parking area. If one or more parking areas with sufficient free capacity are available, the area with the highest priority or lowest time coefficient, respectively, is selected – this corresponds to a type 1 task chain. Otherwise step (2) is carried out.
2. In the second step, an attempt is made to enlarge a free area by relocating vehicles according to the type 2 task chain. If there exist one or more pairs of parking areas that allow the vehicle group to be stored over the given time span, then the pair with highest priority or lowest cumulated time coefficients, respectively, is selected. If this attempt also fails, step (3) is carried out.
3. In the third step, decision is taken to store the vehicle group in the supplementary car park. This assignment corresponds to a type 3 task chain.

The problem representation used by the genetic algorithm MSG follows the concept of the parking area priority list. This means that a chromosome represents a multidimensional priority list which consists of several sub-lists. Each sub-list consists of a permutation vector of parking area identifiers which corresponds to a combination of a port buffer with an outgoing buffer. The decoding procedure is similar to the heuristic HP, however, in contrast to HP a multidimensional priority list is used and relocations of vehicles are excluded. The representation used by MSG is therefore closer to reality.

4.2.4 Deployment Scheduling (Fine Grained Scheduling)

In the fourth phase, the deployment scheduling (fine grained scheduling) for a vehicle group is also performed on the level of individual tasks. For each task of the respective task chain determined for a vehicle group, the PCA submits the order for planning the task to the shift agent of the selected shift. The shift agent executes the order in three steps:

1. First, the operation time of the task which leads to the lowest loss of time is determined. A loss of time occurs if the operation time of the task, which is given in whole time cycles of 15 minutes, is greater than the actual driving time measured in minutes. Time losses may also occur in situations where each driver moves more than one vehicle.
2. For the determined operation time, the starting time of the task is then fixed so that the new local load peak for the drivers caused by the task is as low as possible.
3. In the third step, the task is scheduled taking account of the starting time determined in the second step.

Finally, the shift agent returns the result of the fine-grained scheduling to the PCA for updating the parking reservation list in the next phase. With this activity, the planning process for a vehicle group is finished.

4.2.5 Updating the Parking Area Reservation List

In the fifth phase, the parking area reservation list is updated. For each regular parking area this list contains the number of parking areas that are still free, the partial space reservation for vehicle groups and the start and end of reservation. The parking area reservation list is managed by AAI and must always be updated if a vehicle group is stored on a regular parking area, i.e., a type 1 or type 2 task chain is carried out. In these cases the PCA submits a corresponding order to AAI. The parking area reservation list is used by both of the storage allocation algorithms, HP and MSG.

5. Evaluation of the Multi-Agent Based Planning Approach

For the implementation of the MAS the language Java was used. The MAS was tested on a Pentium IV PC with a cycle frequency of 2.0 GHz. In the following, the test concept and the numerical results derived by series of test runs are described.

For the test a total number of 15 problem instances were generated. All instances are based on the same locations of parking areas. In conformity with practical conditions in the already addressed European seaport the following facilities are assumed: two port buffers for 750 and 1000 vehicles respectively, ten standard parking areas with a capacity of 7000, 6000, 5000, 5000, 4000, 4000, 3000, 3000, 2000, and 1000 vehicles, six outgoing buffers with a capacity of 1000, 1000, 750, 500, 500, and 500 vehicles, and a supplementary car park with sufficient capacity. For the generation of problem instances a planning period of 750 shifts was chosen. According to the practical situation in the mentioned seaport, a total of 400000 vehicles are imported within that period of time. In accordance with the manufacturers, the lengths of stay of the vehicles in the terminal are randomly varied over the problem instances. For every 5 instances a defined length of stay is used. While the average length of stay varies between 100 and 200 shifts, the spread of the averages lies in a range of approximately 10% to 20%. For about 30% of all vehicle groups generated in one problem instance, the departure time is unknown at the time of delivery, so that a departure time estimation is necessary. In conformity with reality, a continuous demand strategy can be considered on the part of the car manufacturers. The seed value z of the random number generator used was varied. The seed values lie in a range of 100 to 500 and for every 3 instances the seed values 100, 200, 300, 400, and 500 are used.

On the basis of these problem instances two series of numerical experiments were carried out, each considering a specific planning situation:

- From the operational point of view of the responsible operations manager, the values of the global objective functions that can be achieved through the application of the MAS and the embedded allocation methods HP and MSG in combination with the learning classifier system are of main interest.
- From a more strategic point of view, the question of how many drivers should permanently be employed is in the foreground.

In the following, the operational and the more strategic aspects of the application of the MAS are considered.

5.1 Results form an Operational Point of View

As to the operational aspect, each of the 15 problem instances was calculated once using the heuristic procedure HP, and the genetic algorithm MSG, each in combination with the described learning classifier system for departure time estimation. The objective function values that were achieved, measured in quarter-hour time cycles, were always averaged over all 15 instances (see Table I). In order to examine the influence of different relations between regular and hired drivers on the solution behaviour of the MAS, different numbers of regular drivers were introduced.

From Table 1 it can be depicted that the genetic algorithm MSG dominates clearly over the heuristic HP for all numbers of regular drivers. The average improvement in the values of the objective function (1) achieved with MSG in comparison with HP ranges from approximately 4.15% for 20 and approximately 6.46% for 100 regular drivers. Similar improvements were calculated for the objective function (2).

Table 1. Averaged objective function values for solution methods HP and MSG.

No. of regular drivers	Solution method	Averaged objective function values Mean total numbers of driver time-cylces (objective function (1))	Mean squares of deviations (objective function (2))
20	HP	1.944.626	3.502.712
	MSG	1.778.445	3.297.325
30	HP	1.967.357	3.373.498
	MSG	1.788.980	3.259.144
40	HP	1.984.200	3.295.305
	MSG	1.783.064	3.148.428
50	HP	2.005.530	3.271.621
	MSG	1.792.501	3.035.176
60	HP	2.007.154	3.143.381
	MSG	1.797.560	2.978.934
70	HP	2.009.307	3.100.632
	MSG	1.802.058	2.907.524
80	HP	1.997.427	3.002.486
	MSG	1.788.750	2.878.044
90	HP	1.980.744	2.959.673
	MSG	1.765.243	2.837.384
100	HP	1.957.906	2.887.345
	MSG	1.751.267	2.801.479

On the basis of the considered demand strategy the integrated learning classifier system shows a robust behavior according to the deviation between the estimated and realized departure time. The deviation values are measured in shifts per estimated vehicle group.

Table 2. Averaged deviation values for solution methods HP and MSG.

No. of regular drivers	Average deviation values shifts per estimated vehicle group HP	MSG
20	4.8	4.6
30	4.4	4.5
40	2.9	3.0
50	2,6	2.6
60	2.8	2.5
70	2.5	2.3
80	3.9	3.5
90	4.4	3.7
100	3.7	4.1

According to Table 2, the deviation is limited on an average of about 3.5 shifts per estimated vehicle group over all problem instances for the MSG allocation method using optimized classifier parameters (cf. [4]). Thereby, the overestimation outweighed the more problematic underestimation clearly. Extensive tests show that the results for departure time estimation are obviously independent from the choice of the allocation method. The achieved average difference of about 0.1% listed in Table 2 can be neglected.

For the heuristic HP the average computing time per problem instance amounts to 15 seconds including the departure time estimation; the corresponding value for MSG amounts to 1450 seconds. These computing times were, however, derived for computational experiments covering 750 shifts or one year and are therefore of no practical relevance. Orders are usually known between one and six weeks beforehand. This means that the MAS could be applied, e.g., every two or three days with the consequence that computing times would be reduced dramatically. The MAS seems therefore to be a suitable tool for supporting the continuous planning of the current terminal operations.

5.2. Results from a Strategic Point of View

From a more strategic point of view, the question of how the total personnel costs can be influenced is of high interest. The reason is that transshipment per seaport terminals is particularly labor-intensive and leads to high personnel costs. The level of the total personnel costs is mainly influenced by regular drivers and by the surcharge rate for hired drivers. In order to show the influence of these critical factors, the sum of the relative costs of regular and hired drivers was calculated for different numbers of regular drivers and for different surcharge rates. The numerical results obtained are shown in Fig.5.

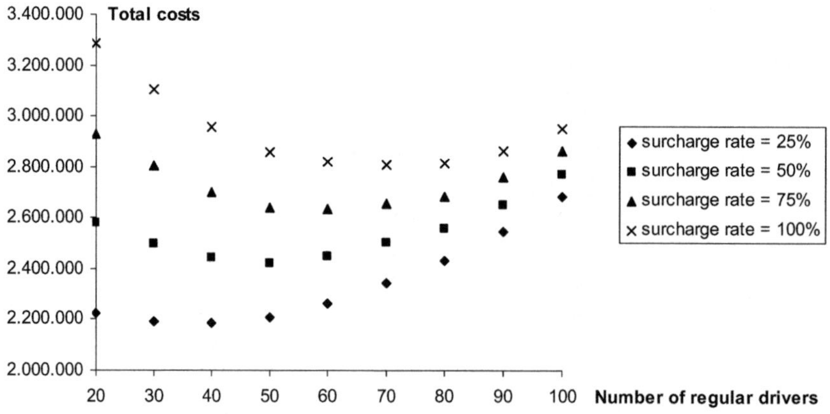

Fig. 5. Total cost curves for different numbers of regular drivers and for different surcharge rates

Idle times of regular drivers are relevant now. Therefore, the relative costs of regular drivers are proportional to their number. The relative costs of hired drivers were determined as the product of the corresponding mean total numbers of driver-time cycles (including idle times of hired drivers) and the cost factors 1.25, 1.5, 1.75, and 2.0 for a surcharge rate of 25%, 50%, 75%, and 100%.

As Fig.5 shows, there exists always a cost optimum. The position of the optimum depends on the level of the cost surcharge per hired driver. With the increase of the surcharge rate from 25% to 50%, 75%, and 100%, the position of the cost optimum is shifted from about 40 to about 50, 60, and 70 regular drivers. This means that an increase in the costs of hired drivers can be compensated by extending the number of regular drivers.

The optimization potential indicated by Fig.5 is of practical relevance. If the number of regular drivers is varied in the given range, then the difference between the minimum and the maximum total costs amounts to 18%, 13%, 11%, and 16% for a surcharge rate of 25%, 50%, 75%, and 100%. These economic dimensions seem to justify the multi-agent based decision support approach presented here from a more strategic point of view, too.

6. Conclusions

To cope with the growing pressure of competition, the automobile terminal operators have to assume arrangements in order to rationalize the underlying workflows and to improve logistic services and customer satisfaction. The modeling of the underlying business processes show that nearly all planning processes are presently carried out manually by using two pure information systems. In order to reduce or eliminate the resulting critical points, a MAS for supporting the planning processes was presented here.

The planning of vehicle transshipment of imported finished vehicles via a seaport automobile terminal focused in this paper includes two main planning tasks: the allocation of parking areas for the temporary storage of vehicles and the allocation of drivers to vehicles that have to be moved in the terminal area. In addition, it may be necessary to estimate the delivery times of vehicles in case this times are unknown at the time of planning. These planning tasks are assigned to three different agent types of a multi-agent system (MAS). A further agent, the coordinator agent, coordinates the activities of the other agents in such a way that the demand for drivers in the planning period is minimized and balanced.

The multi-agent system was tested using 15 problem instances, which were generated in conformity with the situation in a typical European seaport. The test aimed at evaluating the MAS from an operational and, in addition, a more strategic point of view. The test led to the following results:

- The MAS seems to be a suitable tool for supporting the continuous planning of terminal operations.
- Further, the planning results can be improved significantly if a genetic algorithm instead of a heuristic is used for the allocation of parking areas.

- The integrated learning classifier system shows a robust behavior according to the deviation between the estimated and the finally realized delivery times.
- As to the strategic aspect, the total relative costs for regular and hired drivers have an economically relevant minimum, which is shifted towards a higher number of regular drivers for an increasing surcharge rate for hired drivers.

The multi-agent system developed here seems to be suitable for supporting not only current dynamic storage allocation and deployment scheduling, but also the determination of a stock of permanently employed drivers from a more strategic point of view.

References

[1] Brenner, W., R. Zarnekow, and H. Wittig: Intelligente Softwareagenten – Grundlagen und Anwendungen. Springer-Verlag, Berlin u.a. 1998.

[2] Fischer, K., Müller, J. P., and M. Pischel: Cooperative Transportation Scheduling: An Application Domain for DAI. In: Journal of Applied Artificial Intelligence, Special Issue on Intelligent Agents 10, 1996 1, pp. 1-33.

[3] Fischer, T. and H. Gehring: Ein genetischer Algorithmus zur Stellflächenplanung im Rahmen des Fahrzeugumschlagproblems. In: Sebastian, H.-J. and T. Grünert (eds.): Logistik Management, Teubner-Verlag, Berlin/Heidelberg 2001, pp. 365-376.

[4] Fischer, T.: Ein Learning-Classifier-System zur Schätzung von Lagerverweildauern im Rahmen des Fahrzeugimporte, In: Bortfeldt, A., Fischer, T., Homberger, J., Pankratz, G., and R. Strangmeier: Planen, Lernen, Optimieren – Beiträge zu Logistik und E-Learning. Diskussionsbeitrag Nr. 338 des Fachbereichs Wirtschaftswissenschaft, FernUniversität in Hagen, Mai 2003, pp.37–54.

[5] Fischer, T. and H. Gehring: Planning Vehicle Transhipment in a Seaport Automobile Terminal Using a Multi-Agent System. In: Makowski, M., Nakamori, Y. and H.-J. Sebastian: Advances in Complex Systems Modelling. In: European Journal of Operational Research, Volume 166, Issue 3, 2005, pp. 726-740.

[6] Fischer, T.: Multi-Agenten-Systeme im Fahrzeugumschlag – Agentenbasierte Planungsunterstützung für Seehafen Automobilterminals, Deutscher Universitäts-Verlag Gabler Edition Wissenschaft, Wiesbaden 2004.

[7] Holland, J.H. and J.S. Reithmann: Cognitive Systems Based on Adaptive Algorithms. In: Waterman, D.A. and D. Hayes-Roth (eds.): Pattern Directed Inference Systems, Academic Press, New York, 1978.

[8] Jennings, N. R., Faratin, P., Norman, T. J., O'Brien, P., and B. Odgers: Autonomous Agents for Business Process Management. In: International Journal of Applied Artificial Intelligence 14, 2002, pp. 145-189.

[9] Lanzi, P.L., Stolzmann, W., and S.W. Wilson: Learning Classifier Systems – From Foundation to Applications, Springer Verlag, Berlin 2000.

[10] Mattfeld, D. and H. Kopfer: Terminal Operations Management in Vehicle Transhipment. Universität Bremen, Lehrstuhl für Logistik, Technical Report 2001.

[11] Richard, R.A.: Zeroth-Order Shape Optimization Utilizing a Learning Classifier System. Stanford University Press, 1995.

[12] Struve, B.: Agentensysteme zur Modellierung eines Personalplanungsprozesses in einem System- und Softwarehaus. Diplomarbeit, Lehrstuhl Wirtschaftsinformatik, FernUniversität in Hagen 1998.

[13] Zelewski, St.: Multi-Agenten-Systeme – ein neuartiger Ansatz zur dezentralen Feinterminierung von Produktionsaufträgen. In: Wildemann, H. (ed.): Innovationen in der Produktionswirtschaft. Transfer-Centrum-Verlag, München 1998, pp. 133-166.

Commitment Based Sense-and-Respond Framework for Manufacturing Supply Chain

Jun-Jang (JJ) Jeng, Markue Ettl, and Jen-Yao Chung

IBM T.J. Watson Research Center
Yorktown Heights, New York, USA
{jjjeng,mettle,jychung}@us.ibm.com

Abstract. In this paper, we present an agent based framework for building manufacturing supply chain systems for sensing, interpreting, predicting, automating and responding to business activities and the goal is to minimize the latency to make well-informed and prompt business decisions. Ideally, there should be almost zero-latency between the cause and effect of a business decision. This framework is coined as CSR – Commitment Based Sense-and-Respond framework. CSR enables analysis across corporate business processes, notifies the business of actionable recommendations, or automatically triggers business operations, effectively sealing the gap between business intelligence systems and business processes. A CSR based system is an adaptive environment for the continuous monitoring of business performance for discovering business opportunities and unexpected behavior. We will present the foundation of CSR and the CSR reference architecture that takes full advantage of sense-and-respond technology and enables the implementation of distributed, scalable, near real-time manufacturing supply chain management. We discuss the issues and our approach of using CSR for building manufacturing supply chain systems.

1 Introduction

Manufacturing supply chain systems comprise of sophisticated interactions among multiple systems and decision makers, highly complex computation in order to optimize manufacturing plans, and non-trivial integration issues embodied in business data and processes themselves. Traditional manufacturing processes are usually linear and rigid; and once done, will be very hard to change. Modern manufacturing supply chain systems, however, attempt to replace inflexible supply chains with more adaptive supply networks that will result in configurable and more effective management of cross-company manufacturing and logistics to achieve competitive advantage. The ultimate goal of many companies is to build adaptive & "sense and respond" manufacturing supply chain systems flexible enough to sense and respond to changes in their manufacturing environment [2]. A new breed of management platforms and tools coined Sense-and-Respond systems are emerging and commanding corporate attention, forcing businesses to consider alternate platforms with enhanced capabilities for business performance management.

J.-J. Jeng et al.: Commitment Based Sense-and-Respond Framework for Manufacturing Supply Chain, Studies in Computational Intelligence (SCI) 28, 395–417 (2006)
www.springerlink.com

The principal aim of creating a Sense-and-Respond system is to enable and equip an effective decision making process for an enterprise. An adaptive manufacturing supply chain senses the changes of its environment, assesses the situation, executes an analysis, and responds to the changes with a response deemed appropriate. The emergence of e-business technologies has dramatically altered the context in which the decision-making events occur. While the fundamental human and organizational processes remain largely unaffected, e-business has imposed new constraints and demands on the decision maker to provide improved service to the customers:

- The perpetual changes and unpredictably of manufacturing supply chain has made it imperative for decisions to be made in a prompt manner. Process participants must have instant access to information that is pertinent to the current business context.
- Inter-organizational processes require visibility for each involved party, improved alliances and co-operative arrangements and the complete utilization of all resources accessible to the enterprise.
- Result-oriented or outcome-oriented business metrics allow the decision maker to focus on only the relevant information at hand which is germane to their business tasks. Because of the explosion of information widely available on modern manufacturing supply chain systems, decision makers are exposed to a massive and sometimes overwhelming amount of data, which underscores the absolute necessity of filtering out all irrelevant information - a potentially costly and time-consuming task.
- Increased telecommuting has restricted the opportunity to exchange informal information and decision making. The reality is that decisions must frequently be made in real-time to allow a straight-though processing.

These factors suggest that a curtailed focus on the provision of information and analysis tools is insufficient. To maximize the effectiveness, decision support must adopt a more comprehensive view of decision-making. This decision process is often referred to as Sense-and-Respond and was popularized by Stephan Haeckel [1]. In general, the Sense-and-Respond system can be categorized as a system that is continually interacting with its business organizations and supporting organizational agents (software or humans) to make well-informed and prompt decisions. Such a system is capable of autonomous actions in order to satisfy its policy constraints. An adaptive enterprise needs to consider leveraging business processes and systems beyond its own limited boundaries. Yet the current paradigm for delivering quality-of-service to business process applications via the vertical integration of platform-specific components and services does not adequately function in today's loosely coupled business environment.

This paper presents an agent based framework, coined as CSR, for development manufacturing supply chain systems. CSR stands for Commitment Based Sense and-Respond framework. CSR has been exploited to define, develop, and validate manufacturing supply chain system over years. Since the totality of all activities of manufacturing can be viewed as a distributed problem solving process, an agent based approach was intentionally chosen to implement such systems, and CSR is the foundation used to develop such framework. According to Merriam-Webster's collegiate dictionary, commitment is "an agreement or pledge to do something in the future." Commitments can be between trading partners (called external commitments), or between internal parties within a business (called internal commitments). The definition of business commitments captures not only the current stable states ("agreement") but also the future actions ("to do something") and constraints (not to do something); therefore it is an appropriate concept to describe certain types of business relationships and interactions that may require both agreements and actions from participating parties. A set of business commitment establishes the agreement of a management platform to its customers regarding how their artifacts are to be managed in this platform. In our opinion, the concept of commitments nicely fits into the manufacturing supply chain that likely manages multiple trading partners (customers and suppliers) and multiple internal parties (internal distribution centers, multiple factory shops). As will be elaborated in the following sections, the concept of commitment is the corner stone of the CSR architecture. In the manufacturing scenario, commitment can be used for describing management use cases, constraining agent behavior, measuring manufacturing metrics, enforcing desired actions, rendering commitments to other agents and so on.

CSR supports the Sense-and-Respond concept by providing modular software support for each of the four indicated processes. Measurement/Sensing is supported by Managed Objects that interface with business process systems and by Managed Agents which provide data extraction, transformation, and loading capabilities. Interpretation is provided by Situation Agents that identify situations that require attention such as unexpected events and out-of-tolerance metrics. Analysis is performed by Analytics Agents that support risk-based analysis of potential response actions. Such response actions often take the form of optimizations that attempt to achieve enterprise goals subject to current business constraints. Decisions are made by users who interact with Dashboard Portals that display relevant information. Actions are performed by Management Agents that interface with Managed Objects to trigger exception processes in business process systems.

The rest of this paper is organized as follows. Section 2 provides the scenarios considered in this paper as the running example. The conceptual foundation of CSR is described in Section 3. The notion of commitments and CSR are described in Section 4. The CSR architecture and implementations will be presented in Section 5. Related work and concluding remarks will be given at Sections 6 and 7, respectively.

2 Scenarios

This section provides the scenarios of the domain where CSR is aimed for. For the sake of illustration, we are stating an example from the domain of microelectronic manufacturing before presenting a solution based on CSR.

Problem Statement

The core flow for microelectronic manufacturing contains the sub-flows such as:

- From wafers to devices: The wafer is a round thin piece of silicon that looks similar to a CD. The wafers go through an elaborate procedure that has cycle times (time to complete the task) of 1 to 3 months, where millions of circuits are carefully etched onto it. When the wafer is completed, it is transmitted to final test and then cut into small individual rectangular shaped parts called devices. Typically cycle times are few days to 2 weeks.
- From devices to modules: The devices are then placed on a substrate and packaged to create a module which takes between few days to 2 weeks.
- From modules to cards: Modules are then combined together on a card that is the topmost-level component.

Depending on the customer, a microelectronic manufacturer may dispatch wafers, devices, modules, or cards. Within each major manufacturing activity there is series of operations and extensive testing. Operations are grouped into higher level modules such as sectors, work centers, and stages. The actual matching models deployed by the manufacturing planning team work with dynamically established stages as the core manufacturing activity or decision point. The number of levels depends on the specific model.

The key decision points and corresponding decision support systems in the supply chain management (SCM) process can be categorized by the supported supply chain component and its time frame or decision tier.

An SCM process manages the flow of activities from order creation, order processing, order completion to product delivery consists of business processes, control and data flows, and decision structures. An SCM process consists of four steps: demand creation/forecasting, production planning, manufacturing execution, and available to promise (ATP).

SCM decisions in the semiconductor industry typically fall into one of four decision tiers: strategic, tactical, operational, and dispatch (response). The categories are based on the planning horizon, the apparent width of the opportunity window, and the level of precision required in the supporting information.

1. The first decision tier, strategic scheduling, is driven by the time frame or lead time required for business plan, resource acquisition, and new product introduction. Here decision makers are concerned with a set of problems that are three months to seven years into the future. Issues in this tier include general availability of tooling and laborers, manufacturing process changes, changes in or risk assessment of demand for existing product, required or expected incremental improvements in the production process, lead times for additional tooling, manpower and planning.

2. The second tier, tactical scheduling, deals with problems the enterprise encounters in the next week to six months. Issues considered are made of yields, cycle times, and binning percentages, delivery dates estimated for firm orders, available "outs" by time buckets estimated for bulk products, and daily going rates for schedule driven product set.
3. The third tier, operational scheduling, deals with the execution and achievement of a weekly plan such as the shipments are made, serviceability levels to be measured, recovery actions to be taken. Tools typically use in support of daily activities are material resource planning (MRP), decision support, recovery models, prioritization techniques and deterministic forward schedulers.
4. The fourth tier, real-time response system, addresses the problems of the next hour to a few weeks by responding to conditions as they emerge in real time and accommodate variances from availability assumed by systems in the plan creation and commitment phases. Usually, analytics agents are used to render response based upon commitments, business policies, and business rules.

CSR is expected to support the aforementioned decision tiers. CSR dispatches scheduling decisions that are concerned with the control of the actual manufacturing flow or logistics; and instructs the operator the next steps to achieve current manufacturing commitments. CSR can also make decisions concerning trade offs between running test lots for a change in an existing product or a new product and running regular manufacturing lots, lot expiration, prioritizing late lots, positioning preventive maintenance downtime, production of similar product to reduce setup time, down stream needs, simultaneous requests on the same piece of equipment.

The following use case describes the scenarios for continual demand-driven build plan and inventory optimization mainly focusing on the operational scheduling tier and real-time response decision tier. End-of-quarter revenue targets (per module family) are released/updated after the meetings among business line managers and executives. A business line manager (BLM) has a pre-determined set of module families for which he/she has financial responsibility and, therefore, whose actual revenue (accumulated so far) and revenue outlook (for remaining weeks in the current quarter) he/she is interested in tracking against the revenue target of the current quarter. Whether the progression of the accrued revenue is on target or above target or below target is determined by the system using a wineglass model [3]. The BLM has concerns for the "below target" situations and may choose to set up corresponding personal alerts for them. The above-mentioned monitoring capability is provided through a business dashboard for microelectronic manufacturing. We use the term CSR to indicate the system that will fulfill the requirements delineated by the use cases.

Main Course

1. On the ith day of the current week, a BLM selects a set of saleable part numbers (they are built from the same common wafer and they are under the BLM's financial responsibility) to view the future weekly actual sales and planned demand quantities through CSR monitoring portal.
2. Detection of Situation and Alert:
   a. CSR issues an alert showing the current sales quantities of some selected saleable part numbers in the nth week are out of their bands. A band is defined as an n-week-ahead statistically acceptable interval for the current sales quantity.
   b. CSR recommends adjusting the planned demand quantities and safety stock requirements for the nth week.
3. Recommendation:
   a. CSR invokes demand planning module and inventory planning module to provide recommended demand quantities and safety stock requirements for the nth week.
   b. CSR disaggregates the weekly demand quantities into daily demand quantities.
   c. CSR recommends altering daily build plan in order to optimally match new daily demand statements, thus achieve high serviceability, and minimize manufacturing and inventory costs.
   d. CSR invokes MRP explosion and implosion module to generate optimal daily build plan, including common wafer start quantity and manufacturing release quantity in each part number level within the BOM chart associated with the selected saleable part numbers. The optimal build plan minimizes manufacturing requirements as well as resulted inventory levels, hence minimizes the manufacturing cost and inventory cost.
4. Prediction and Risk Assessment:
   a. CSR predicts the to-be manufacturing cost, inventory cost, and service level associated with the selected saleable part numbers based on new demand statements and new build plan.
   b. CSR predicts the as-is manufacturing cost, inventory cost, and service level associated with the selected saleable part numbers based on new demand statements and old build plan.
   c. CSR concludes the financial and serviceability benefits of applying newly recommended demand statements, safe stock requirements, and build plan.
5. Decision Making by BLM:
   a. The BLM summarizes the financial and service benefits and reports to the strategic management team (BLE's and executive's) for their approvals.
   b. Upon the approval, the BLM releases the new build plan to sites for manufacturing execution. The BLM releases the new demand statements to procurement team to alter buy plan.

A CSR Solution

CSR has been built to support the aforementioned decision tiers. CSR dispatches scheduling decisions that are concerned with the control of the actual manufacturing flow or logistics; and instructs the operator the next steps to achieve current manufacturing commitments. CSR can also make decisions concerning trade offs between running test lots for a change in an existing product or a new product and running regular manufacturing lots, lot expiration, prioritizing late lots, positioning preventive maintenance downtime, production of similar product to reduce setup time, down stream needs, simultaneous requests on the same piece of equipment.

Figure 1 gives an overview of the functions for the CSR system for microelectronics manufacturing. The functions are indicated in Figure 1 as triangles with a number and include: 1) the integration and monitoring of events, 2) the generation and measurement of key performance indicators (KPI's), 3) the detection of business situations and exceptions, 4) the optimization of a business situation, 5) the performance prediction of optimization alternatives, 6) the risk assessment of the optimization alternatives, 7) the automatic recommendation of a decision, 8) the actual decision making, and finally 9) the response and invocation of actions in the operational target environment. The CSR system is built around three major functional units: a data management solution supporting real-time data, a management agent system, and an enterprise dashboard.

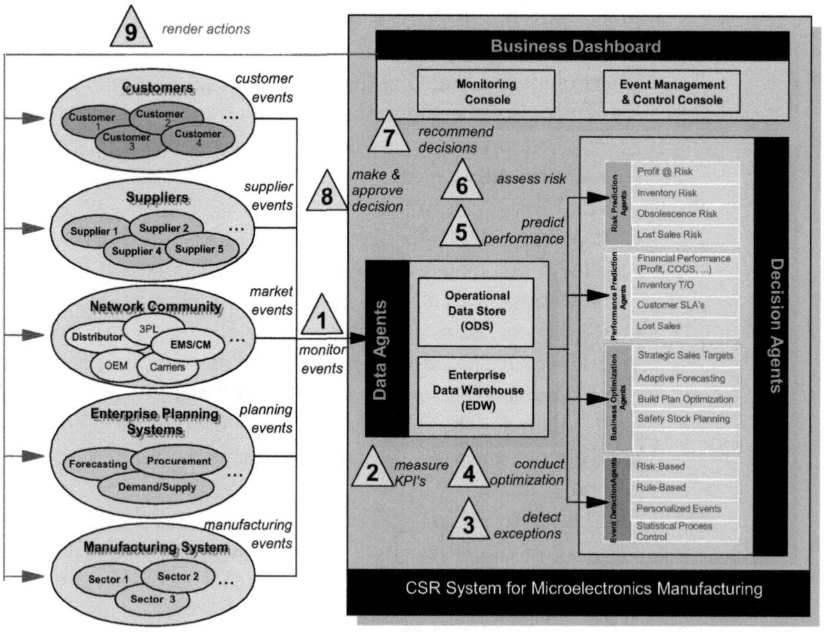

Figure 1: Sense & Respond in the Microelectronics Scenario

The CSR data repositories are used to store business event trails and performance data that originate from enterprise applications and advanced planning systems. Business metrics are updated in real time as new events are posted. The CSR management agents provide business intelligence and analytics to improve the performance of the enterprise. It consists of processes and algorithms that enable real time detection and responses to event conditions. Responses can be as simple as sending an alert to a dashboard, or as complex as loading information into analytical models and changing the operating plan based on the output of the analytics.

Information feedback from the operation of enterprise applications is critical to ensure visibility and control over the business processes enabled by such applications. For the CSR system we provide application adapters that facilitate the collection of enterprise applications, from customer relationship management, to procurement, to advanced planning and scheduling, to the manufacturing floor. The volume of business events, such as a release of a purchase order, receipt of a sales order, shipment of a customer order, or the completion of a manufacturing lot, demand high levels of data integrity and transaction processing. This is fully supported by a flexible and extensible grid-based software infrastructure.

The system's ability to identify potential out of tolerance situations, whether to unexpected fluctuations in supply and demand, or emerging customer, partner, and supplier needs, is enabled by analytical exception detection agents. These agents utilize standardized or configurable measurements to observe business events; for example to ensure that enterprise revenue goals are being accomplished. The policies are managed pro-actively. Alert messages inform business process owners in advance if a new trend is emerging and actions must be taken.

Finally, a CSR system provides a suite of domain-dependent optimization, performance prediction, and risk assessment agents that make exception management even more effective. The agents adopt existing cost structures and business process flexibility, and recommend optimized business policies and actions that drive business performance to higher levels of productivity, efficiency, and financial predictability.

An example of a typical use case for continuous demand-driven build plan and inventory optimization in the domain of microelectronic manufacturing can be described as follows. End-of-quarter revenue targets (per module family) are released/updated after the meetings among business line managers and executives. A business line manager has a pre-determined set of module families for which she has financial responsibility and, therefore, whose actual revenue (accumulated so far) and revenue outlook (for remaining weeks in the current quarter) she is interested in tracking against the revenue target of the current quarter. The actual demand, revenue, revenue outlook and inventory levels are key metrics for the business line managers for making adjustments in the manufacturing process. The following scenario illustrates a scenario how the business line manager utilizes the CSR system.

- The CSR system receives events from various source systems from the supply chain. Some of these events impact the inventory levels or revenue metrics for the manufactured modules (such as "order placed" or "order cancelled" events). The CSR system continuously updates the actual revenue, the revenue outlook and inventory levels (step 1 and 2).
- Whether the progression of the accrued revenue is normal or below target is determined by the CSR system using a wineglass model [3]. In the case where the revenue is below target, the CSR system automatically detects such a situation and issues an alert showing the current sales quantities of some selected saleable part numbers in the nth week are out of their bands (step 3).
- The CSR system recommends adjusting the planned demand quantities and safety stock requirements for the nth week. As next step, it invokes a demand planning module and inventory planning module to analyze demand quantities and safety stock requirements for the nth week (step 4).
- It further recommends altering the daily build plan in order to optimally match new daily demand statements, thus high serviceability, and minimize manufacturing and inventory costs. By doing so, it also shows the effects and risks of all suggested alternatives for changing the build plan (step 5 and 6).
- Finally, the business line manager looks at the suggestions of the CSR system and makes a final decision for improving the build plan (step 7 and 8).
- The CSR system immediately revises the actual build plan in the ERP system (action) and continues the monitoring of the performance indicators with the updated build plan (step 9).

3 Framework

In general, CSR can be categorized as a system that is continually interacting with its managed substrate, and that is capable of autonomous actions in this substrate in order to meet its management commitments. A platform for CSR takes inputs from the managed substrate, and produces actions that affect it. As such, CSR interacts directly or indirectly with the situated entities in the managed substrates. Examples of situated entities include business processes, business organizations, managed resources, and business systems, where

- Business processes are the first-class citizens that can be observed, measured, analyzed and managed. In the example of PLM, managed business processes can be supply-chain management, customer relationship management, enterprise resource planning and so on.
- Business organizations refer to the participating parties of the business processes and BPM systems which are managed by CSR. Business organizations can come from many roles: enterprises, business analysts,

BPM system administrators, BPM integrators and the business executives who are interested in knowing the status of business processes.

- Management Policies refer to the management contracts established between CSR and virtual organizations. An example is that the maximum cycle time of some supply chain management process shall not be greater than 48 hours. If the agreement was violated, certain amount of penalty will be exerted.
- Business Systems comprise manageable entities that are situated in the substrate. Resource's manageability defines information that is useful for managing a resource and details the aspects of the resource including the instrumentation which allows CSR to interact with it. There have been many standards of defining manageability at various levels. Through instrumentation, a resource is turned into managed resource because its state can be perceived, aggregated, analyzed and modified through the standard interfaces provided by the instrumentation layer that is located between CSR and its environment.

Horizontal Decomposition

The functionality of CSR can be decomposed either horizontally or vertically. Figure 2 shows that CSR is decomposed horizontally into three pillars: perception, evaluation, and actuation. (1) Perception pillar receives the data and events from the CSR substrate. (2) Evaluation pillar processes the perceived information. (3) Actuation pillar renders management directives to the managed entities within the substrate.

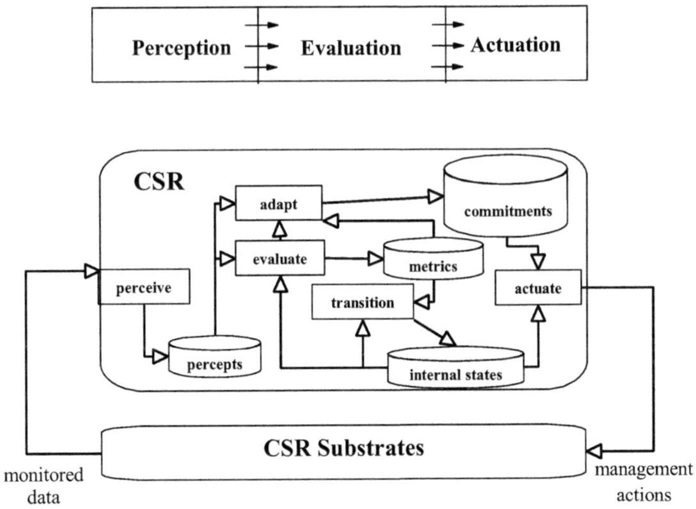

Figure 2: Horizontal Decomposition

Conceptually, the Evaluation pillar can be further decomposed into three sub-functions: measure, transition, and adapt; and four local data stores: percepts, metric stores, control states, and commitment store. The percepts store contains the perceived values from the CSR substrate. The function measure computes the metrics according to the values of percepts and stores them into the metrics store. The control states represent the current situation of CSR as a whole. Control states are different from the states perceived from the substrate.

In general, control states capture the status of CSR as a whole, and the environment states capture the status of the managed entities within the CSR substrate. The function transition changes the control state of CSR according to the current control state, the local commitments, and the resultant metrics. The function adapts changes the local management commitments according to the existing control state and the business policies. The commitments store contains the local management commitments that the CSR enforces on the managed entities of the substrates, business organizations, business systems.

Vertical Decomposition

CSR can be also decomposed vertically into three layers: Reactive Management, Deliberate Management and Reflective Management. Figure 3 illustrates the vertical decomposition, where solid lines represent the flows of management directives, and the dotted lines represent the flows of management events.

- Reactive management layer responds to the management events quickly and directly through scripted business process models. A notable example of the reactive management is deterministic workflow management where workflow models are defined at the build time and executed at the run time. Another example is the alarm system that will notify the system administrator if some managed resource is suffering from severe performance problems, and demanding immediate attention.
- Deliberate management layer performs managerial tasks that require more reasoning and more complicated computation. It is not uncommon that CSR needs to provide decision support capability so more intelligent management directives can be derived towards managed resources. An example of such managerial tasks is the business processes with the ability of Sense-and-Respond [1]. Another example of deliberate management is mapping QoS metrics from IT-level into business process level and vice versa. An event such as "disk failure" may mean little out of business context. However, it may imply a loss of gigantic capital for a business organization if it has causal relationship with critical business process performance such as financial trading. The mapping rules in the layer of deliberate management should capture this relationship to prevent business loss.

- Reflective management layer enables CSR to maintain information about itself and use this information to remain extensible and adaptable. Reflexive management layer performs meta-management directives unto the lower management layers and managed entities. A meta-management directive is a higher sphere of control such as adapting the management commitments, modifying measurement and analysis algorithms in the deliberative management layer, or changing the alarm rules in the reactive management layers. As such, CSR can have detailed knowledge of the managed resources, current status of managed business processes and business systems, the ultimate capacity in the inventory, performance expectation, and all connections to other systems to manage itself. Therefore, through reflective management mechanism, CSR achieves the goals of both 2nd order management and autonomic computing.

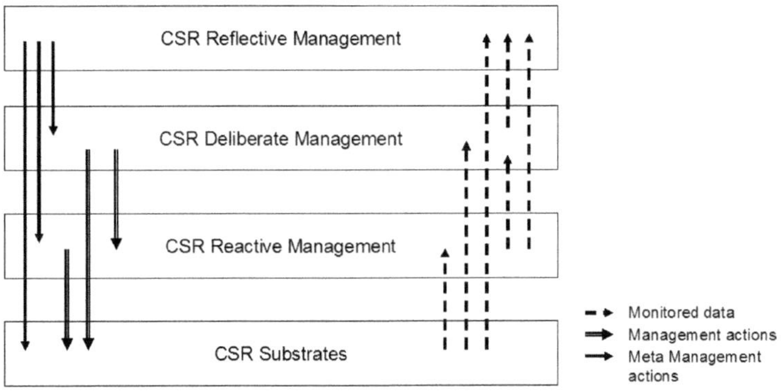

Figure 3: Vertical Decomposition of CSR

Mesh Decomposition

Horizontal and vertical decompositions can be combined to form mesh decomposition as shown in Figure 4. Mesh model is used as a formal modelling tool for cognitive architecture [10]. We found that Mesh model is very suitable to illustrate different architectural aspects of management spaces for CSR. Figure 4 defines nine regions, ($S_{i,j}$ $1 \leq i,j \leq 3$), called management spaces (or m-spaces) and their potential interactions. Only legitimate flows, both management directives and management events, are allowed to be transmitted between m-spaces. The entities in the CSR substrates emit management events and receive management directives. Evaluation is completely an internal processing inside CSR without interaction with its substrate. The meta-management directives are rendered only through the actuation pillar. Within CSR, management events can be generated or transformed between layers upwardly and management directives can be rendered downwardly.

Figure 4 shows two typical management scenarios based upon the decomposed m-spaces: Flow A is a common management scenario: management events are perceived by the reactive management layer, the evaluation is performed by the deliberate management layer, and the actuation is activated through the reactive management layer. This management scenario is called Ω model of management flow, which is the most common scenario among all. Flow B is a meta-management scenario where management events are delivered all the way to the reflective layer, evaluated, eventually some meta-management directives are delivered through the actuation pillar in the reflective management layer. An example of such scenario is data mining of the business data residing in the substrate. The reactive management layer extracts the data, the deliberate management layer transforms and loads the data, and the reflective management layer analyzes the data and makes/suggests business decisions accordingly.

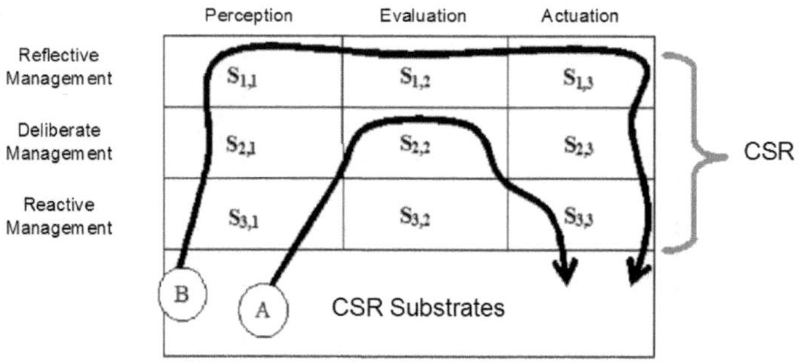

Figure 4: Mesh Decomposition of CSR

We refer to members of a management space as management agents or m-agents shortly, by which we mean autonomous actors interacting with each other, and with the substrate. Such an m-agent can be a software component, with its own state and thread of control, or it might be a proxy for human users interacting with CSR with some interface. Mesh decomposition model actually define a couple of m-agent classes, each of which embodies certain rules of engagement that its situated m-agents must abide. For example, the m-agents of m-space $S_{2,3}$ only receive management events from m-space $S_{2,2}$ and must comply the management commitments imposed by the m-agents in m-space $S_{1,3}$. Consequently, the CSR conceptual framework entails regulated m-agents, management events, managerial functions, and management directives.

4 Commitment Based Manufacturing Supply Chain

In the domain of microelectronic manufacturing, CSR interacts with multiple participating parties such as business line managers who make decisions, building superintendent who conducts the day to day scheduling, production planning team who determine intermediate target outs, manufacturing team who make the final determination of what is built and when. Moreover, CSR needs to interact with multiple systems such as:

- Central planning engine (CPE) that helps decision makers determine how to best meet prioritized demand without violating temporal, asset, or capacity constraints.
- The optimal manufacturing resource planning (OMRP) tool that assists decision makers define detailed instructions about what manufacturing activities must be accomplished and when they must be completed.
- The available to promise (ATP) tool that enables an organization to dynamically reallocate projected supply in response to incremental changes in the demand statement (new orders arriving, orders being filled, and order changes or cancellations) according to business policy guidelines, identify projected shortfalls with respect to committed orders, and provide real-time order commits and status.
- The demand management (DM) tool that help decision makers coordinate demand estimates from different sources such as orders, sales rep forecasts, customer forecasts, internal demand, and marketing forecasts in logical step by step

Therefore, the management of business processes and systems is critical for CSR to succeed. There is a need for a set of management policies that work as "contract" among business processes and systems constraining their behavior so as to guarantee the interests of all the stakeholders. The contracts are formalized as "commitments" in CSR. A commitment can be either exogenous or endogenous, where the former aimed for defining contracts among decision makers and the latter for internal agents. Commitments define the constraints that would follow certain courses of actions, or to hold certain agreed and trusted situations manifested by the entities in the CSR substrates. CSR based system composed of management agents, which are aimed to fulfill the commitments deployed to CSR models.

A commitment is a sextuple (T, E, M, S, D, A) where

- T is the target manufacturing supply chain system to be controlled.
- E is the set of events emitting from T.
- M is the set of metrics that are used to measure the behavior and performance of T.
- S is the set of situations (exceptions) that are defined over M.
- D is the set of decision rules that will be triggered when one or more situations in S are evaluated to be true.
- A is the set of potential actions that will be render to T according to the evaluation outcome of decision rules in D.

A commitment is established among the stakeholders of a manufacturing supply chain including the decision makers from different levels of decision tiers. Commitments are realized and enforced by specialized agents named commitment rings. Each commit ring consumes a commitment and performs required functionality. Behaviorally, there are five functions embedded in a commitment ring: Sense, Detect, Analyze, Decide and Effect. Figure 5 shows the functional structure of a commitment ring.

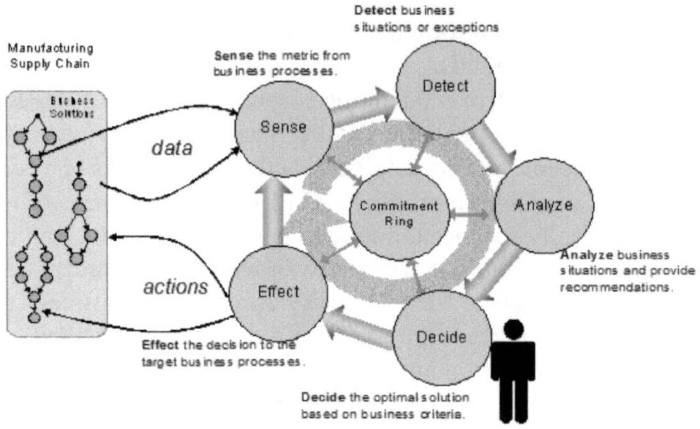

Figure 5: The Commitment Ring

- "Sense" is the functional stage where a commit ring interacts with target manufacturing supply chain and provides data extraction, transformation, and loading capabilities for the sake of preparing qualified data that is to be further monitored and analyzed. This stage produces business metrics, e.g., the cycle time of a business process of supply chain management.
- "Detect" is the stage of detecting business situations and/or exception occurring in the manufacturing supply chain. An example of situation detection could be higher-than-expected business process cycle time in certain supply chain.
- "Analyze" is the stage when a commitment ring performs analytics such as risk-based analysis of resolving business exceptions. The output of this stage often comes with recommendation of potential resolutions to decision makers. An example is new set of build plans or modified inventory policies for the business process that is problematic.
- "Decide" is the stage where a commitment ring (on the behalf of decision makers) will make decision about what to respond to business situations. Note that decision makers can be either humans or automated agents. The mission of a commitment ring is to help decision makers to make the right decision at the right time to the right targets.
- "Effect" is the stage where a commitment ring carries out certain actions in order to enforce the decisions made through the Decide stage. Actions can be of many forms. The simplest kind of action is alerting interested parties

about the decisions. More complicated ones can be involved sophisticated invocations of business processes.

A commitment ring represents a scalable mechanism of realizing sense-and-respond capabilities at different decision tiers. A commitment ring consists of nodes and links. A commitment node is a basic service that enables transformation from input data to output data based on its capabilities and the pre-defined policies. A commitment link transmits data with specific types from one node to another node. A commitment node can have multiple instances of input and output links. Therefore, it can process multiple input requests concurrently. The number of commitment nodes in a commitment ring is subject to the actual requirements. Commitment rings can be used as a simple modeling vehicle of integrating sense-and-respond capabilities for manufacturing supply chain.

Commitment rings provide the means of building highly configurable and adaptive integration platform for BPM solutions. In our example, we have come up with 5 typical sense-and-respond service nodes in a commitment ring: (1) event processing service that takes raw data and produce qualified data to be further processed; (2) metric generation service that receives the qualified data and produced metrics; (3) situation detection service that analyzes incoming metrics and raise situations if needed; (4) action planning service that is triggered by situations and creates an action plan in order to resolve the situation; and (5) action rendering service that takes a group of actions from action planning service and actually renders them to the target business solutions. A commitment node can process multiple input data requests based on the functionality to which it is aimed.

Multiple commitment rings form a commitment net in that each commitment ring becomes a node and interactions among commitment rings constitute the links. While commitment rings capture the monitoring and control patterns of specific business situations (or exceptions), commitment net represents the pattern of communicating autonomous commitment rings in order to capture a global behavior of monitoring and control across manufacturing supply chain systems. Hence, a commitment net realizes the sense-and-respond capabilities for a business organization (enterprise). Commitment rings collaborate with one another and aggregate into higher granularities. The structure of commitment nets can represent contractual (committal) bindings between business organizations (enterprises) and typically result in information exchange between business organizations (enterprises). Figure 6 illustrates a potential structure of commitment net formed by commitment rings and the interactions among them.

A key goal of commitment net is to provide ubiquitous sense-and-respond services for target manufacturing supply chain systems. Furthermore, the commitment net, is a dynamic and adaptive agent where the availability and state of these services and resources are constantly changing. The pool of currently available sense-and-respond services is represented as a graph where the node represents services and the links, can be modeled as potential interactions.

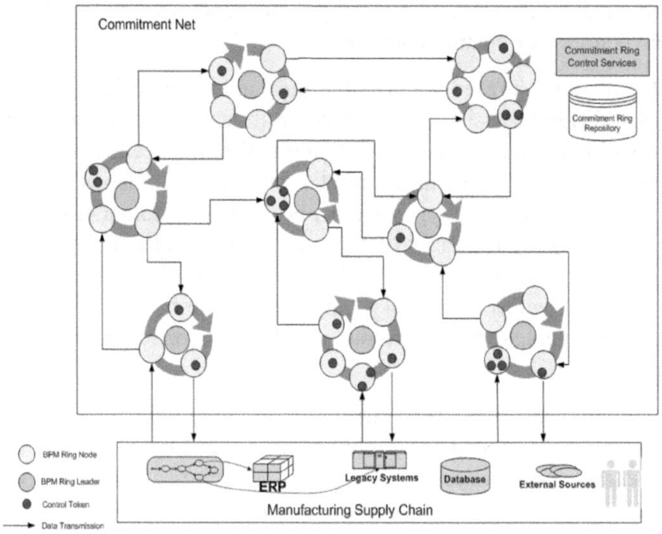

Figure 6: Commitment Nets

The execution paths generated from commitment net based on constrains and goals defined in the commitment requirement actually manifest the Sense-and-respond *capabilities* of a manufacturing supply chain system on the sense-and-respond ability. As described in previous section, commitment rings are applied to multiple levels of emprise abstraction: strategy, operation, execution, and implementation. Each layer consists of corresponding commitment rings that are specialized in monitoring and controlling specific layer of enterprise resources. Sense-and-respond capabilities can be defined either horizontally or vertically. Horizontal Sense-and-respond capability is an execution path that consists of commitment rings exclusively of a specific layer, e.g. the strategic sense-and-respond capability. On the other hand, the vertical sense-and-respond capability is an execution path which contains the commitment rings across different layers. In the diagram, it is also indicated that some commitment rings are for processing external events and some for internal events among commitment rings.

In the example of microelectronic manufacturing, the CSR system interacts with multiple participating parties such as business line managers who make decisions, building superintendent who conducts the day-to-day scheduling, production planning team who determines intermediate target outs, and manufacturing team who makes the final determination of what is built and when. Moreover, CSR needs to interact with multiple systems such as: (1) Central planning engine (CPE) that helps decision makers determine how to best meet prioritized demand without violating temporal, asset, or capacity constraints. (2) The optimal manufacturing resource planning (OMRP) tool that assists decision makers define detailed instructions about what manufacturing activities must be accomplished and when they must be completed.

(3) The available to promise (ATP) tool that enables an organization to dynamically reallocate projected supply in response to incremental changes in the demand statement (new orders arriving, orders being filled, and order changes or cancellations) according to business policy guidelines, identify projected shortfalls with respect to committed orders, and provide real-time order commits and status. (4) The demand management (DM) tool that helps decision makers coordinate demand estimates from different sources such as orders, sales rep forecasts, customer forecasts, internal demand, and marketing forecasts in logical step by step.

Each commitment is realized by an agent. An agent can commit a commitment to another agent. For example, an ATP agent expects an Inventory agent to have inventory level less than certain quantified value. If the expectation is violated, some actions will be taken based upon corresponding bound response that is defined in the ATP agent. A commitment can be applied to several agents simultaneously, i.e., those agents share the same commitment. We can thus define the locus of control for the commitment by defining a set containing the agents committing to it. In CSR, it is assumed that an agent will always communicate its commitments truthfully with the other agents in response to queries and actions. Each agent may have one more commitments. Agents commit to one another through the resource bindings. Since agents can commit to one another, the "commit-to" relationships form a commitment network with agents as nodes and commitment as the edges. For example, the Sales agent is committed to sending an alert to the Demand Agent when the current sales quantities of some selected saleable part numbers in the *n*th week are out of their bands. The Demand Agent is committed to the Presentation Agent by providing a list of recommendation on adjusting the planned demand quantities and safety stock requirements for the *n*th week. However, the Demand Agent cannot work by itself. The Demand Planning Agent and the Inventory Planning Agent commit themselves to the Demand Agent such that they promise to recommend the Demand Agent demand quantities and safety stock requirements for the *n*th week.

Network of Commitments (NoC) is a generalized notion of Chains of Commitment [4]. NoC is used to define the triggering points for evaluation, the control and data flows, the monitoring commitments, and the (situation) detection commitments. Since a agent can *commit* and *un-commit* commitments dynamically, the applications built on CSR become extremely configurable and can be adaptive to the enterprise needs, e.g., Balanced Score Card, by simply modifying the definitions of CSR through configuration management tools. From the viewpoints of any stakeholder associated with the aforementioned microelectronic manufacturing processes, commitments themselves provide a means to define the quality of services for the CSR system itself because the whole system will become more visible, controllable and configurable. Figure 7 depicts a NoC in CSR for the use case scenario described previously. Sales Agent detects the "out-of-band" situation and notifies its committed agent, i.e., Demand Agent, which will consequently notify Recommendation Agent and Risk Assessment Agent in sequence to obtain recommended build plan(s) and necessary assessment such as inventory cost, manufacturing cost and SLA measurement. Note that commitment relationship may imply either event/situation flows or data flows between commitment-related agents, and the actions to be taken really depend on the definition of the involved commitments, i.e., on the expectations, actions and responses that are delineated in the committed agents. Commitments that are established among participating parties of CSR systems are called *exogenous commitments*. On the other hand, commitments are named *endogenous commitments* if they are mainly exploited to enforce the behavior of target systems. The aforementioned commitments are all endogenous commitments. However, Presentation Agent has an exogenous commitment to the decision maker that represents the ultimate user of the manufacturing system.

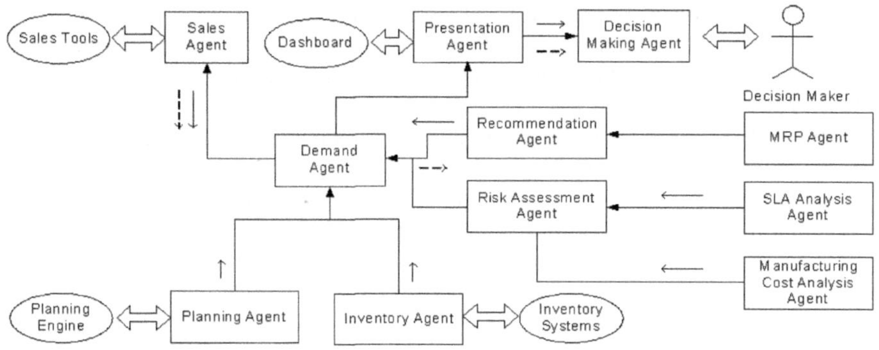

Figure 7: Network of Commitments (NoC)

The aforementioned scenario leads to the fact that an agent must be adaptive to dynamic information requirements and resource limitations via adjusting its perceptual strategy. We introduce a new modeling concept called *smart distance* [5] that allows an agent to obtain all the distance values, based upon pre-defined mutual commitments, for example, connecting to all the Inventory Agents with inventory less than tolerant level and establishing a shared plan [6] to increase the inventory. A CSR system can be a complex system that consists of interwoven dynamic and heterogeneous agents. Figure 7 shows various kinds of agents that realize the management relationships (reflective, deliberate, reactive) in the scenario. The springs are used to represent the smart distances between agents as adjusted by contextual information. There are forces that will influence the distances between two agents. Whenever there are needs to change the strategy, the user inputs the updated strategy into the corresponding reflective agent. The reflective agent analyzes the updated strategy and forms a set of commitments. These commitments will be sent to the "related" agents in the deliberate or reactive layers. Here, the relatedness is defined by the value of the distance between two agents. When receiving a new commitment, an agent will adapt its behavior so as to meet the commitment requirements. Figure 8 illustrates the potential interactions among agents through smart distances. For example, the strategic agent R1 forms an agent hierarchy consisting of two deliberate agents (D1, D2) and five reactive agents (A1, A2, A3, A4, A5). The whole hierarchy constitutes a new agent with R1 as its façade agent. Agents in the same hierarchy are close to one another in terms of smart distances. Agent hierarchies are created and dismissed dynamically because the smart distance is a dynamic function over time and business context. The smart distances enable the establishment of commitment relationships among agents dynamically. For example, initially, the manufacturing system (referring to the example described in the scenario) consists only of a pool of unorganized resource agents. Upon arrival of a sales event, a Sales Agent is created if it does not pre-exist. The smart distance between this Sale Agent and the Demand Agent is reduced. Hence, the Sales Agent begins to be aware of the Demand agents, and starts to negotiate with them regarding the provision of certain manufacturing operations. In the mean time, the distance between Demand Agent and Planning Agent is also decreased and also makes both aware of each other. They also start the negotiation. During the negotiation process, the Demand agent requires specific properties from the operation, such as high quality or high throughput, when the Planning Agents try to optimize their utilization. At the end of the negotiation, the Planning Agents combine to form the agreed manufacturing line and the Planning Agent initiates the creation of Product Agents. Finally, an ad-hoc network of commitments is formed.

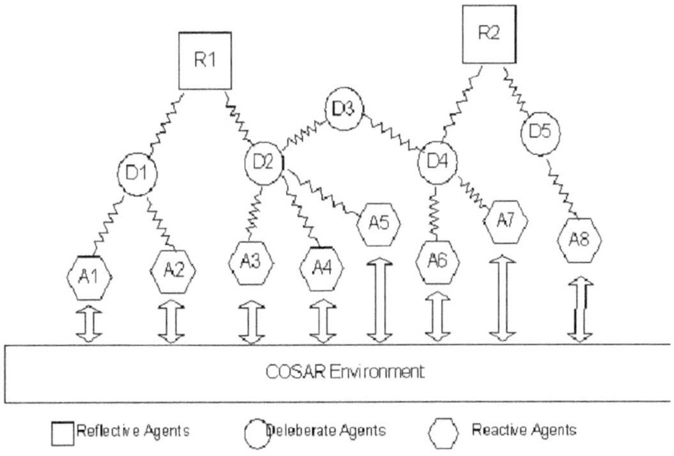

Figure 8: An Example of Smart Distances

5 Architecture

Figure 9 shows a reference architecture for agent based CSR systems. The architecture includes 5 groups of agents: 1) Management Agents for analytical processing and decision making, 2) Data Agents for the real-time transformation of business activity events, 3) Commitments and Distances, 4) Presentation Agents to display desired metrics and analytical results to users, and 5) Probe Agents sense information out of the BAM environment, and renders actions to manageable business activities and systems. The core functions are realized by the management agents where specialized agents are created for different purposes. Monitoring Agents sense the business situations from the message bus and dispatching them to appropriate agents for further processing. Response Agents render actions back to the target business activities and systems via synchronous calls or asynchronous messages. Hence, Monitoring and Response Agents realize the depicted agent based functions *perception* and *actuation*, respectively. The agent function *evaluation* is implemented by three types of agents: Reactive Agents, Business Intelligence (BI) Agents, and Autonomic Agents.

Figure 9 also shows the agents (shown as round boxes) that are managed by the layer of Data Agents: 1) Event Agents, 2) ETL Agents, and 3) Evaluation Agents. The purpose of Event Agents is to receive events from various event sources and to unify the different event formats from various business event sources. The events will be sent from Event Agents to subscribed ETL Agents, which have the commitments from corresponding event Agents. The ETL Agent processing logic can include the calculation of process metrics and storing these metrics in the process data store or the process warehouse. ETL Agents can also publish the process metrics to Evaluation Holons that have subscribed themselves to the *metric type.* The main purpose of the process information factory is to provide a data foundation for a process-driven decision support system to monitor and improve the business activities continuously. It is a global process information repository, which enables BI Agents and process analysts to access comprehensive information on business processes very quickly, at different aggregation levels, from different and multidimensional points of view, over a long period of time, using a huge historic data basis prepared for analyzing purposes to effectively support the management of business processes.

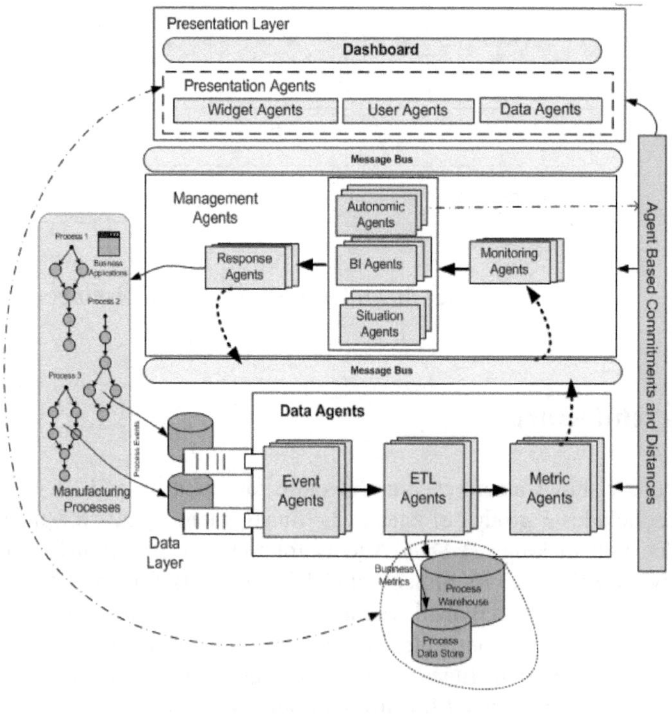

Figure 9: Agent-Based CSR System

6 Related Work and Discussion

The Supply Chain Operation Reference (SCOR) Model provided by the Supply Chain Council defines inter-organizational business processes and their workflows [7]. The SCOR Model contains the metrics for operational control and best practices of supply chain design. SCOR identifies key performance indicators (KPIs) in five process areas: plan, source, make, deliver, and return. However, it does not have standards for regulating actions when some KPI-based thresholds are exceeded. CSR, however, addresses this issue by defining management commitments for both manufacturing metrics and management actions. Chains of Commitments architecture was proposed by Ervin et al [4]. The domain of microelectronic manufacturing that CSR is aimed for also belongs to the arena of supply chain management. Nevertheless, CSR is focused on providing mechanism and concrete architecture enabling the representation, evaluation, and enforcement of management commitments. CSR is an agent-based framework where agents can commit themselves to one another to form a network of commitments that can be used to drive management scenarios.

Minsky and Ungureanu [8] described a mechanism called law-governed interaction (LGI), which is designed to satisfy three principles: (1) coordination policy needs to be coordinated; (2) the enforcement needs to be decentralized; and (3) coordination policies need to be formulated. CSR satisfies all of the LGI principles. LGI uses decentralized controllers co-located with agents. LGI does not address the issues such as commitments-driven management processes. Similarly, Verdicchio and Colombetti also incorporate commitments into the supply chain domain [9]. They are concerned with the issue of information sharing among agents and proposed a data structure for commitments that can be used for the communication of agent-based framework for the management of a supply chain. However, CSR is concerned with the holistic infrastructure that can be used for commitment-based management in the domain of microelectronic manufacturing.

7 Conclusion

In this paper, we have presented the concepts of CSR that enables commitment-based management for microelectronic manufacturing processes. Based on such concepts, an agent-based system has been developed and presented in this paper. Instead of using traditional integration approaches, CSR takes commitment as the driving factor of managing manufacturing processes, systems, and data. The consequence is that we have obtained an adaptive framework and architecture for microelectronic manufacturing. With CSR, business line manager has better control over the microelectronic manufacturing system since it is highly configurable and easier to work on commitments instead of disintegrated policies.

CSR can be definitely applied to domains beyond microelectronic manufacturing, e.g., executive decision support systems, business process management system etc. Since the formation of "Network of Commitment" would be a costly and time-consuming task, it is worthy to conducting the NoC simulation. It is important to consider many factors during the simulation such as "level of commitment", "penalty of failing to make the commitment", commitment escalation.

References

1. Haeckel, S.H., and Slywotzky, A.J. *Adaptive Enterprise: Creating and Leading Sense-and-Respond Organizations*, Harvard Business School Publisher, August, 1999.
2. Lin, G., Buckley S., Cao H., Caswell N., Ettl M., Kapoor S., Koenig L., Katircioglu K., Nigam A., Ramachandran B., and Wang K., The Sense-and-Respond Enterprise, IBM researchers develop integrated SAR system of global value chain optimization, OR/MS Today, April 2002, available at http://www.lionhrtpub.com/orms/orms-4-02/frvaluechain.html
3. Wu, L.S.Y., Hosking, J.R.M., and Doll, J.M., "Business Planning Under Uncertainty: Will We Attain Our Goal?," IBM Research Report RC 16120, Sep. 24, 1990, Reissued with corrections Feb. 20, 2002.
4. Ervin, R., "Chains of Commitment Software Architecture," *ACM SIGecom Exchanges* Special Issues on Chains of Commitment, http://www.acm.org/sigs/sigecom/exchanges/volume_3_(02)/3.1-Ervin.pdf, 2002.
5. Ye, Y., Boies, S., Huang, P., and Tsotsos, J., (2001). Agents-Supported Adaptive Group Awareness: Samrt Distance and WWWaware. IEEE *Transactions on Systems, Man, and Cybernetics. Part A: Systems and Humans*, Vol. 31, No. 5. pp: 369-379.
6. Grosz, B., Hunsberger, L., and Kraus, S. , Planning and Acting Together. *AI Magazine*, 20(4):23-34. 1999.
7. Supply-Chain Operations Reference Model, Supply Chain Council, http://www.supply-chain.org, Version 3.1, March, 2000.
8. Minsky, N.H., and Ungureanu V., "Law-Governed Interaction: A Coordination and Control Mechanism for Heterogenous Distributed Systems," ACM Transaction on Software Engineering and Methodology, Vol. 9, No. 3, July, 2000, Pages 273-305.
9. Verdicchio, M. and Colombetti, M., "Commitments for Agent-based Supply Chain," *ACM SIGecom Exchanges* Special Issues on Chains of Commitment, http http://www.acm.org/sigs/sigecom/exchanges/volume_3_(02)/3.1-Verdicchio.pdf, 2002.
10. Sloman, A., Varieties of Affect and the CogAff Architecture Schema, in *Proceedings of AISB'01 Symposium on Adaptive Agents and Multi-agent Systems*, University of York, United Kingdom, March 21-24, 2001.

A Multi-Agent Approach to Supply Chain Management in the Chemical Industry

Rajagopalan Srinivasan*, Mukta Bansal, I.A. Karimi

The National University of Singapore,
Department of Chemical and Biomolecular Engineering,
4 Engineering Drive 4, 10 Kent Ridge Crescent
Singapore 117576
*Email: chergs@nus.edu.sg

Abstract. Supply chains in the chemical industry typically span across several corporations and involve several departments within each. The functioning and characteristics of the entities and the intra-enterprise and inter-enterprise interactions have to be explicitly considered in decision-making. In this chapter, we describe an agent-based model of a refinery's supply chain. Software agents emulate the entities such as procurement, sales, operations, storage and logistics departments of the refinery as well as the suppliers, logistics service providers, and oil-exchanges. These agents model the embedded business policies and thus mimic the different business processes of the enterprise. Uncertainties are captured by stochastic elements embedded in the agents. The dynamics of the supply chain is emulated by discrete event simulation of the agent-based model. The application of the supply chain model and simulation in decision-making is illustrated here. Different business processes and supply chain configurations are evaluated based on their effect on entity-specific as well as supply chain wide key performance indicators. This enables well-rounded decisions related to both the structure and parameters of the supply chain.

Keywords: Refinery supply chain, business decision support, simulation, disruptions

Notations

Abbreviations

MADE	Multi-Agent Development Environment
PRISMS	Petroleum Refinery Integrated Supply chain Modeler and Simulator
RFQ	Request-For-Quote
RRFQ	Reply-to-Request-For-Quote
SCM	Supply Chain Management
3PL	3rd Party Logistics provider

R. Srinivasan et al.: *A Multi-Agent Approach to Supply Chain Management in the Chemical Industry*, Studies in Computational Intelligence (SCI) **28,** 419–450 (2006)
www.springerlink.com

Variables

$C(j)$	amount of crude needed to meet demand in jth procurement cycle (kbbl)
$\tilde{C}(j)$	amount of crude to procure in jth procurement cycle based on forecasted demand (Kbbl)
$D(k)$	processing cost for crude k ($/kbbl)
$E(i)$	forecasted product price for product i ($/kbbl)
G	crude cut
$H(n)$	shipment of crude on day n (kbbl)
$\tilde{H}(n)$	shipment of crude scheduled to arrive on day n (kbbl)
i	product
j	procurement cycle
k	crude
N	number of products
n	day
$P(k)$	profit of crude k ($)
$Q(i)$	forecasted product quantity for product i (kbbl)
R	transportation cost ($)
$S(n)$	total stock of crude on day n (kbbl)
$\tilde{S}(n)$	planned stock of crude on day n (kbbl)
$T(n, j)$	throughput on day n and for jth procurement cycle (Kbbl/day)
$\tilde{T}(n, j)$	planned throughput for day n and for jth procurement cycle (kbbl/day)
$\hat{T}(n)$	backlog order for day n (kbbl)
$U(k)$	cost of crude k ($/kbbl)
$y(i,k)$	yield of product i for crude k

Parameters

A	planning horizon (days)
B	simulation horizon (days)
F	length of procurement cycle (days)
J	number of procurement cycles
T_{min}	minimum throughput of refinery (kbbl/day)
T_{max}	maximum throughput of refinery (kbbl/day)
W	safety stock (kbbl)

1 Introduction

The chemical industry is one of the world's largest manufacturing industries, producing more than 50,000 chemicals and formulations. Starting from raw materials such as oil, coal, gas, water, air and minerals, the chemical industry produces a vast array of substances that form the basis for almost every other manufacturing activity. It operates on a global scale; it exists in nearly every country in the world, and contributes 7% of global income and accounts for 9% of international trade.

Supply chains in the electronics, automobile and other industries have received much attention in literature. Although some of these lessons can be partly extended to the chemical industry, supply chains in the chemical and process industry have distinctive features and require special attention. As an example, consider a petroleum refinery supply chain.

1.1 Petroleum Refinery Supply Chain

Figure 1 shows a schematic of a typical petroleum refinery supply chain. Refining is a complex process that transforms crude oil into valuable products such as gasoline, heating oil, and jet fuel, as well as petrochemical intermediates, which are further processed to produce fertilizers, plastics, synthetic fibers, detergents, etc. A refinery supply chain begins with the production of crude oil and gas from either ground fields or offshore platforms. After pretreatment and storage, these are transported via Very Large Crude Carriers (VLCCs) and Liquefied Natural Gas (LNG) tankers to various refineries around the world. The petroleum refinery converts these into a variety of intermediate bulk chemicals that are used as feedstock in petrochemical plants as well as fuels for aviation, ground transport, electricity generation, etc. Thus, the supply chain has at least three distinct centers of manufacturing, namely the oil/gas fields & platforms, the petroleum refineries, and the petrochemical plants. Each of these manufacturing entities is in turn surrounded by a host of logistics services for storage, transportation, distribution, packaging, etc.

Oil products are distributed to customers via various modes that depend on the distance, the nature of products, and demand quantities. The main oil products leave the refinery in bulk loads. Large consumers like petrochemical manufacturers may be supplied directly from the refinery via pipelines, rail, road, or sea. Smaller customers are generally supplied via storage and distribution centers known as terminals or depots. These disparate entities make the task of supplying the right product and the right quantity to the right customer at the right time with the right quality and service a very complex endeavor.

The long refinery supply chain that spans the globe suffers from long transportation times (for example, it takes 4-6 weeks to ship crude oil from the Middle East to a refinery in Asia). Further, the price of crude oil, the basic raw material for the refinery, is very volatile even on a daily basis; the demands and the prices for the products are also highly variable. These confound production planning, scheduling and supply chain management. As one example, higher than forecasted demand for products can lead to market opportunities for the refiner that can be exploited if adequate stock of crude is available at hand; however a lower than forecasted demand would lead to high inventory costs that can significantly erode refinery profits. Determining the safety stock levels for crude oil is therefore tricky. Similarly, numerous products and their variants can be produced from a crude by suitably utilizing the complex manufacturing process consisting of a highly interconnected system of reactors, separators and blenders. However, the yields of the different products from different crudes are different as are the operating costs for each combination. Given forecasted demands and prices for the products, the process of determining the right mix of crudes has to account for these as well as the landed cost of the crude that includes the purchase cost as well as the costs involved in moving it to the refinery. The fluctuation in the costs, demands and prices on a daily-basis necessitates frequent and speedy re-evaluations of numerous supply chain alternatives. Each evaluation should account for the complex relationships between the raw materials, operating units, and products to arrive at a feasible and optimal solution.

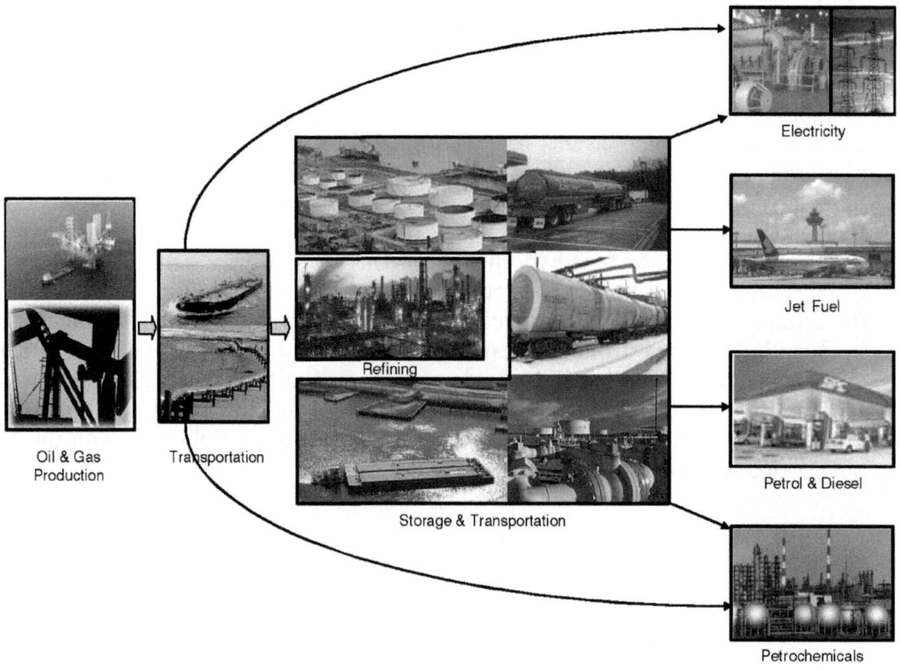

Fig. 1. Schematic of petroleum refinery supply chain

1.2 Distinguishing Features of Chemical Supply Chains

It is evident from the above example that, while chemical supply chains show some similarity with other supply chains, they have many unique features as well. These can be summarized as due to:

1. Complex Nature of Chemical Industry
A primary feature of chemical supply chains is the huge variety of non-discrete, immiscible, incompatible, non-substitutable, huge-volume products, each of which has its own unique characteristics. The concepts of "discrete parts" and "assembly" do not exist in chemical manufacturing. The industry is highly capital-intensive with long and divergent supply chains with recycle loops that simply do not exist in other supply chains. The industry is the biggest consumer of itself and many of its businesses are high-volume and low-margin. Huge inventories that are critical to the continuity and profitability; need for safety-first; sociopolitical uncertainties, and environmental regulations; and extensive trading are other key features of the chemical industry that set them apart from the other manufacturing industries.

2. Fluctuations in Oil Price:
Volatility in crude oil poses a tremendous challenge to manage the chemical supply chain. OPEC, the Organization of Petroleum Exporting Countries, has significant influence on the price of crude oil as its members control a great portion of the world's oil supply. The price of oil strongly influences the price of petrochemical products. The efficiency of chemical supply chain is dependent on the fluctuations in oil prices. The variation in the oil price may disrupt the supply chain and leads to risk management of chemical supply chain.

3. Intricate Manufacturing Process:
The manufacturing complexity of the chemical industry poses a challenge to the efficient management of the supply chain. There are many factors to account for in a complex manufacturing process such as choice of raw materials, selection of towers and pipes, selection of storage units, hazardous nature of chemicals, etc. Chemical process manufacturing plants are expensive to build and maintain and are designed for specific production modes. Due to these factors, manufacturing plant is not flexible to reconfigure according to the dynamics of supply chain.

4. Complex Transportation Process and Large Inventory:
The chemical industries transport huge amount of chemicals all over the world. They are transported by either land or sea with maritime transport as the workhorse. This makes the transportation process very slow. Further, the hazardous nature and huge volumes of chemicals necessitate the use of highly expensive and sophisticated transport equipment and storage facilities that require complex and expensive cleaning procedures and maintenance, and result in long lead times. The slow transportation induces high in-transit inventory, which have to be accounted for during inventory management. Logistics costs in the chemical industry could be as high as 20% of the purchase cost. Variability of transport times make it necessary to have safety stock at the company ends to ensure that customer services would not be affected by any disruptions to in-transit inventory.

5. Environmental Regulations:
As most chemicals are hazardous, there are stringent regulatory compliances imposed on transporting it on land and sea. Environmental regulations relate to pollution during manufacturing and transport. In an effort to protect the environment, specific standards exist for packaging, labeling, distribution and transport of chemicals. For example, certification of vessels is a widely prevalent requirement.
This chapter describes an agent-based approach to model and simulate chemical supply chains. Relevant literature is reviewed in Section 2. Various platforms/environments for agent-based modeling are also discussed. Section 3 provides an overview of the Multi-Agent Development Environment (MADE), a special-purpose modeling environment developed using Gensym's G2 Expert System shell. The salient features of the refinery crude supply chain are described in Section 4. The supply chain decision support system called Petroleum Refinery Integrated Supply Chain Modeler and Simulator (PRISMS) implemented using MADE is presented in Section 5. In Section 6, we illustrate the application of PRISMS to support effective supply chain management.

2. Literature Review

Supply chain management has gained much attention in recent years as businesses feel the pressure of increased competition. In literature, mathematical and operations research models have been most popular for supply chain studies. In the following, we review the literature related to managing chemical supply chains.
The **design of a supply chain** is a strategic decision addressing the location and capacities of production units and distribution centers, transportation links between them, as well as the modes of transportation. Supply chain design is a difficult task because (1) the sub-systems are intrinsically complex, (2) there are many interactions among the sub-subsystems, and (3) external factors such as demand uncertainties intricately affect performance. Dynamic modeling of the supply chain is an essential requirement for such studies. Perea, Grossmann, Ydstie & Tahmassebi (2000) apply ideas of process dynamics and control for supply chain management. Their model accounts for the flow of information and materials and provides insights to trade-offs between various performance indicators. Tsiakis et al. (2001) considered the design of a multi-product, multi-echelon supply chain network comprising of a number of manufacturing sites at locations fixed a priori, a number of warehouses and distributions centers at locations to be selected from a set of choices, and fixed customer zones. The design problem is modeled as a mixed-integer linear program whose objective is to minimize the total annualized cost of the network, taking into account both infrastructure and operating costs. Uncertainty in product demand is handled using a scenario-planning approach where a set of scenarios are constructed representative of both optimistic and pessimistic situations.
Papageorgiou et al. (2001) describe an optimization-based approach to addresses the related problem commonly faced by the pharmaceutical industry – selecting one or products to be introduced from a set of potential products and jointly planning site production capacity. The overall problem is formulated and solved using a mixed-

integer linear programming (MILP) model that considers many aspects specific to the pharmaceutical sector such as product lifetime constraints, scale-up, and qualification. Guillen et al. (2005) consider the design of a supply chain consisting of several production plants, warehouses and markets, and the associated distribution systems. Uncertainty in the production scenario is represented as a set of scenarios with given probabilities of occurrence. The design problem is then formulated as a multi-objective optimization to maximize profit and customer satisfaction while minimizing the financial risk. Pareto optimal design alternatives that represent the trade-off among the different objectives are generated rather than a unique solution.

Another key issue in supply chain management is the optimization of **supply chain behavior**. Given a supply chain and demands from customers over a time horizon, the goal is to calculate the optimal material acquisition plan, the production schedule, the ordering policies between distribution nodes and the inventory levels for the distribution network such that profit is maximized and customer service is close to a desired level. Perea-Lopez et al. (2003) formulate this as a multi-period mixed-integer linear programming (MILP) problem and use a rolling horizon Model Predictive Control approach to solve it in the face of disturbances. Gupta et al. (2000, 2003) use two-stage stochastic modeling framework coupled with a chance constraint programming approach. Supply chain decisions are classified into production and logistics decisions. In their approach, production decisions are made before the demand is known (first stage) while the logistics decisions are delayed. The latter are made in the second stage to handle evolving uncertainty in the product demand. Lababidi et al. (2004) developed an optimization model for the supply chain of a petrochemical company operating under uncertain operating and economic conditions. They also use a two-stage method to solve the stochastic model. In the first stage, decisions are made regarding the production volumes of different products for every planning period. In the second stage, decisions are made regarding the volume shipped to the distribution center, demand losses, backlog orders, and product inventories. Jung et al. (2004) used a simulation-based optimization approach for determining the safety stock level to meet a desired level of customer satisfaction. Choudhury et al. (2004) developed a decision matrix to allocate production plans to different manufacturing locations to minimize the total cost including manufacturing costs, transportation cost, tax rates, inventory carrying costs, warehousing costs, etc. Singhvi et al. (2004) focused on production planning in the supply chain and developed a graphical procedure motivated by pinch technology for production decision making.

With the increasing push towards collaboration, fair **profit sharing between supply chain entities** becomes an important issue. Gjerdrum et al. (2001) developed a strategy for deciding transfer prices along with production resource utilization, production levels, inventory levels, and flows so as to maximize the profit levels of the separate enterprise that comprise the supply chain. Chen et al. (2003) constructed a multi-product, multistage, and multi-period production and distribution-planning model to achieve multiple objectives such as maximizing the profit of each participant enterprise, customer service level, and safe inventory level. The model is formulated as a multi-objective mixed-integer nonlinear programming (MOMINLP) problem and fuzzy-set theory is used to attain a compromise solution among all participant companies of the supply chain.

Complex interaction between entities and the multi-tiered structure of supply chains obviate analytical models that can accurately capture the dynamics of entire supply chains. Agent-based systems are a promising alternative to supply chain modeling and simulation. Software agents exhibit the following properties (Woolridge, 2002):

- **Reactivity:** Ability to respond to changes that occur in their environment
- **Social Ability:** Ability to interact i.e., cooperate, co-ordinate, and negotiate with other agents to meet their objectives
- **Pro-activeness:** Ability to take initiative to satisfy their objectives

These qualities make agents ideal for modeling and analysis of supply chains, where collaboration, intelligence, and mobility are essential. Further, agent-based technologies support concurrent and distributed decision-making that is also an integral element of supply chain management (Bond & Gasser, 1988). As will be evident from Sections 3 and 5, agent-based approaches are also versatile and can easily capture qualitative and transactional events in the supply chain in addition to the quantitative aspects that traditional modeling approaches are best suited for.

Agent-based approaches have therefore received some attention in the chemical supply chain context as well. Flores, Wang & Goltz (2000) describe an ongoing effort in developing an integrated framework for supporting supply chain management of process industries. Retailers, warehouse, plants, and raw material are modeled as a network of co-operative agents, each performing one or more supply chain functions. Julka, Srinivasan and Karimi (2002a; b) propose a unified, flexible and scalable framework for modeling, monitoring and managing supply chains. Their framework has two basic elements: object modeling of supply chain flows and agent modeling of supply chain entities. Their framework uses three classes of agents: (1) emulation; (2) query and (3) project agents. Emulation agents model the supply chain entities such as manufacturers, logistics providers, etc. Query agent handle queries from the user and assist in supply chain analysis. Project agent performs the tasks needed to perform the study or solve the problem. Aldea et al. (2004) present case studies illustrating the application of multi-agent systems to diverse problems in the chemical process industry.

An agent toolkit is a development environment that provides modelers (agent builders) with a sufficient level of abstraction to allow them to implement intelligent agents with desired attributes, features and rules (Serenko & Detlor, 2002). Agent environments such as JADE, JATlite, and Zeus have been commonly developed in Java. In the last few years, several toolkits have been developed with special attention to interoperability and compatibility. The Foundation for Intelligent Physical Agent (FIPA) reference model has emerged as a standard for developing agent platforms. We have developed a new environment called G2 Multi-Agent Development Environment (G2-MADE) for agent modeling chemical supply chains. G2-MADE is developed using Gensym's expert system shell, G2, a commonly used platform for real-time decision support systems in the chemical industry. G2 object-oriented graphical environment and open-connectivity provide an ideal environment for chemical supply chain applications. As described next, G2-MADE contains the essential building blocks for modeling supply chains. Models, simulators, and decision support systems for any supply chain can therefore be developed without significant programming effort.

3 G2-Multi-Agent Development Environment

G2-MADE is an integrated environment to design, develop, debug, simulate and deploy agents. It supports the development of scalable multi-agent applications capable of running on a single machine or on a distributed network. The main components of MADE are SCAgent, SCMessage, SC-Activity and Host

SCAgent: A SCAgent is an autonomous, multi-threaded object having the ability to specify characteristics of various supply chain entities. Agent is specialized according to the intended role in supply chain for example, supplier agent, logistics agent, 3PLs etc. The behavior of a SCAgent is described in the form of Grafcets which is embedded inside the agents (explained below). Each SCAgent has a network-wide unique name and communicates with other SCagents through messages. G2-MADE has a "Yellow Pages" facility thorough which SCAgents can find other SCAgents with specific properties.

SCMessage: SCAgents communicate with each other by exchanging objects, called SCMessages. All SCMessages share common attributes including the originating (source) SCAgent, departure time, Recipient SCAgent, and time of receipt. In addition, a SCAgent can use customized messages that are subclasses of SCMessage. These subclasses are suitable for representing specific types of supply chain communications such as Request-for-Quote (RFQ). A SCMessage can be sent by one SCAgent to another (including itself) in general, or a specific activity within a SCAgent.

Activity: An activity defines a specific behavior of a SCAgent. At any time, a SCAgent may perform multiple activities of the same or different types (i.e., multi-threaded agents). A SCMessage sent to an agent may initiate a new activity or continue a dialog with an ongoing activity. Once an activity is initiated, messages can be sent specifically to it. In G2-MADE, Grafcets are used to specify the activity of SCAgents (David and Alla, 1992). A Grafcet is as a graphical programming language that is widely used for specifying process control actions in the chemical industry. A Grafcet is a graph that consists of two types of nodes – steps and transitions, as shown in Figure 2. A step represents a state, phase, or mode that can be active or inactive. Associated with a step are actions that are performed when a step is active. In G2-MADE, actions can be specified using G2's full-scale procedural language. Some examples of actions include sending message, calculating profits, selecting operable crudes from a set, etc. A transition signifies a change from one state to another. Each transition represents a condition that is necessary for the change to occur and for execution to move to the next step. An example of a transition condition is "Wait for quotations from at least three suppliers". In G2-MADE, simple transition conditions can be expressed in the form of G2-rules while complex ones can be coded as G2 procedures. The flow of control during the activity of an agent is thus defined by the Grafcet. A Grafcet can contain parallel threads of actions (see below), thus agents can perform parallel tasks. G2-MADE contains standard Grafcet templates for each class of agents; these can be modified by the developer to bestow a specific behavior to a SCAgent. A new activity can be initiated by the receipt of a specific message class as defined in the SCAgent handler.

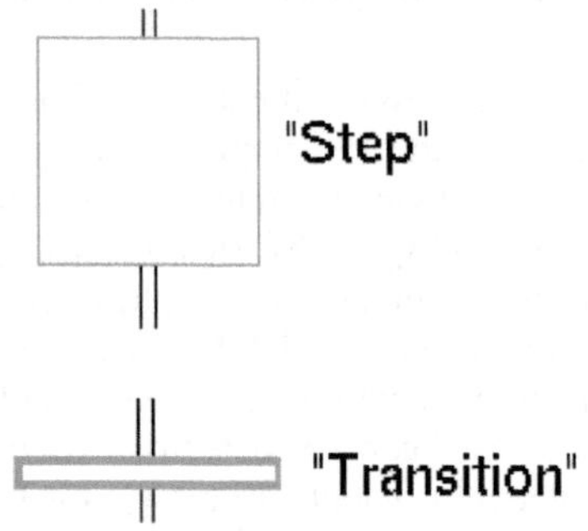

Fig. 2. Steps and Transitions are used to develop a Grafcet that specifies the activities of a SCAgent

Agent Handler: The agent handler determines how a SCAgent will respond to messages. The agent handler thus accepts or refuses the message, routes accepted messages to the appropriate activity, and initiates new activities as necessary. Messages that can initiate a new activity are specified in the agent handler by the developer.

Host: Every multi-agent application contains a host. Every agent, when created, registers with the host. The host serves as the post office and is responsible for message delivery. All SCMessages are routed through the host. The host sorts all outstanding messages according to the requested destination time of the message, locates the destination SCAgent and delivers the message to the agent handler of the addressee SCAgent. As described above, the message may be processed by the addressee in one of two ways – it will either start a new activity or continue an ongoing one.

The activities involved in message exchange are illustrated in Figure 3 using the example of a Producer and Supplier agent. Producer agent seeks to procure material "A". If the Producer agent knows the identity of a suitable supplier, then it can directly send a purchase-order message to the supplier. If a suitable supplier is not known beforehand, the Producer agent can look up the Yellow Pages directory to find out all supplier agents who sell "A". The Producer can then send some or all of them RFQ messages. All messages will go through the host and would be delivered to the agent handler of the destination agent.

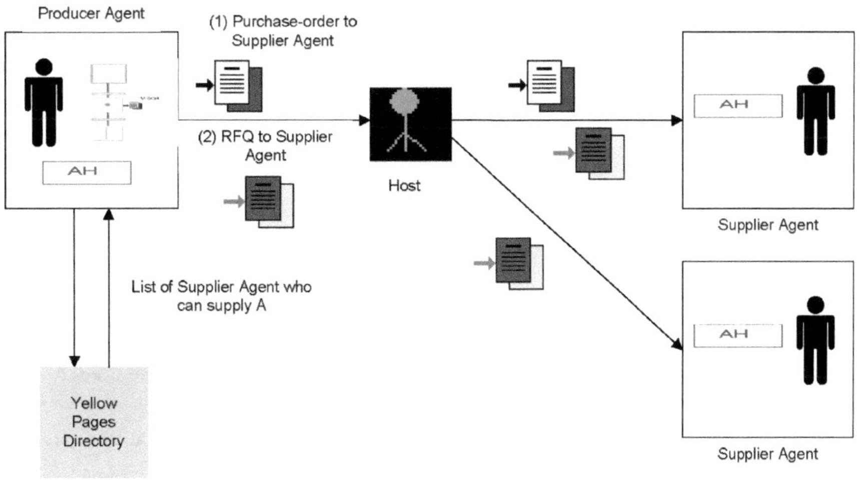

Fig. 3. Message Passing in G2-MADE

G2-MADE-Scheduler: G2-MADE comprises a message scheduler based on discrete-event simulation. Whenever a message to initiate a new activity is received, the scheduler creates a new instance of the SCAgent's Grafcet and starts it execution. Figure 4 shows a Grafcet that models the activity of the Supplier in the above example. Whenever a Supplier agent receives a new message, this Grafcet will be instantiated and execution will begin from the main thread at the step marked M. Based on the class of message, the appropriate thread T1 or T2 will be executed. If the message is of class RFQ, then the condition embedded in the first transition of T1 (notated as "Receive RFQ details) will be satisfied and the next step ("Send Quotation") will be executed. Similarly, a message of class Purchase-order will result in the execution of thread T2 and the task to "Send confirmation to procurement" will be performed. The "End of thread" transition signals completion of the activity and execution of the Grafcet instance will be terminated. Any complex, multi-threaded supply chain activity can be modeled using this formalism.

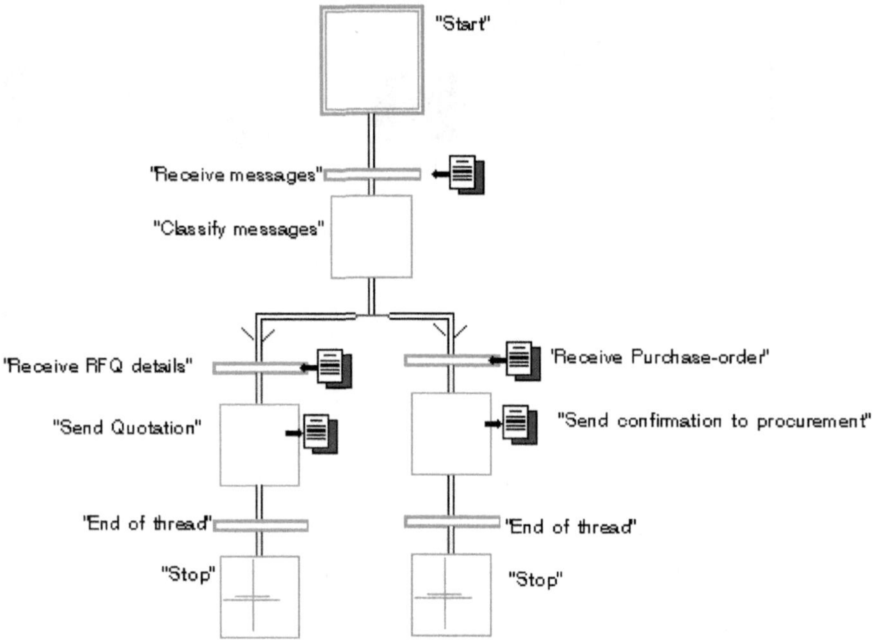

Fig. 4. Grafcet for Supplier agent

In the next section, we describe the model used for refinery supply chain management.

4 Refinery Supply Chain Management

Crude procurement is one of the most important supply chain activities in the refinery and has a direct impact on refinery profits. Large buffers of crude degrade the economics of the refinery due to the high inventory cost; insufficient crude would lead to crude stock-out situations that necessitate unit shutdowns – both should therefore be avoided. Further, crude procurement is a complex activity that requires interaction and closed coordination between several departments in the refinery as well as third parties. It therefore serves as a suitable illustration to explore the benefits of agent-based supply chain management.

A brief overview of the internal departments of the refinery is given below.

Procurement: Coordinates the crude procurement process. It retrieves crude availability and decides which crude to purchase and in what quantity. To do this, it needs information about crude availability, refinery targets, and logistics.

Sales: Provides product prices and demands, both current and forecasted

Operations: Decides which crude and how much to process every day
Storage: Manages the crude inventory and releases crude to operations.
Logistics: Arranges transport of crude from the oil supplier terminal to the refinery
In addition to the refinery, oil suppliers and third party logistics providers (3PLs) are important players in the refinery supply chain. 3PLs arrange for the transportation of the crude from the oil supplier's terminal to the refinery.
The crude procurement process varies from refinery to refinery; the following is one popular approach. The major events during crude procurement are shown in Figure 13. The entire crude procurement process can be divided into three sub-processes: crude selection and purchase; crude transportation, delivery and storage; and crude refining. Each of the sub-processes is explained below.
Crude Selection and Purchase
The crude selection and purchase is normally done at fixed intervals called *procurement cycles*. The purchase of crude is done significantly in advance of the time the crude will be processed – this duration is known as the *planning horizon*, and notated as A. The following activities are performed in each procurement cycle:

1. At the beginning of the procurement cycle, the procurement department requests the estimated demands during the *target week* (at the end of the planning horizon) of the various petroleum products from the sales department.
2. The sales department subsequently sends forecasted prices and demands of products to the procurement department.
3. The procurement department also acquires the list of crudes available for purchase in the petroleum exchange. Based on the characteristics of each crude (crude assays and cuts), its price, the forecasted product demands, and their prices, the procurement department calculates the profit margin, (also called the *netback value*, for each crude. The procurement department shortlists the most profitable crudes and sends the list, called the *crude basket*, to the operations department.
4. The operations department confirms the operability of the crudes in the crude basket based on plant constraints and previous experience and returns the *refined crude basket* to the procurement department.
5. The procurement department compiles the pickup location and time for the crudes in the refined crude basket and requests the logistics department for estimates of transportation costs.
6. The logistics department invites various 3PLs to bid for the contract of transporting each crude from its pickup terminal to the refinery. 3PLs bid for the contracts; their bids contain the transportation costs, demurrage terms and cost, etc.
7. The logistics department processes the bids and transportation cost estimates sent to procurement department based on the best bid.
8. The procurement department calculates the net profit on each crude (including the transportation cost) in the refined crude basket and finally selects the crudes to be purchased. Purchase orders are sent to the supplier. On receiving confirmation from the oil supplier, the logistics department is informed.

9. The logistics department in turn awards the contract for transporting the crude to the lowest cost 3PL. After receiving confirmation from the 3PL, the logistics department forwards the transport details to the procurement and storage departments.

Crude Transportation, Delivery, and Storage

The delivery and storage sub-process are given below:

10. Depending upon the date of pickup, the 3PL dispatches the ship to the pickup terminal.
11. The oil supplier starts loading crude. On completion, the tanker starts its journey to the refinery and informs the storage department of its expected arrival date.
12. The storage department checks the jetty schedule and arranges for the timely unloading of the crude. Demurrage charges may be levied by the 3PL if the tanker has to wait for a long period for the jetty to become available. On arrival at the refinery, the tanker informs the storage department. After berthing and approval from the storage department, crude unloading begins.

Crude Refining

This is the actual processing in the refinery. The refinery is a continuous process that runs 24x7 at a throughput that can be specified daily in the range $[T_{\min}\ T_{\max}]$. The task of selecting the mix of crudes to process is carried out daily:

13. The operations department decides the crude mix to run based on the present process conditions, crude stock, and the day's production targets (received from the sales department). It requests the storage department to release the required crude.
14. The storage department releases the required amounts to the operations department and updates the inventory database.

5 Agent Modeling of Refinery Supply Chain

The crude procurement process described above can be modeled using the multi-agent paradigm (Julka et al., 2002a; b). The resulting model has been implemented in G2-MADE and a decision support tool called Petroleum Refinery Integrated Supply chain Modeler and Simulator (PRISMS) developed. In PRISMS, agents are used to model the departments of refinery as well as the external entities. The hierarchy of SCAgents in refinery supply chain model is shown in Figure 5.

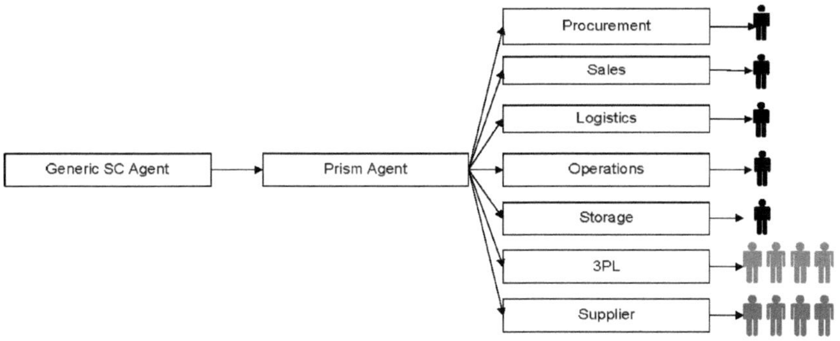

Fig. 5. Hierarchy of Agent Classes in PRISMS

The Generic-SCM-Agent is the superior class of all agents in G2-MADE. The Prism Agent is the superior class of all agents in PRISMS. It has seven subclasses, each customized to reflect the activities of a different entity – the Procurement, Sales, Logistics, Operations, and Storage departments of the refinery and the 3PLs and Supplier. One or more instances of these agent classes are used to model the structure and functioning of the refinery's supply chain. Messages are exchanged between the agents to emulate flow of information and chemicals in the refinery supply chain. Figures 6–12 describe the activities of the different agents involved in crude procurement. The procurement and logistics agents have one thread while the storage, sales, operations, 3PL, and supplier agents are multi-threaded.
The procurement agent receives a message from the clock agent (not described here) at the beginning of each procurement cycle. This message initiates the procurement process for that cycle. As shown in Figure 11, the procurement agent sends a message (Message M_P^1) to the sales department asking for the market data, which is the first step in the crude selection and purchase process as explained in Section 4. The sales agent has two threads (marked T1 and T2) as depicted in Figure 6. Any message requesting for 'market data' will activate thread T1 and the sales agent will respond with the market data (Message M_s^1) to procurement agent, as described in Step 2 of crude selection and purchase. Using this, the procurement agent calculates the profit and amount of crude to procure based on equations (1) and (2):

$$P(k) = \sum_{i=1}^{N} (y(i,k) * E(i)) - U(k) - D(k) \ . \qquad (1)$$

$$\tilde{C}(j) = (\sum_{i=1}^{N} Q(i) / G) / N - S + W \ . \qquad (2)$$

The procurement agent then sends the crude basket to the operations agent (Message M_P^2, Step 3). The operations agent has two threads as shown in Figure 8. The message with the crude basket will activate thread T1 and the operations agent will refine it and return refined crude basket as Message M_o^1 to the procurement agent (Step 4). The procurement agent then requests the logistics agent for transportation costs (Message M_p^3, Step 5) which is provided as Message M_L^2. The procurement agent recalculates the profit including the transportation cost.

$$P(k)=\sum_{i=1}^{N}(y(i,k)*E(i))-U(k)-D(k)-R \tag{3}$$

The procurement agent then selects the crude to purchase and sends the purchase order to supplier (Message M_p^4, Step 8). On receiving the purchase order, the supplier agent, whose Grafcet is shown in Figure 7, sends a confirmation (Message M_{SP}^1) to the procurement agent. The latter then sends the purchase details to the logistics agent (Message M_p^5) and waits for the transportation details (Message M_L^4). Finally, the transportation details are forwarded to storage and supplier agents (Message M_p^6).

In addition to the above tasks, some agents have other tasks to be performed on a regular basis. These are spawned by messages from the clock-agent. The sales agent performs the routine task of calculating the actual demands at regular intervals based on customer orders. Similarly, the Operations agent (Figure 8) decides the crude mix to run daily based on the crude stock and product demand (Step 13). Equations (4) and (5) are used in this thread to calculate planned throughput $\tilde{T}$ and actual throughput T in every procurement cycle.

$$\tilde{T}(n,j)=\tilde{C}(j)/F\ . \tag{4}$$

$$T(n,j)=\begin{cases}0 & \text{if } S(n)<T_{\min}\\ \min(C(j)/F+\hat{T}(n-1),S(n),T_{\max}) & \text{otherwise}\end{cases} \tag{5}$$

$$\hat{T}(n+1)=\max[(T'(n)-T(n)),0]\ \text{where}\ T'(n)=C(j)/F+\hat{T}(n-1) \tag{6}$$

The Operations agent sends Message M_O^2 daily to the storage department to release the crude necessary for the day's processing. The storage agent (Figure 9) has six threads. On getting Message M_O^2, storage agent releases the crude and updates the stock inventory based on equation (7).

$$S(n)=S(n-1)-T(n,j)+H(n)\ . \tag{7}$$

The Grafcet of the logistics agent is shown in Figure 10. When Message M_P^3 is received from Procurement agent to arrange for transportation, the logistics agent sends a RFQ (Message M_L^1) to the 3PLs registered with it. The 3PL agents reply (see Figure 11) with a RRFQ (Message M_{PL}^1) The logistics collates the information from these RRFQ and sends the best price for transportation to the Procurement agent. The logistics agent also sends a message (M_L^3) to the winning bidder, who replies with the transportation details. These are forwarded by the logistics department to the procurement agent via Message M_L^4.

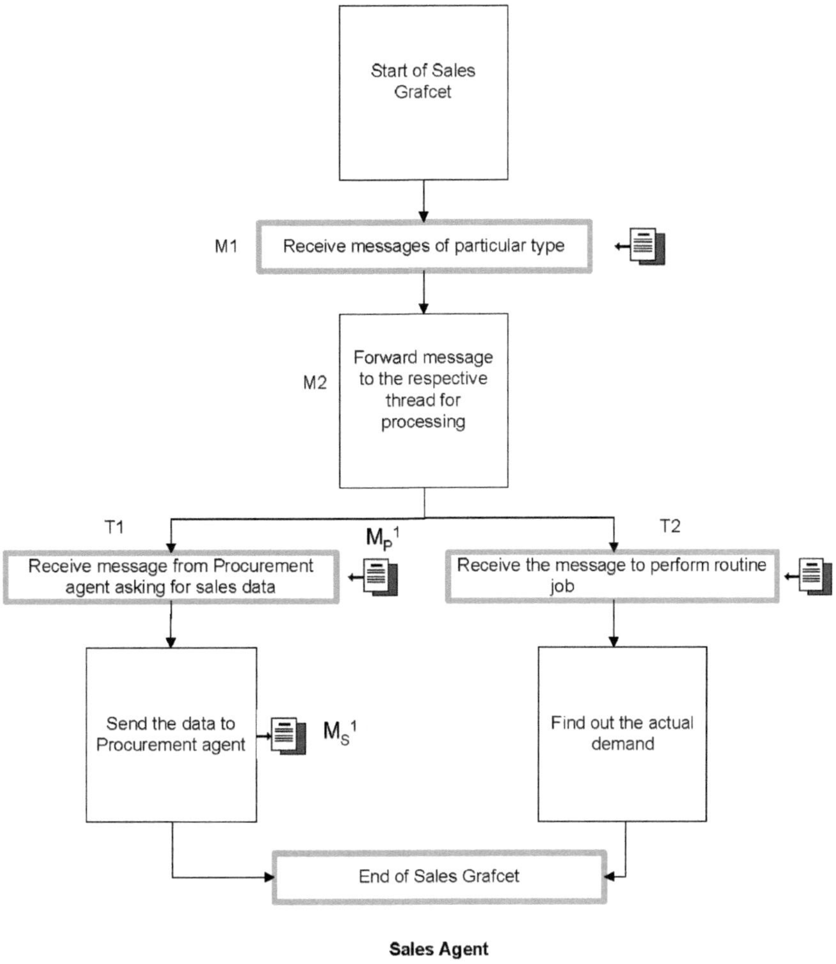

Fig. 6. Grafcet for Sales agent in PRISMS

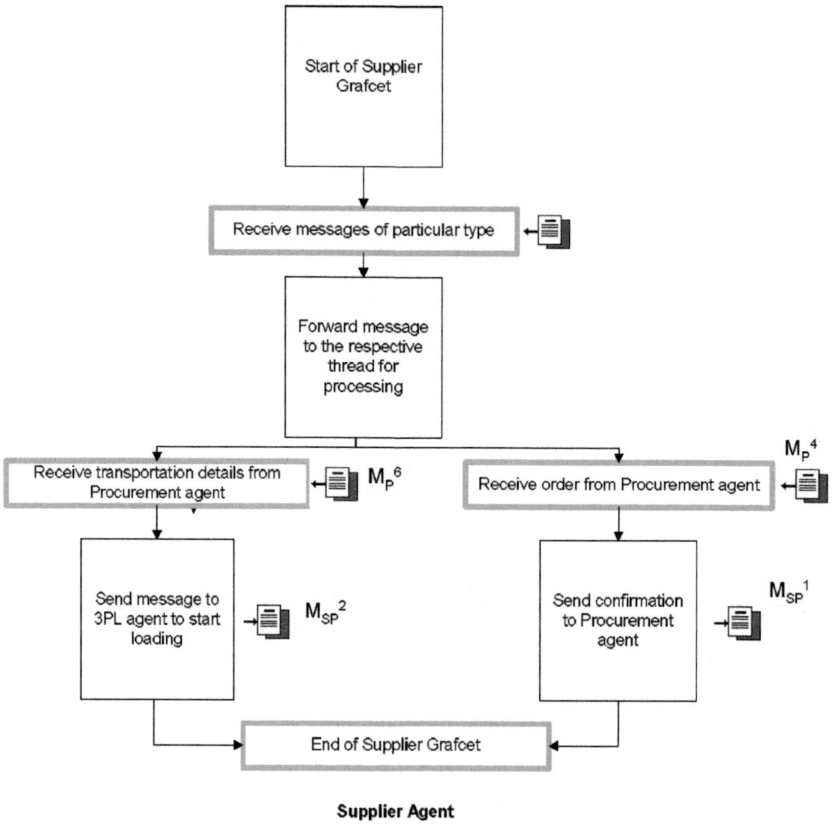

Fig. 7. Grafcet for Supplier agent in PRISMS

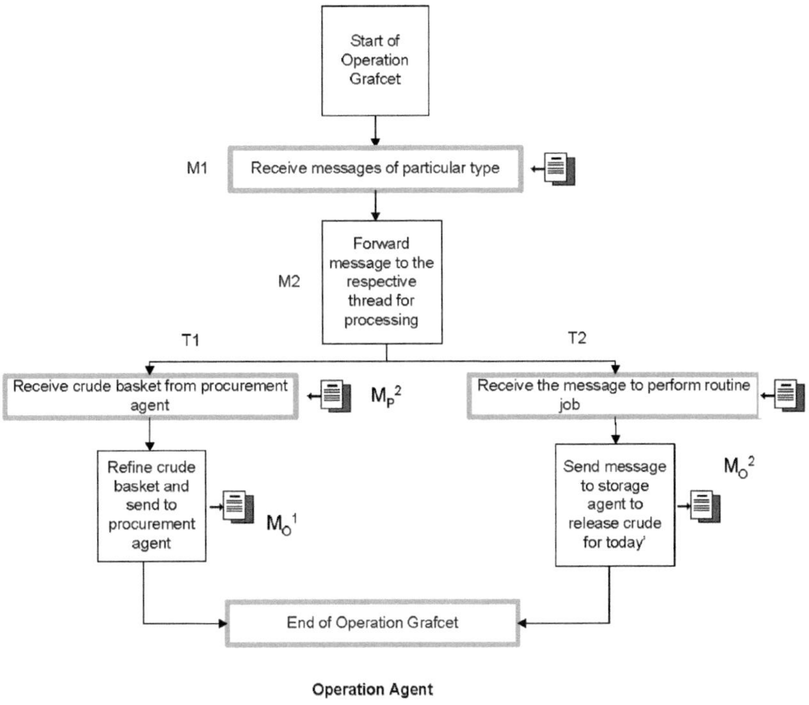

Fig. 8. Grafcet for Operation agent in PRISMS

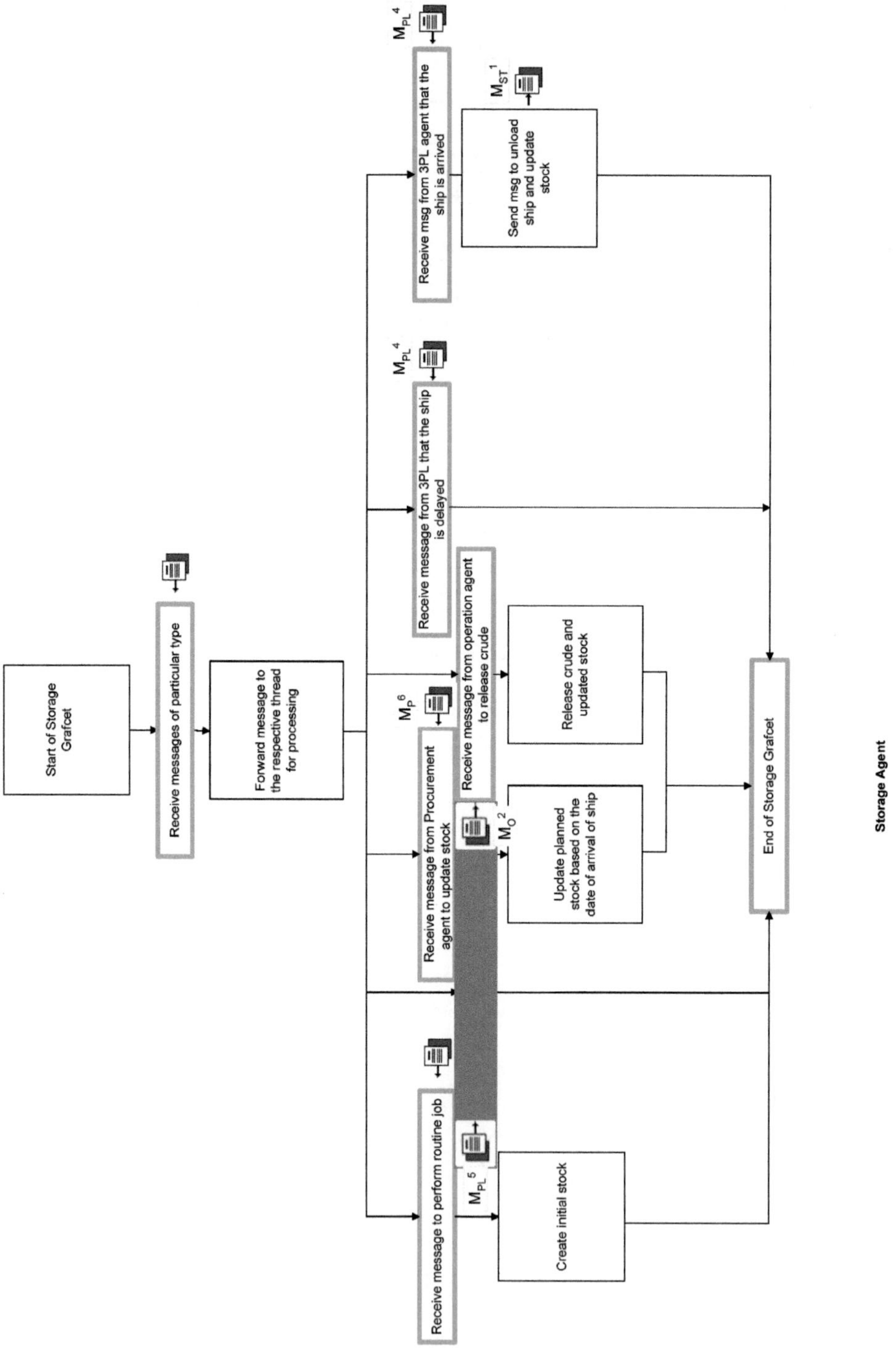

Fig. 9. Grafcet for Storage agent in PRISMS agent

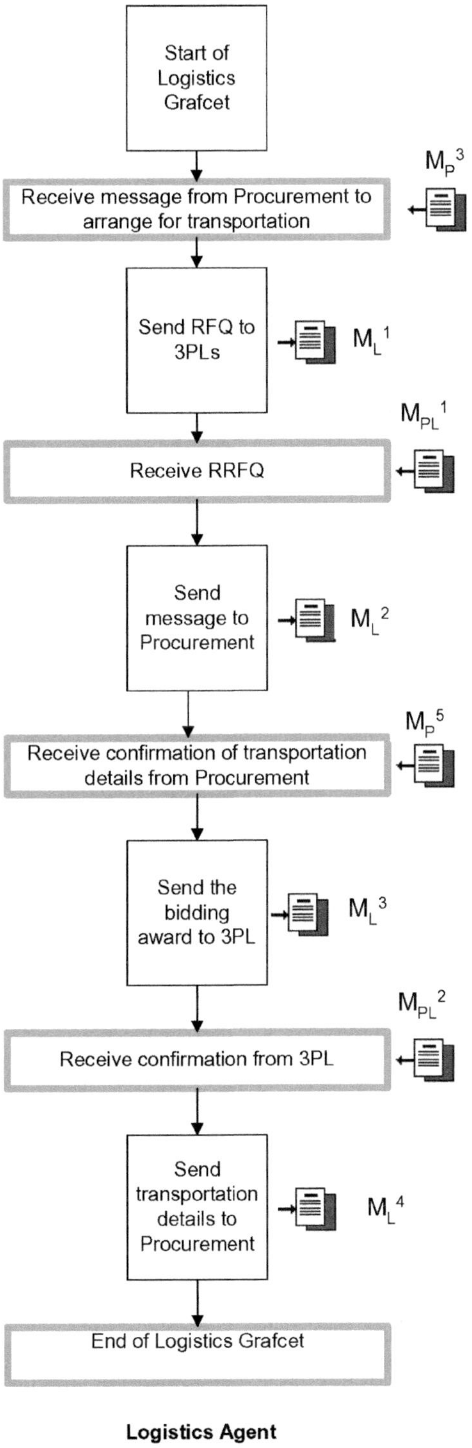

Fig. 10. Grafcet for Logistics agent in PRISMS

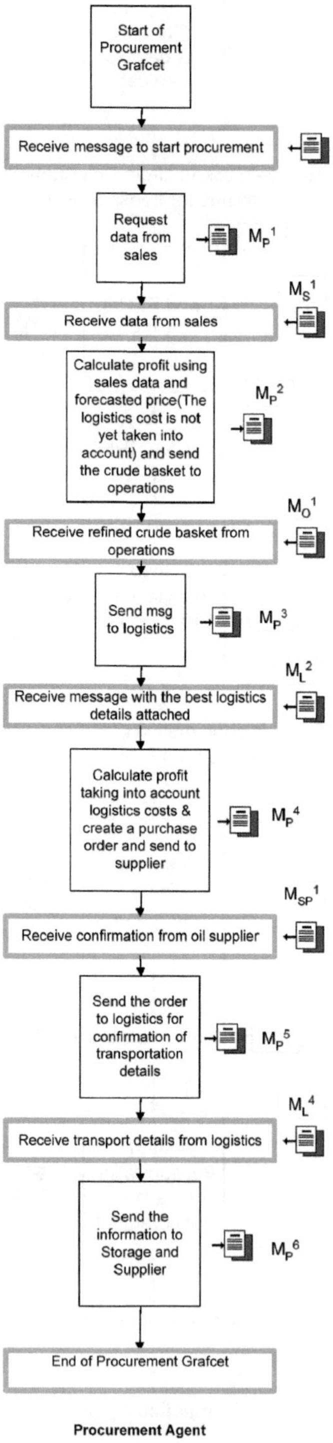

Fig. 11. Grafcet for Procurement agent in PRISMS

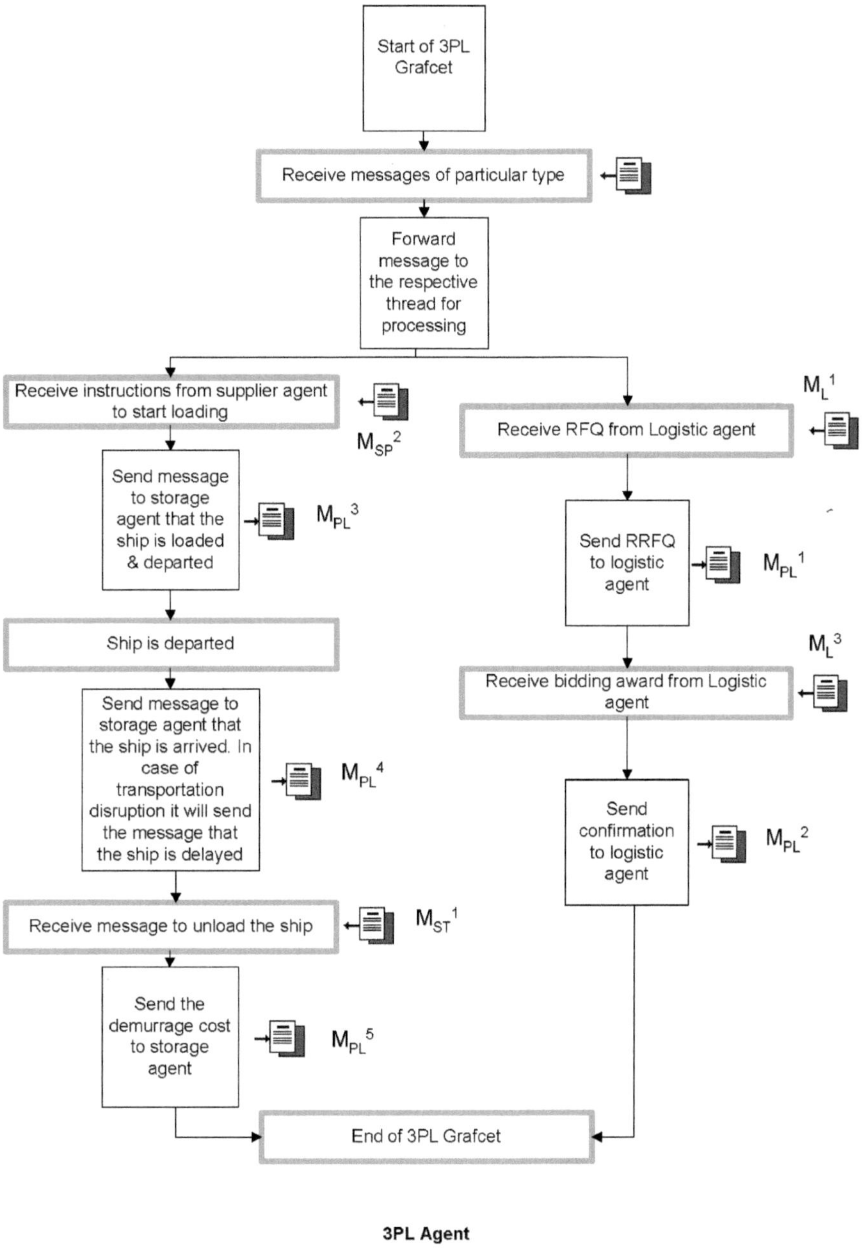

Fig. 12. Grafcet for 3PL agent in PRISMS

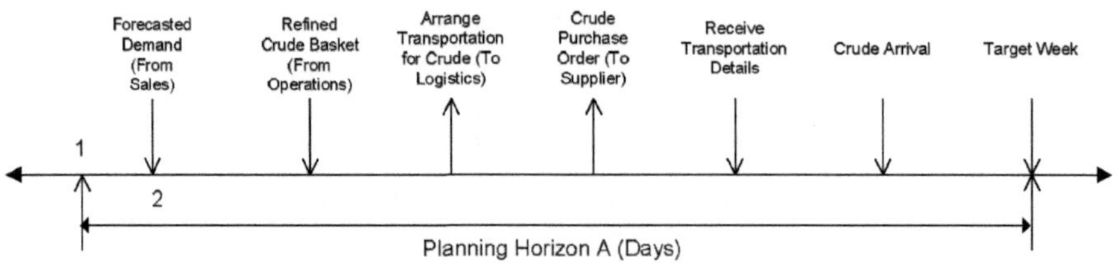

Fig. 13. Supply Chain Events during a Procurement Cycle

In the next section, we illustrate how this agent-based model of the refinery supply chain can be used for decision support.

6 Case Studies

Consider a simplified refinery with the following characteristics:

1. The refinery makes seven products.
2. Only one crude is procured in each procurement cycle.
3. The refinery operates on a pull mechanism, i.e. throughput is calculated based on demand. The Sales agent generates demand forecast and actual demand data stochastically.
4. A safety stock (W) of crude is maintained so that inventories do not fall below a pre-specified minimum level.
5. During normal operation, a max of 5% difference may exist between forecasted and actual demand. If the difference is larger, a disruption is considered to have occurred in the supply chain.

The following parameters values are used for the refinery and its supply chain operation:

Minimum throughput of refinery	T_{min}	70 kbbl/day
Maximum throughput of refinery	T_{max}	120 kbbl/day
Planning horizon	A	50 days
Simulation horizon	B	105 days
Length of procurement cycles	F	7 days
Number of procurement cycles	J	10
Safety stock	W	150 kbbl

In the following studies, results are shown from Day 35 of the simulation when the refinery and the supply chain has reached steady state after initialization.

6.1 Study 1: Normal Scenario

In this study, the normal operation of the supply chain is illustrated. Figure 14 shows the planned stock versus actual stock of refinery for ten procurement cycles. In the first procurement cycle, products have to be delivered on the 50th day. In this cycle, crude is delivered by ship on the 35th and 42nd day, consequently stock levels go up on these days. Inventory level trend down on other days due to production. The same saw-tooth trend occurs in other procurement cycles as well. Table 1 shows the projected and actual demand for the first ten procurement cycles. Figure 16 plots the crude procured to fulfill the forecasted demand for each procurement cycle. To fulfill the demand for the first procurement cycle 702 kbbl of crude is procured. The actual demand is the same as the planned demand for the first and second procurement cycles. Therefore, the planned and actual inventory profiles match. In subsequent procurement cycles, forecast and real demands differ, resulting in the CDU throughput differing from the original plan (See Figure 15). In the 4th – 6th procurement cycles, actual demand is less than the forecasted demand, so the actual throughputs are lower and stocks higher than planned. This effect is carried forward to the 7th procurement cycle. The refinery throughput changes to meet actual demands. Similar responsiveness of the supply chain can be seen in other procurement cycles as well.

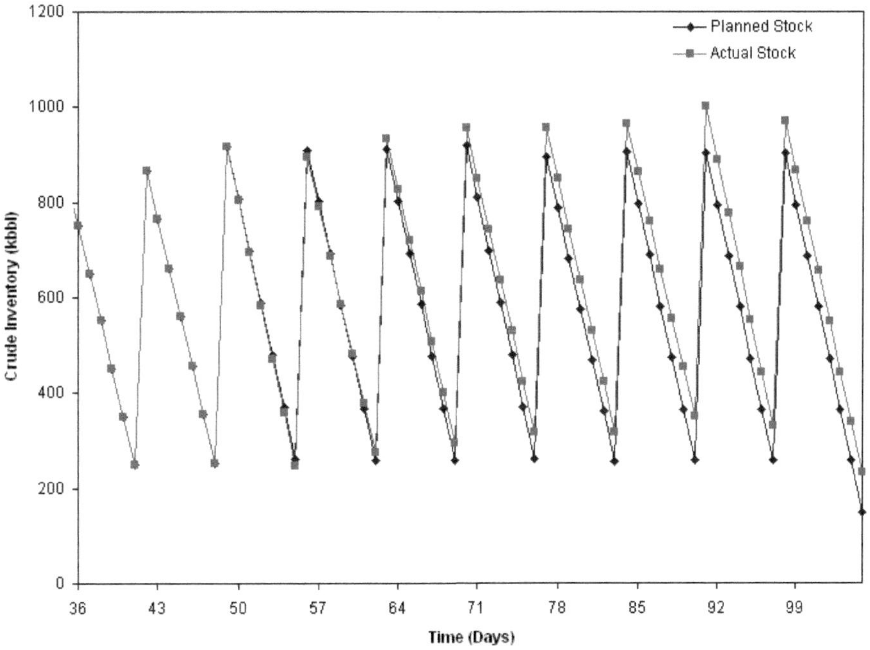

Fig. 14. Crude Inventory profile over simulation horizon

Table 1. Forecasted and actual crude demand in the first 10 procurement cycles

Procurement Cycle	Delivery Date	Crude Procured based on forecasted demand (kbbl)	Crude needed to meet the actual demand (kbbl)
1	50	702	702
2	57	717	717
3	64	768	783
4	71	759	721
5	78	761	745
6	85	769	746
7	92	744	744
8	99	754	717
9	106	752	782
10	113	752	737

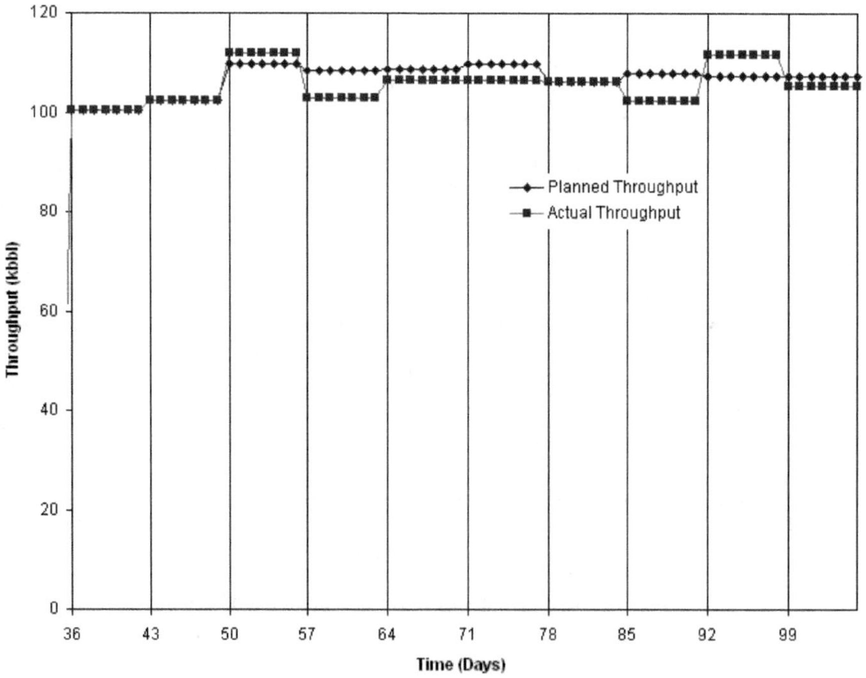

Fig. 15. Actual versus Planned throughput over simulation horizon

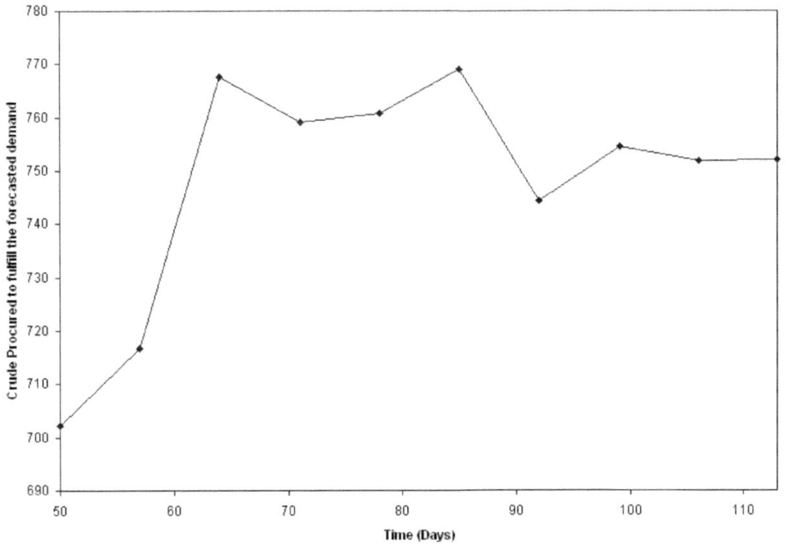

Fig. 16. Crude procurement in each procurement cycle

6.2 Study 2: Transportation Disruption

One use of the supply chain model above is to understand the effect of disruptions on the supply chain. The effect can be studied in terms of impact on various performance indicators such as inventory, refinery operation, demand fulfilled, etc. In this case, transportation disruption – an important and frequent disruption in the supply chain – is considered. The disruption is introduced through a stochastic increase in the crude transportation time. For example, a ship scheduled to arrive on day 42nd is delayed at sea and arrives on the 48th day instead. Because of the ship delay, there is stock out in the refinery from the 44th to 48th day when even the safety stock is used up. Figure 17 plots the planned versus actual stock. Stock falls to 51.3 kbbl at the end of day 43, which is inadequate to operate the refinery even at minimum throughput. As seen in Figure 18, the throughput over this period goes to zero and the refinery unit has to be shutdown. This would result in the inability to meet demands and customer dissatisfaction. When the delay ship arrives on the 48th day, the inventory level goes up. The crude for the third procurement cycle also arrives on the 49th day and the stock becomes much higher than planned. Throughput is increased to maximum (120kbbl/ day) to meet demands.

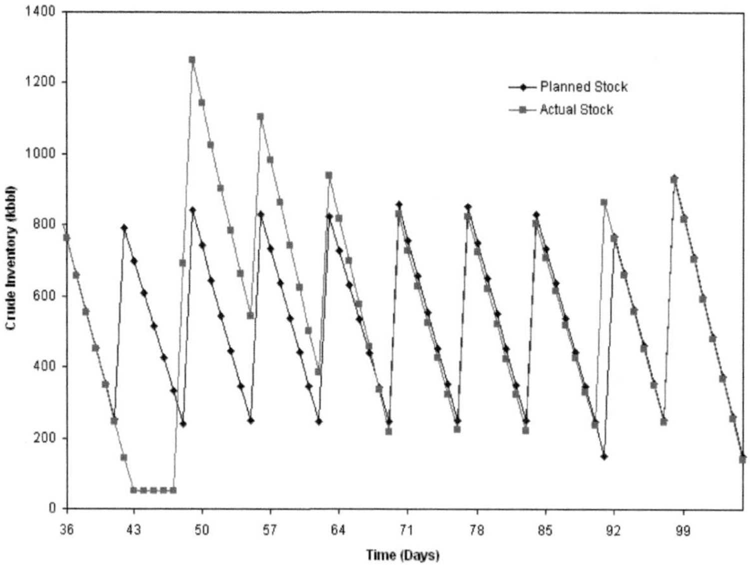

Fig. 17. Crude inventory in case of transport disruption

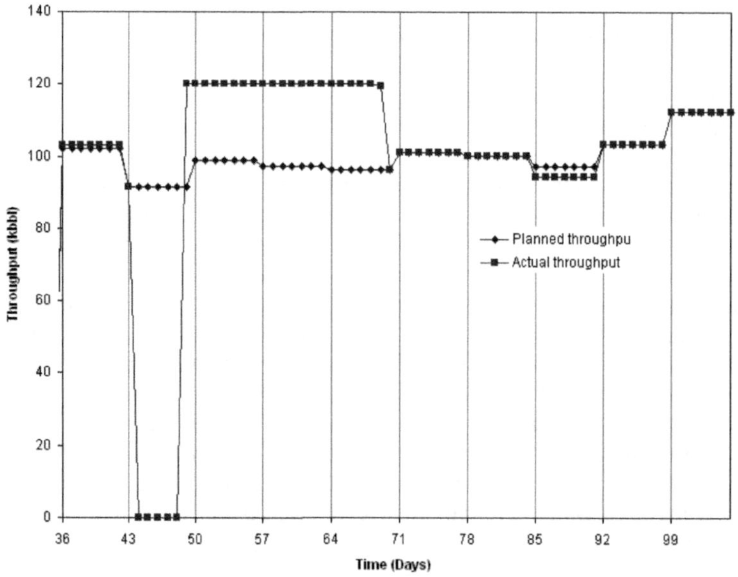

Fig. 18. Actual versus Planned throughput in case of transport disruption

6.3 Study 3: Demand High

The normal operations of the refinery are designed to handle a 5% difference between projected and actual demand. Order fulfillment can be expected to be 100% in these cases since the refinery keeps a safety stock of 150 kbbl to handle small demand increases. In this study, we evaluate the effect of larger demand increases. Figure 19 shows the order fulfillment (%) along with uncertain demand. The small demand variations during cycles 1-6 are absorbed completely and order fulfillment remains at 100%. A large increase in demand during cycles 7–9 however leads to a drop in fulfillment to 64% – 69% since adequate crude inventories are not available. These lead to missed market opportunities, which could have been exploited if the supply chain can be made more nimble.

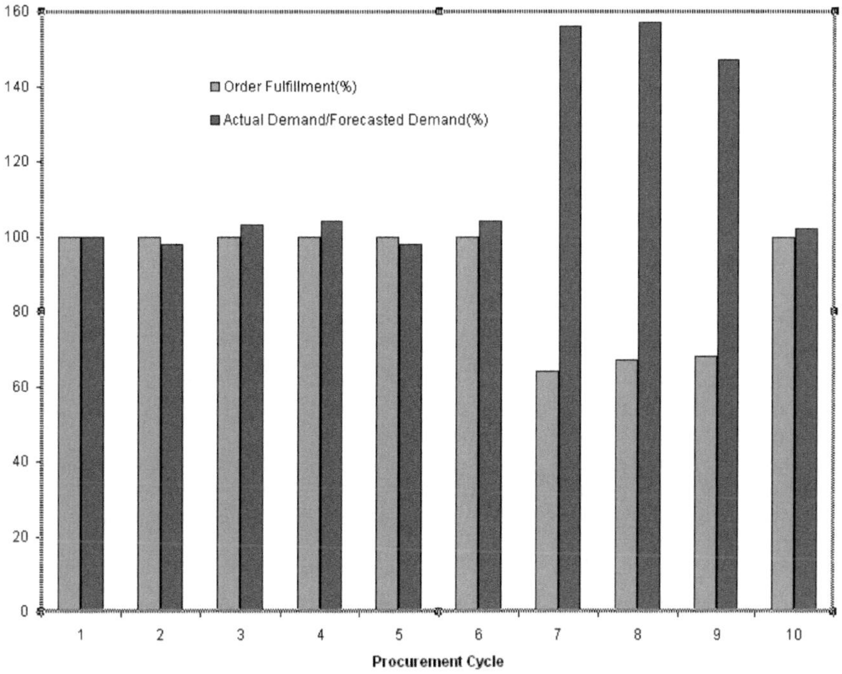

Fig. 19. Order Fulfillment in case of increase in demand

7 Discussion

Chemical Industries supply chain are unlike other industries supply chain due to specific features such as longer chains, complex transportation process, large inventory, complex manufacturing process, etc. These features of the chemical supply chain pose challenges to its effective management. The multi-agent approach is suitable for modeling the behavior of such supply chains as it can capture the dynamics and complexity of the supply chain and in a comprehensive and extendable fashion. The G2-MADE illustrates one easy to use framework to model the functions and activities within a supply chain. A model and simulation developed using the agent-based approach, such as PRISMS, can then be used to study the dynamics of the supply chain in its normal as well as disrupted states.

Unhindered and timely material, information, and finance flow between different entities of supply chain is important. Blockage in any of these would lead to undesirable events like process shutdown, financial loss, under-supply or over-supply, etc. Hence, there is a greater need for risk and disruption management. The agent-based refinery supply chain model described above can be extended to provide decision support during disruptions. An agent-based disruption management system should be capable of detecting abnormal situations before they occur, diagnose the root cause, and propose corrective actions as required (Bansal et al, 2005). The agents that model the department can be endowed with additional capabilities (by including suitable threads to their Grafcets) to measure entity-specific key performance indicators (KPIs). These KPIs can also be monitored by the agents by comparing their day-to-day values against pre-planned limits. Alarms can be generated when a sustained deviation in any KPI is detected. Corrective agents can be proposed and scheduled into the supply chain operation as necessary.

References

1. Aldea, A., Bañares-Al antra, R., Jiménez, L., Moreno, A., Martínez, J., Riaño, D. (2004). The scope of application of multi-agent systems in the process industry: three case studies. Expert Systems with Applications 26 (2004) 39 -47.
2. Bansal, M., Adhitya, A., Srinivasan, R., Karimi, I., (2005). An Online Decision Support Framework for Managing Abnormal Supply Chain Events, Presented in the European Symposium on Computer Aided Process Engineering – 15, Barcelona, Spain, May 29 – Jun, 2005.
3. Beamon, B. M. (1998). Supply chain design and analysis: Models and methods. International Journal of Production Economics 55 (3), 281.
4. Bond, A. H., & Gasser, L. (1988). An analysis of problems and research in DAI. Readings in distributed artificial intelligence. San Mateo, CA: Morgan Kaufman.
5. Chen, C., Wang, B., Lee, W., (2003). Multiobjective Optimization for a Multienterprise Supply Chain Network. Industrial Engineering Chemistry Research 2003, 42, 1879-1889.
6. Choudhary, A., Tiwari, M., Mukhopadhyay, S., (2004). Application of an analytical network process to strategic planning problems of a supply chain cell: case study of a pharmaceutical firm. Production Planning & Control Vol. 15, No. 1, Jan 2004, 13-26.
7. Cohen, M. A. & Lee, H. L. (1998). Strategic analysis of integrated production-distribution systems: Models and methods. Operations Research 36 (2), 216.
8. Cohen, M. A. & Moon, S. (1990). Impact of production scale economies, manufacturing complexity, and transportation costs on supply chain facility networks. Journal of Manufacturing and Operations Management 3, 269.
9. David, R., and Alla, H., (1992), Petri nets and Grafcet : tools for modelling discrete event systems, Prentice-Hall.

10. De Souza, R., (2004). Logistics: Complexity, challenges and opportunities, Logistics & SCM, Inaugural Issue.
11. European Commission – Chemicals Introduction, http://www.europa.eu.int/comm/energy_transport/atlas/htmlu/chemintro.html
12. Fishhaut, Eric., (2003), Petrochemicals firms take a stand against oil price volatility. GlobalView Software, May.
13. FIPA Agent Management Specification. Foundation for Intelligent Physical Agents, (2002), http://www.fipa.org/specs/fipa0023/
14. Flores, R. Garcia., Wang, X.Z. & Goltz, G. E. (2000). Agent-based information flow for process industries' supply chain modelling. Computers & Chemical Engineering 24, 1135-1141
15. Gaonkar, R., Viswanadham, N., (2004). A Conceptual and Analytical Framework for the Management of Risk in Supply Chains. Proceedings - IEEE International conference on Robotics and Automation, Volume 2004, Issue 3, 2004, Pages 2699-2704.
16. Gjerdrum, J., Shah, N., Papgeorgiou, L., (2001). Transfer Prices for Multienterprise Supply Chain Optimization. Industrial Engineering Chemistry Research, 40, 1650-1660.
17. Guillén, G., Mele, F., Bagajewicz, M., Espuña, A., Puigjaner, L., (2005). Multiobjective supply chain design under uncertainty. Chemical Engineering Science 60, 1535-1553.
18. Gupta, A., Maranas, C., Mcdonald, C., (2000). Mid-term supply chain planning under demand uncertainty: customer demand satisfaction and inventory management. Computers and Chemical Engineering 24 (2000), 2613-2621.
19. Gupta, A., Maranas, C., (2003). Managing demand uncertainty in supply chain planning. Computers and Chemical Engineering 27 (2003) 1219-1227.
20. Julka, N., Srinivasan, R., Karimi, I., (2002). Agent-based supply chain management-1: Framework. Computers and Chemical Engineering 26 (2002) 1755-1769.
21. Julka, N., Karimi, I., Srinivasan, R., (2002). Agent-based supply chain management-2: a refinery application. Computers and Chemical Engineering 26 (2002) 1771-1781.
22. Jung, J., Blau. G., Pekny, J., Reklaitis, G., Eversdyk, D., (2004). A simulation based optimization approach to supply chain management under demand uncertainty. Computers and Chemical Engineering 28 (2004) 2087-2106.
23. Karimi, I., Srinivasan, R., Viswanadham, N. (2000). Chemical Supply Chains Supply Chain Mapping and Tool Development. The Logistic Institute-Asia Pacific, Chemicals Supply Chain Research, WP: TLI-AP/00/09, May 2000.
24. Kimbrough, S. O., Wu, D. J. & Zhong, F. (2002). Computers play the beer game: can artificial agents manage supply chains? Decision Support Systems 33 (2002) 323-333.
25. Lababidi, H., Ahmed, M., Alatiqi, I., Al-Enzi, A. (2004). Optimizing the Supply Chain of a Petrochemical Company under Uncertain Operating and Economic Conditions. Industrial Engineering Chemistry Research 2004, 43, 63-73.
26. Lambert, D. M. & Cooper, M.C. (2000). Issues in supply chain management. Industrial Marketing Management 29 (1), 65.
27. Lee, H. L. & Billington, C. (1993). Material management in decentralized supply chains. Operations Research 41 (5), 835.
28. Mishra, M., R. Srinivasan, and I. A. Karimi, (2003). Managing disruptions in refinery supply chain using an agent-based decision support system, Presented in the AIChE annual meeting, San Francisco, Nov 16-21.
29. Papageorgiou, L., Rotstein, G., Shah, N., (2001). Strategic Supply Chain Optimization for the Pharmaceutical Industries. Industrial Engineering Chemistry Research 2001, 40, 275-286.
30. Perea, E., Grossmann, I., Ydstie, E., Tahmassebi, T., (2000). Dynamic modeling and classical control theory for supply chain management. Computers and Chemical Engineering 24, 1143-1149.
31. Perea-López, E., Ydstie, E., Grossmann, I., (2003). A model predictive control strategy for supply chain optimization. Computers and Chemical Engineering 27 (2003) 1201-1218.

32. Qi, Xiangtong., Bard, J.F., Yu, G., (2004). Supply Chain coordination with demand disruptions. Omega, Volume 32, Issue 4, 2004, Pages 301-312.
33. Serenko, A., Detlor, B., (2002), Agent Toolkits: A general overview of the market and an assessment of instructor satisfaction with utilizing toolkits in the classroom.
34. Sheffi, Y., (2001). Supply Chain Management under the Threat of International Terrorism. International Journal of Logistics Management 12 (2001) 1-11.
35. Sheffi, Y., Rice, J.B., Fleck, J.M., Caniato, F., (2003). Supply Chain Response to Global Terrorism: A situation scan, Presented in the EurOMA POMS Joint International Conference, Cernobbio.
36. Singhvi, A., Madhavan, k., Shenoy, U., (2004). Pinch analysis for aggregate production planning in supply chains. Computers and Chemical Engineering 28 (2004) 993-999.
37. Swaminathan, J. M., Smith, S. F. & Sadeh, N. M. (1998). Modeling Supply Chain Dynamics: A Multiagent Approach. Decision Sciences Volume 29 Number 3.
38. Tsiakis, P., Shah, N., Pantelides, C., (2001). Design of multi-echelon supply chain networks under demand uncertainty. Industrial Engineering Chemistry Research 2001, 40, 3585-3604.
39. Weiss, G. (1999). Multiagent systems. A modern approach to distributed artificial intelligence. Cambridge, MA: MIT Press
40. Whitfield, Mark (2004). A Stronger link. European Chemical News, September 20-26, 2004
41. Williams, J. F. (1983). A hybrid algorithm for simultaneous scheduling of production and distribution in multi-echelon structures. Management Science 29 (1), 77.
42. Wilson, M.C., (2002), Transportation Disruptions in the Supply Chain: Simulator as a Decision Support Tool, Proceedings of the 31st Annual Logistics Educators Conference, 2002.
43. Woolridge, M., (2002), An Introduction to MultiAgent Systems, John Wiley & Sons.
44. Wooldridge, M. & Jennings, N. (1995). Intelligent agents: theory and practice. Knowledge Engineering Review 10 (2), 115.
45. Xu, Minghui., Qi, Xiangtong., Yu, G., Zhang, H., Gao, Chengxiu.,(2003) The Demand Disruption Management Problem for a Supply Chain with Nonlinear Demand Functions. Journal of Systems Science and Systems Engineering, Volume 12, No.1 Pages 1-16.

Batch number: 10762627

Printed by Printforce, the Netherlands